Planetary Science: Emerging Concepts and Applications

Planetary Science: Emerging Concepts and Applications

Editor: Joe Carry

www.callistoreference.com

Callisto Reference,
118-35 Queens Blvd., Suite 400,
Forest Hills, NY 11375, USA

Visit us on the World Wide Web at:
www.callistoreference.com

ISBN: 978-1-63239-881-9 (Hardback)

The publisher's policy is to use permanent paper from mills that operate a sustainable forestry policy. Furthermore, the publisher ensures that the text paper and cover boards used have met acceptable environmental accreditation standards.

Printed in the United States of America.

Cataloging-in-Publication Data

Planetary science : emerging concepts and applications / edited by Joe Carry.
    p. cm.
Includes bibliographical references and index.
ISBN 978-1-63239-881-9
1. Planetary science. 2. Solar system. 3. Planetary systems. I. Carry, Joe.
QB601 .P53 2017
523.4--dc23

Table of Contents

Preface

This book traces the progress of planetary science and highlights some of its key concepts and applications. Celestial bodies like planets, moons, gas giants, micrometeoroids, etc. are extensively studies under planetary science. It also involves the study of planetary systems like the solar system. Over the years, this field has evolved from comparison only astronomy and earth science to several other sub-fields like planetary geology, atmospheric science, exoplanetology, etc. This text strives to provide a fair idea about this discipline and to help develop a good understanding of planetary science and its related fields. The various studies that are constantly contributing towards advancing technologies and evolution of planetary science are examined in detail. Those in search of information to further their knowledge will be greatly assisted by this book. It will help the readers in keeping pace with the rapid changes in this field.

Significant researches are present in this book. Intensive efforts have been employed by authors to make this book an outstanding discourse. This book contains the enlightening chapters which have been written on the basis of significant researches done by the experts.

Finally, I would also like to thank all the members involved in this book for being a team and meeting all the deadlines for the submission of their respective works. I would also like to thank my friends and family for being supportive in my efforts.

Editor

Frictional velocity-weakening in landslides on Earth and on other planetary bodies

Antoine Lucas[1,2,†], Anne Mangeney[1,3] & Jean Paul Ampuero[2]

One of the ultimate goals in landslide hazard assessment is to predict maximum landslide extension and velocity. Despite much work, the physical processes governing energy dissipation during these natural granular flows remain uncertain. Field observations show that large landslides travel over unexpectedly long distances, suggesting low dissipation. Numerical simulations of landslides require a small friction coefficient to reproduce the extension of their deposits. Here, based on analytical and numerical solutions for granular flows constrained by remote-sensing observations, we develop a consistent method to estimate the effective friction coefficient of landslides. This method uses a constant basal friction coefficient that reproduces the first-order landslide properties. We show that friction decreases with increasing volume or, more fundamentally, with increasing sliding velocity. Inspired by frictional weakening mechanisms thought to operate during earthquakes, we propose an empirical velocity-weakening friction law under a unifying phenomenological framework applicable to small and large landslides observed on Earth and beyond.

[1] Équipe de sismologie, Institut de Physique du Globe de Paris, Sorbone Paris Cité, Université Paris Diderot, UMR 7154 CNRS, 1 rue Jussieu, 75238 Paris cedex 05, France. [2] Division of Geological and Planetary Sciences, California Institute of Technology, 1200 E. California Blvd, Pasadena, California 95125, USA. [3] Équipe ANGE INRIA, Laboratoire Jacques-Louis Lions, UPMC Paris 6, 4 place Jussieu, Case 187 75252 Paris cedex 05, France. † Present address: Laboratoire Astrophysique, Instrumentation et Modélisation (AIM), CNRS-UMR 7158, Université Paris-Diderot, CEA-Saclay, Gif-sur-Yvette 91191, France. Correspondence and requests for materials should be addressed to A.L. (email: lucas@ipgp.fr).

A valanches, debris flows and landslides are key components of mass transport at the surface of the Earth. They have also been observed on other planetary bodies of our Solar System, from the interior planets to the icy moons of Saturn as well as small bodies such as the asteroid Vesta[1]. On Earth these mass wasting processes feed rivers with solid materials and thus participate in the evolution of the landscape. Moreover, catastrophic landslides constitute a significant hazard for life and property. The great diversity of natural gravitational flows in terms of volumes involved (from a few cubic metres to hundreds of cubic kilometres), flowing materials (for example, soil, clay, rocks, water ice and mixtures of different materials with or without the presence of a fluid phase such as gas or water), environment (for example, different gravitational acceleration on different planetary bodies, underlying topography), triggering mechanisms (seismic, volcanic, hydrological and climatic external forcing) and physical processes involved during the flow (for example, fragmentation[2] and erosion/deposition[3,4]) hinder a unified view of these phenomena. Therefore, despite the great number of experimental, theoretical and field studies, the behaviour of natural landslides is still poorly understood. This prevents accurate estimation of their maximum travel distance (also known as runout distance), the area covered by the flows and of their velocity, which are crucial components of hazard assessment for these catastrophic events. Calculating these quantities requires the use of appropriate laws describing energy dissipation during the landslide. Much research has been carried out in this direction, ranging from basic energy balance considerations of a sliding block to sophisticated models of granular flows over real topography including a variety of friction laws[5-10]. Readers can refer to a broad review already published[11].

On the basis of the energy balance of a rigid sliding block, the Heim's ratio[5] (that is, the ratio of the drop height H to the runout distance $\Delta L'$), is commonly used as an estimate of the effective friction coefficient. This ratio shows a clear decrease with increasing volume, raising two types of questions: (1) What is the physical meaning of the Heim's ratio? Does it really represent the effective friction during the landslide? If not, is the suggested volume dependence only an artefact of the use of the Heim's ratio instead of the effective friction? (2) Is there a more appropriate way to quantify the effective friction? Does it really decrease with increasing volume? If so, what is causing this volume dependence?

Here we address these questions with a focus on dense, catastrophic and rapid landslides. We first compile data on landslides from the literature and from our own analysis of remote-sensing data and confirm the common observation that the Heim's ratio decreases with increasing volume. We then develop an analytical relation between flow properties and the effective friction coefficient in a continuum model of granular column collapse over an inclined plane, and show that it generally differs from the Heim's ratio. On the basis of this analytical result, we develop consistent estimates of the effective friction coefficient of real landslides across the Solar System and find that friction decreases with increasing volume. We further confirm this scaling by a more complete approach, conducting simulations that account for topography and mass deformation to determine the friction coefficient that best reproduces the observed deposits of each landslide. These numerical simulations provide also estimates of landslide velocity, which indicate that the effective friction decreases with increasing velocity. We finally propose a single velocity-weakening friction law, inspired from concepts in earthquake physics (for example, flash heating of granular contacts), and incorporate it in our numerical models to show that it reproduces well the ensemble of observations of small to large landslides on Earth and other planetary surfaces.

Results

Heim's ratio for a large landslide data set. We gathered a large collection of data on landslides on Earth (82) and other planetary bodies (89), including data from the literature and new data (42) obtained using existing digital topography models (DTM) or DTMs that we built ourselves for this purpose. Some examples are illustrated in Fig. 1. Our DTMs, based on the most accurate available imagery and state-of-the-art photogrammetry analysis tools (see Methods), all together provide the best data set presently available on planetary landslides (see Supplementary Table 1). Despite the great complexity and diversity of these landslides and the large dispersion in the data, a general relation between Heim's ratio and volume is clearly identified (Fig. 2a, see Supplementary Note 1 for the discussion on the origin of the Heim's ratio). Essentially, $H/\Delta L' \sim 0.4$–0.7 for volumes smaller than $10^6\,\mathrm{m}^3$ and drops to values < 0.1 for volumes $> 10^9\,\mathrm{m}^3$. This has been previously observed and interpreted as a decrease of the effective friction coefficient as the volume increases[5,8,12,13]. The practical implication is that large landslides run over distances much longer than expected from the usual values of friction coefficient measured in laboratory experiments (~ 0.6–0.7). In this sense, large landslides are said to present high mobility or long runout distances[5,6,8]. However, as already long discussed in the literature, this interpretation is not as straightforward as it seems[8,9,11,12,14,15]. Confusion between the two independent questions formulated in the introduction has led some to refute the decrease of effective friction with volume because of the questionable meaning of the Heim's ratio[15]. Its limitations as a measure of the effective friction have long been recognized in the literature, based on the fact that its definition does not involve the displacement of the centre of gravity[16], the spreading of mass or the role of the topography[8,9,12,14,15]. Regardless of the meaning of the Heim's ratio, a possible volume dependence of the effective friction coefficient remains to be properly established by taking into account the effects of mass deformation and topography, as will be demonstrated here.

Effective friction from analytical description. To advance our understanding of the mechanics of catastrophic landslides beyond the energy balance of a sliding block, and in particular to take into account the deformation of the mass, we consider here the collapse of a granular column over an inclined plane. This situation is simple enough to be amenable to an analytical solution, leading to a closed-form relationship between the Heim's ratio and the other parameters involved in the problem (friction coefficient, bed slope and initial dimensions of the mass). On the basis of an analytical solution of the one-dimensional thin-layer depth-averaged equations of mass and momentum conservation with a Coulomb friction law[17-19], we derive the following relationship between the effective friction coefficient μ_{eff} and the initial thickness of the released mass H_0, the slope $\tan\theta$ and the distance travelled by the front along the slope ΔL (see Supplementary Note 2 for the complete derivation and Fig. 2a for the definition of the parameters. Different geometries are illustrated in Supplementary Fig. 1):

$$\mu_{\mathrm{eff}} = \tan\delta = \tan\theta + \frac{H_0}{\Delta L}. \qquad (1)$$

The analytical solution also shows that the Heim's ratio is

$$\frac{H}{\Delta L'} = \tan\theta + \frac{1}{\cos^2\theta\left(\frac{2k}{\tan\delta - \tan\theta} + \frac{L_0}{H_0} - \tan\theta\right)}, \qquad (2)$$

where L_0/H_0 is the inverse of the initial aspect ratio and k an empirical coefficient (for example, with $k = 0.5$, the results of granular collapse experiments are quantitatively reproduced[4,20]).

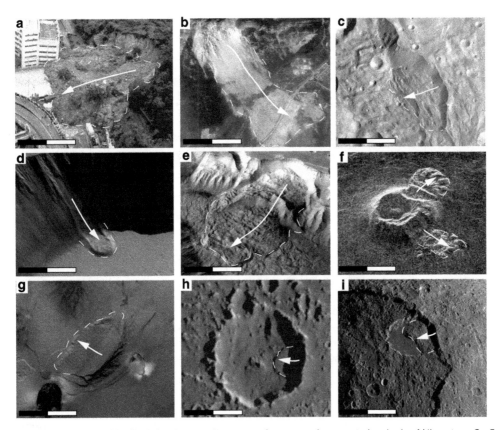

Figure 1 | Examples of landslides observed in the Solar System. Sizes range from tens of metres to hundreds of kilometres. On Earth: (**a**) Fei Tsui, Hong Kong (scale bar, 30 m); (**b**) Frank Slide, Canada (scale bar, 1 km). On Vesta (**c**) in the South pole region (scale bar, 80 km). On Mars: (**d**) Olympus Mons (scale bar, 2 km) and (**e**) Tithonium Chasma (scale bar, 10 km). On Venus: (**f**) in the Navka Region (scale bar, 25 km). On Jupiter's moons: (**g**) Euboea Montes on Io (scale bar, 100 km). (**h**) Inside Callisto's crater (scale bar, 10 km). On Saturn's moons: (**i**) Malun Crater on Iapetus (scale bar, 100 km). Deposits and sliding direction are highlighted with dashed white lines and arrows, respectively. Information on these landslides are reported in Supplementary Table 1.

Comparing equations (1) and (2), we observe that Heim's ratio results from a complex interplay between different quantities and differs significantly from the effective friction (details are provided in Supplementary Note 2). This is indeed confirmed by our observations (Fig. 3a). Equation (1) provides a more consistent estimate of the effective friction coefficient found from observations. We apply it to our data set and find that the effective friction clearly decreases with increasing volume whatever the planetary environment (Fig. 2c and see the subsection concerning application to real data in Supplementary Note 2). Combining laboratory and natural scale data (Fig. 2c), we find that the friction decreases for volumes larger than $\sim 10^3 \, \mathrm{m}^3$.

Effective friction from numerical modelling. The analytical approach does not account for the complex effect of the three-dimensional (3D) topography, which plays a key role in the landslide dynamics and deposits. Furthermore, the value of the effective friction coefficient derived from the analytic solution is mainly derived from the data on runout distance. We therefore assess here the ability of this empirical friction coefficient to explain the full extension of the deposits by simulating some landslides using the *SHALTOP*[20–22] numerical model that takes into account the real 3D topography. Although this model does not account for all the complex aspects of natural flows, such as the presence of heterogeneous materials (nature and size) or a fluid phase (gas or liquid) and the physical processes potentially acting during the flow (for example, fragmentation[2], erosion/

deposition[3,4]), it represents a significant advance compared with simple analytical solutions or scaling laws because of its ability to account for topography effects and mass deformation.

For each landslide, we used *SHALTOP* simulations to determine by trial and error the friction coefficient μ_s that best reproduces the deposit area inferred by satellite imagery analysis[9,10] (see Supplementary Fig. 6). We found that μ_s was very similar (within 3°) to the effective friction coefficient μ_{eff} obtained from the analytical solution (Fig. 3b). Their difference is within the range induced by the uncertainty of the estimate of the initial scar geometry[10]. This is an important result for the calibration of numerical models, indicating that an estimate of the friction coefficient needed to simulate real landslides can be readily obtained from equation (1).

Many numerical studies have found friction coefficients similar to ours for landslides on Earth[23–25]. Discrete element modelling of large Martian landslides requires similar values of the friction coefficient[26], showing that, whatever the type of model used, low friction is required to explain observations.

Numerical simulations also provide an estimate of the landslide velocity. The maximum velocity is known to be overestimated in depth-averaged thin layer models by only about 20% and the mean velocity is correctly estimated[20,27,28]. We calculate the mean velocity from the simulations by averaging the velocity in space and time over the whole landslide surface and duration. Velocities in our simulations agree with those reported for terrestrial landslides[25] (Fig. 4a). Dispersion in velocity values is expected given that topographic slopes over which landslides occur can vary from 0° to 40° locally as well as variousness in

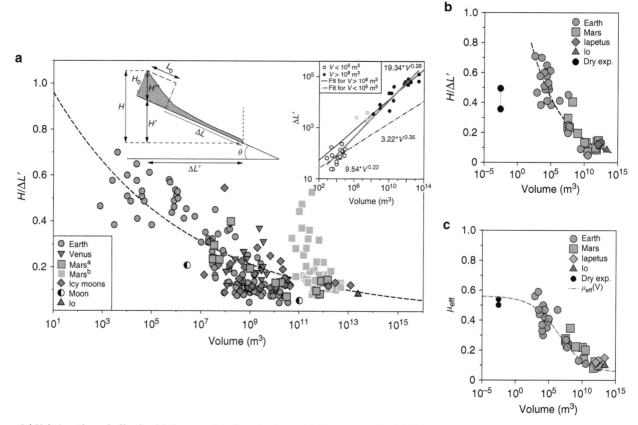

Figure 2 | Heim's ratio and effective friction as a function of volume. (**a**) Heim's ratio ($H/\Delta L'$) for different planetary bodies: terrestrial data come from literature[8,50,62] (Blue bullet); Martian data (Mars[a]) come from the literature[10] and this study after stereo extraction from CTX and HRSC images (orange square); Martian data (Mars[b]) from previous work[63] (pink square); data for the Moon (half black circle) and icy moons including Rhea, Callisto and Iapetus (grey lozenge) are from the literature[36,64]; Io data are from the previous work[65] (purple triangle); Venus data come from a former study[66] (orange triangle). The dashed line is the best fit of Terrestrial, Martian[b] and Io data for which DTMs or accurate field data[62] are available (symbols with thick outlines): $H/\Delta L' = 1.2 \times V^{-0.089}$. Owing to the absence of accurate DTMs, volume of landslides on the Moon, Venus and icy moons are estimated from the empirical relationship $\Delta L'(V)$ derived from terrestrial data shown in the right inset. Metrics are defined by the sketch. (**b**) ($H/\Delta L'(V)$) and (**c**) ($\mu_{eff}(V)$) for constrained landslides on Earth, Mars, Iapetus and Io. Experimental results[4] (black plain circle). The dashed red curve is the prediction from the final velocity-dependent law fitted to the terrestrial values including experimental results. All the values of panels **b** and **c** are reported in Supplementary Table 1. The data follows a trend $\mu_{eff} \sim V^{-0.0774}$ for volumes larger than $\sim 10^3$ m^3. The scatter of the data for $\mu_{eff}(V)$ in **c** is significantly smaller than for $H/\Delta L'(V)$ in **b** (see statistics in Supplementary Table 4). For the sake of clarity, error bars are not shown but are approximately the size of each symbol on the horizontal axis and twice this size on the vertical axis.

materials involved. Experimental work[4] has shown that the mean velocity is ~ 2.5 times greater on a slope of 22° than that on a flat bottom. Although new constraints on landslide velocities on Earth are emerging from seismological observations[28,29] and from a few rock avalanches that have been observed and/or filmed *in situ*[25], the simulation approach adopted here is the only way to quantitatively recover landslide velocities for remote and past events on Earth and beyond.

We find that the landslide velocity increases with the volume of the released mass (Fig. 4a). Hence, the volume dependence of μ_{eff} can be interpreted as a velocity dependence of friction, which is a more usual representation to investigate frictional weakening in solid mechanics and earthquake physics. Figure 4b shows that the effective friction decreases as a function of velocity. This general trend, observed for all landslides studied here, suggests a common mechanism that induces frictional weakening at increasing flowing velocities.

Quantification of frictional weakening. While several processes may lead to friction reduction under certain conditions[2–4,8,12,24,30–37] (see Supplementary Discussion), it is not

clear if the observations can be explained by a single process or if different processes act at different scales or in different environments. To clarify this, we need to quantify these potential processes and assess their compatibility with observations and environmental conditions. As a first step in this direction, our aim is to identify empirical friction laws that provide a unifying framework to explain the landslide scaling observations across all planetary environments (for example, low gravity, airless environment and absence of water table). The challenge will then be to quantitatively compare this empirical friction law to the mechanical behaviour derived from a specific mechanism.

Here we investigate quantitatively the compatibility between our observations of catastrophic landslides and two weakening mechanisms that have been introduced in very different contexts and for which constitutive relationships are available in a form that enables comparison with our results on real landslides. These relationships are (i) a friction law controlled by flash heating proposed to explain frictional weakening during earthquakes[32,38] and (ii) a rheology law proposed to describe laboratory granular flows[39].

Frictional weakening with increasing slip velocity has been invoked to explain key features of earthquakes[40] and has been

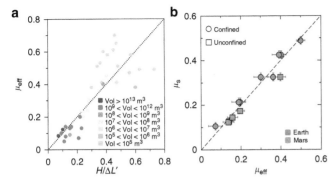

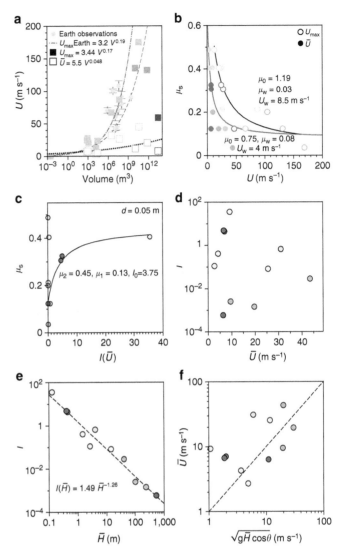

Figure 3 | Heim's ratio, analytical effective friction and friction derived from simulations. (**a**) Comparison between μ_{eff} and $H/\Delta L'$. Each landslide is represented by a circle the colour of which is related to its volume. The dotted line represents $x = y$, where x and y are the abscissa and ordinate, respectively. (**b**) The friction coefficient μ_s calibrated in the numerical modelling to reproduce the observed extension of the deposits is comparable to the effective friction coefficient μ_{eff} estimated from natural data using the simple analytical formula (1). Error bars in panel **b** represent the uncertainties in the estimation of μ_{eff} and uncertainties of μ_s calibrated by trial and error. For the sake of clarity, error bars are not shown on panel **a** but are approximately twice the size of each symbol.

observed in laboratory experiments of fast sliding on rocks and gouge materials[41]. A micro-mechanical process that can lead to dramatic velocity-weakening is flash heating[32,38]. The real contact between two rough solid surfaces generally occurs over a small fraction of their nominal contact area, on highly stressed micro-contacts (asperities). Slip produces frictional heating at the micro-contact scale. If slip is fast enough to prevent heat dissipation by conduction, the micro-contact experiences a significant transient temperature rise that activates thermal effects such as melting, dehydration and other phase transformations. This reduces the local shear strength of the micro-contact and leads to a macroscopic decrease of the friction coefficient as a function of sliding velocity. Laboratory observations of velocity-weakening have been previously interpreted within the framework of flash heating for rock sliding[38,42] and granular flows[43]. Owing to its microscopic scale and effects, application of flash heating to a very rough, irregular and/or heterogeneous surface where material in contact can be different and the shear zone not well defined is still valid and thus suitable for landslides. Conceivably, flash heating may occur at the inter-granular contact level during a landslide, a point that we will discuss more quantitatively below.

The applicability of these concepts to icy moons[44] is essentially speculative as the behaviour of icy grounds on planetary bodies remains a very unconstrained problem. Theoretical and experimental studies[45–47] have shown that similar behaviour can be observed for water ice at low temperatures and for rocky materials, suggesting that flash heating might occur on ice[48]. However, quantitative differences between planetary bodies similar to the variability observed between different rocky materials on Earth is expected. Because this is a very unconstrained topic, our approach here is only a first-order analysis[36].

Basic flash weakening theory[32,38] yields the following compact form to describe the steady-state friction coefficient as a function of slip velocity U: if $\|U\| > U_w$

$$\mu(U) = \frac{\mu_o - \mu_w}{\|U\|/U_w} + \mu_w, \tag{3}$$

otherwise,

$$\mu(U) = \mu_o, \tag{4}$$

Figure 4 | Frictional and dynamic properties from simulations. (**a**) Mean (that is, spatially and temporally integrated) and maximum velocities calculated in the simulation of real landslides over 3D topography as a function of the volume. Colours correspond to Earth in blue, Mars in orange and Iapetus in grey. Circle symbols represent measurements of maximum velocity (and their respective uncertainties) from terrestrial landslides reported in the literature[25]. The red dash-dotted curve is the best fit for all terrestrial data (simulations and reported values). Black dash-dotted lines are the best fit of the maximum and mean velocities, respectively, obtained from simulations of landslides on Earth, Mars and Iapetus. (**b,c**) The calibrated friction coefficient μ_s as a function of (**b**) the flow velocity and (**c**) the inertial number calculated from equation (5). Equation (8) is fitted to the maximum velocity (values in black) and to the mean velocity (values in red). (**d,e**) Inertial number as a function of (**d**) the mean velocity and (**e**) the mean thickness. (**f**) Mean velocity as a function of $\bar{v}_f = \sqrt{g\bar{H}\cos\theta}$, where $\bar{H}$; is the mean thickness during the flow and θ the mean slope (dashed line is equality). The best fits are displayed except for **d** and **f**. Colour scale indicating volumes, that is, cool colours for small volumes, warm colours for large volumes as for Fig. 3a except for panel **a** for which colours correspond to the planetary body as mentioned in the legend. For the sake of clarity, error bars are not shown but are approximately the size of each symbol.

where μ_o and μ_w are the static and thermally weakened friction coefficients, respectively, and U_w is a characteristic velocity for the onset of dramatic weakening, controlled by competition between frictional heating and heat conduction (see Supplementary Discussion). Even though natural landslides could span a broad range of parameters μ_o, μ_w and U_w, we found that equations (3) and (4) with a single set of parameters are consistent with the whole set of friction coefficients and velocities obtained in our simulations of real landslides (Fig. 4b). The best fitting parameters are $\mu_o = 0.75$, $\mu_w = 0.08$ and $U_w = 4\,\mathrm{m\,s^{-1}}$.

In order to investigate the effect of velocity-weakening friction on landslide dynamics, we introduced the friction law (3) with the global best fitting parameters in the *SHALTOP* numerical model. We then simulated natural landslides over 3D topography for several examples on Earth, Mars and Iapetus following previous works[9,10]. Without any further calibration, we were able to reproduce the observed runout distance and the shape of the deposit with errors below 12%. We then determined a set of parameters that provided the best global agreement between simulations and observations for all the tested cases: $\mu_o = 0.84$, $\mu_w = 0.11$, $U_w = 4.1\,\mathrm{m\,s^{-1}}$ (details on the simulations are given in Supplementary Methods and Supplementary Figs 8 and 9). Results from a simulation with velocity-dependent friction are shown in Fig. 5. The friction coefficient fluctuates in space and time during the landslide: fast flowing regions experience friction as low as μ_w while slow flowing regions experience friction as high as μ_o. The runout distances are then reproduced with an error <8%. The morphology of the deposit is also well reproduced. This shows that a single set of friction parameters can reproduce first- and second-order features of landslides with volumes varying by up to 14 orders of magnitude.

Note that when the threshold in equation (3) is not used in the simulation, the deposit area is affected by <12% and the runout distance by <8%. The distribution of the velocity and hence the thickness is slightly affected, especially at the beginning of the event ($t < 0.3t_f$), but solutions tend to converge towards the end of the simulation. The mean velocities calculated with the velocity-weakening friction law (3) are similar to the velocities obtained using a constant friction law with the friction coefficient μ_s fitted to reproduce the deposit. However, strong differences in the local

velocity field and thickness are observed between these two types of simulations.

Because of the large variability of the materials involved in the different landslides, we also varied the parameters of equation (3) to maximize the agreement between simulation and observations for each example individually. The parameters that best reproduce the deposits of studied landslides fall in the following ranges: $\mu_o \in [0.5–0.96]$, $\mu_w \in [0.08, 0.16]$ and $U_w \in [0.8–5.2]\,\mathrm{m\,s^{-1}}$. These values are comparable to the range derived from laboratory experiments[32] (where $\mu_o \in [0.6–0.88]$; $\mu \in [0.12–0.16]$ and $U_w \in [0.1–0.3]\,\mathrm{m\,s^{-1}}$). The main difference is observed for U_w, which might be explained by the different conditions prevailing in landslides, that is, granular material, lower confinement stress (e.g. see also Supplementary Tables 2 and 3).

Note that small-scale laboratory experiments of granular collapse are well reproduced by finite-element numerical models using a Coulomb friction law with a constant friction coefficient[49]. Experimental studies[4] show that granular collapse over beds with slope angles up to 22° have a maximum velocity of about $2\,\mathrm{m\,s^{-1}}$, that is, $<4\,\mathrm{m\,s^{-1}}$, so that equation (3) predicts a constant $\mu = \mu_o$. The characteristics of natural rockfalls from 1 to $10^3\,\mathrm{m}^3$ can be reproduced with a constant friction coefficient[50]. If we use our scaling law between volume and initial thickness of the released mass (see Supplementary Equation (20) in Supplementary Note 2 and Supplementary Note 1), a volume of $10^3\,\mathrm{m}^3$ would correspond to $H_0 \simeq 2.8\,\mathrm{m}$. Granular collapse experiments[4,51] show that the maximum velocity is a function of $\sqrt{gH_0 \cos\theta}$. As a result, on slope angles $\theta \simeq 30°$ typical for these rockfalls, the threshold velocity would be higher than $4.8\,\mathrm{m\,s^{-1}}$. All these simple calculations are in very good agreement with the empirical parameters of the friction laws (3) and (4).

Rheological laws for granular materials have been proposed in which the friction coefficient depends on the so-called inertial number[39,52,53],

$$I = \frac{\dot{\gamma}d}{\sqrt{P/\rho_s}} \qquad (5)$$

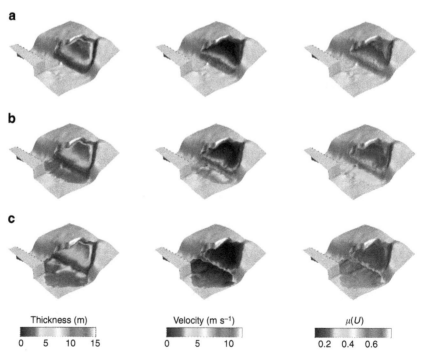

a

b

c

Thickness (m)	Velocity (m s⁻¹)	μ(U)
0 5 10 15	0 5 10	0.2 0.4 0.6

Figure 5 | Fei Tsui (Hong Kong) landslide is simulated with the velocity-weakening friction law. Thickness, velocity and friction coefficient are shown at three different times during the flow (that is, (**a**) 1 s, (**b**) 3 s and (**c**) 9 s). Friction weakening derived from equation (8) with $\mu_o = 0.84$, $\mu_w = 0.11$ and $U_w = 4.1\,\mathrm{m\,s^{-1}}$ has been used here. The example refers to Fig. 1a.

where $\dot{\gamma}$ is the shear strain rate, d the particle diameter, P the confining stress and ρ_s the particle density. Assuming hydrostatic pressure $P = \rho_s g H \cos\theta$ and $\dot{\gamma} \simeq U/H$, where g is the gravitational acceleration, H the thickness during the flow, θ the mean slope along the landslide path and U the mean velocity, we obtain

$$I \simeq \frac{Ud}{H\sqrt{gH\cos\theta}} \tag{6}$$

If we replace the velocity and thickness by their values averaged over space and time for each landslide, we find that, in the simulations with a constant friction coefficient, μ_s increases as I increases (Fig. 4c).

On the basis of laboratory experiments, the following friction law for dense granular flows has been introduced[39]:

$$\mu(I) = \frac{\mu_2 - \mu_1}{(1 + I_0/I)} + \mu_1 \tag{7}$$

The parameters that best fit our simulation results are $\mu_1 = 0.13$, $\mu_2 = 0.45$ and $I_0 = 3.75$, assuming $d = 0.05\,\mathrm{m}$ (Fig. 4c). For reference, the following parameters from laboratory experiments have been obtained[39]: $\mu_1 = 0.36$, $\mu_2 = 0.62$ and $I_0 = 0.279$. We implemented equation (7) in the *SHALTOP* model. Simulations adopting the best global fitting parameters obtained from Fig. 4c failed to reproduce the observed deposits. By tuning the parameters for each case individually, we were able to satisfactorily reproduce the runout distance but not the morphology of the deposits.

A multiscale phenomenological friction law for granular flows. We will now consider the similarities between the two friction laws described above (equations (3) and (7)). The flash heating friction law (equation (3)) features velocity-weakening that is a function of $1/U$ at large velocities. Granular shear laboratory experiments with specified thicknesses generally present velocity-strengthening behaviour consistent with equation (7). However, in our simulations the flow thickness is unconstrained and we find that, in contrast to what is first expected from equation (6), the inertial number I does not increase systematically with velocity U (Fig. 4d). We find that I is mainly controlled by $\bar{H}$; (Fig. 4e). This is related to a very rough scaling found here between $\bar{U}$ and $\sqrt{g\bar{H}\cos\theta}$ (Fig. 4f). Assuming the same relationship for non-averaged values of U and H, taken literally, would lead to $I \propto 1/H$ or, equivalently, to $I \propto 1/U^2$. As a result, the granular friction law (equation (7)) also leads to velocity-weakening, with an asymptotic behaviour of the form $1/U^2$ at high velocities. Note that $\mu \propto U^{-0.5}$ has been reported for water ice[47]. This suggests that velocity-weakening with asymptotic behaviour of the power-law form $1/U^n$ may be compatible with landslide observations and our current analysis favours an exponent n closer to 1 than to 2. Although both friction laws lead to qualitatively similar (power-law) velocity-weakening behaviour, quantitative differences in their velocity-weakening exponent make the flash heating law suitable for reproducing landslides, over a wide range of volumes and planetary environments, but not the granular flow law.

Our study suggests that a universal velocity-weakening friction law can describe small to large catastrophic landslides occurring in natural environments on Earth and other planetary bodies. Essentially, the friction coefficient is required to vary from around $\mu_1 = 0.7$ (a friction angle of $\delta \simeq 35°$) for low velocity flows to values as small as $\mu_r = 0.1$ ($\delta_r = 5.7°$) for rapid flows. The nature of the materials involved is expected to add scatter to these values. Typically, spherical beads used in laboratory experiments[39] have friction coefficients of around 0.65, while friction coefficients of natural rocks may reach 0.6–0.7 and even 0.7–0.75 for low-

temperature water ice[45] relevant for icy moons. On the basis of the foregoing discussion, we propose the following friction law:

$$\mu(U) = \frac{\mu_o - \mu_w}{(1 + \|U\|/U_w)} + \mu_w, \tag{8}$$

with $\mu_o = 0.84$, $\mu_w = 0.11$ and $U_w = 4\,\mathrm{m\,s^{-1}}$. Including a sharp velocity threshold as in equations (3) and (4) does not significantly affect the results of our landslide simulations. Even though this friction law is similar to the one derived from flash heating theory, the physical weakening process controlling landslides could be different. Indeed, the value of U_w found here is an order of magnitude higher than in solid friction experiments[54]. While this can be interpreted as a macroscopic velocity distributed over a granular shear zone a few tens of particles thick, there is evidence that inter-granular slip activity is highly intermittent[55] resulting in inter-particle velocities comparable to or faster than the macroscopic velocity[32].

An intuitive explanation of this velocity-weakening is that, whatever the scale, higher velocities increase the fluctuations in granular flows and may locally decrease the volume fraction, possibly decreasing frictional dissipation and enabling more complex flows (vortices and so on). Indeed, recent discrete element modelling of dry granular flows highlights the appearance of possible regimes dominated by large-scale vortices, significantly reducing the effective friction[56], although this remains to be observed in laboratory experiments. Different mechanisms may also be responsible for this friction weakening, owing to the great complexity of natural landslides that involve very different material properties (for example, composition and strength), environment variables (for example, air pressure and gravity) and physical processes at play (for example, fluid/grain interactions and erosion/deposition processes). Rock friction experiments showing a similar velocity-weakening behaviour have also been associated with a multitude of physical processes[54] (note that equation (8) can be used to fit these experiments, see Supplementary Fig. 7). Finally, we were able to obtain a predictive curve from the velocity-dependent friction by returning the relationship between velocity and volume (see $U_{\max_{\mathrm{Earth}}}$ on Fig. 5a) into equation (8) as shown in Fig. 2c. The best fit on the terrestrial data (that is, including experimental results) is obtained for $\mu_o = 0.56$, $\mu_w = 0.05$, $U_w = 4$ and $U = 0.17V^{0.21}$ with a coefficient of determination $R^2 = 0.90$. Nevertheless, variations of these parameters are expected for each individual landslide as mentioned previously.

Discussion

Using a large range of well-constrained data on landslides observed on different planets, an analytical solution of dry granular flows and numerical modelling over realistic 3D topography, we have demonstrated that the classically used Heim's ratio is not equal to the effective friction acting on landslides. We propose a more accurate way to quantify this effective friction based on field observations or remote-sensing data. As previously observed for the Heim's ratio and despite the large variety of landslide environments examined here, a clear decrease of the effective friction with increasing volume of the released mass is found while accounting for the deformation of the sliding mass and the 3D topography.

The novelty in our study is the correlation between the sliding velocity and the volume. The observed decrease of the friction coefficient with volume can be interpreted as velocity-dependent frictional weakening. Comparing numerical models of natural landslides on real 3D topography to field data on their deposits, we find an empirical relationship between the effective friction and the flow velocity. This relationship is surprisingly similar to a friction law derived for weakening by flash heating. The resulting friction coefficient varies from high values, up to $\mu_o = 0.7$–0.8, at low velocities (that is, small volumes) to very small values, down to $\mu_w < 0.1$, for rapid flows (that is, large volumes). Simulations of natural landslides based on this empirical friction law, with a single set of parameters, reproduce the observed landslide deposits with good accuracy over a very broad range of volumes and contexts. In contrast, we find that a friction law derived from laboratory experiments of granular flows is not compatible with natural observations.

Although our analysis cannot determine the physical origin of frictional weakening in landslides, we propose here a velocity-weakening friction law capable of describing, under a unifying phenomenological framework, the behaviour of small to large landslides observed on different planetary bodies.

Methods

Analytical solution. The analytical solution was developed on the basis of various studies[17–19] and describes the collapse over an inclined plane of a granular mass of effective friction coefficient $\mu_{eff} = \tan\delta$, where δ is the effective friction angle of the granular material. This solution is derived from the one-dimensional thin-layer depth-averaged equations of mass and momentum conservation with a Coulomb friction law, using the method of characteristics[17,19] (see Supplementary Notes 1 and 2).

Landslide identification and topographic reconstruction. For geomorphic measurements and numerical simulations, the initial scar geometry, the shape of the initial released mass and the bottom topography were reconstructed from observations using remote-sensing data in order to identify the landslide deposits in optical and elevation data. Landslides were identified using imagery provided by the Planetary Data System from the Mars Reconnaissance Orbiter, Mars Express, Cassini and Galileo missions. We produced the digital elevation models by photogrammetry based on rigorous sensor models from which ephemeris (that is, SPICE kernels) were extracted from the USGS ISIS software distribution[57] (Integrated Software for Imagers and Spectrometers) and the NASA Ames StereoPipeline[58] for data from the Cassini and Galileo missions. For the Mars Reconnaissance Orbiter cameras (HiRISE and CTX), DTM extraction was carried out on our photogrammetry workstation using the SOCET SET commercial software suite from BAE Systems. Bundle adjustments were performed for all of our examples in order to minimize errors due to uncertainties on SPICE kernels. In order to provide a realistic bottom topography, we reconstructed the initial mass released using geographic information system applications as already described[10,59] (see Section 4.2.2 of previous study[10]) and performed previously[9,10,27,59].

Numerical simulations. The simulations were performed using the *SHALTOP* model[20,21] based on the thin layer approximation for the depth-averaged equations of mass and momentum conservation with a Coulomb friction law. This model has been successfully applied to the modelling of laboratory experiments and natural examples[9,10,20,21,27–29,50,60]. Comparison with experiments of granular collapse and with discrete element models[20,61] shows that thin-layer models adequately reproduce the deposit and the overall dynamics for aspect ratios $a < 1$. This is the case for all the landslides for which we have a DTM and detailed field data (see Supplementary Table 1). Although the driving forces are overestimated in thin layer depth-averaged models, leading to overestimates of the initial velocity by up to 20% for aspect ratios $a \simeq 1$ the deposits are correctly reproduced. Recent simulation of the seismic waves generated by the flow along the topography and comparison with seismic records suggest that the landslide dynamics are also well reproduced by these numerical models[28,29,50].

For each landslide, we calibrated the effective friction coefficient of the simulation, $\mu_s = \tan\delta$, in order to best fit the deposit area observed from the field and/or satellite imagery analysis[9,10,27–29,60]. Note that using only one empirical parameter (μ_s), these models reproduce the whole deposit extension and even its general mass distribution quite well over a wide range of conditions (Supplementary Fig. 6). Here, we also integrated the new rheological laws into the model as shown in Fig. 5 and Supplementary Fig. 8.

Sensitivity of the results. The impact of the DTM resolution has already been discussed in former studies[10,27]. Reducing the spatial resolution of the DTM by a factor of two leads to an error of about 3° on the friction angle. When necessary, the topographic grid is oversampled in order to converge in terms of spatial resolution. In order to reconstruct the pre-failure and bottom topographies, we used the method previously developed[10] which has been shown to be efficient and lead to small error on the derived friction coefficient from numerical modelling (see Section 4.2 of ref. 10). As shown previously, the runout distance is weakly affected by the geometry of the initial scar geometry[10]. This leads to uncertainties

in the estimation of the friction angle δ of about 3° (Fig. 3b). The DTM resolutions are good enough to not affect our results in this 1-sigma band.

References

1. Otto, A. *et al.* Mass-wasting features and processes in Vesta's south polar basin Rheasilvia. *J. Geophys. Res. Planets* **118**, 2279–2294 (2013).
2. Davies, T. R., McSaveney, M. J. & Hodgson, K. A. A fragmentation spreading model for long-runout rock avalanches. *Can. Geotech. J.* **36**, 1096–1110 (1999).
3. Mangeney, A., Tsimring, L. S., Volfson, D., Aranson, I. S. & Bouchut, F. Avalanche mobility induced by the presence of an erodible bed and associated entrainment. *Geophys. Res. Lett.* **34**, L22401 (2007).
4. Mangeney, A. *et al.* Erosion and mobility in granular collapse over sloping beds. *J. Geophys. Res.—Earth Surf.* **115**, F03040 (2010).
5. Heim, A. *Bergsturz und Menschenleben* (Fretz and Wasmuth, 1932).
6. Dade, W. B. & Huppert, H. E. Long-runout rockfalls. *Geology* **26**, 803–806 (1998).
7. Pouliquen, O. Scaling laws in granular flows down rough inclined planes. *Phys. Fluids* **11**, 542–548 (1999).
8. Legros, F. The mobility of long-runout landslides. *Eng. Geol.* **63**, 301–331 (2002).
9. Lucas, A. & Mangeney, A. Mobility and topographic effects for large valles marineris landslides on Mars. *Geophys. Res. Lett.* **34**, L10201 (2007).
10. Lucas, A., Mangeney, A., Mège, D. & Bouchut, F. Influence of the scar geometry on landslide dynamics and deposits: application to martian landslides. *J. Geophys. Res.—Planets* **116**, E10001 (2011).
11. Pudasaini, S. P. & Hutter, K. *Avalanche Dynamics: Dynamics of Rapid Flows of Dense Granular Avalanches* (Springer, 2007).
12. Davies, T. R. Spreading of rock avalanche debris by mechanical fluidization. *Rock Mech.* **15**, 9–24 (1982).
13. Hayashi, J. N. & Self, S. A comparison of pyroclastic flow and debris avalanche mobility. *J. Geophys. Res.* **97**, 9063–9071 (1992).
14. Campbell, C., Clearly, P. W. & Hopkins, M. Large-scale landslide simulations: Global deformation, velocities and basal friction. *J. Geophys. Res.* **100**, 8267–8283 (1995).
15. Staron, L. & Lajeunesse, E. Understanding how volume affects the mobility of dry debris flows. *Geophys. Res. Lett.* **36**, L12402 (2009).
16. Hungr, O. Mobility of rock avalanches. *Report of the National Research Institute for Earth Science and Disaster Prevention, Tsukuba, Japan* **46**, 11–20 (1990).
17. Mangeney, A., Heinrich, P. & Roche, R. Analytical solution for testing debris avalanche numerical models. *Pure Appl. Geophys.* **157**, 1081–1096 (2000).
18. Kerswell, R. R. Dam break with Coulomb friction: a model of granular slumping? *Phys. Fluids* **17**, 057101 (2005).
19. Faccanoni, G. & Mangeney, A. Exact solution for granular flows. *Int. J. Num. Anal. Meth. Goemech.* **37**, 1408–1433 (2013).
20. Mangeney-Castelnau, A. *et al.* On the use of Saint-Venant equations to simulate the spreading of a granular mass. *J. Geophys. Res.* **110**, B09103 (2005).
21. Mangeney, A., Bouchut, F., Thomas, N., Vilotte, J.-P. & Bristeau, M. O. Numerical modeling of self-channeling granular flows and of their levee-channel deposits. *J. Geophys. Res.* **112**, F02017 (2007).
22. Farin, M., Mangeney, A. & Roche, O. Fundamental changes of granular flow dynamics, deposition and erosion processes at high slope angles: insights from laboratory experiments. *J. Geophys. Res.* doi:10.1002/2013JF002750 (2013).
23. Hungr, O., Dawson, R., Kent, A., Campbell, D. & Morgenstern, N. R. Rapid flow slides of coal-mine waste in British Columbia, Canada. *GSA Rev. Eng. Geol.* **15**, 191–208 (2002).
24. Hungr, O. & Evans, S. G. Entrainment of debris in rock avalanches: an analysis of a long runout mechanism. *Bull. Geol. Soc. Am.* **116**, 1240–1252 (2004).
25. Sosio, R., Crosta, G. B. & Hungr, O. Complete dynamic modeling calibration for the Thurwieser rock avalanche (Italian Central Alps). *Eng. Geol.* **100**, 11–26 (2008).
26. Smart, K. J., Hooper, D. & Sims, D. Discrete element modeling of landslides in Valles Marineris, Mars. *AGU Fall Meeting.* Abstract P51B–1430 (2010).
27. Lucas, A., Mangeney, A., Bouchut, F., Bristeau, M. O. & Mège, D. in *Proceedings of the 2007 International Forum on Landslide Disaster Management* (eds Ho, K. & Li, V.) (International Forum on Landslide Disaster Management, Hong Kong, 2007).
28. Favreau, P., Mangeney, A., Lucas, A., Crosta, G. & Bouchut, F. Numerical modeling of landquakes. *Geophys. Res. Lett.* **37**, L15305 (2010).
29. Moretti, L. *et al.* Numerical modeling of the Mount Steller landslide flow history and of the generated long period seismic waves. *Geophys. Res. Lett.* **39**, L16402 (2012).
30. Melosh, H. J. Acoustic fluidization: a new geologic process? *J. Geophys. Res.* **84**, 7513–7520 (1979).
31. Erismann, T. Mechanisms of large landslides. *Rock Mech.* **12**, 15–46 (1979).

32. Rice, R. J. Heating and weakening of faults during earthquake slip. *J. Geophys. Res.* **111**, B05311 (2006).

33. Goren, L. & Aharonov, E. Long runout landslides: the role of frictional heating and hydraulic diffusivity. *Geophys. Res. Lett.* **34**, L07301 (2007).

34. De Blasio, F. V. Landslides in Valles Marineris (Mars): a possible role of basal lubrication by sub-surface ice. *Planet. Space Sci.* **59**, 1384–1392 (2011).

35. Viesca, R. & Rice, J. R. Nucleation of slip-weakening rupture instability in landslides by localized increase of pore pressure. *J. Geophys. Res.* **B3**, 2156–2202 (2012).

36. Singer, K., McKinnon, W., Schenk, P. & Moore, J. Massive ice avalanches on Iapetus caused by friction reduction during flash heating. *Nat. Geosci.* **5**, 574–578 (2012).

37. Pudasaini, S. P. & Miller, S. A. The hypermobility of huge landslides and avalanches. *Eng. Geol.* **157**, 124–132 (2013).

38. Beeler, N. M., Tullis, T. E. & Goldsby, D. L. Constitutive relationships and physical basis of fault strength due to flash heating. *J. Geophys. Res.* **113**, B01401 (2008).

39. Jop, P., Forterre, Y. & Pouliquen, O. A constitutive law for dense granular flows. *Nature* **441**, 727–730 (2006).

40. Heaton, T. H. Evidence for and implications of self-healing pulses of slip in earthquake rupture. *Phys. Earth. Planet. Inter.* **64**, 1–20 (1990).

41. Tsutsumi, A. & Shimamoto, T. High velocity frictional properties of gabbro. *Geophys. Res. Lett.* **24**, 699–702 (1997).

42. Goldsby, D. L. & Tullis, T. E. Flash heating leads to low frictional strength of crustal rocks at earthquake slip rates. *Science* **334**, 216–218 (2011).

43. Kuwano, O. & Hatano, T. Flash weakening is limited by granular dynamics. *Geophys. Res. Lett.* **38**, L17305 (2011).

44. Lucas, A. Slippery sliding on icy Iapetus. *Nat. Geosci.* **5**, 524–525 (2012).

45. Beeman, M., Durham, W. B. & Kirby, S. H. Friction of ice. *J. Geophys. Res.* **93**, 7625–7633 (1988).

46. Miljković, K., Mason, N. J. & Zarnecki, J. C. Ejecta fragmentation in impacts into gypsum and water ice. *Icarus* **214**, 739–747 (2011).

47. Kietzig, A.-M., Hatzikiriakos, S. G. & Englezos, P. Physics of ice friction. *J. Appl. Phys.* **107**, 081101 (2010).

48. Kennedy, F. E., Schulson, E. M. & Jones, D. E. The friction of ice on ice at low sliding velocities. *Philos. Mag. A* **80**, 1093–1110 (2000).

49. Crosta, G. B., Imposimato, S. & Roddeman, D. Numerical modeling of 2-D granular step collapse on erodible and nonerodible surface. *J. Geophys. Res.* **114**, F03020 (2009).

50. Hibert, C., Mangeney, A., Grandjean, G. & Shapiro, N. Slope instabilities in the Dolomieu crater, Réunion Island: from seismic signals to rockfall characteristics. *J. Geophys. Res.* **116**, F04032 (2011).

51. Roche, O., Montserrat, S., Niño, Y. & Tamburrino, A. Experimental observations of water-like behavior of initially fluidized, unsteady dense granular flows and their relevance for the propagation of pyroclastic flows. *J. Geophys. Res.* **113**, B12203 (2008).

52. Ancey, C., Coussot, P. & Evesque, P. A theoretical framework for granular suspension in a steady simple shear flow. *J. Rheol.* **43**, 1673–1699 (1999).

53. Savage, S. The mechanics of rapid granular flows. *Adv. Appl. Mech.* **24**, 289–366 (1984).

54. Di Toro, G. *et al.* Fault lubrication during earthquakes. *Nature* **471**, 494–498 (2011).

55. da Cruz, F., Emam, S., Prochnow, M., Roux, J. N. & Chevoir, F. Rheophysics of dense granular materials: discrete simulation of plane shear flows. *Phys. Rev., E* **72**, 021309 (2005).

56. Brodu, N., Richard, P. & Delannay, R. Shallow granular flows down flat frictional channels: steady flows and longitudinal vortices. *Phys. Rev.* **E87**, 022202 (2013).

57. Torson, J. & Becker, K. ISIS—a software architecture for processing planetary images. *LPSC* **XXVIII**, 1443–1444 (1997).

58. Moratto, Z. M., Broxton, M. J., Beyer, R. A., Lundy, M. & Husmann, K. Ames Stereo Pipeline, NASA's open source automated stereogrammetry software. *Lunar and Planetary Science Conference.* **41**, Abstract 2364 (2010).

59. Lucas, A. *Dynamique des Glissements de Terrain Par Modélisation Numerique et Télédétection: Application aux Exemples Martiens* (Thèse de doctorat, IPGP, 2010).

60. Pirulli, M. & Mangeney, A. Result of back-analysis of the propagation of rock avalanches as a function of the assumed rheology. *Rock Mech. Rock Eng.* **41**, 59–84 (2008).

61. Mangeney, A., Staron, L., Volfsons, D. & Tsimring, L. Comparison between discrete and continuum modeling of granular spreading. *WSEAS Trans. Math.* **2**, 373–380 (2006).

62. King, J. P. Natural terrain landslide study—The natural terrain landslide inventory. *GEO Report No. 74. Hong Kong SAR: Geotechnical Engineering Office* (1999).

63. Quantin, C., Allemand, P. & Delacourt, C. Morphology and geometry of Valles Marineris landslides. *Planet. Space Sci.* **52**, 1011–1022 (2004).

64. Moore, M. *et al.* Mass movement and landform degradation on the icy Galilean satellites: results of the Galileo nominal mission. *Icarus* **140**, 294–312 (1999).

65. Schenk, P. & Bulmer, M. H. Thrust faulting, block rotation and large-scale mass movements at Euboea Montes, Io. *Science* **279**, 1514–1517 (1998).

66. Malin, C. M. Mass movements on Venus: preliminary results from Magellan cycle 1 observations. *J. Geophys. Res.* **97**, 16337–16352 (1992).

Acknowledgements

The authors would like to thank J. Rice, N. Cubas, N. Lapusta, F. Passelègue, A. Schubnel, N. Brantut, M. Lapotre, J. Melosh, L. Moretti, A. Valance, P. Richard, R. Delannay, N. Brodu, O. Pouliquen, K. Miljković and O. Aharonson for interesting discussions and feedback. The authors also thank the French Space Agency (CNES) for its support. The research was funded by French ANR PLANETEROS and LANDQUAKES, by the Terrestrial Hazard Observation and Reporting (THOR) Center at Caltech and by Campus Spatial Grant from Université Paris-Diderot.

Author contributions

A.L. collected the data, ran the numerical simulations, derived the digital terrain models, produced the figures and supplementary information ancillary materials and played a major role in the quantitative analysis. A.M. supervised the numerical analysis. A.L. and A.M. carried out the analytical development. J.P.A. developed the perspective on friction weakening processes from earthquake science. All the authors shared ideas, contributed to the interpretation of the results and to writing the manuscript.

Additional information

2

Tracing the fate of carbon and the atmospheric evolution of Mars

Renyu Hu[1,2], David M. Kass[1], Bethany L. Ehlmann[1,2] & Yuk L. Yung[1,2]

The climate of Mars likely evolved from a warmer, wetter early state to the cold, arid current state. However, no solutions for this evolution have previously been found to satisfy the observed geological features and isotopic measurements of the atmosphere. Here we show that a family of solutions exist, invoking no missing reservoirs or loss processes. Escape of carbon via CO photodissociation and sputtering enriches heavy carbon (^{13}C) in the Martian atmosphere, partially compensated by moderate carbonate precipitation. The current atmospheric $^{13}C/^{12}C$ and rock and soil carbonate measurements indicate an early atmosphere with a surface pressure <1 bar. Only scenarios with large amounts of carbonate formation in open lakes permit higher values up to 1.8 bar. The evolutionary scenarios are fully testable with data from the MAVEN mission and further studies of the isotopic composition of carbonate in the Martian rock record through time.

[1] Jet Propulsion Laboratory, California Institute of Technology, Pasadena, California 91109, USA. [2] Division of Geological and Planetary Sciences, California Institute of Technology, Pasadena, California 91125, USA. Correspondence and requests for materials should be addressed to R.H. (email: renyu.hu@jpl.nasa.gov).

The evolution of the atmosphere of Mars is one of the most intriguing problems in the exploration of the solar system. Presently Mars has a very thin 6-mbar atmosphere in equilibrium with polar caps and regolith. Yet, both morphological and mineralogical evidence suggests that the climate of Mars more than 3 billion years ago was warmer and wetter than the present[1]. The atmospheric conditions conducive to this environment are still uncertain. A denser CO_2 atmosphere may have facilitated early warm and wet surface conditions, at least locally and transiently at high orbital obliquities[2].

The pressure of the early Martian atmosphere has not yet been constrained by models of atmospheric evolution, due to uncertainties in the planet's early outgassing history, atmospheric escape and carbonate precipitation[3,4]. To transition from a thicker early atmosphere to the thin current atmosphere, carbon needs to be removed by either escape to space[5,6] or deposition near the surface as carbonates[7]. Recent models of the upper atmosphere of Mars suggest that < 300 mbar of CO_2 has escaped to space since the late heavy bombardment (LHB)[8,9], and current Mars exploration has only found local evidence of carbonate deposits[10]. Neither mechanism, alone or coupled, fully accounts for the 'missing' CO_2, if a multi-bar early atmosphere is assumed.

Here we combine the recent Mars Science Laboratory (MSL) isotopic measurements of Mars' atmosphere, and orbital remote sensing and in situ measurements of Mars' surface composition to place hard constraints on Mars' atmospheric evolution. The isotopic signature of carbon offers a unique tracer for the atmospheric evolution of Mars because CO_2 is the major constituent of Mars' atmosphere, and because carbon is not incorporated into surface minerals except for carbonates. Our study is driven by the following three recent and important constraints from in situ and remote sensing observations.

Carbon isotope signature of Mars' atmosphere. Early data analyses from the Phoenix lander showed an isotopically light atmosphere but were influenced by a calibration artefact[11,12]. Telescopic measurements had a large uncertainty of 20‰ and were subject to telluric contamination[13]. The Sample Analysis at Mars (SAM) instrument suite on MSL has reported the most precise isotopic measurements of atmospheric CO_2 to date: $\delta^{13}C = 46 \pm 4$ (ref. 14), measured by both the tunable laser spectrometer and the quadrupole mass spectrometer, which shows that the current Martian atmosphere is enriched in ^{13}C than the Martian mantle (Fig. 1). $\delta^{13}C$ is defined as the relative enhancement of the ratio $^{13}C/^{12}C$ with respect to a reference standard (VPDB), reported in parts per thousand (‰):

$$\delta^{13}C \equiv \frac{(^{13}C/^{12}C)_{Sample} - (^{13}C/^{12}C)_{VPDB}}{(^{13}C/^{12}C)_{VPDB}} \times 1,000 \quad (1)$$

where $(^{13}C/^{12}C)_{VPDB} = 0.0112372$ (ref. 15).

Carbonates formed in the Amazonian era. Orbital remote sensing indicates the Martian dust contains 2–5 wt% of carbonate[16]. Phoenix-evolved gas experiments have measured up to 6 wt% carbonate in soil of the northern plains[17]. MSL's evolved gas experiments in Gale Crater found ~ 1 wt% carbonate at the Rocknest aeolian deposit[18]. Certain young, large geologic units on Mars, including the southern polar layered deposits and the Medusae Fossae formation may contain up to 10 m global equivalent of dust[19,20]. On the basis of these measurements, we estimate an upper limit of carbonate formation during the Amazonian Era to be 7 mbar of CO_2, corresponding to global presence in the upper 10 m of soil at an abundance of ~ 2 wt%.

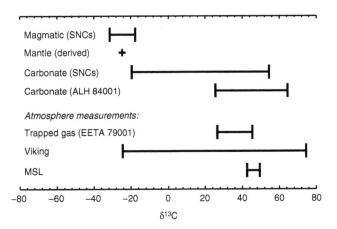

Figure 1 | Summary of $\delta^{13}C$ measurements of Mars. The $\delta^{13}C$ value of the magmatic component of SNC meteorites are used to derive the $\delta^{13}C$ value of the Martian mantle[31]. The carbonates in SNC meteorites have highly variable $\delta^{13}C$ values and the carbonates in ALH 84001 (formed ~ 3.9 Ga before the present) are generally enriched in ^{13}C than the Martian mantle[11]. The rest of the measurements are for modern Mars' atmosphere[14,51,52].

Carbonates formed in the Noachian and Hesperian era. Carbonate-bearing rocks have been discovered in various Noachian terrains by orbital remote sensing[10] and in one rock formation at Gusev Crater in situ[21]. The largest contiguous exposure of carbonate-bearing rocks in Nili Fossae covers 15,000 km², is a few tens of metres thick, and may host up to 12 mbar of CO_2 (refs 22,23). Deep crustal carbonate rocks may also exist, exposed in several impact craters[24]. However, carbonates are not widespread on Mars and are rare compared with other secondary minerals like hydrated silicates and sulfates[25]. There is no inherent difference in the detectability between phyllosilicate and carbonate from an infrared spectroscopy methodological perspective, and the planet Mars has been globally sampled. Thus, the difference is most simply explained as a real difference in abundance. On the basis of this fact, an upper bound of the amount of carbonates is 5 wt% in the volume of crust interrogated by remote sensing, constrained by an upper bound of carbonate non-detectability from infrared absorption features. Assuming 500 m depth, this upper bound corresponds to an equivalent atmospheric pressure of 1.4 bar. Similarly, 1 wt% of carbonates in the crust, more plausibly non-detectable in remote sensing and rover-based analyses, would correspond to 0.3 bar of CO_2.

Driven by these three constraints and together with a newly identified mechanism (photodissociation of CO) that efficiently enriches the heavy carbon isotope, we find a group of plausible atmospheric evolution solutions that can indeed satisfy the current atmospheric pressure and isotopic signatures, and the amount of carbonate deposition, invoking no missing reservoirs or loss processes. We therefore derive new quantitative constraints on the atmospheric pressure of Mars through time, extending into the Noachian, ~ 3.8 Gyr before the present. The atmospheric $\delta^{13}C$ data and the carbonate content in rock and soil indicate an early atmosphere with a surface pressure < 1 bar. Only scenarios with large amounts of carbonate deposition in open-water systems permit higher values up to 1.8 bar.

Results

Isotope fractionation in CO photodissociation. Photodissociation of CO is the most important photochemical source of escaping carbon atoms from Mars, responsible for ~ 90%

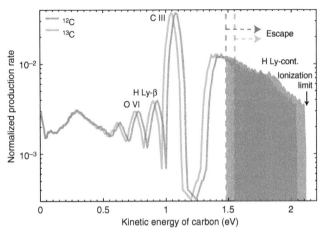

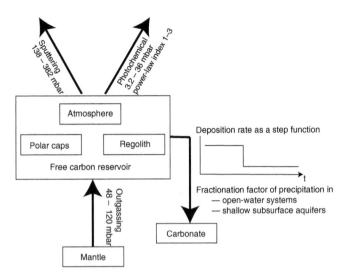

Figure 2 | Energy distribution of carbon atoms produced by photodissociation of CO. The critical energy for each isotope to escape is shown in dashed lines for comparison. The energy distribution is calculated using the current solar minimum spectrum for an exobase at 200 km and the measured CO photodissociation cross-section[26]. The grey and red areas indicate the fraction of ^{12}C and ^{13}C that escape, respectively, which is 0.40 for ^{12}C and 0.24 for ^{13}C. The fractionation ratio ^{12}C/^{13}C via CO photodissociation is thus 0.6. Using early Sun proxies[47] or assuming higher exobases gives quantitatively similar results.

photochemical loss[9]. Its fractionation factor, however, has never been evaluated. Here we show that CO photodissociation on Mars has a fractionation factor of 0.6 and is a highly efficient mechanism to enrich ^{13}C of the atmosphere.

In a CO photodissociation, energy from the incident photon, in excess of the bond dissociation energy, is imparted to carbon and oxygen atoms as kinetic energy. We use the solar spectrum and the cross-section of CO photodissociation as a function of wavelength to calculate the kinetic energy distribution of carbon isotopes (Fig. 2). The significant fractionation effect of CO photodissociation is mainly due to two effects: first, conservation of momentum determines that ^{13}C takes a lesser fraction of the excess energy than ^{12}C in each photodissociation event; and second, ^{13}C requires more energy to escape from the gravity of Mars. The excess energy needs to be >2.6–2.9 eV to produce escaping carbon atoms (that is, the escape threshold energy is 1.5 eV for ^{12}C and 1.6 eV for ^{13}C, and the corresponding energy for non-escaping ^{16}O by conservation of momentum is 1.1–1.3 eV). Because the bond dissociation energy of CO is 11.2 eV, escaping carbon atoms can only be produced by photons more energetic than 13.8 eV, that is, the solar Lyman continuum. In this regime, the cross-section of CO photodissociation does not have strong lines[26]. Furthermore, the branching ratios of CO photodissociation do not affect the fractionation factor, because only the channel that produces ground-state atoms leads to escape of any carbon atoms. If one of the dissociation products is in its excited state (for example, C(^{1}D) or O(^{1}D)), the produced carbon atom will not have enough kinetic energy to escape for any photon less energetic than the ionization threshold (83.5 nm). Thus, the fractionation factor of 0.6 is not sensitive to the evolution of the solar extreme ultraviolet (EUV) spectrum.

Martian atmosphere evolution scenarios. With the newly calculated fractionation factor for CO photodissociation, we construct a model to trace the history of δ^{13}C of a free-carbon reservoir with carbonate deposition and atmospheric escape as the two sinks and magmatic activity as the sole source (Fig. 3).

Figure 3 | A box model for long-term exchanges between the carbon reservoirs on Mars. The model traces a single free-carbon reservoir that includes the atmosphere, CO_2 in polar ice, and adsorbed CO_2 in the regolith, with magmatic outgassing as the source, and carbonate deposition and atmospheric escape as the sinks. The escape mechanisms include pick-up ion sputtering and photochemical escape.

The free-carbon reservoir includes all reservoirs exchangeable on short timescales, that is, the atmosphere, CO_2 in polar ice and adsorbed CO_2 in the regolith. The carbon history of Mars has been extensively studied with reservoir models[3–7,27–30], but none of the previous models have included photochemical escape as a major mechanism that enriches atmospheric ^{13}C.

We model the evolution of this reservoir starting from 3.8 Ga before the present, that is, the mid/late Noachian after any impact-enhanced loss during the LHB, beginning with an atmospheric δ^{13}C value equal to that of mantle-degassed CO_2 derived from the magmatic component of the SNC meteorites (shergottites, nakhlites, chassignites)[31] (other starting values produce similar results; see the next section). The modelled initial reservoir size is calculated from the sum of the current reservoir size, the total removal by atmospheric escape and the total amount of carbonate formation, minus the total outgassed. For outgassing, we adopt the volcanic emplacement rates from thermal evolution models and photogeological estimates, which are in agreement (48–120 mbar)[32,33]. For sputtering, we adopt the three-dimensional (3D) Monte Carlo simulations[8] as the standard escape rates, and the total sputtered is 138–382 mbar, depending on the age dependency of the solar flux. We adopt the photochemical escape rate calculated for the present-day solar flux[9], and scale up the rate for earlier, more intense solar Lyman continuum with a power law:

$$F_{\mathrm{pr}} = 7.9 \times 10^{23} F_{\mathrm{Lyc}}^{a} \qquad (2)$$

where F_{pr} is the photochemical escape flux in particles per s, a is the power-law index and F_{Lyc} is the solar Lyman continuum flux in units of the current solar Lyman continuum flux. The power-law index is a free parameter in the model, because existing calculations of the photochemical escape rate only study current solar conditions[9,34]. It is however reasonable to explore a power-law index ranging between 1 and 3, because an increasing solar EUV flux would lead to increasing mixing ratios of CO, CO$^+$ and electrons in the thermosphere, and these multiple factors could contribute to increasing the photochemical escape rate of carbon. The total photochemical loss for this range is 3.2–36 mbar.

The rate of carbonate formation is simply assumed to be a step function, characterized by an early carbonate formation rate, a late carbonate formation rate and a time of transition. Reality would, of course, be a more gradual transition. For the effects on the carbon isotopic ratio of the atmosphere, two carbonate formation scenarios are considered. One scenario is precipitation in open-water systems (for example, lake and ponds) that have good isotopic communication with the atmosphere. The carbonate formed in this way is $\sim 10‰$ enriched than the atmosphere[15]. The other scenario is precipitation in shallow subsurface aquifers that are semi-isolated and have poor isotopic communication with the atmosphere, that is, there is no influx of gas to replace carbonate precipitated in rock pores. Carbonate is formed by evaporation of water originally derived from the surface, and the water can be enriched by up to $\sim 50‰$ after 99% of the original volume evaporated. The carbonate formed in this way is thus up to $\sim 60‰$ enriched relative to the atmosphere. The shallow subsurface aquifer scenario has been suggested to explain the high $\delta^{13}C$ values of the carbonates in the Martian meteorite ALH 84001 (ref. 35). Some carbonate formation may also proceed in subsurface, closed aquifers from carbon-bearing gases sourced from hydrothermal fluids, but these would not influence the atmospheric reservoir's evolution.

We undertake a million-model approach to quantify the relationship between the amount of carbonate formation and the escape rate, and derive constraints on the early surface pressure. We explore the power-law index of the photochemical escape rate from 1 to 3, the time of transition from high carbonate formation to low carbonate formation ranging from 3.0 to 3.5 Ga, the amount of early carbonate formation from 0.001 to 10 bar, the amount of late carbonate formation from 0.01 to 7 mbar and the amount of early volcanic outgassing from 48 to 120 mbar. We also consider the uncertainties in how the solar EUV flux varies with age, which affect the total removal of sputtering and photochemical processes (see Methods). We performed ~ 50 million simulations using combinations of parameters for both carbonate formation scenarios, and show the combinations of parameters that produce the $\delta^{13}C$ value in the 1-σ range measured by MSL in Fig. 4. The range of early surface pressure—including the atmosphere, the absorbed carbon in the regolith and the polar caps—is also shown in Fig. 4.

Most scenarios permitted by the measurements of the current atmospheric $\delta^{13}C$ and the surface carbonate content have an early surface pressure < 1 bar. Owing to the ^{13}C enrichment effect, even a small amount of atmospheric escape via CO photo-dissociation can drive the atmospheric $\delta^{13}C$ value to the present-day value measured by MSL. Therefore, the isotopic data themselves do not require massive atmospheric loss, and existing known escape mechanisms are fully consistent with all evidence from measured isotopic values and carbonate abundance. In fact, the enrichment from 3.8 billion years ago to present is so significant that it must be compensated by Noachian/Hesperian carbonate deposition, because carbonate formation and out-gassing during the Amazonian are low (see the next section for a discussion on Amazonian volcanic outgassing). Figure 4 quan-tifies this compensation: a higher photochemical escape flux implies a greater amount of early carbonate deposition; but if more carbonates precipitated in shallow subsurface aquifers that were locally enriched, a lesser volume would be required. If all carbonate precipitated in highly enriched shallow subsurface aquifers as ALH 84001, the upper bound of carbonate formation is 0.5 bar, yielding an upper bound of the surface pressure of 0.9 bar. A surface pressure > 1 bar is only permitted if the power-law index is > 2 and most carbonates formed in open-water systems. That would also require more carbonates, not yet detected by rovers and orbiters. The upper bound on the amount

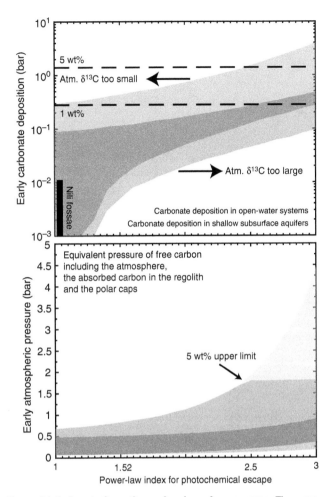

Figure 4 | Carbonate formation and early surface pressure. These are constrained by the current atmospheric $\delta^{13}C$ value measured by the MSL[14]. In both panels the red area shows the permitted range when carbonate deposition occurred in shallow subsurface aquifers and the blue area shows the permitted range when carbonate deposition occurred in open-lake systems. The amount of early carbonate formation must be commensurate with the amount of photochemical escape to produce the measured $\delta^{13}C$ of the current Mars' atmosphere. A lesser amount of carbonate formation is required if the fractionation factor is greater. The early surface pressure is constrained by both the current atmospheric (Atm.) $\delta^{13}C$ value and the upper limit of 1.4 bar (5 wt%) for the early carbonate formation, shown in dark blue versus light blue colour in the bottom panel.

of carbonates allowed by the geologic record (5 wt% everywhere globally in the top 500 m, or 1.4 bars of CO_2) thus determines an overall maximum early atmospheric pressure of 1.8 bar. The two carbonate formation scenarios examined are endmembers, and any solution between the two is viable and results in initial atmosphere values between the two cases.

Figure 5 shows four standard scenarios that lead to a present-day $\delta^{13}C$ value consistent with the MSL measurement. The scenarios are chosen for a power-law index of 2 for the photochemical escape rate. Calculations of the photochemical escape rate for present-day low solar activity and high solar activity indicate a power-law index of 2–2.4 (refs 9,34), although the range of the EUV flux in these calculation (a factor of 2) is smaller than the range from the present day to 3.8 Ga. Two observations can be made from the evolutionary tracks of these scenarios. First, photochemical processes are the main processes that enrich ^{13}C, although sputtering is the main atmospheric escape process. The amount removed by sputtering is ~ 30 times

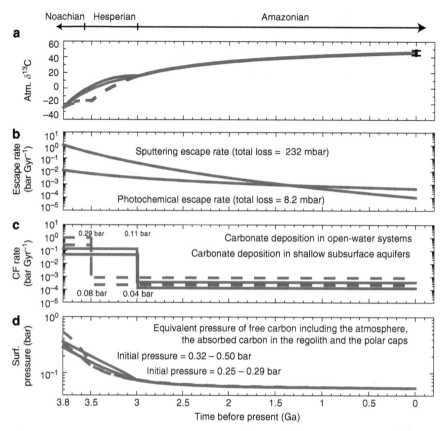

Figure 5 | Standard scenarios of carbon evolution on Mars since the LHB at 3.8 Ga that arrive at present-day $\delta^{13}C$ values. (a) The evolution of the atmospheric (Atm.) $\delta^{13}C$ value, in comparison with the MSL measurement shown by the error bar. **(b,c)** The evolution of the escape rate and the carbonate formation rate, respectively. All scenarios have the same escape rate, corresponding to a power-law index of 2 for the photochemical escape rate. The solid lines are the scenarios where carbonate deposition persisted through the Hesperian Era and the broken lines are the scenarios where carbonate deposition only occurred during the Noachian Era. The blue lines are the scenarios where carbonate deposition occurred in shallow subsurface aquifers and the red lines are the scenarios where carbonate deposition occurred in open-water systems. For each scenario, the amount of early carbonate formation (CF) is determined by the $\delta^{13}C$ value. **(d)** The evolution of the surface (Surf.) pressure. The definition of the age boundaries is taken from a recent crater-density study[53]. The escape rate in the second panel can be converted to atom per s by $1\,\mathrm{bar\,Gyr^{-1}} = 1.7 \times 10^{27}$ atom per s.

Table 1 | Jacobian values for how the amount of sputtering escape (M_{SP}), photochemical escape (M_{PH}), carbonate deposition (M_{CD}) and volcanic outgassing (M_{VO}) would affect the present $\delta^{13}C$.

Early carb. (bar)	Late carb. (bar)	Transition time (Ga)	Index	$\partial\delta^{13}C/\partial\ln(M_{SP})$	$\partial\delta^{13}C/\partial\ln(M_{PH})$	$\partial\delta^{13}C/\partial\ln(M_{CD})$	$\partial\delta^{13}C/\partial\ln(M_{VO})$
Carbonate deposition in open-water systems							
0.29	<0.007	3.5	2.0	74.0	78.9	− 46.3	− 0.34
0.11	<0.007	3.0	2.0	66.3	65.1	− 54.2	− 0.46
Carbonate deposition in shallow subsurface aquifers							
0.08	<0.007	3.5	2.0	89.9	82.4	− 62.6	− 0.57
0.04	<0.007	3.0	2.0	75.3	77.5	− 63.3	− 0.92

We determine the Jacobian values of each parameter by calculating the $\delta^{13}C$ for a variation of 10% of each parameter from the standard scenarios shown in Fig. 5. The columns 'Early carb' and 'Late carb' indicate the amount of carbonate deposition in the Noachian and Hesperian, and that in the Amazonian, respectively.

greater than the amount removed by photochemical processes, but their effects on the carbon isotopic ratio are comparable (Table 1). Second, if carbonate formation persisted through the Hesperian period, the required total amount of carbonates would be less than if carbonate formation only occurred during the Noachian. This is because a unit mass of carbonate formation has a greater impact on the final $\delta^{13}C$ value if it occurs later in the history. With the uncertainties in the time of transition and in the history of the solar EUV flux, the amount of CO_2 deposited as carbonate would be 20 mbar–0.7 bar, corresponding to 2–100

deposits of the size of Nili Fossae, or up to 3 wt%, if distributed globally. The early surface pressure is constrained to be 0.1–0.5 bar for carbonates formed in shallow subsurface aquifers, and the upper limit can be extended up to 1 bar for carbonates formed in open-water systems (Fig. 4).

Finally our results show that carbonate formation from the late Noachian to the Hesperian period is not required when the power-law index is <1.5 and the amount of sputtering is at the lower end of the reasonable range. These are the most 'carbonate conservative' scenarios fully consistent with the isotopic data, which do not require carbonate deposits beyond Nili Fossae and imply an early surface pressure of <0.3 bar.

Sensitivity to the starting $\delta^{13}C$ and the outgassing history. To understand the effects of the starting atmospheric $\delta^{13}C$ value and the volcanic outgassing rates to our results, we performed additional sets of simulations that assume the $\delta^{13}C$ value at 3.8 Ga before present to be −35 and −15‰, and simulations that assume higher volcanic outgassing rates or longer volcanic outgassing period than the standard scenarios. The results of these simulations are shown in Fig. 6.

If the starting atmosphere has a lower value for $\delta^{13}C$, fewer carbonate rocks would be required for a fixed photochemical escape rate. For example, for a power-law index of 2, the minimum amount of carbonate formation is 0.05 bar if the starting $\delta^{13}C$ is −25‰, 0.02 bar if the starting $\delta^{13}C$ is −35‰ and 0.09 bar if the starting $\delta^{13}C$ is −15‰ (the left panel of Fig. 6). The fractional variation of the required amount of carbonate deposition is significant when the carbonate deposition amount is small. When the amount of carbonate formation is >0.1 bar, we find the sensitivity to the initial $\delta^{13}C$ becomes much less significant. In general, the uncertainty in the amount of carbonate formation introduced by the starting $\delta^{13}C$ value is <0.1 bar (Fig. 6) and so is the uncertainty in the estimate of the early surface pressure.

We adopt the model by Grott et al.[32] for the outgassing rate history. Two endmember scenarios for outgassing are suggested, one is termed 'global melt' scenario and the other is termed 'mantle plume' scenario. The difference between the 'global melt' scenario and the 'mantle plume' scenario is that the latter has the volcanic outgassing flux more evenly distributed throughout the Hesperian period and extended into the Amazonian period (Fig. 7). For the standard models, we assume the 'global melt' scenario for the oxygen fugacity one order of magnitude higher than the iron-wustite buffer (IW + 1) and an efficiency of $\eta = 0.4$ (for a total outgassing amount of 48 mbar). Increasing the efficiency to $\eta = 1$ (for a total outgassing amount of 120 mbar) results in quite minimal changes (Fig. 6).

Without changing the total outgassing rate, but more evenly distributing it over the Hesperian (that is, the 'mantle plume' scenario) would cause a decrease in the required amount of carbonate formation for a power-law index of ∼1.5 (the right panel of Fig. 6). Prolonging the volcanic outgassing period

decreases the minimum amount of required carbonate formation, because both processes lower the $\delta^{13}C$ value, and because outgassing during the late Hesperian has a greater impact on the final $\delta^{13}C$ value than that during the early Hesperian. To summarize, either increasing the total outgassing rate or prolonging the outgassing period lead to minor changes to our standard models, and our results appear to be relatively insensitive to the volcanic outgassing rates varied in a wide range.

For completeness, we test the evolution scenarios by both increasing the total outgassing rate and prolonging the outgassing activity. Specifically, we assume the mantle plume scenario of Grott for outgassing, at an oxygen fugacity of IW + 1 and a degassing efficiency of 0.4. The total outgassing amount since 3.8 Ga would then be 420 mbar, in which 350 mbar would be outgassed between 3.8 and 3.0 Ga and 70 mbar would be outgassed between 3.0 Ga and present. This is compared with the standard models in which 48 mbar would be outgassed between 3.8 and 3.0 Ga, and essentially 0 would be outgassed after 3.0 Ga.

For this kind of volcanic outgassing history, the planet must have started to build-up the atmosphere from close to zero pressure at ∼3 Ga before the present, to arrive at the appropriate present-day size of free carbon (that is, 54 mbar). This is a result of simple mass budget balance owing to the fairly significant volcanic outgassing source after 3.0 Ga and insignificant mass loss due to non-thermal escape. If the planet had an atmosphere at ∼3 Ga, its current atmosphere would be more massive than 7 mbar.

To fit the current $\delta^{13}C$ value, the solution would be substantially different from our standard scenarios, in that the final $\delta^{13}C$ value is no longer sensitive to any evolutionary events before 3.0 Ga, including the early carbonate formation rate, and that the photochemical escape rate becomes the sole factor that controls atmospheric evolution since 3.0 Ga. In particular, we find that the power-law index would have to be >3.7 to provide enough fractionation during the Amazonian and lead to the present-day $\delta^{13}C$ value consistent with the MSL measurement. This solution cannot represent the evolution of planet Mars, because it requires Mars to have no or minimal atmosphere 3.0 Ga before the present, and an extremely large power-law index for the photochemical escape rate, both of which are unlikely. We therefore suggest that the solutions with substantial outgassing during the Amazonian period is unlikely for planet Mars.

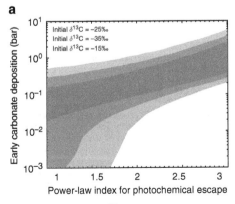 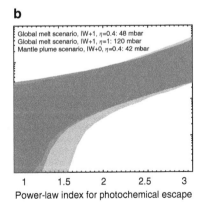

Figure 6 | Sensitivity of the results to the initial $\delta^{13}C$ value and the volcanic outgassing rates. The same as the upper panel of Fig. 4, but assuming $\delta^{13}C$ to be −35 and −15‰ at the beginning of the modelled period (**a**), or assuming different volcanic outgassing models (**b**). The coloured areas show the permitted range of the amount of carbonate deposition, assuming deposition in open-lake systems. Different colours correspond to different sensitivity studies, as labelled on the figure. The sensitivity results for the scenarios assuming deposition in shallow subsurface aquifers are similar. The constraints are relatively insensitive to the initial $\delta^{13}C$ value or the volcanic outgassing rates in reasonable ranges.

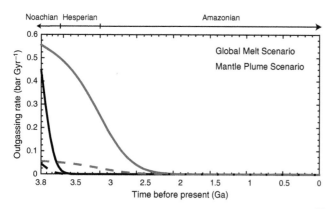

Figure 7 | Volcanic outgassing rates. We adopt the model by Grott *et al.*[32]. Cases of oxygen fugacity of IW + 1 and IW for an efficiency of 0.4 are shown in solid and dashed lines, respectively.

Discussion

In this work we show that a large 'missing' carbon reservoir is unnecessary unless the volcanic outgassing rate was very high. Rather, starting from a few hundred mbar to about 1 bar of CO_2 after the LHB, sputtering and moderate carbonate formation is able to reduce the atmospheric pressure, and subsequent photochemical escape is able to appropriately enrich ^{13}C with known processes. The surface pressure constrained by isotopic modelling is also consistent with an upper bound provided by analyses of impact crater size[36]. The uncertainty in the surface pressure is dominated by the uncertainties in the photochemical and sputtering escape rates, as well as the geological settings of early carbonate formation. Our results thus highlight the crucial importance of a reliable understanding of non-thermal escape processes on Mars. How the photochemical escape rates scale with the solar EUV flux plays a key role, and is yet to be studied. The Mars Atmosphere and Volatile EvolutioN (MAVEN) mission will provide data to calibrate current non-thermal escape models and improve the extrapolation to early Mars.

The evolutionary tracks shown in this paper can be compared with carbonates in SNC meteorites that formed in the past 1 billion years and have varied $\delta^{13}C$ (ref. 10). Their formation from atmosphere-sourced carbon, mixed with magma-sourced carbon, is fully consistent with our evolution scenarios. Importantly, our model provides a methodology for determining even more precisely the past atmospheric escape flux, testable by examination of isotopic ratios in Martian carbonates. Coupled measurements of isotopic ratios and isotopologues of Noachian to Early Amazonian deposits, measured *in situ* or in meteorites and returned samples, would uniquely distinguish the timing of major carbonate formation, its geologic setting, as well as the amount of carbonate deposition that balanced photochemical escape.

Methods

A box model for the atmospheric evolution. We adopt the box model of Kass[37] to trace the evolution of the free-carbon reservoir and its $\delta^{13}C$ value. We make major changes to the model, including: using the latest calculations of sputtering rates; adding photochemical escape as a mechanism of non-thermal escape and calculating the fractionation factor of CO photodissociation; using the latest estimate of the outgassing history of Mars that has taken into account the newly measured solubility of carbon in a more reduced mantle melt; considering the atmosphere, polar caps and the regolith to be one exchangeable reservoir for free carbon, and using the latest measurements of their masses for the current size; and modifying the implementation of the atmospheric collapse according to recent 3D early Mars climate simulations and allowing the model to be directly constrained by the geologic record.

Free-carbon reservoir. The current Martian atmosphere is at or near equilibrium with polar caps and regolith. We define 'free carbon' as the atmosphere, CO_2 ice in polar caps and CO_2 absorbed by the regolith. All free-carbon species are treated as

a single reservoir, assuming they exchange with each other over geological times and they do not significantly fractionate relative to each other in this exchange. Laboratory measurements indicate that the carbon in CO_2 does not fractionate during condensation, an unusual phenomenon probably due to the effect of isotopic substitution on the inter-molecular binding energy of the condensed phase[38]. Recent measurements also show that regolith absorption results in a $\sim 1\permil$ enrichment of ^{13}C in the gaseous phase[39]. Therefore, condensation or absorption of even large amounts of CO_2 does not change $\delta^{13}C$ in the atmosphere.

Detailed characterization of Mars and relevant laboratory studies in recent years have provided good constraints on the size of this free-carbon reservoir. We adopt a current free-carbon reservoir of 54 mbar (7 mbar atmosphere, 2 mbar polar caps[40], 5 mbar subsurface polar deposits[41] and 40 mbar regolith absorption[42]) as a fixed parameter in our baseline models.

The advantage of combining these reservoirs into a single reservoir is that we do not have to explicitly trace the evolutionary history of the surface temperature, which is primarily a function of the surface pressure, the solar luminosity and the orbital obliquity[2]. The surface temperature is the dominant factor that controls the partitioning of CO_2 among the atmosphere, the regolith and the polar caps[3]. This way, we can focus on evaluating the relationship between non-thermal escape that enriches ^{13}C and carbon deposition that depletes ^{13}C.

Outgassing. Outgassing is the primary source of new carbon into the free-carbon reservoir. The carbon in the mantle is released into the atmosphere through volcanic emplacement of mantle material and outgassing from the magma. The outgassing flux can be estimated from the history of volcanic activity, the estimated intrusive emplacements and the carbon content of the magmas. The history of volcanic activity of Mars has been estimated photogeologically by determining the ages of volcanic units of the planet's surface[33] and theoretically by modelling the thermal history of the planet[32]. Photogeological estimates based on Viking data suggest extrusive magma of 67×10^6 km^3 from the late Noachian to present[33], which corresponds to a total outgassing of 0.5–50 mbar for complete degassing of 10–1,000 p.p.m. dissolved CO_2. In addition, for a ratio of intrusive to extrusive magma of 8.5:1, similar to Earth, the total outgassing can range from 2.4 to 240 mbar assuming 40% outgassing efficiency for the intrusive magma. No quantitative photogeological estimates have been published after, but recent observations using high-resolution imagery have suggested Martian volcanism started earlier and the decay in intensity was more rapid than previously thought[43].

The latest models of Mars' thermal evolution history appear to fully cover the uncertainties in the flux and timing of volcanic outgassing[32]. In the 'global melt' scenario of Grott, the planet cools fast, and most volcanic outgassing concentrates in the pre-Noachian and Noachian periods. For this scenario, the total outgassing rate during our modelled period (from 3.8 Ga to present) would be 48 mbar at an oxygen fugacity of IW + 1 and an outgassing efficiency of 40%. At IW + 1, $\sim 1,000$ p.p.m. CO_2 can be dissolved in the magma, and this amount scales with the oxygen fugacity exponentially[44]. In the 'mantle plume' scenario of Grott, the planet cools more slowly, and volcanic outgassing persists throughout the Hesperian period and extends into the Amazonian period. For this scenario, the total outgassing rate during our modelled period would be 470 mbar. The photogeological estimates are within the range defined by these two endmember scenarios.

For this study, we adopt the volcanic outgassing flux modelled by Grott, as

$$P_{\text{outgassing}} = 10^{\text{IW}}\eta A[\tanh(t/a)^{\alpha}]^{1/\beta} \qquad (3)$$

where $P_{\text{outgassing}}$ is the cumulative outgassed partial pressure at t Myr after formation and η is the outgassing efficiency. For a 'global melt' scenario the parameters are $A = 252.45$ mbar, $a = 719.89$ Myr, $\alpha = 3.6206$ and $\beta = 6.7809$, and for a 'mantle plume' scenario the parameters are $A = 224.39$ mbar, $a = 1505.1$ Myr, $\alpha = 2.7606$ and $\beta = 3.3600$.

We assume volcanic outgassing has been mainly in the form of CO_2, rather than more reduced forms of carbon, during the modelled period. The speciation of volcanic outgassing on Mars is mainly controlled by the oxygen fugacity of the source magma. It has been recently shown, experimentally, that outgassing would be in CO_2 when the oxygen fugacity is >IW − 0.55 (ref. 45). Petrologically primitive SNC meteorites show that the Martian mantle has an oxygen fugacity between IW and IW + 1 (ref. 44). For this range, volcanic outgassing would be in CO_2. However, the early Martian mantle may have been more reduced than the sources of SNC meteorites, since the meteorite ALH 84001 formed 3.9 Ga before the present shows an oxygen fugacity as low as IW − 1 (ref. 44). But, if this has been the case, the total outgassing rate would be insignificant (that is, < 12 mbar for an oxygen fugacity of IW − 1 for the most optimistic estimate of volcanic emplacement). We therefore suggest that the possibility of an early reduced mantle and consequently CO or even CH_4 outgassing during the modelled period has little impact on our model or results.

Pick-up ion sputtering and photochemical escape. During the model period, hydrodynamic escape has ceased, impact delivery and removal of volatiles are limited and the dominant atmospheric escape processes are pick-up ion sputtering and photochemical processes.

Pick-up ion sputtering is a process by which oxygen ions in Mars' upper atmosphere are picked up by the solar wind magnetic field and then collide with the molecules and atoms in the upper atmosphere sputtering them away[6,8]. This process may have been quite efficient in removing carbon from early Mars when the solar wind was much stronger than present. We expect sputtering occurred during the entire modelled period. Detection and mapping of crustal magnetic anomalies over the Mars surface implies the Martian magnetic field should have ceased before the formation of Hellas or the rise of Tharsis, because the interiors of these basins or most volcanic edifices lack magnetic remanence[46]. Therefore, the magnetic field does not affect our study. We adopt the 3D Monte Carlo simulations[8] as the standard values of the sputtering rate, fitted to this functional form

$$F_{sp} = \exp\left(-0.462\ln(F_{EUV})^2 + 5.086\ln(F_{EUV}) + 53.49\right) \qquad (4)$$

where F_{sp} is the sputtering escape flux in particles per s and F_{EUV} is the solar EUV flux in units of the current solar EUV flux.

The evolution of the solar EUV flux has been derived by observing young solar-like stars[47]. We adopt $F_{EUV} \propto t^{-1.23 \pm 0.1}$, where t is the age. The uncertainty of the index corresponds to 20% uncertainty in the flux at 3.8 Ga, determined from the measurement errors of the observations that are used to derive this flux index[47,48]. The total amount of atmosphere sputtered for the range of $F_{EUV} \propto t^{-1.23 \pm 0.1}$ is 232^{+150}_{-94} mbar, or 138–382 mbar, calculated using equation (4).

The carbon sputtered will be between 15 and 40‰ (depending on the epoch) lighter than the source atmosphere. The sputtering process itself does not fractionate the atmosphere, because the carbon atoms that actually escape all have sufficient energy to escape regardless of the isotope. But the sputtering occurs at altitudes well above the homopause[6,8], where each species takes on its mass-dependent scale height. Thus, the atmosphere being sputtered is lighter than the total atmosphere and the net effect of the sputtering process is to enrich the atmosphere.

The fractionation factor α of the sputtering process can be calculated by

$$\alpha = \exp\left(\frac{-g\,\Delta m\,\Delta z}{kT}\right), \qquad (5)$$

where g is Mars' gravitational acceleration, Δz is the distance from the homopause to the source altitude of escaping carbon atoms, Δm is the mass difference between the isotopes, k is the Boltzmann constant and T is the mean temperature of the upper atmosphere. We adopt the source altitude from Kass[6] and calculate the fractionation factors to be 0.96–0.98. A later calculation appears to suggest a higher source altitude, by tens of kilometres[8]. The impact of this difference to the sputtering fractionation factor is minimal, as the fractionation factor of sputtering is close to unity in all cases, that is, sputtering is inefficient in enriching ^{13}C in the atmosphere compared with photochemical processes. However, it is efficient in removing atmospheric mass.

The main photochemical processes that generate escaping carbon atoms are photodissociation of CO and dissociative recombination of CO^+ and CO_2^+, and the escape rates are calculated for the present-day solar EUV conditions[9]. The escaping carbons are mainly produced by Lyman continuum photons (Fig. 2), and the appropriate way to scale up the rates with the solar Lyman continuum flux (F_{Lyc}) is unclear, so this is a free parameter in the model. We adopt $F_{Lyc} \propto t^{-0.86 \pm 0.1}$ based on observations of young solar-like stars[47,48]. Again, the uncertainty of the index corresponds to 20% uncertainty in the flux at 3.8 Ga, determined from the measurement errors of the observations used to derive this index[47,48]. The total amount of photochemical escape for the range of $F_{EUV} \propto t^{-0.86 \pm 0.1}$ is $3.4^{+0.4}_{-0.2}$ mbar for a power-law index of 1, $8.2^{+2.2}_{-1.6}$ mbar for a power-law index of 2 and 24^{+12}_{-8} mbar for a power-law index of 3, calculated using equation (2).

Photochemical processes fractionate the atmosphere in a much more efficient way than sputtering. This is because sputtering, when it occurs, is usually too energetic to have any isotopic effects, whereas the fate of photochemical products is sensitive to their masses. The fractionation factor of dissociative recombination for ^{13}C versus ^{12}C is ~ 0.8 (ref. 49), and that of CO photodissociation is ~ 0.6 (main text). Both depend weakly on the exobase location and fractionate more than sputtering (> 0.95). Since about 90% of the photochemical escape flux is via CO photodissociation[9], the composite fractionation ratio is $0.8 \times 10\% + 0.6 \times 90\% = 0.62$.

Carbonate deposition. Since we do not explicitly trace the evolution of the atmospheric temperature and pressure, carbonate deposition has to be approximated. We assume the carbon deposition rate to be a step function, characterized by a constant relatively high early carbonate formation rate, a constant relatively low late carbonate formation rate and a transition time. We allow the transition time to vary between 3.5 to 3.0 Ga before the present, covering the scenarios where carbonate deposition persisted through the Hesperian Era, and the scenarios where carbonate deposition only occurred during the Noachian Era. Assuming the step function minimizes the number of free parameters and allows straightforward comparison with geologic evidence. In reality the carbonate formation rate would have been variable on small timescales due to transient existence of liquid water on the surface or in the subsurface. The cumulative amount of carbonate deposition during the Noachian and Hesperian, that during the Amazonian, and the transition

time are three tuning parameters, and the first two are independently constrained by geologic record.

The standard temperature-dependent formulation of the carbon fractionation in carbonate formation indicates that carbonate precipitates in open-water systems at 0 °C is ~ 13‰ heavier than the source atmosphere. A temperature of 0 °C was selected because most of the carbonate rocks are expected to have formed in very cold water (since it is hard to warm the atmosphere even that much). However, the carbonates in the Martian meteorite ALH 84001 formed 3.9 Ga before the present and have high ^{13}C values[10]. One of the interpretations of the stable isotopic features of the carbonates of ALH 84001 is that they formed in an evaporative environment localized enriched in heavy carbon and oxygen, that is, shallow subsurface aquifers[35]. Therefore, we assume $\alpha_{carbonate} = \alpha_{precipitation} + \alpha_{evaporation}$, where $\alpha_{precipitation} = 13$‰ is the fractionation at precipitation and $\alpha_{evaporation}$ varies from 0 to 50‰, which corresponds evaporation of 0–99.2% of water before precipitation as calculated by the Rayleigh distillation formula. The upper bound of $\alpha_{evaporation}$ corresponds to ALH 84001. This treatment is coarse because the acidity of the residual solution will change as evaporation progresses[35]. Another possible way to increase $\alpha_{carbonate}$ above $\alpha_{precipitation}$ is that the source CO_2 was produced photochemically from a methane-rich background atmosphere[50]. For our purpose, it is not necessary to distinguish these possible causes for the large fractionation factor of carbonate formation suggested by ALH 84001.

References

1. Fassett, C. I. & Head, J. W. Sequence and timing of conditions on early Mars. *Icarus* **211**, 1204–1214 (2011).
2. Forget, F. *et al.* 3D modelling of the early martian climate under a denser CO_2 atmosphere: temperatures and CO_2 ice clouds. *Icarus* **222**, 81–99 (2013).
3. Haberle, R. M., Tyler, D., McKay, C. P. & Davis, W. L. A model for the evolution of CO_2 on Mars. *Icarus* **109**, 102–120 (1994).
4. Manning, C. V., McKay, C. P. & Zahnle, K. J. Thick and thin models of the evolution of carbon dioxide on Mars. *Icarus* **180**, 38–59 (2006).
5. Jakosky, B. M., Pepin, R. O., Johnson, R. E. & Fox, J. Mars atmospheric loss and isotopic fractionation by solar-wind-induced sputtering and photochemical escape. *Icarus* **111**, 271–288 (1994).
6. Kass, D. M. *Change in the Martian Atmosphere.* Ph.D. Thesis (California Institute of Technology, 1999).
7. Kahn, R. The evolution of CO_2 on Mars. *Icarus* **62**, 175–190 (1985).
8. Leblanc, F. & Johnson, R. E. Role of molecular species in pickup ion sputtering of the Martian atmosphere. *J. Geophys. Res. Planets* **107**, 5010 (2002).
9. Groeller, H., Lichtenegger, H., Lammer, H. & Shematovich, V. Hot oxygen and carbon escape from the Martian atmosphere. *Planet. Space Sci.* **98**, 93–105 (2014).
10. Niles, P. *et al.* Geochemistry of carbonates on Mars: implications for climate history and nature of aqueous environments. *Space Sci. Rev.* **174**, 301–328 (2013).
11. Niles, P. B., Boynton, W. V., Hoffman, J. H., Ming, D. W. & Hamara, D. Stable isotope measurements of Martian atmospheric CO_2 at the Phoenix landing site. *Science* **329**, 1334–1337 (2010).
12. Niles, P. B. *et al.* in *Lunar and Planetary Science Conference*, vol. 45, 2573 (The Woodlands, TX, 2014).
13. Krasnopolsky, V. A., Maillard, J. P., Owen, T. C., Toth, R. A. & Smith, M. D. Oxygen and carbon isotope ratios in the Martian atmosphere. *Icarus* **192**, 396–403 (2007).
14. Webster, C. R. *et al.* Isotope ratios of H, C, and O in CO_2 and H_2O of the Martian atmosphere. *Science* **341**, 260–263 (2013).
15. Faure, G. *Principles and Applications of Inorganic Geochemistry* (Macmillan Coll Div, 1991).
16. Bandfield, J. L., Glotch, T. D. & Christensen, P. R. Spectroscopic identification of carbonate minerals in the Martian dust. *Science* **301**, 1084–1087 (2003).
17. Sutter, B. *et al.* The detection of carbonate in the martian soil at the Phoenix Landing site: a laboratory investigation and comparison with the Thermal and Evolved Gas Analyzer (TEGA) data. *Icarus* **218**, 290–296 (2012).
18. Leshin, L. A. *et al.* Volatile, isotope, and organic analysis of Martian Fines with the Mars Curiosity Rover. *Science* **341**, 3 (2013).
19. Zuber, M. T. *et al.* Density of Mars' south polar layered deposits. *Science* **317**, 1718–1719 (2007).
20. Watters, T. R. *et al.* Radar sounding of the Medusae Fossae Formation Mars: equatorial ice or dry, low-density deposits? *Science* **318**, 1125–1128 (2007).
21. Morris, R. V. *et al.* Identification of carbonate-rich outcrops on Mars by the Spirit rover. *Science* **329**, 421–424 (2010).
22. Ehlmann, B. L. *et al.* Orbital identification of carbonate-bearing rocks on Mars. *Science* **322**, 1828–1832 (2008).
23. Edwards, C. S. & Ehlmann, B. L. Carbon sequestration on mars. *Geology* **43**, 863–866 (2015).
24. Michalski, J. R. & Niles, P. B. Deep crustal carbonate rocks exposed by meteor impact on Mars. *Nat. Geosci.* **3**, 751–755 (2010).
25. Ehlmann, B. L. & Edwards, C. S. Mineralogy of the Martian surface. *Annu. Rev. Earth Planet. Sci.* **42**, 291–315 (2014).

26. Chan, W., Cooper, G. & Brion, C. Absolute optical oscillator strengths for discrete and continuum photoabsorption of carbon monoxide (7200 ev) and transition moments for the x 1 + a 1 system. *Chem. Phys.* **170,** 123–138 (1993).

27. Jakosky, B. M. Mars volatile evolution: evidence from stable isotopes. *Icarus* **94,** 14–31 (1991).

28. Pepin, R. O. Evolution of the Martian atmosphere. *Icarus* **111,** 289–304 (1994).

29. Jakosky, B. & Phillips, R. Mars' volatile and climate history. *Nature* **412,** 237–244 (2001).

30. Gillmann, C., Lognonn, P., Chassefiere, E. & Moreira, M. The present-day atmosphere of Mars: where does it come from? *Earth Planet. Sci. Lett.* **277,** 384–393 (2009).

31. Wright, I., Grady, M. M. & Pillinger, C. Chassigny and the nakhlites: carbon-bearing components and their relationship to Martian environmental conditions. *Geochim. Cosmochim. Acta* **56,** 817–826 (1992).

32. Grott, M., Morschhauser, A., Breuer, D. & Hauber, E. Volcanic outgassing of CO_2 and H_2O on Mars. *Earth Planet. Sci. Lett.* **308,** 391–400 (2011).

33. Greeley, R. & Schneid, B. D. Magma generation on Mars: amounts, rates, and comparisons with Earth, Moon, and Venus. *Science* **254,** 996–998 (1991).

34. Fox, J. L. $Co_2 +$ dissociative recombination: a source of thermal and nonthermal c on mars. *J. Geophys. Res. Space Phys.* **109,** A08306 (2004).

35. Halevy, I., Fischer, W. W. & Eiler, J. M. Carbonates in the Martian meteorite Allan Hills 84001 formed at (18 ± 4 c) in a near-surface aqueous environment. *Proc. Natl Acad. Sci. USA* **108,** 16895–16899 (2011).

36. Kite, E. S., Williams, J.-P., Lucas, A. & Aharonson, O. Low palaeopressure of the Martian atmosphere estimated from the size distribution of ancient craters. *Nat. Geosci.* **7,** 335–339 (2014).

37. Kass, D. M. *Change in the Martian Atmosphere.* PhD thesis, (California Institute of Technology, 1999).

38. Eiler, J. M., Kitchen, N. & Rahn, T. A. Experimental constraints on the stable-isotope systematics of CO_2 ice/vapor systems and relevance to the study of Mars. *Geochim. Cosmochim. Acta* **64,** 733–746 (2000).

39. Rahn, T. & Eiler, J. Experimental constraints on the fractionation of $^{13}C/^{12}C$ and $^{18}O/^{16}O$ ratios due to adsorption of CO_2 on mineral substrates at conditions relevant to the surface of Mars. *Geochim. Cosmochim. Acta* **65,** 839–846 (2001).

40. Smith, D. E. *et al.* Time variations of Mars' gravitational field and seasonal changes in the masses of the polar ice caps. *J. Geophys. Res. Planets* **114,** E05002 (2009).

41. Phillips, R. J. *et al.* Massive CO_2 ice deposits sequestered in the south polar layered deposits of Mars. *Science* **332,** 838–841 (2011).

42. Zent, A. P. & Quinn, R. C. Simultaneous adsorption of CO_2 and H_2O under Mars-like conditions and application to the evolution of the Martian climate. *J. Geophys. Res. Planets* **100,** 5341–5349 (1995).

43. Werner, S. C. The global Martian volcanic evolutionary history. *Icarus* **201,** 44–68 (2009).

44. Hirschmann, M. M. & Withers, A. C. Ventilation of CO_2 from a reduced mantle and consequences for the early Martian greenhouse. *Earth Planet. Sci. Lett.* **270,** 147–155 (2008).

45. Wetzel, D. T., Rutherford, M. J., Jacobsen, S. D., Hauri, E. H. & Saal, A. E. Degassing of reduced carbon from planetary basalts. *Proc. Natl Acad. Sci. USA* **110,** 8010–8013 (2013).

46. Lillis, R. J., Frey, H. V. & Manga, M. Rapid decrease in Martian crustal magnetization in the Noachian era: implications for the dynamo and climate of early Mars. *Geophys. Res. Lett.* **35,** L14203 (2008).

47. Ribas, I., Guinan, E. F., Güdel, M. & Audard, M. Evolution of the solar activity over time and effects on planetary atmospheres. I. High-energy irradiances (1-1700 Å). *Astrophys. J.* **622,** 680–694 (2005).

48. Claire, M. W. *et al.* The evolution of solar flux from 0.1 nm to 160 μm: quantitative estimates for planetary studies. *Astrophys. J.* **757,** 95 (2012).

49. Fox, J. L. & Hac, A. Velocity distributions of C atoms in CO^+ dissociative recombination: implications for photochemical escape of C from Mars. *J. Geophys. Res. Space Phys.* **104,** 24729–24737 (1999).

50. Galimov, E. On the phenomenon of enrichment of Mars in ^{13}C: a suggestion on the reduced initial atmosphere. *Icarus* **147,** 472–476 (2000).

51. Carr, R. H., Grady, M. M., Wright, I. P. & Pillinger, C. T. Martian atmospheric carbon dioxide and weathering products in SNC meteorites. *Nature* **314,** 248–250 (1985).

52. Nier, A. O. & McElroy, M. B. Composition and structure of Mars' upper atmosphere: results from the neutral mass spectrometers on Viking 1 and 2. *J. Geophys. Res.* **82,** 4341–4349 (1977).

53. Werner, S. & Tanaka, K. Redefinition of the crater-density and absolute-age boundaries for the chronostratigraphic system of mars. *Icarus* **215,** 603–607 (2011).

Acknowledgements

Support for this work was provided by NASA through Hubble Fellowship grant #51332 awarded by the Space Telescope Science Institute, which is operated by the Association of Universities for Research in Astronomy, Inc., for NASA, under contract NAS 5-26555. The research was carried out at the Jet Propulsion Laboratory, California Institute of Technology, under a contract with the National Aeronautics and Space Administration.

Author contributions

R.H. modelled the fractionation factor of photochemical escape, developed the million-model approach, simulated the evolution scenarios and wrote the manuscript; D.M.K. provided the framework of the evolution model and modelled the fractionation factor of sputtering; B.L.E. assembled the measurements of carbonate content in rock and soil, and provided geological constraints on scenarios; Y.L.Y. provided the insight into the evolution of stellar radiation and escape rates; all authors interpreted the results and commented on the manuscript.

Additional information

Competing financial interests: The authors declare no competing financial interests.

3

Ubiquity of Kelvin–Helmholtz waves at Earth's magnetopause

Shiva Kavosi[1,*] & Joachim Raeder[1,*]

Magnetic reconnection is believed to be the dominant process by which solar wind plasma enters the magnetosphere. However, for periods of northward interplanetary magnetic field (IMF) reconnection is less likely at the dayside magnetopause, and Kelvin-Helmholtz waves (KHWs) may be important agents for plasma entry and for the excitation of ultra-low-frequency (ULF) waves. The relative importance of KHWs is controversial because no statistical data on their occurrence frequency exist. Here we survey 7 years of *in situ* data from the NASA THEMIS (Time History of Events and Macro scale Interactions during Substorms) mission and find that KHWs occur at the magnetopause ~19% of the time. The rate increases with solar wind speed, Alfven Mach number and number density, but is mostly independent of IMF magnitude. KHWs may thus be more important for plasma transport across the magnetopause than previously thought, and frequently drive magnetospheric ULF waves.

[1] Department of Physics and Space Science Center, University of New Hampshire, 8 College Road, Durham, New Hampshire 03824, USA. * These authors contributed equally to this work. Correspondence and requests for materials should be addressed to J.R. (email: J.Raeder@unh.edu).

The magnetosphere is filled with plasma from the iono-sphere and the solar wind (SW). Although ionosphere plasma can be easily traced by its composition, the entry pathways of SW plasma are much less clear. Early models suggested magnetic reconnection at the dayside magnetopause (MP) and subsequent convection into the tail as the main path[1,2]. However, later research showed that the SW plasma density in the tail often maximizes during times of northward interplanetary magnetic field (IMF), forming the Cold Dense Plasma Sheet. As magnetic reconnection does not occur during northward IMF at the dayside MP, and because of the Cold Dense Plasma Sheet plasma properties, other entry mechanisms must be involved in the entry process. High-latitude magnetic reconnection near the cusps[3,4], impulsive penetration[5], gradient drift[6], particle diffusion[7] and the Kelvin–Helmholtz instability (KHI)[8-10] have been suggested as viable mechanisms operating during northward IMF. Although cusp reconnection and Kelvin–Helmholtz waves (KHWs) have received much attention recently[3,11], the relative importance of all these processes remains unknown[7]. Although KHWs may play an important role as a SW plasma entry mechanism, they are also considered drivers of magnetosphere ultra-low-frequency (ULF) waves[12,13], which in turn strongly affect the radiation belts[14].

KHWs have been studied extensively using *in-situ* data and simulations. Event studies have shown that KHWs occur at times at the MP, and have revealed some of their basic properties[10,11,15,16]. Both magneto-hydrodynamic (MHD) and kinetic simulations, mostly in a simplified two-dimensional geometry, have shown how KHWs can roll up and mix SW plasma with magnetosphere plasma[17,18]. More recently, global MHD simulations of the magnetosphere have shown the development of KHWs[19]. Despite the progress in understanding KHWs properties and their effect on transport, little is known about their occurrence rate. Linear theory[20] suggests that KHWs are most unstable at high-flow shear, for example, high solar wind speed, and when the IMF is nearly northward, for example, nearly parallel to the magnetosphere field. As these conditions occur rarely together, KHWs have often been considered infrequent events.

Because satellite orbital dynamics makes it impossible to monitor the MP over long-time periods, in the past only intermittent observations of the MP were available. However, the THEMIS (Time History of Events and Macro scale Interactions during Substorms) mission, originally designed to study substorms[21], has almost ideal equatorial orbits to study KHWs. With orbit apogees between 12 RE (Earth radii) and 30 RE, the spacecraft frequently cross the MP flanks during the spring and fall seasons, as the orbits undergo precession around the Earth.

We survey the THEMIS data to obtain a database of MP crossings, and identify crossings where KHWs are present. The statistical analysis shows that KHWs occur ~19% of the time regardless of the solar wind conditions. We find that the KHWs occurrence rate increases with solar wind speed, Alfven Mach number and number density, but is mostly independent of IMF magnitude. The occurrence rate increases with IMF cone angle and maximizes at zero IMF clock angle. We find that KHWs also occur at higher rate than expected for southward IMF. We conclude that KHWs may thus be more important for plasma transport across the MP than previously thought, and frequently drive magnetospheric ULF waves.

Results

Occurrence rate and IMF dependence. The duration of MP encounters can last from minutes to hours. To obtain occurrence rates, we divided each encounter into 5-min intervals. Each interval is classified as KHW or not, and tagged with ancillary data, such as time-shifted SW and IMF data. Our database (Supplementary Data 1) consists of ~11,500 5-min samples, covering ~960 h dwell time at the MP. The samples are nearly evenly divided between northward (~500 h) and southward (~460 h) IMF conditions.

We find that about half of the crossings show waves or quasi-periodic variations, but not all of them are KHWs. Figure 1 shows the KHWs occurrence rate as a function of IMF clock angle and IMF cone angle. As a function of clock angle, the occurrence rate is ~35% for near northward IMF, near 20% if the IMF lies in the equatorial plane, and about 10% for southward IMF. The fact that KHWs occur during southward IMF at a significant rate is not expected; because it is generally thought that magnetic reconnection dominates over KHI during such conditions and prevents KHWs growth. The IMF cone angle dependence is as expected from the linear dispersion relation of KHWs[20], which predicts that the instability maximizes when the magnetic field on either side of the shear layer is close to collinear, which occurs for ~90° cone angle. The overall occurrence rate of KHWs is ~19% regardless of solar wind and IMF conditions. This is a substantially higher rate than the linear dispersion relation would suggest.

Solar wind parameter dependence. Figure 2a shows occurrence percentage of KHWs (orange bins) and the corresponding

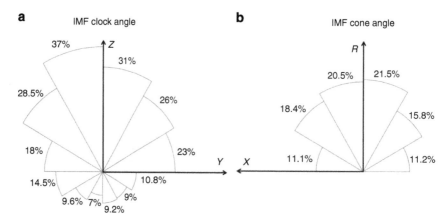

Figure 1 | KHW occurrence rate as a function of IMF clock angle and cone angle. The clock angle is defined as $atan(B_y/B_z)$, and the cone angle is defined as $acos(B_X/B)$. X points towards the Sun, Y points duskward, Z points north, and $R = (Y^2 + Z^2)^{1/2}$. KHW occurrence maximizes for northward IMF, but is still significant during southward IMF (**a**). The IMF is more effective generating KHW when it is oriented perpendicular to the Sun–Earth line (**b**).

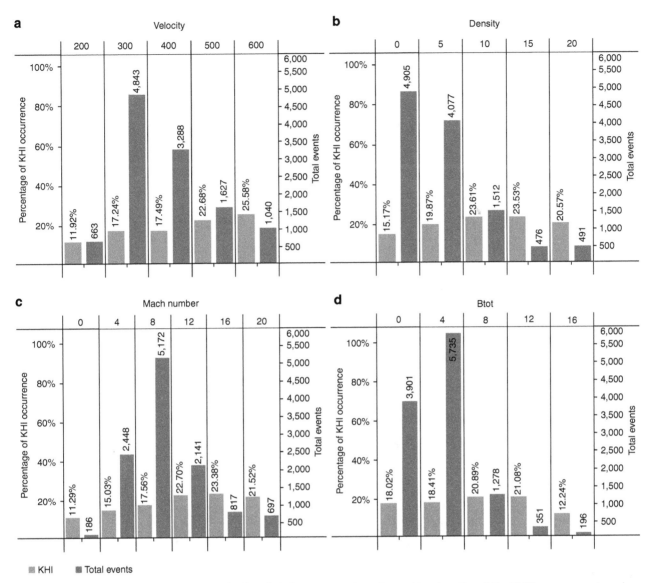

Figure 2 | Occurrence rate of KHW as a function of solar wind plasma parameters. Orange bins show the relative KHW occurrence rate and grey bins show the number of 5-min KHW intervals in that bin. The panels show, respectively, the dependence on the solar wind velocity (**a**), the solar wind density (**b**), the solar wind Mach number (**c**) and the IMF magnitude. The parameter dependence is mostly as expected from the Kelvin-Helmholtz dispersion relation, but the significant occurrence rate at low velocity (<300 km s^{-1}) is not expected.

number of 5-min intervals (grey bins) as a function of solar wind speed. The latter is shown to assess the statistical significance of the data. As expected, the occurrence frequency increases with solar wind speed. However, the occurrence of KHWs at very low solar wind speed is unexpected. There appears to be no low-speed cutoff for KHW; KHW are still observed at 270 km s^{-1} solar wind speed. The KHWs dependence on solar wind density (Fig. 2b) is weak. At low densities, there is a positive correlation, which tapers out for densities that are larger than 10 cm^{-3}. There is also a positive correlation with the solar wind Alfven Mach number (Fig. 2c), which also tapers out at high (>16) Mach numbers. The IMF magnitude (Fig. 2d) appears only to have an effect for unusual high values of more than 16 nT. It is tempting to compare the KHWs dependencies with the dispersion relation for KHWs; however, the solar wind parameters are not the same as the plasma and field parameters on the magnetosheath side of the MP flanks. In particular, the solar wind is slowed down by the bow shock and then re-accelerates along the flanks of the magnetosphere. Therefore, the magnetosheath velocity is generally slower than the solar wind, but it is also possible that the draped

IMF accelerates the magnetosheath plasma to speeds larger than the solar wind speed. However, the trends shown in Fig. 2 are in agreement with linear theory, in particular the increase of the KHW rate with solar wind speed, and the apparent suppression of KHWs for strong IMF, which was predicted by theory[22].

Discussion

Linear MHD theory predicts that KHWs are most unstable when the magnetic field on either side of the shear layer is perpendicular to both the flow direction and the direction of the velocity gradient. Furthermore, the growth rate increases with flow shear. Thus, it has commonly been assumed that KHWs at Earth's MP are restricted to times of nearly northward IMF and high solar wind speed. This would make them rare events with little importance for magnetospheric dynamics. Although the dispersion relation does not distinguish between northward and southward IMF, that is, whether the magnetic field is parallel or anti-parallel across the shear layer, it was commonly assumed that during southward IMF periods magnetic reconnection would

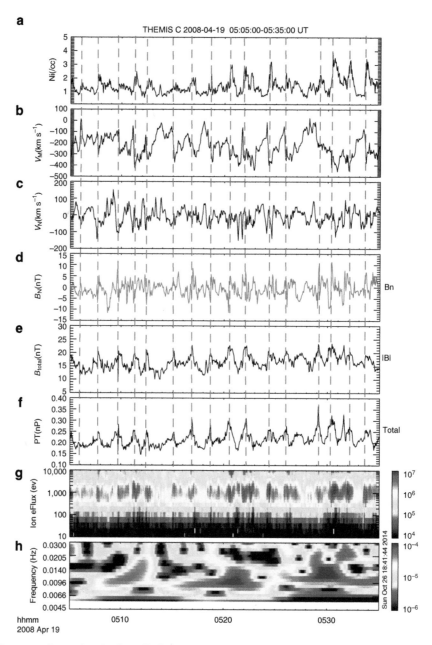

Figure 3 | Kelvin–Helmholtz waves observed at the dawn flank magnetopause by THEMIS C on 19 April 2008. The panels show, from top to bottom: (**a**) the ion number density, (**b**) the M component of ion velocity, (**c**) normal component of the ion velocity vector, (**d**) normal component of magnetic field, (**e**) total magnetic field, (**f**) total (magnetic plus ion) pressure, (**g**) omnidirectional ion energy spectrogram and (**h**) wavelet spectra of the total pressure. The wave period is approximately 100 s.

dominate over KHW generation. However, recent reports have shown that KHWs may also occur during southward IMF conditions[23,24], but these were case studies that give no indication as to whether these were singular events or whether they would occur more common. We find that the southward KHW events in our database are generally shorter (20 min on average) than those observed during northward IMF (45 min on average). The relative short duration of KHW events during southward IMF may explain why only few such events have been reported.

Refined theoretical analysis that has taken into account the finite width of the shear layer and its structure has further narrowed the parameter range under which KHWs should occur[25]. However, the true occurrence rate of KHWs remained uncertain, and many researchers assumed they were rare events. Statistical studies have long been hampered by the lack of suitable data. Although Pc5 waves observed on the ground are often

associated with KHWs, they may also have other causes and thus provide no suitable statistics. On the other hand, *in situ* observations are restricted by satellite orbital dynamics. Before THEMIS, most missions had orbits that would preclude frequent KHW observations, or the missions were too short to obtain sufficient data for statistical studies. THEMIS has for the first time provided a sufficiently large database of MP crossings in the equatorial plane, together with suitable instrumentation, to allow for the study presented here.

Our results show that KHWs are much more ubiquitous and occur under most SW and IMF conditions. We confirm the presence of KHWs even during southward IMF conditions, in line with recent event studies[23,24]. During northward IMF, KHWs occur frequently, and particularly also during periods of very low solar wind speed. Theoretical models suggest that the growth rate diminishes for small flow shear and that there even may be a

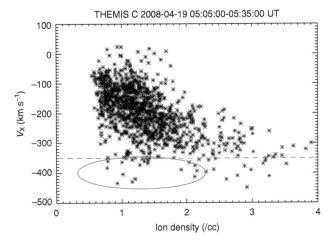

THEMIS C 2008-04-19 05:05:00-05:35:00 UT

Figure 4 | Scatter plot of the ion velocity V_X component versus ion density. The data are from THEMIS C for the 19 April 2008 event. Each symbol represents one of the samples, which were taken at 5 min cadence. Negative V_X values indicate anti-sunward flow. The plot confirms that this event consists of rolled-up Kelvin-Helmholtz vortices, because a fraction of the low-density magnetospheric plasma, indicated by the red ellipse, flows faster than the magnetosheath plasma (< -350 km s^{-1}, shown by the dashed red line).

cutoff velocity. By contrast, we find only weak velocity dependence and no indication of a cutoff.

It is not clear why the data are difficult to reconcile with linear theory, but the most likely reason seems to be that the dispersion relations are based on the assumption of a simply structured shear layer, that is, either a jump-like discontinuity or a smooth transition of finite thickness. In reality, however, the MP often has a complicated boundary layer structure, which generally does not match these assumptions. The presence of such boundary layers makes it thus difficult to test the dispersion relations, because single spacecraft observations generally do not reveal their structure. This is evident, for example, in the event shown in Fig. 3. During this KHW event, neither the magnetosheath plasma velocity nor the plasma density is well defined, and both vary over a large range of values.

Because KHWs can facilitate the entry of SW plasma into the magnetosphere[9,10], they may be more important for the magnetosphere mass budget than previously thought. It has been shown that Kelvin–Helmholtz vortices can efficiently mix plasma. The breaking vortex greatly increases the area through which plasma can diffuse, and it also creates sharp gradients that make it possible for particles to cross from the magnetosheath to the magnetosphere because of finite Larmor radius effects. There is also evidence that magnetic reconnection may occur on small scales within the vortices[9], which would also enable plasma transport across the boundaries. However, quantifying the importance of KHWs for mass transport across the MP remains to be done.

KHWs are also thought to be significant drivers of magnetospheric ULF waves, which in turn can energize the particles in Earth's radiation belts[12–14]. However, such ULF waves can also be generated through other different processes, such as solar wind buffeting or drift-resonant instabilities[26]. Thus, a reassessment of the importance of KHWs for radiation belt dynamics may be necessary.

Methods

Event selection. We surveyed data from 2007 to 2013, when the THEMIS spacecraft frequently crossed the MP during the dawn and dusk orbital phases[27]. We examined the plasma and magnetic field data to catalogue MP crossings with the motivation to identify KHWs. The magnetic field measurements were provided by the FGM instrument[28] and plasma measurements were from the ESA spectrometer[29]. Figure 3 shows a typical example of a crossing where KHWs were present. We show the THEMIS magnetic field and velocity components in boundary normal coordinates (L,M,N)[30] to facilitate the characterization of the oscillations. Different regions and waves were most easily identified in the ion energy spectra, shown in Fig. 3g). Oscillations occurred, where the probe C observed alternately hot magnetosphere plasma and cold magnetosheath plasma. The oscillations were also visible in the magnetic field normal component, B_N (Fig. 3d), the M and N components of the smoothed velocity (Fig. 3b and c), the ion number density (Fig. 3a) and the total pressure (magnetic plus ion pressure; Fig. 3f). The oscillations of the velocity normal component, V_N, show that the MP moved back and forth. As the spacecraft orbital velocity was slow compared with V_N, the oscillations must have been caused by MP surface waves. However, not all MP surface waves are the result of KHI. Other mechanisms can also lead to the excitation of surface waves, such as dynamic pressure variations in the solar wind or magnetosheath[31], non-steady MP reconnection that can generate bulges in the MP or Flux Transfer Events (FTEs)[30]. We thus needed to discriminate all MP wave observations against FTEs and buffeting of the magnetosphere by the solar wind. We inspected solar wind data for every event, where possible, to confirm that the event was not preceded by rapid or periodic SW pressure changes[31–33] that may have caused buffeting. Such events only produce a single bipolar B_N and are thus easily distinguishable from surface waves and ruled out by our requirement of at least four wave periods. They can also often be ruled out by their irregular structure, because KHWs are to large degree monochromatic wave trains. In the initial linear stage of the KHWs, the MP can be approximated by a planar surface, and thus there are no significant total pressure variations across the MP[11,34,35]. An example of such linear KHWs (Supplementary Fig. 1) shows that KHWs in the linear stage can be easily distinguished from FTEs by the absence of bipolar B_N signatures, and by the absence of maxima in |B| and the total pressure.

Discrimination between KHW and FTEs. As KHWs in the nonlinear stage have some similar characteristics as FTEs, such as bipolar B_N and possibly similar wave periods of a few minutes, they may be difficult to differentiate. The properties of FTEs are well known[36,37]. FTEs are magnetic flux ropes whose magnetic signatures include a distinctive bipolar excursion in the magnetic field component B_N normal to the MP surface, either enhancements or crater-like variations of the magnetic field strength at the event centre[38–40], and a deflection of the tangential B_L and B_M components, as shown in Supplementary Fig. 2. The bipolar FTE signature is brief (0.5–2 min) and sequences of FTEs are separated by longer periods of quiet, typically 3–8 min (refs 36–41), which is summarized in Supplementary Table 1, whereas KHWs are continuous wave trains. In addition, the total (thermal and magnetic) pressure in a FTE typically maximizes at the centre of the event[37] as can be seen in Supplementary Fig. 3. By contrast, in a KHW in the nonlinear stage, that is, within a rolled-up vortex the total pressure is expected to have a minimum at the centre and a maximum at the edge of the vortex[11,34,35,42,43]. The pressure minimum occurs because the centrifugal force of the rotating vortex pushes the plasma radially outward, as depicted in Supplementary Fig. 4. This figure also shows that there is a density jump at the edges of the vortices where the pressure should maximize. Figure 3 and Supplementary Fig. 5 show examples of nonlinear KHWs where the density jumps from magnetospheric to magnetosheath values indeed closely coincide with total pressure maxima at the edges of the vortices.

The above criteria, which are summarized in Supplementary Table 2 were not always sufficient to differentiate FTEs from KHWs. Therefore, we also exploited the fact that a rotating KHW vortex accelerates the plasma. When the KHW enters the nonlinear phase, at some distance r_c from the vortex centre, the centrifugal force $\rho V\varphi^2/r_c$ should be nearly equal for both denser and less dense media, where ρ is the plasma mass density and $V\varphi$ is the azimuthal flow velocity; otherwise the vortex would disintegrate[44,45]. Thus, the less dense part of the vortex rotates faster than the denser part. Such low density, accelerated magnetosphere plasma can be exposed in a ρ (or number density N) versus V_X scatter plot, where the KHW or vortex, exhibits a distinct pattern[45–47]. This is demonstrated in Supplementary Fig. 6, where simulations have been used to create the expected scatter plot patterns. Figure 4 shows such a V_X–N scatter plot generated from THEMIS C observations of the KHW example presented in Fig. 3. Concurrent Themis B observations in the magnetosheath showed plasma with ~350 km s^{-1} velocity and ~8 cm^{-3} density. Figure 4 shows that for part of the low-density (<4 cm^{-3}) boundary layer ions, $|V_X|$ is larger than that of the high-density magnetosheath ions ($V_X \sim -350$ km s^{-1}), which is due to the vortex rotation and is not expected for a FTE. Supplementary Fig. 7 shows that no such signature occurs for linear KHWs, that is, the case shown in Supplementary Fig. 1. Supplementary Fig. 8 shows the V_X–N scatter plot for the FTEs presented in Supplementary Fig. 3. The pattern is clearly different from that produced by KHWs in the nonlinear stage, such as the one shown in Fig. 4, and can be used to distinguish them. However, this method could only be used for cases with northward IMF, because with southward IMF low-density, high-speed flows can also result from reconnection[11]. Supplementary Fig. 9 shows a case of KHWs during southward IMF. In this case, we ruled out FTEs because there are no distinct pressure maxima, and neither are there maxima or crater-like structures in the magnetic field magnitude. Instead, the magnetic field magnitude shows distinct minima, which would not be present at FTEs.

Supplementary Fig. 10 shows a unique case where FTEs and KHWs occur back-to-back. The corresponding scatter plot (Supplementary Fig. 11) shows that during the FTE interval, the tangential flows are mostly less than $100 \, \text{km s}^{-1}$, as opposed to the faster tangential flows during the KHW interval. However, the normal flow component (V_N in Supplementary Fig. 10) is much larger for the FTEs than for the KHWs. THEMIS A and THEMIS D also observed this event (not shown here). THEMIS A was located at (3.2, − 10.0,3.4) at the beginning of the interval, that is, very close to THEMIS E. It observed essentially the same signatures as THEMIS E. THEMIS D, on the other hand, was located at (3.9, − 9.8,3.7), that is, further on the magnetosheath side, and only observed the FTEs, but not the KHWs. This implies that the fluctuations in each sub-interval are of different nature. In particular, the amplitude of the KHW must be smaller than the size of the FTEs.

Whenever an event still remained ambiguous, we considered it not to be a KHW.

References

1. Dungey, J. W. Interplanetary magnetic field and the auroral zones. *Phys. Rev. Lett.* **6**, 47–48 (1961).
2. Pilipp, W. G. & Morfill, G. Formation of plasma sheet resulting from plasma mantle dynamics. *J. Geophys. Res.* **83**, 5670–5678 (1978).
3. Øieroset, M. *et al.* THEMIS multi-spacecraft observations of magnetosheath plasma penetration deep into the dayside low-latitude magnetosphere for northward and strong By IMF. *Geophys. Res. Lett.* **35**, L17S11 (2008).
4. Li, W. *et al.* Cold dense magnetopause boundary layer under northward IMF: Results from THEMIS and MHD simulations. *J. Geophys. Res.* **114**, A00C15 (2009).
5. Woch, J. & Lundin, R. Signatures of transient boundary layer processes observed with Viking. *J. Geophys. Res.* **97**, 1431–1447 (1992).
6. Olson, W. P. & Pfitzer, K. A. Magnetospheric responses to the gradient drift entry of solar wind plasma. *J. Geophys. Res.* **90**, 10823–10833 (1985).
7. Wing, S., Johnson, J. R. & Fujimoto, M. Timescale for the formation of the cold-dense plasma sheet: A case study. *Geophys. Res. Lett.* **33**, L23106 (2006).
8. Terasawa, T. *et al.* Solar wind control of density and temperature in the near-Earth plasma sheet: WIND/GEOTAIL collaboration. *Geophys. Res. Lett.* **24**, 935–938 (1997).
9. Nykyri, K. & Otto, A. Plasma transport at the magnetospheric boundary due to reconnection in Kelvin-Helmholtz vortices. *Geophys. Res. Lett.* **28**, 3565 (2001).
10. Hasegawa, H. *et al.* Transport of solar wind into Earth's magnetosphere through rolled-up Kelvin–Helmholtz vortices. *Nature* **430**, 755–758 (2004).
11. Hasegawa, H. Structure and dynamics of the magnetopause and its boundary layers. *Monogr. Environ. Earth Planets* **1**, 71–119 (2012).
12. Walker, A. D. M. The Kelvin-Helmholtz instability in the low-latitude boundary layer. *Planet. Space Sci.* **29**, 1119–1133 (1981).
13. Rae, I. J. *et al.* Evolution and characteristics of global Pc5 ULF waves during a high solar wind speed interval. *J. Geophys. Res.* **110**, A12211 (2005).
14. Elkington, S. R. A review of ULF interactions with radiation belt electrons, *Magnetospheric ULF Waves: Synthesis and New Directions. Geophys. Monogr. Ser* **169**, 177–193 (2006).
15. Otto, A. & Fairfield, D. H. Kelvin-Helmholtz instability at the magnetotail boundary: MHD simulation and comparison with Geotail observations. *J. Geophys. Res.* **105**, 21175–21190 (2000).
16. Foullon, C. *et al.* Evolution of Kelvin-Helmholtz activity on the dusk flank magnetopause. *J. Geophys. Res.* **113**, A11203 (2008).
17. Cowee, M., Winske, D. & Gary, S. P. Two-dimensional hybrid simulations of super diffusion at the magnetopause driven by Kelvin-Helmholtz instability. *J. Geophys. Res.* **114**, A10209 (2009).
18. Matsumoto, Y. & Seki, K. Formation of a broad plasma turbulent layer by forward and inverse energy cascades of the Kelvin–Helmholtz instability. *J. Geophys. Res.* **115**, A10231 (2010).
19. Claudepierre, S. G. *et al.* Solar wind driving of magnetospheric ULF waves: Field line resonances driven by dynamic pressure fluctuations. *J. Geophys. Res.* **115**, A11202 (2010).
20. Talwar, S. P. Hydromagnetic stability of the magnetospheric boundary. *J. Geophys. Res.* **69**, 2707–2713 (1964).
21. Angelopoulos, V. *et al.* First results from the THEMIS mission. *Space Sci. Rev.* **141**, 453–476 (2008).
22. Miura, A. Anomalous transport by magnetohydrodynamic Kelvin-Helmholtz instabilities in the solar wind magnetosphere interaction. *J. Geophys. Res.* **89**, 801–818 (1984).
23. Yan, G. Q. *et al.* Kelvin-Helmholtz vortices observed by THEMIS at the dusk side of the magnetopause under southward interplanetary magnetic field. *Geophys. Res. Lett.* **41**, 4427–4434 (2014).
24. Hwang, K. J. *et al.* Kelvin-Helmholtz waves under southward interplanetary magnetic field. *J. Geophys. Res.* **116**, A08210 (2011).
25. Gratton, F. T. *et al.* Concerning a problem on the Kelvin-Helmholtz stability of the thin magnetopause. *J. Geophys. Res.* **109**, A04211 (2004).
26. Yeoman, T. K. & Wright, D. M. ULF waves with drift resonance and drift-bounce resonance energy sources as observed in artificially induced HF radar backscatter. *Ann. Geophys.* **19**, 159–170 (2001).
27. Sibeck, D. G. & Angelopoulos, V. THEMIS science objectives and mission phases. *Space Sci. Rev.* **141**, 35–59 (2008).
28. Auster, H. U. *et al.* The THEMIS fluxgate magnetometer. *Space Sci. Rev.* **141**, 235–264 (2008).
29. McFadden, J. P. *et al.* The THEMIS ESA Plasma Instrument and in-flight calibration. *Space Sci. Rev.* **141**, 277–302 (2008).
30. Russell, C. T. & Elphic, R. C. Initial ISEE magnetometer results: magnetopause observations. *Space Sci. Rev.* **22**, 681–715 (1978).
31. Sibeck, D. G. A model for the transient magnetospheric response to sudden solar wind dynamic pressure variations. *J. Geophys. Res.* **95**, 3755–3771 (1990).
32. Lockwood, M. Flux-transfer events at the dayside magnetopause: transient reconnection or magnetosheath dynamic pressure pulses? *J. Geophys. Res* **96**, 5497–5509 (1991).
33. Fairfield, D. H. *et al.* Upstream pressure variations associated with the bow shock and their effects on the magnetosphere. *J. Geophys. Res.* **95**, 3773 (1990).
34. Miura, A. Self-organization in the two-dimensional Kelvin- Helmholtz instability. *Phys. Rev. Lett.* **83**, 1586–1589 (1999).
35. Hasegawa, H. *et al.* Kelvin-Helmholtz waves at the Earth's magnetopause: multiscale development and associated reconnection. *J. Geophys. Res.* **114**, A12207 (2009).
36. Elphic, R. C. Observations of flux transfer events: Are FTEs flux ropes, islands, or surface waves? *Phys. Magn. Flux Ropes Geophys. Monogr. Ser* **58**, 455–471 (1990).
37. Elphic, R. C. Observations of flux transfer events: A review. *Phys. Magnetopause Geophys. Monogr. Ser* **90**, 225–233 (1995).
38. Farrugia, C. J. *et al.* Field and flow perturbations outside the reconnected field line region in flux transfer events: Theory. *Planet. Space Sci.* **35**, 227–240 (1988).
39. Southwood, D. J., Farrugia, C. J. & Saunders, M. A. What are flux transfer events? *Planet. Space Sci.* **36**, 503–508 (1988).
40. Wang, Y. *et al.* Initial results of high-latitude magnetopause and low-latitude flank flux transfer events from 3 years of Cluster observations. *J. Geophys. Res.* **110**, A11221 (2005).
41. Paschmann, G. *et al.* Plasma and magnetic field characteristics of magnetic flux transfer events. *J. Geophys. Res.* **87**, 2159–2168 (1982).
42. Miura, A. Compressible magnetohydrodynamic Kelvin-Helmholtz instability with vortex pairing in the two-dimensional transverse configuration. *Phys. Plasmas.* **4**, 2871–2885 (1997).
43. Nakamura, T. K. M. *et al.* Decay of MHD-scale Kelvin-Helmholtz vortices mediated by parasitic electron dynamics. *Phys. Rev. Lett.* **92**, 145001 (2004).
44. Takagi, K. *et al.* Kelvin-Helmholtz instability in a magnetotail flank-like geometry: three-dimensional MHD simulations. *J. Geophys. Res.* **111**, A08202 (2006).
45. Hasegawa, H. *et al.* Single-spacecraft detection of rolled-up Kelvin-Helmholtz vortices at the flank magnetopause. *J. Geophys. Res.* **111**, A09203 (2006).
46. Taylor, M. G. G. T. *et al.* Spatial distribution of rolled-up Kelvin-Helmholtz vortices at Earth's dayside and flank magnetopause. *Ann. Geophys.* **30**, 1025–1035 (2012).
47. Plaschke, F., Taylor, M. G. G. T. & Nakamura, R. Alternative interpretation of results from Kelvin-Helmholtz vortex identification criteria. *Geop. Res. Lett.* **41**, 244–250 (2014).

Acknowledgements

This research was supported by NASA THEMIS subcontract SA405826326 to UCB, and NASA grant NNX11AJ11G. We acknowledge the THEMIS ESA and FGM instrument teams for the use of their data.

Author contributions

J.R. initiated the study and guided the graduate student S.K., who screened the data and obtained the statistics. Both authors discussed the methods, results and implications at all stages.

Additional information

4

Cosmochemical fractionation by collisional erosion during the Earth's accretion

Asmaa Boujibar[1,†], Denis Andrault[1], Nathalie Bolfan-Casanova[1], Mohamed Ali Bouhifd[1] & Julien Monteux[1]

Early in the Solar System's history, energetic collisions of differentiated bodies affected the final composition of the terrestrial planets through partial destruction. Enstatite chondrites (EC) are the best candidates to represent the primordial terrestrial precursors as they present the most similar isotopic compositions to Earth. Here we report that collisional erosion of >15% of the early Earth's mass can reconcile the remaining compositional differences between EC and the Earth. We base our demonstration on experimental melting of an EC composition at pressures between 1 bar and 25 GPa. At low pressures, the first silicate melts are highly enriched in incompatible elements Si, Al and Na, and depleted in Mg. Loss of proto-crusts through impacts raises the Earth's Mg/Si ratio to its present value. To match all major element compositions, our model implies preferential loss of volatile lithophile elements and re-condensation of refractory lithophile elements after the impacts.

[1] Laboratoire Magmas et Volcans, Université Blaise Pascal, CNRS UMR-6524, 5 rue Kessler, 63000 Clermont-Ferrand, France. † Present address: Astromaterials Research and Exploration Science, NASA Johnson Space Center, 2101 Nasa Parkway, Houston, Texas 77058, USA. Correspondence and requests for materials should be addressed to A.B. (email: asmaa.boujibar@nasa.gov).

Large-scale melting occurred broadly in the first stages of planetary accretion[1]. The impressive homogeneity of stable isotopic compositions of large bodies suggests extensive melting and formation of magma oceans[1]. However, while large-scale melting can efficiently erase isotopic heterogeneities, low degrees of partial melting are a primary cause of chemical segregation. The possibility to form proto-crusts by low degrees of melting of chondritic material is evidenced by the discovery of felsic achondrites (GRA-06128/9; ref. 2). Early differentiation of planetary embryos was also recently suggested by a study that determined the initial content of the short-lived radionuclide [26]Al in angrites[3]. Therefore, early crusts could form by different processes, such as fractional crystallization of a magma ocean or the migration of silicate melts over networks of veins and dikes[4]. Despite the formation mechanism of these proto-crusts, their occurrence on accreting bodies should have played a major role in the final planetary composition, because energetic episodes of accretion have eroded the planetary surfaces[5-8].

The chemical composition of the building blocks that accreted to form the Earth remains controversial. Our planet shows remarkable isotopic similarities with enstatite chondrites (EC), especially with those of the EH type[9], for the elements whose isotopes do not fractionate during core segregation (O, Ca, N, Mo, Ru, Os, Cr, Ni, Ti and Sr). However, EC and the Earth present important chemical differences. First, EC are so reduced that their silicate phases are free of FeO, which differs from the present-day silicate mantle with 8 wt% FeO. This issue may eventually be solved by internal oxidation processes for some of the Fe metal[9,10]. A second issue is that EC show a Mg/Si weight ratio lower than 1 ($\sim$0.63; ref. 11), differing substantially from that of the Earth's upper mantle ($\sim$1.1; ref. 12). The Mg/Si ratio of bulk Earth (BE) could be slightly overestimated if the lowermost mantle is bridgmanitic (that is, perovskitic) rather than pyrolitic[13]. Furthermore, the Earth shows higher abundances of refractory lithophile elements (RLE) and lower concentrations of moderately volatile elements (as the alkali elements Li, Na, K and Rb) compared with EC and all other chondrites (Fig. 1).

In this study, we test whether collisional erosion of early crusts can explain the chemical divergence between the BE and EC. Based on experiments on the melting properties of synthetic EC at pressures between 1 bar and 25 GPa, we show that early

differentiated planetary bodies can develop silica- and alkali-enriched crusts. Loss of these crusts through impact erosion can ultimately increase the Mg/Si ratio of the planetary bodies to match the current Earth ratio. In addition, to further increase the budget in RLE and accurately reproduce the terrestrial concentrations of the major and minor elements, impact erosion must be accompanied with preferential loss of volatile lithophile elements and re-condensation of RLE.

Results

Partial melting of enstatite chondrites. We experimentally investigated the composition of melts produced by low degrees of melting of synthetic EC powders, at low oxygen fugacity (3.6 to 1.8 log units below the iron/wustite buffer), at different pressure conditions expected for melt segregation in partially molten planetary embryos (see Methods section and Supplementary Fig. 1). Our pseudo-eutectic melts are all characterized by high concentrations of SiO_2, Al_2O_3 and Na_2O, and low MgO contents (Fig. 2). The change with pressure of the low-degree melt composition agrees with that previously reported at 1 bar (ref. 14). The most striking features are the increase of MgO (Fig. 2b) and decrease of SiO_2 (Fig. 2a) and Al_2O_3 (Fig. 2c) with pressure. The disappearance of clinopyroxene at 16 GPa and garnet at 24 GPa

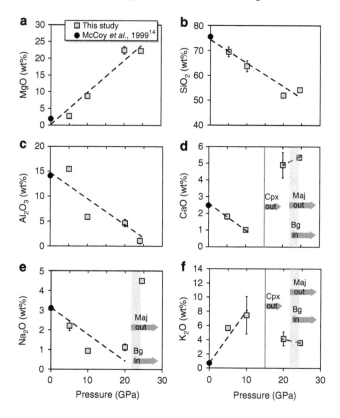

Figure 2 | Chemical composition of pseudo-eutectic melts synthesized at high pressures. We report the compositions of melts generated by partial melting of an EC-like mantle at pressures between 5 and 25 GPa (yellow squares). Our data set plots are in agreement with previous results performed at 1 bar (ref. 14) (dark circles). Whereas MgO content (**a**) increases with pressure in the silicate melts, the concentrations of SiO_2 (**b**) and Al_2O_3 (**c**) decrease. Sharp changes of the liquid compositions occur when the sample encounters phase transformations (shown with the grey arrows), in particular, with the disappearance of clinopyroxene (Cpx) at 15 GPa (vertical lines in **d** and **f**) and simultaneous disappearance of majorite (Maj) and appearance of bridgmanite (Bg, the $MgSiO_3$ perovskite; vertical grey areas in **d-f**) at 24 GPa.

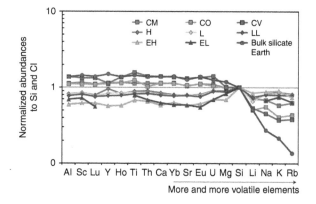

Figure 1 | Major chemical differences between the chondrites and the bulk silicate Earth (BSE). The abundances are normalized to Si and CI chondrites[11,12]. From left to right, the lithophile elements are reported with decreasing their 50% condensation temperature. The depletion of volatile elements in the BSE results from the erosion of crusts enriched with incompatible elements and the subsequent loss of the most volatile elements.

(ref. 15) induces major CaO (Fig. 2d), Na_2O (Fig. 2e) and K_2O (Fig. 2f) enrichment of the liquid, respectively. Altogether, pseudo-eutectic liquids show compositions between rhyolitic and trachy-andesitic in the range of pressures investigated. In partially molten planetesimals, such melts should ascend relatively easily towards the planetary surface due to their low melt densities, even for low degrees of partial melting[4]. As the melts can re-equilibrate during their ascent to the surface, those produced at shallower depths are more likely to stay unaltered and produce proto-crusts enriched in incompatible elements. The most appropriate melts for the formation of proto-crusts should then be those produced at relatively low pressures and degrees of partial melting.

Change of proto-Earth composition with collisional erosion. We now evaluate how collisional erosion of proto-crusts made of these pseudo-eutectic liquids would affect the chemical composition of an EH-type planetary embryo[11]. First, we calculate the Mg/Si ratio of a planetary body after removal of pseudo-eutectic melts generated at average pressures of 1 bar to 25 GPa and compare it with the BE[12], with the hypothesis that $\sim$7 wt% Si is present in the Earth's core[9,16]. By increasing the amount of crustal erosion, the Mg/Si ratio of the depleted EH planetoid increases towards the present-day BE ratio of $\sim$0.9, owing to the high SiO_2 content of the melts (Fig. 3a). The lower the pressure of melt-solid equilibration, the higher the SiO_2 content in the melt, and therefore the crustal erosion should be the less extensive. The BE Mg/Si ratio can be achieved by accretion of EH chondrites and erosion of a crust of 15–18% of the planetary mass for solid-melt equilibrium at pressures below 10 GPa. This amount of crustal erosion is comparable to the loss of highly incompatible elements required to explain the mantle budget in volatile elements[17], and is comparable to the amount of mass lost during hit and run simulations[18]. As the formation of such SiO_2-rich crust necessitates a low degree of partial melting

(at a level of 5–7%), removing proto-crust to a level of 15–18 wt% of the planetary mass requires repeated processes of proto-crust formation and collisional erosion. Repeated partial melting of the EH-type mantle would constantly produce an SiO_2-rich proto-crust, owing to the fixed pseudo-eutectic composition. The nature of planetary accretion itself provides the necessary energy to melt planetesimals and erode them or even disrupt them (see below). Simulations indicate that an accreting proto-planet should experience $\sim 10^5$ collisions[18].

Removal of proto-crust also affects the Al/Si, Ca/Si and Na/Si ratios (Fig. 3b and Supplementary Fig. 2). Interestingly, the misfits between Mg/Si, Al/Si, Ca/Si and Na/Si ratios in BE and in our model of EH-type planetoid depleted by crustal erosion show a clear correlation with the condensation temperatures of the elements of interest[19] (Fig. 3b). Rather than being fortuitous, this trend can be understood in terms of differential re-condensation of the elements after collisional erosion. Our model implies that early differentiation of a planetary embryo forms a silica-rich crust (Fig. 4a,b) that is subsequently eroded and vaporized by energetic impacts (Fig. 4c). The eroded material is then chemically fractionated with a preferential condensation of the refractory elements relative to the volatiles (Fig. 4d). This explains the previously reported marked consequences on

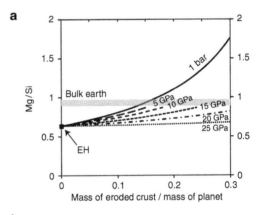

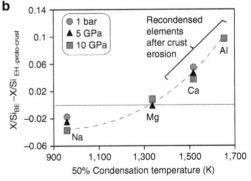

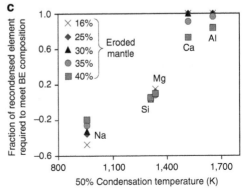

Figure 3 | Change by collisional erosion of the proto-planetary chemical composition. (**a**) Change by collisional erosion of the planetary Mg/Si ratio by removal of a differentiated crust with a composition equivalent to that of pseudo-eutectic melts produced at pressures of 1 bar and 5, 10, 15, 20 and 25 GPa. Removal by collisional erosion from the model EH composition results in an increase of the Mg/Si ratio of the planetesimal. The grey area represents the Mg/Si ratio of the BE with 7 wt% Si dissolved into the core[9,16]. Erosion of a crust of >15% of the planet mass is required to reconcile the Mg/Si ratio of an EH-type planet with the present-day BE[12]. (**b**) Correlation between residual misfits between BE and the EH proto-planet X/Si for X = Na, Ca and Al as a function of the 50% condensation temperature of the elements[19]. After the mass of eroded crust is adjusted to meet the Mg/Si ratios of BSE, there is a residual misfit for the abundances of other major elements (Supplementary Fig. 2). The correlation between BE enrichments and the condensation temperatures of the different elements suggests chemical fractionation during the processes of vaporization of the planetary surface, with re-condensation of the eroded material on the planetary surface (Fig. 4). (**c**) Degree of chemical fractionation required by our model, for different amounts of collisional erosion. Here we consider that the 15% of the crust (produced in the 0–5-GPa range) required to match the Mg/Si ratio of the BE and 16–40% of the planetary mantle are eroded by the impacts. For a total erosion of 31–45% of the planetary mass, for example, the actual Na/Si, Mg/Si, Ca/Si and Al/Si ratios of BE are reproduced when 100% of Al and Ca, 10% of Mg and 5% of Si are re-condensed on the planetary surface, which is in agreement with their condensation temperature. A negative value for Na denotes the fact that the residual mantle of our EH-type model (after collisional erosion) would still contain high Na contents compared with BE. In this case, additional Na volatilization from the residual mantle is required.

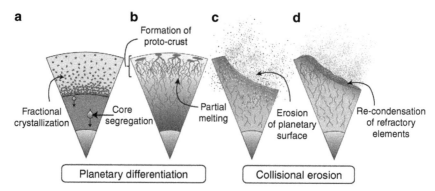

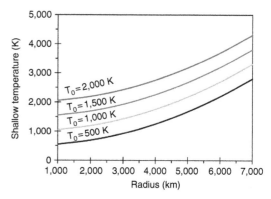

Figure 4 | Schematic model of chemical fractionation by collisional erosion. (a,b) Early heating leads to the rapid segregation of Fe-rich metal into the core. It is accompanied with the formation of a deep magma ocean[1] (**a**) and/or small-scale partial melting and formation of complex networks of veins and dikes[4] (**b**) which allows fast transfer of melts to the surface[4]. Both can yield compositional stratification of the mantle and the formation of an SiO_2-rich proto-crust. (**c**) The repeated collisions induce erosion of the proto-crust enriched in incompatible elements, as well as part of the planetary mantle. (**d**) Within the fraction of material volatilized by meteoritic impacts, re-condensation of refractory elements is favoured compared with the volatile elements.

the thermal history of Earth that could result from the loss of incompatible and refractory heat-producing elements, such as U and Th[5,20,21].

In fact, the Earth's budget in RLE can be reached by a number of different erosion and re-condensation models. The lowest amount of erosion (15–18 wt% of proto-crust to meet the right Mg/Si ratio) requires re-condensation of 100% of the refractory elements (Ca and Al) on the planetary surface (Fig. 3c). For higher amounts of planetary surface erosion (15–18 wt% proto-crust plus a mantle fraction), the models require less chemical fractionation during vaporization or re-condensation. For 40% mantle erosion, re-condensation of 84% of Al, 73% of Ca, 10% of Mg and 5% of Si is necessary (Fig. 3c and Table 1) to produce a planetary composition similar to Earth. Comparable fractionation of major elements with Ca- and Al-rich condensates depleted in Mg and Si was reported in the most primitive unequilibrated Semarkona ordinary chondrites (LL3.0; ref. 22). Also, such a degree of chemical fractionation during re-condensation may not be necessary if the composition of the parent body deviates from the EH chondrites (see below). Then, non-equilibrium processes, such as solar wind, could have favoured the loss of volatile elements from the gravitational field of the proto-planet. In addition, the volatile elements could have been partially atmophile due to hot planetary surfaces, which could have contributed to the volatile and refractory fractionation.

The major sources of energy provided by planetary accretion. During planetesimal growth, the thermal state of a proto-planet depends on its initial heating caused by the decay of short-lived radionuclides such as [26]Al and [60]Fe (ref. 23), on its accretionary history and on how potential energy is dissipated during the iron/silicates segregation. Temperatures in excess of 2,000 K could have been reached within the first 2–3 million years after the formation of the first solids of the Solar System (calcium–aluminium-rich inclusions, CAI)[24]. The early radioactive heating can therefore cause both segregation of a metal-rich core and silicate melting on planetary embryos that have quickly accreted (to attain a radius of 30 km within the first 3 million years; see Fig. 2 in Yoshino et al.[24]). Kleine et al.[25] showed that high temperatures of early asteroids are in agreement with the young ages of iron meteorites, OC, CO and CR chondrites and the peak temperatures recorded by the same meteorites (see in ref. 25).

In addition to this radioactive heating, in the shallow parts of the impacted planet, impact heating (ΔT), superimposed to a sufficiently hot proto-planetary interior (with an initial

Figure 5 | Shallow post-impact temperature as a function of the target radius R. Different pre-impact temperatures are illustrated (from $T_0 = 500$ to 2,000 K). We consider that (i) surface gravity is equal to $g = 4/3\,\pi\,G\,\rho\,R$ (G is the gravitational constant and ρ is the density of the target), (ii) kinetic energy of the impactor is controlled by the escape velocity of the impacted body (that is, $v_{impactor} = \sqrt{(2gR)}$), (iii) the impactor and impacted body have the same densities, and (iv) only 30% of the incoming kinetic energy is converted into heat.

temperature of T_0) can lead to temperatures ($T = T_0 + \Delta T$) much larger than the vaporization temperature of silicates ($\sim 1,300\,°C$ at 0.001 bar; see Fig. 5 above). This heating is localized in a spherical region called the isobaric core, just beneath the impact site (for example, ref. 26). By making the conservative assumptions that (i) kinetic energy of the impactor is controlled by the escape velocity of the impacted body, (ii) impactor and target body have the same densities and (iii) only 30% of the incoming kinetic energy is converted into heat, a simple energy balance indicates that ΔT (K) $= 4.7 \times 10^{-5} \times R^2$ (km) (refs 26,27).

Assuming that the impact occurs with the escape velocity of the impacted body is a conservative assumption in the sense that it is the minimum impact velocity; larger impact velocities can substantially increase the post-impact temperature below the surface. For instance, some episodes of hypervelocity impacts have probably played a major role during the accretion of the terrestrial planets. For example, a recent model suggests that Jupiter and Saturn have likely migrated inwards and then outwards early in the Solar System history[28], leading to strong resonance and high eccentricities of the planetesimals. As the resonance with Jupiter can significantly increase the impact

Table 1 | Resulting compositions with dual processes of collisional erosion and fractional re-condensation.

	Bulk Earth	Assumed core composition[44]	EH-like planet			EH-like planet with adjusted core size		
			0% eroded mantle	25% eroded mantle	40% eroded mantle	0% eroded mantle	25% eroded mantle	40% eroded mantle
Mg	15.6		13.8	12.8	10.8	15.9	15.8	15.6
Al	1.6		1.0	1.1	1.1	1.2	1.4	1.6
Si	16.6	7	15.0	13.7	11.5	17.2	16.9	16.6
Ca	1.7		1.1	1.2	1.2	1.3	1.5	1.7
Na	0.2		0.2	0.2	0.1	0.2	0.2	0.2
K	0.0		0.0	0.0	0.0	0.0	0.0	0.0
O	30.5	4	27.9	25.8	21.9	30.5	30.5	30.5
Fe	31.4	82	38.0	41.9	49.6	31.4	31.4	31.4
Co	0.1	0.1	0.1	0.1	0.1	0.1	0.1	0.1
Ni	1.6	5	2.3	2.5	3.0	1.6	1.6	1.6
S	0.6	2	0.6	0.6	0.7	0.6	0.6	0.6

We compare the actual bulk Earth composition (first column) (calculated based on bulk silicate Earth[12] composition and cosmochemical estimates of core composition[44] (second column)) with the EH-like Earth (third column) after erosion of (i) 15% of a crust composed of pseudo-eutectic melts produced at 0–5 GPa and (ii) 0–40% of its mantle. The erosion yields a significant depletion in the lithophile elements (Mg, Si, O, Na, Al and Ca) compared with the siderophile elements (Fe, Ni, Co and S). By adjusting the core size to the actual terrestrial core size (by fixing the concentrations of the siderophile elements to that of the Bulk Earth), the composition of our EH-like planetary model can reach values very close to that observed in the present-day Earth.

velocities[29], the kinetic energy, available during the impact processes on growing bodies, may have enhanced the vaporization and the excavation of their shallow material. This model, with marked migration of Jupiter and Saturn, is one suggested model of the dynamical evolution of the Solar System. Other models may not produce such high velocity collisions, but they would have a longer period during which frequent collisions could occur.

Vaporization following collisional erosion from large impacts is widely accepted[30]. The abundance of impacts is supported by observations of the iron meteorites, which represent fragments of cores of differentiated bodies, which are now destroyed. These iron meteorite parent bodies are as old as the CAI, which are the oldest objects of the Solar System, that formed during the first million year after T_0 (ref. 31). Thus, collisional stripping operated in the very early stages of the formation of the Solar System.

Discussion

Our experimental results show that surface erosion modifies more efficiently the planetary composition of smaller planetary bodies owing to lower internal pressures (Fig. 3a). This suggests a more efficient chemical fractionation early in the accretion history. As a matter of fact, silica- and alkali-rich compositions (andesitic to rhyolitic and trachyitic) have been observed in various asteroidal materials: bulk achondrites (GRA-06128/09), silicate enclaves in iron meteorites, glass inclusions or igneous clasts in aubrites, glass inclusions in the howardite, eucrite and diogenite meteorites, clasts in the ureilites. The oldest sampled materials of the Solar System with a granitic composition are only 5–30 Myr younger than the formation of the Solar System (see ref. 32 for a review). This indicates the formation of silicic crust nuclei very early in the inner Solar System.

A widespread Si isotopic dichotomy is observed between planetary/asteroidal bodies and the chondrites. There is a correlation between the enrichment in heavy Si (^{30}Si) with both (i) the depletion of the moderately volatile elements (K) and (ii) the depletion of Si relative to Mg for the Earth, the Moon, Vesta, Mars, the angrites (basaltic meteorites) parent body and the chondrites[33]. As suggested by Pringle et al.[33], this correlation could be caused by early impact-induced evaporative loss, occurring as early as the formation of angrites, 2 Myr later than the formation of the oldest currently known objects of the Solar

System[34,35] (Supplementary Note 1). Thus, the Si isotopic measurements fully support our model of early collisional erosion. It has also been proposed that energetic episodes of impacts can induce significant chemical fractionation of RLE in EL chondrites[36].

In contrast to the other planetary differentiation processes, such as core segregation or magma ocean crystallization, an early impact erosion affects the composition of both planetary bodies and the remaining unaccreted material, in a complementary way. Owing to a higher surface/volume ratio compared with the planetesimals, the chondritic material that was left over from the accretion may have preferentially reacted with gases produced by vaporization of the eroded crusts, which are enriched in SiO_2 and volatile lithophile elements. This can explain the enrichments in Na, S and Si observed at the edges of chondrules contained in carbonaceous and ordinary chondrites[37–39]. EC are the most alkali- and silica-rich chondrites[11,40] (Fig. 1), but in contrast to the carbonaceous chondrites, this enrichment is global and not limited to chondrules (Supplementary Note 2). This is likely owing to the fact that EC have experienced important thermal metamorphism that usually erases the chemical gradients. Several studies have analysed the behaviour of alkali elements in chondrules in order to retrieve the conditions under which the chondrules form. The majority of these studies concluded that very high dust densities with high solid/gas ratios are required in the nebula to enrich the edges of chondrules in Na[37]. This was previously attributed to aqueous alteration[41]. In light of our results, we propose that the destruction and vaporization of eroded proto-crusts can also contribute to this elevation of the dust/gas ratio of the nebula in the formation regions of chondrules.

The EH meteorites that are sampled today could have suffered an SiO_2 enrichment compared with their parent bodies, owing to interaction with crusts eroded from primordial planetesimals. This would have drifted their composition away from that of the Earth, compared with the original building blocks (Fig. 1). This common original material could have had a higher Mg/Si ratio and lower content in alkali elements than the EH chondrites. Consequently, our quantitative model of collisional erosion developed in this study (for example, Fig. 3) can be considered as the most extreme one, as less proto-crust erosion would be required to meet the actual BE chemical composition. Altogether, we show that collisional erosion could readily explain the major chemical divergences between the EH chondrites and Earth. It reinforces the model of the EH-like Earth[9], nevertheless adding to this model the idea that the common starting material could have been slightly depleted in SiO_2 and volatile lithophile elements in comparison with the EH chondrites.

Methods

Preparation of starting materials. The starting material was composed of 68 wt% silicate and 32 wt% metal with a chemical composition equivalent to the average composition of EC[9] (Supplementary Table 1). While we are exclusively interested in the silicate properties, the presence of metals helps reproduce the EH-Earth conditions more precisely and buffer the oxygen fugacity. The silicate fraction was composed of ultra-pure oxides and carbonates (SiO_2, MgO, Al_2O_3, TiO_2, Cr_2O_3, MnO, Na_2CO_3, $CaCO_3$ and K_2CO_3) that were finely ground together, decarbonized and then dried overnight at 1,000 °C. The metal powders were composed of a fine mixture of ultra-pure Fe, Ni, Si, Mn, Co and/or FeS. All samples were S free, except for 5 wt% S in the metal of sample no. 104. Silicate and metal fractions were intimately mixed to obtain a homogeneous chondritic powder, that was kept constantly in a vacuum oven to avoid hydration.

Details of the experimental conditions. We conducted experiments at 5, 10, 20 and 25 GPa, with temperatures ranging from 1,380 to 1,900 °C using Kawai-type 1,000-t and 1,500-t multi-anvil presses. We heated the experiments to temperatures at or slightly above the solidus, that is, between 1,380 and 1,900 °C, for oxygen fugacities between -2.1 and -3.6 log units below the IW buffer (Supplementary Table 2). Assemblies were composed of Cr-doped MgO octahedra pressure media with edges of 18, 14 and 10 mm, coupled with tungsten carbide cubes with 11, 6 and 4 mm truncations, respectively. Pressure calibrations of both 1,000-t and 1,500-t presses were previously described[42]. The sample powder was loaded into a graphite capsule that was surrounded by MgO sleeves to prevent the sample pollution. High temperatures were achieved using an $LaCrO_3$ furnace and a ZrO_2 sleeve thermal insulator. Temperature was measured using a $W_5Re/W_{26}Re$ thermocouple in experiment nos. 1,223 and 174, and estimated using the relation between temperature and electrical power for experiment nos. 1,216 and 104. Pressure and temperature uncertainties are estimated to be ~ 0.5 GPa and 100 °C, respectively.

To achieve homogeneous samples without relict of the starting powders, we first heated the samples above the liquidus temperature[43] for a couple of minutes. Temperature was then rapidly reduced (with a rate of 100 K in <20 s) to a temperature just above the reported solidus. The sample was then equilibrated between 30 min and 3 h. This procedure helps grain growth and allows segregation of relatively large pools of melt, which are usually difficult to collect with low degrees of partial melting[43].

Chemical compositions of the coexisting phases. The microstructure of recovered samples was observed using a scanning electron microscope (Supplementary Fig. 1). Phase relations and chemical compositions were determined using a CAMECA SX100 electron probe micro-analyser (Supplementary Table 1). We used an accelerating voltage of 15 kV, an electron beam defocused to 2–20 μm and a current of 15 nA, except for the chemical analyses of bridgmanites that were analysed with 2 nA. As standards, we used pure metals and silicates.

References

1. Greenwood, R. C., Franchi, I. A., Jambon, A. & Buchanan, P. C. Widespread magma oceans on asteroidal bodies in the early Solar System. *Nature* **435**, 916–918 (2005).
2. Day, J. M. D. *et al.* Early formation of evolved asteroidal crust. *Nature* **457**, 179–182 (2009).
3. Schiller, M., Connely, J. N., Glad, A. C., Mikouchi, T. & Bizzarro, M. Early accretion of protoplanets inferred from a reduced inner solar system ^{26}Al inventory. *Earth Planet. Sci. Lett.* **420**, 45–54 (2015).
4. Wilson, L. & Keil, K. Volcanic activity on differentiated asteroids: a review and analysis. *Chemie der Erde - Geochemistry* **72**, 289–321 (2012).
5. O'Neill, H. S. C. & Palme, H. Collisional erosion and the non-chondritic composition of the terrestrial planets. *Phil. Trans. R. Soc. A* **366**, 4205–4238 (2008).
6. Asphaug, E., Agnor, C. B. & Williams, Q. Hit-and-run planetary collisions. *Nature* **439**, 155–160 (2006).
7. Leinhardt, Z. M. & Stewart, S. T. Collisions between gravity-dominated bodies. I. Outcome regimes and scaling laws. *Astrophys. J.* **745**, 79 (2012).
8. Davison, T. M., O'Brien, D. P., Ciesla, F. J. & Collins, G. S. The early impact histories of meteorite parent bodies. *Meteor. Planet. Sci.* **48**, 1894–1918 (2013).
9. Javoy, M. *et al.* The chemical composition of the Earth: enstatite chondrite models. *Earth Planet. Sci. Lett.* **293**, 259–268 (2010).
10. Wade, J. & Wood, B. J. Core formation and the oxidation state of the Earth. *Earth Planet. Sci. Lett.* **236**, 78–95 (2005).
11. Wasson, J. T. & Kallemeyn, G. W. Composition of chondrites. *Phil. Trans. R. Soc. Lond.* **A325**, 535–544 (1988).
12. McDonough, W. F. & Sun, S. S. The composition of the Earth. *Chem. Geol.* **120**, 223–253 (1995).
13. Murakami, M., Ohishi, Y., Hirao, N. & Hirose, K. A perovskitic lower mantle inferred from high-pressure, high-temperature sound velocity data. *Nature* **485**, 90–94 (2012).
14. McCoy, T., Dickinson, T. L. & Lofgren, G. Partial melting of the Indarch (EH4) meteorite: a textural, chemical, and phase relations view of melting and melt migration. *Meteor. Planet. Sci.* **34**, 735–746 (1999).
15. Liebske, C. & Frost, D. J. Melting phase relations in the MgO–$MgSiO_3$ system between 16 and 26 GPa: implications for melting in Earth's deep interior. *Earth Planet. Sci. Lett.* **345-348**, 159–170 (2012).
16. Gessmann, C. K., Wood, B. J., Rubie, D. C. & Kilburn, M. R. Solubility of silicon in liquid metal at high pressure: implications for the composition of the Earth's core. *Earth Planet. Sci. Lett.* **184**, 367–376 (2001).
17. Halliday, A. N. The origins of volatiles in the terrestrial planets. *Geochim. Cosmochim. Acta* **105**, 146–171 (2013).
18. Bonsor, A. *et al.* A collisional origin to Earth's non chondritic composition? *Icarus* **247**, 291–300 (2015).
19. Lodders, K. Solar System abundances and condensation temperatures of the elements. *Astrophys. J.* **591**, 1220–1247 (2003).
20. Warren, P. H. A depleted, not ideally chondritic bulk Earth: The explosive-volcanic basalt loss hypothesis. *Geochim. Cosmochim. Acta* **72**, 2217–2235 (2008).
21. Jackson, M. G. & Jellinek, A. M. Major and trace element composition of the high $^3He/^4He$ mantle: Implications for the composition of a nonchondritic Earth. *Geochem. Geophys. Geosyst.* **14**, 2954–2976 (2013).
22. Nagahara, H., Kita, N. T., Ozawa, K. & Morishita, Y. Condensation of major elements during chondrule formation and its implication to the origin of chondrules. *Geochim. Cosmochim. Acta* **72**, 1442–1465 (2008).
23. Urey, H. C. The cosmic abundances of potassium, uranium and thorium and the heat balances of the Earth, the Moon and Mars. *Proc. Natl Acad. Sci. USA* **41**, 127–144 (1955).
24. Yoshino, T., Walter, M. J. & Katsura, T. Core formation in planetesimals triggered by permeable flow. *Nature* **422**, 154–157 (2003).
25. Kleine, T. *et al.* Hf-W chronology of the accretion and early evolution of asteroids and terrestrial planets. *Geochim. Cosmochim. Acta* **73**, 5150–5188 (2009).
26. Pierazzo, E., Vickery, A. M. & Melosh, H. J. A reevaluation of impact melt production. *Icarus* **127**, 408–423 (1997).
27. Monteux, J., Coltice, N., Dubuffet, F. & Ricard, Y. Themo-mechanical adjustment after impacts during planetary growth. *Geophys. Res. Lett.* **34**, L24201 (2007).
28. Walsh, K. J., Morbidelli, A., Raymond, S. N., O'Brien, D. P. & Mandell, A. M. A low mass for Mars from Jupiter's early gas-driven migration. *Nature* **475**, 206–209 (2011).
29. Weidenschilling, S. J., Davis, D. R. & Marzari, F. Very early collisonal evolution in the asteroid belt. *Earth Planets Space* **53**, 1093–1097 (2001).
30. Canup, R. M. & Asphaug, E. Origin of the Moon in a giant impact near the end of the Earth's formation. *Nature* **412**, 708–712 (2001).
31. Amelin, Y., Krot, A. N., Hutcheon, I. D. & Ulyanov, A. A. Lead isotopic ages of chondrules and calcium-aluminum-rich inclusions. *Science* **297**, 1678–1683 (2002).
32. Bonin, B. Extra-terrestrial igneous granites and related rocks: a review of their occurrence and petrogenesis. *Lithos* **153**, 3–24 (2012).
33. Pringle, E. A., Moynier, F., Savage, P. S., Badro, J. & Barrat, J.-A. Silicon isotopes in angrites and volatile loss in planetesimals. *Proc. Natl Acad. Sci. USA* **111**, 17027–17032 (2014).
34. Baker, J. A., Bizzarro, M., Witting, N., Connelly, J. & Haak, H. Early planetesimal melting from an age of 4.5662 Gyr for differentiated meteorites. *Nature* **436**, 1127–1131 (2005).
35. Bouvier, A. & Wadhwa, M. The age of the solar system redefined by the oldest Pb-Pb age of a meteoritic inclusion. *Nat. Geosci.* **3**, 637–641 (2010).
36. Rubin, A. E., Huber, H. & Wasson, J. T. Possible impact-induced refractory-lithophile fractionations in EL chondrites. *Geochim. Cosmochim. Acta* **73**, 1523–1537 (2009).
37. Libourel, G., Krot, A. N. & Tissandier, L. Role of gas-melt interaction during chondrule formation. *Earth Planet. Sci. Lett.* **251**, 232–240 (2006).
38. Alexander, C. M. O. & Grossman, J. N. Alkali elemental and potassium isotopic compositions of Semarkona chondrules. *Meteor. Planet. Sci.* **40**, 541–556 (2005).
39. Marrocchi, Y. & Libourel, G. Sulfur and sulfides in chondrules. *Geochim. Cosmochim. Acta* **119**, 117–136 (2013).
40. Krot, A. N., Keil, K., Goodrich, C. A., Scott, E. R. D. & Weisberg, M. K. in *Treatise on Geochemistry.* (ed. Davis, A. M.) Vol. 1 (Elsevier, 2003).
41. Alexander, C. M. O., Grossman, J. N., Ebel, D. S. & Ciesla, F. J. The formation conditions of chondrules and chondrites. *Science* **320**, 1617–1619 (2008).
42. Boujibar, A. *et al.* Metal–silicate partitioning of sulphur, new experimental and thermodynamic constraints on planetary accretion. *Earth Planet. Sci. Lett.* **391**, 42–54 (2014).

43. Tsuruta, K. & Takahashi, E. Melting study of an alkali basalt JB-1 up to 12.5 GPa: behavior of potassium in the deep mantle. *Physi. Earth Planet. Inter.* **107,** 119–130 (1998).

44. Allègre, J. A., Poirier, J. P., Humler, E. & Hofmann, A. W. The chemical composition of the Earth. *Earth Planet. Sci. Lett.* **134,** 515–526 (1995).

Acknowledgements

We thank B. Bourdon, M. Boyet, S. Charnoz, M. Javoy, E. Kaminski, R. Mishra, S. Keshav, D. Laporte, G. Libourel, H. Martin, Y. Marrocchi, M. Toplis, K. Righter and M. Zolensky for their fruitful discussions and various members of the Laboratoire Magmas et Volcans staff for essential help. This project received funds from INSU-CNRS, Université Blaise Pascal, Région Auvergne and the ANR 'Oxydeep' project. It is a LabEx Clervolc 170 contribution.

Author contributions

A.B. conceived the study, conducted the experiments and acquired the data. A.B., D.A., N.B.-C. and J.M. modelled, interpreted and discussed the results. The manuscript was prepared by A.B., D.A., N.B.-C., M.A.B. and J.M.

Additional information

5

The deuterium/hydrogen distribution in chondritic organic matter attests to early ionizing irradiation

Boris Laurent[1], Mathieu Roskosz[1,*], Laurent Remusat[2,*], François Robert[2], Hugues Leroux[1], Hervé Vezin[3], Christophe Depecker[1], Nicolas Nuns[4] & Jean-Marc Lefebvre[1]

Primitive carbonaceous chondrites contain a large array of organic compounds dominated by insoluble organic matter (IOM). A striking feature of this IOM is the systematic enrichment in deuterium compared with the solar hydrogen reservoir. This enrichment has been taken as a sign of low-temperature ion-molecule or gas-grain reactions. However, the extent to which Solar System processes, especially ionizing radiation, can affect D/H ratios is largely unknown. Here, we report the effects of electron irradiation on the hydrogen isotopic composition of organic precursors containing different functional groups. From an initial terrestrial composition, overall D-enrichments and differential intramolecular fractionations comparable with those measured in the Orgueil meteorite were induced. Therefore, ionizing radiation can quantitatively explain the deuteration of organics in some carbonaceous chondrites. For these meteorites, the precursors of the IOM may have had the same isotopic composition as the main water reservoirs of the inner Solar System.

[1]UMET, Université Lille 1, CNRS UMR 8207, Villeneuve d'Ascq F-59655, France. [2]IMPMC, CNRS UMR 7590, Sorbonne Universités, Université Pierre et Marie Curie, IRD, Muséum National d'Histoire Naturelle, CP 52, 57 rue Cuvier, Paris 75231, France. [3]LASIR, Université de Lille 1, CNRS UMR 8516, Villeneuve d'Ascq F-59655, France. [4]Institut M.E. Chevreul, Université de Lille 1, CNRS, FR 2638, Villeneuve d'Ascq F-59655, France. *These authors contributed equally to this work. Correspondence and requests for materials should be addressed to M.R. (email: mathieu.roskosz@univ-lille1.fr).

Carbonaceous chondrites result from the accretion of materials formed during the early stages of the Solar System's history. These meteorites typically contain a large amount (up to 4 wt%) of organic matter, which was formed by abiotic processes[1] and occurs as small grains in the fine-grained matrix[2] (from a few hundred nanometres to a few micrometres in diameter). The insoluble organic matter (IOM) consists of small aromatic units, which are highly substituted and are connected by short and branched aliphatic chains[3,4]. The presence of monoradicals and diradicals is a major structural feature of extraterrestrial IOM, and they are rarely found in terrestrial counterparts[5]. Another striking feature of this IOM is the systematically large enrichment in deuterium compared with the solar hydrogen reservoir and the Jovian atmosphere[6,7]. The bulk hydrogen isotopic composition of the IOM also contrasts significantly with terrestrial signatures ($\delta D = 972 \pm 2$‰ for the (Ivuna-like) carbonaceous chondrite Orgueil and up to 3,527‰ for the (Renazzo-like) carbonaceous chondrite LEW85332, relative to standard mean ocean water (SMOW)[8]). In addition, the δD of the IOM of Orgueil is heterogeneous at the molecular scale[9]: the isotopic ratio depends on the hydrogen position in the macromolecule (for example, the type of C–H bond). Several mechanisms, occurring either in the interstellar medium[10–12], the protosolar nebula[9,13] or later on in parent asteroids[14], have been proposed to explain the isotopic signature of the IOM. These mechanisms include ion-molecule reactions at low temperatures in molecular clouds[15,16], water reduction and subsequent isotopic exchange with organic molecules during aqueous alteration[14], or a synthesis from ice grains catalysed by the ultraviolet irradiation[17].

In the protosolar nebula, charged particles (mainly protons and electrons) forming the solar wind are streamed along magnetic lines through the protoplanetary disk, and they constantly collide with micrometre-sized dust particles. The interactions of these charged particles with matter are dominated by an ionizing mechanism similar to the effects of energetic photons (ultraviolet and X-rays). Ionizing irradiation refers to an interaction with the electron shells of the target atoms. After such interaction, the electronic structure of the target is at least temporarily affected. The main consequences for the target are ionization and breaking of chemical bonds, but they do not include direct ejection of atoms from the target. During the T-Tauri stage of the Sun's evolution, such irradiation was very efficient because the solar flux was larger than at present[18] by three orders of magnitude[19]. Protons may interact with the nuclei of atoms, causing structural modification and sputtering, but these effects are restricted to the surfaces of solids; the penetration depth of protons is limited to a few tens of nanometres for an energy typical of solar wind particles (for example, within the range of 1–10 keV). For electrons, damage results mainly from the interactions with the electron shells of the target atoms, and the contribution of nuclear interactions (such as sputtering) is negligible. However, electrons induce ionization at depth that can cause chemical, structural and isotopic modifications. Recently, significant D-enhancements were produced by irradiating organic macromolecules with the electron beams generated by transmission and scanning electron microscopes (SEM)[20–22]. Such ionizing irradiation was also shown to produce monoradicals and diradicals[22]. However, these studies were unable to derive intramolecular fractionation factors (α) in relation to the different types of C–H bonds present in the IOM and their possible links with the α values of the chondritic IOM[5]. Consequently, no direct comparison could be made between the experimental results and the D/H signatures measured in the IOM.

Here, we present the effects of electron irradiation on the D/H fractionation of organic precursors containing different functional groups to derive typical intramolecular fractionation factors. Electrons were selected as ionizing radiation because their ionizing efficiency and their penetration depth are both higher than other candidates (ultraviolet photons and protons). Consequently, electron irradiations are easier to carry out in the laboratory, irradiated materials are easier to characterize by conventional spectroscopic and isotopic methods, and derived properties are bulk properties rather than surface properties. For comparison, we have also performed some ultraviolet irradiation experiments (see Supplementary Note 1 for details). The results suggest that the nature of the incoming radiation may not have a determining effect.

Results

Experimental strategy. Despite being structurally distinct from the IOM, polymers are suitable analogues. Polymers and IOM share the same functional groups (for example, aliphatic chains and aromatic cycles). Yet, polymers consist of long chains composed of many repeated units (monomers), whereas the IOM does not show any significant repetition of any pattern. Three polymers, polyethylene terephthalate (PET, $(C_{10}H_8O_4)_n$), polystyrene (PS, $(C_8H_8)_n$) and polyethylene (PE, $(C_2H_4)_n$)), were chosen to derive the intramolecular fractionation factors between the moieties identified in the IOM. Each polymer exhibits specific and distinct combinations of different types of C–H bonds. In PS, five out of eight of H atoms are bonded to aromatic carbon atoms (hereafter referred to as 'ring'), two out of eight to aliphatic carbon (CH_2) and one out of eight to benzylic carbon (CH). Linear aliphatic chains of CH_2 characterize PE, whereas half of the C–H bonds are aliphatic and half are aromatic in PET. The bulk initial hydrogen isotope compositions of these three polymers, assumed to be homogeneous, were close to SMOW and were typical of manufactured organic products (PET: -33 ± 8‰, PS: -30 ± 10‰ and PE: -76 ± 7‰).

Thin films (ranging from 900 nm up to 10 µm) were irradiated in a SEM at 30 keV and room temperature. The electron doses (that is, the amount of deposited energy per unit volume) ranged from 1.6×10^{24} to 1.4×10^{25} eV cm^{-3}. These doses correspond to typical exposure durations as short as a few hundred years, according to the current understanding of the dynamics of protoplanetary disks[18,19,23] and taking into account the fact that direct ionizing irradiation is essentially effective on a thin layer of the disk surface (see the Methods section). Samples were then characterized by using Fourier–Transform Infrared Spectroscopy (FTIR), Time of Flight Secondary Ion Mass Spectrometry (ToF-SIMS) and nano Secondary Ion Mass Spectrometry (NanoSIMS; see the Methods section).

Effects of electron irradiation. Electron irradiation of the three polymers affects their structures in a similar way. First, for each polymer, the intensity of the infrared signal decreases relative to all of the C–H groups, namely CH_2 (1,465 cm^{-1}), CH (2,890 cm^{-1}) and ring (3,030 cm^{-1}), at similar rates (Fig. 1). This indicates that all C–H bonds are simultaneously affected by irradiation. Hence, the decrease of the signal of the functional C–H groups is independent of the dissociation energy of individual C–H bonds. Second, these infrared features and the decrease of the $H^+/C_2H_4^+$, measured by ToF-SIMS, follow a comparable trend for the three polymers (Fig. 2). All of these structural and compositional data can be described by a first-order rate equation. The dose constant (K_E) derived from this formalism describes the rate dependence of the structural evolution for each analogue versus energy deposition. For the three polymers, the K_E values are comparable (4×10^{-25} cm^3 eV^{-1} for PE and PET, 5×10^{-25} cm^3 eV^{-1} for PS, also see the Methods section). This indicates that both C–H

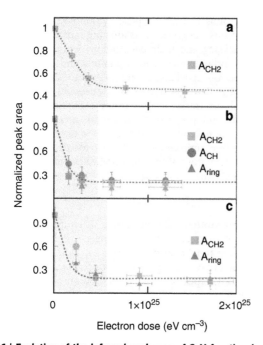

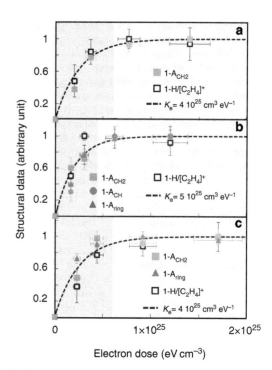

Figure 1 | Evolution of the infrared peak area of C–H functional groups. All of the experiments were performed at 30 keV and 300 K. This evolution is documented for (**a**) polyethylene (PE), (**b**) polystyrene (PS) and (**c**) polyethylene terephthalate (PET). For each sample, the intensity of the infrared signal related to CH_2 (1,465 cm^{-1}), CH (2,890 cm^{-1}) and ring (3,030 cm^{-1}) are considered. The peak areas (namely A_{CH2}, A_{CH}, A_{ring}) are normalized to the peak area of non-irradiated starting materials. On the three panels, the blue region represents the plateau, whereas structural, compositional and isotopic changes are still functions of the electron dose in the grey area. The doses required to reach 90% of the structural plateau delimit the blue and grey regions of the plots.

Figure 2 | Structural and compositional evolution of the three polymers. All of the experiments were performed at 30 keV and 300 K. A combination of ToF-SIMS ($H^+/C_2H_4^+$) and FTIR data (the peak areas A_{CH2}, A_{CH}, A_{ring}) are presented for (**a**) polyethylene (PE), (**b**) polystyrene (PS) and (**c**) polyethylene terephthalate (PET). Data are all normalized between 0 and 1, allowing a direct comparison of different physical quantities. All evolutions are relative to the properties of the non-irradiated precursor. The coefficient K_E is the electron dose constant obtained from a first-order rate equation (see the Methods section for details). On the three panels, the blue region represents the plateau, whereas structural, compositional and isotopic changes are still functions of the electron dose in the grey area. The doses required to reach 90% of the structural and chemical plateaus delimit the blue and grey regions of the plots. Data for PET are from ref. 22.

bond break-up and hydrogen loss from the polymer films occurred at comparable rates for all polymers.

More importantly, the strong correlation between the structural, compositional and isotopic evolutions previously established for the PET film[22] is confirmed for all samples. The aforementioned first-order rate equation and K_E dose constant derived from structural and compositional data were used to describe the isotopic data set. No additional adjustments were required to produce remarkable fits to the data (Fig. 3). As the electron dose increases, the δD of each polymer increases until it reaches a plateau value (see the Methods section). Here, the major feature is that each polymer exhibits a different isotopic plateau, despite their comparable rates of structural and compositional evolution. These plateau values correspond to the maximum bulk isotopic compositions (relative to the non-irradiated starting materials), with δD values of $463 \pm 113‰$, $315 \pm 30‰$ and $272 \pm 60‰$ for PE, PS and PET, respectively. This is a consequence of the nature and the relative proportions of the different C–H groups present in the starting material.

Discussion

The three studied polymers have different D/H plateau values. The nature and the relative proportions of the different C–H groups therefore control the extent of the bulk isotopic signature under ionizing irradiation. From a plateau value, the individual isotopic signatures of aliphatic, aromatic and benzylic C–H bonds of the irradiated polymers can be calculated by mass balance, taking into account the relative proportions of these groups in the three different polymers. A single set of values is determined

thanks to three independent mass balance equations (see the Methods section): δD (vs SMOW) values of $81 \pm 128‰$, $463 \pm 113‰$ and $1,192 \pm 173‰$ for aromatic, aliphatic and benzylic C–H groups, respectively. Assuming local equilibrium, the intramolecular fractionation factors (for example, $\alpha_{A-B} \approx (\delta D_A + 1,000)/(\delta D_B + 1,000)$) associated with the irradiation process were derived: $\alpha_{benzylic-aromatic} = 2.03 \pm 0.29$ and $\alpha_{aliphatic-aromatic} = 1.35 \pm 0.19$. These values are in remarkable agreement with the intramolecular fractionation factors determined from the IOM of the Orgueil meteorite (Table 1), for which the measured fractionation factors are $\alpha_{benzylic-aromatic} = 1.96 \pm 0.05$ and $\alpha_{aliphatic-aromatic} = 1.35 \pm 0.05$[9]. Furthermore, our results are also in agreement with the fractionation factors characterizing the isotopic equilibrium between organic molecules and a deuterium plasma (D_3^+)[24]. This match may suggest that the ionizing irradiation induces a local isotopic equilibration between the irradiated solid and the hydrogen-rich plasma. This assertion is further sustained by the lack of significant isotopic exchange between organic insoluble macromolecules and H_2, H_2O or other neutral volatile species under ambient conditions[24]. Our study shows that the heterogeneous distribution of H isotopes at the molecular scale in Orgueil IOM is likely a result of an ionizing irradiation process on different types of C–H groups.

Another excellent agreement is found between the δD (relative to SMOW) derived for each C–H group of irradiated samples and the δD measured in the IOM of Orgueil. Hence, if the isotopic signature of Orgueil was shaped by irradiation, the bulk isotopic composition of the precursor of the IOM before irradiation was very similar to the composition of our starting material, being very close to SMOW. The composition of the water trapped in Orgueil is also commensurate with SMOW[14]. Therefore, our result suggests that both the precursor of the IOM and the water in Orgueil may have had the same initial isotopic composition as most of the water of the inner bodies. More generally, this new fractionating process could explain the systematic D-enrichment measured in the IOM relative to the water present in all carbonaceous chondrites[14]. Such a contrasting behaviour between water and organic matter would be consistent with the poor

efficiency of the disk ionization in enhancing the D/H of the solar water reservoirs[25].

In the protoplanetary disk, other solar ionizing particles, such as energetic photons and protons, may also have contributed to shaping the structural and isotopic signature of the IOM. Preliminary results of the structural and isotopic evolutions of a PET film irradiated by ultraviolet radiation suggest that the nature of the incoming ionizing particle may not be the pivotal factor. For a given deposited energy, the D/H and the nature of the radicals formed are comparable for ultraviolet and electron irradiation (Fig. 3, Methods section and Supplementary Fig. 4). However, each particle has its own characteristic penetration depth and ionizing efficiency. In this respect, electrons appear to be relatively versatile and are able to modify the target over several micrometres in depth. Still, our results call for a detailed comparison of the effects of these different forms of ionizing radiation at the molecular scale. Such a comparison could help to better understand the molecular processes responsible for the deuteration of organic precursors. These mechanisms are not yet clearly identified and may either operate during the interaction with the incoming particles or during the rearrangement/recombination of the material. The fact that, for a given deposited energy, ultraviolet and electron irradiation induce comparable D-enrichments suggests that fractionation may be primarily produced during recombination rather than during the interactions between a given type of particle and the organic precursors. However, this assertion must be experimentally confirmed because it would lead, in turn, to a more comprehensive description of the D/H evolution in organic matter in the protoplanetary disk regions exposed to ionizing radiation from the protosun. Finally, a better understanding of the irradiation-driven fractionation of other light elements (such as nitrogen, carbon and oxygen) should also shed new light on the nature and evolution of the extraterrestrial IOM and water reservoirs, and it might even extend our findings to CR chondrites that show strong enrichments in both heavy nitrogen and deuterium.

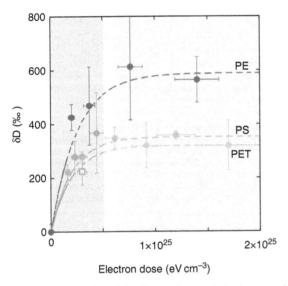

Figure 3 | Isotopic evolution of the three polymers. Isotopic compositions (δD) for polyethylene (PE—brown symbols), polystyrene (PS—green symbols) and polyethylene terephthalate (PET—blue symbols) are presented relative to the non-irradiated starting materials. All of the experiments were performed at 30 keV and 300 K. Dashed lines represent the calculated trends given by the first-order rate equations. The same K_E values were used for isotopic and structural evolutions. The blue region represents the plateaus, whereas structural, compositional and isotopic changes are still functions of the electron dose in the grey area. The doses required to reach 90% of the isotopic plateaus delimit the blue and grey regions of the plots (these doses are different for each analogue but sit within the white intermediate region of the plot). An additional datum (blue open square), corresponding to the isotopic signature of a PET film irradiated by ultraviolet, is also provided. It plots along with other data collected on PET films irradiated by electrons. The calculations of both electron and ultraviolet doses are detailed in the Methods section.

Methods

Sample preparation and irradiation. The PS samples were prepared by spin-coating deposition on a silicon wafer. Spin coating is a centrifugal force-driven method for the deposition of thin films on a flat substrate. The deposition was achieved by the rotation of the substrate, on which a small quantity of the solubilized film is spread. The PS was solubilized in toluene. For a speed of 2,000 r.p.m. and a concentration of 10 wt% PS in toluene, a final thickness of 1.3 µm was obtained. After the deposition, samples were stored under secondary vacuum conditions for several hours to ensure complete evaporation of toluene and polymerization of the film. The PE samples were derived from a 10-µm-thick film (GoodFellow Cambridge Limited) and were mounted to a brass sample holder using conductive copper tape. The PET consisted of a 900-nm-thick film that was biaxially oriented and semi-crystalline (from Good Fellow Co.), and it was deposited on a brass sample holder.

Electron irradiation experiments were carried out on a Hitachi S4700 field-emission gun scanning electron microscope at 30 keV (at the Electron Microscope Facility, University Lille 1). Electron fluences were measured with a Faraday cup. The onset of the melting of PE is typically lower than 80 °C. After irradiation, the film did not show any indication of thermal degradation. We can thus estimate the temperature of the samples under the electron beam to be between 20 and 80 °C. After irradiation experiments and between the different analytical steps, all samples were stored under a flow of dry argon in the dark.

Ultraviolet irradiations were performed on PET films. They were performed with a polychromatic ultraviolet lamp (Hamamatsu irradiation system) equipped with a band-pass filter at 239 nm. At this wavelength, the coupling between ultraviolet and the PET film was especially efficient[26]. The nominal energy of the lamp was 300 µW cm^{-2}. However, the filter absorbed 80% of the incident energy at 239 nm so that the actual energy of the lamp during these experiments was estimated to be 60 µW cm^{-2}.

Determination of the energy deposited (electron and ultraviolet doses). The modifications induced by electron irradiation are directly proportional to the energy deposited in the sample[22]. The fluence (that is, the number of electrons per surface unit) is an important parameter in the determination of the energy

Table 1 | Intramolecular isotopic signatures and equilibrium fractionation factors.

	δD_{Benz}	δD_{Aliph}	δD_{Arom}	$\alpha_{Benz\text{-}Arom}$	$\alpha_{Aliph\text{-}Arom}$
This study	1,192 ± 173	463 ± 113	81 ± 128	2.03 ± 0.29	1.35 ± 0.19
Remusat et al.[9]	1,250 ± 50	550 ± 50	150 ± 50	1.96 ± 0.05	1.35 ± 0.05
Robert et al.[24]				1.99 ± 0.38	1.22 ± 0.39

This table provides experimental isotopic signatures for benzilic (δD_{Benz}), aliphatic (δD_{Aliph}) and aromatic (δD_{Arom}) C–H functional groups determined from the irradiated samples. Derived equilibrium fractionation factors between these groups are also reported ($\alpha_{Benz\text{-}Arom}$ and $\alpha_{Aliph\text{-}Arom}$). Measurements on the Orgueil (CI) chondrite[9] and results of the experimental exchange experiments[24] are reproduced for comparison.

deposited by electrons in the target samples. Electron fluences were measured with a Faraday cup with a precision of $\pm 15\%$ (Supplementary Table 1). For irradiation performed at 30 keV, the energy is not entirely dissipated in the film, as the path followed by the electrons is longer than the thicknesses of the PE and PS samples (penetration depth up to 17.2 μm at 30 keV, see Supplementary Fig. 1). In addition, we calculated the energy deposition in the first few hundred nanometres, which is the typical thickness analysed by NanoSIMS for the D/H measurements. The energy deposited in the film directly depends on the electronic stopping power (that is, the linear rate of the energy transfer between the incoming particle and the target). Here, the characteristic electron path and the stopping power are taken from the ESTAR database[27] (Supplementary Table 1). To take into account both the fluence and the stopping power, all of the isotopic and structural data are presented as a function of the *dose (E)*:

$$E = \sigma_E . F \qquad (1)$$

where E is the electron dose (eV cm^{-3}), σ_E is the electronic stopping power (MeV cm^{-1}) and F is the fluence (e$^-$ cm^{-2}).

A comparable approach was used for ultraviolet irradiation to directly compare the effects of ultraviolet and electron irradiation as a function of the energy deposited on the PET film. Here, an absorption coefficient of 2.5×10^5 cm^{-1} was selected for photons with a wavelength of 239 nm (ref. 26). The calculated deposited energy after 26 h of irradiation is 3.1×10^{25} eV cm^{-3} within the first 100 nm of the PET films, which is the typical thickness analysed by NanoSIMS.

Sample analysis. The evolution of the molecular structure of samples during irradiation was studied by infrared spectroscopy (FTIR) using the imager spotlight 300 from Perkin-Elmer at UMET, University Lille 1. Transmission spectra were collected by averaging 32 scans in the 750–4,000 cm^{-1} range. A linear baseline correction was applied to each spectrum between 750 and 3,700 cm^{-1}. The peak areas were evaluated using Gaussian deconvolution (the commercial Peakfit software by Systat software, Inc. was used for this purpose). For the PS film, the infrared spectra were complicated by the presence of interference fringes. These fringes typically result from the thin sample (1.3 μm) synthetized by spin-coating deposition. The substrate and the film act as a plane–parallel interface from which interference patterns (sinusoidal waves) arise. These interferences (a sinusoidal contribution) were removed from the general spectra using a conventional subtraction method.

For all irradiated samples, a clear signal of adsorbed water is observed at the surface of the films (Supplementary Fig. 2). Furthermore, in the case of PE, the infrared spectra in the range 750–1,750 cm^{-1} present additional optical features that could not be attributed to any vibrational modes, based on the available literature. These features and the spectral contribution of adsorbed water make the analysis of structural changes very complicated and necessarily qualitative. For this reason, we only focused on the normalized evolution of the areas of the main characteristic peaks for each polymer but not on unattributed peaks or those originated from surface artefacts because of adsorbed water. In the case of the PE samples, the typical bands of adsorbed water were first subtracted from the spectra before quantification of the peak areas (Supplementary Fig. 2).

All of the D/H analyses were performed on the Cameca NanoSIMS 50 at MNHN Paris, France. The surfaces of the samples were gold-coated (20 nm thick) before analysis, and an electron-flooding gun was used during analysis. The sample surface was rastered by a primary Cs$^+$ beam over areas of 8×8 μm^2 divided into 64×64 pixels at a raster speed of 2 ms per pixel. The isotopes were collected from the inner 5.6×5.6 μm^2 area in multicollection mode with a beam current of 16 pA. Secondary ions of H- and D- were collected by electron multipliers with a dead-time of 44 ns. The mass spectrometer was set to 4,000 mass resolving power. Before each analysis, a 400-pA primary beam was rastered over a 10×10 μm^2 surface area over a period of 50 s to remove gold coating and surface contamination. This procedure was particularly important for removing the sample layers containing adsorbed water. For such polymers, this presputtering typically removes ~ 50 nm of material before the analysis starts. The instrumental mass fractionation was corrected by measuring several times a day the unirradiated films used as internal standards. Turning to the instrumental drift, a daily measurement of a natural kerogen type III showed a reproducibility of $\pm 30\%$. A typical analysis consisted of bracketing measurements of irradiated areas by nearby unirradiated zones. Error bars reported in this study are 1σ errors based on the quadratic sums of standard deviations on replicates performed on non-irradiated (three to four analysis) and irradiated areas (two to four analysis). For each polymer and before irradiation, the δD relative to the SMOW was determined by conventional gas source mass spectrometry (performed at CRPG, Nancy) and was found to be $-33 \pm 8\%$, $-30 \pm 10\%$ and $-76 \pm 7\%$ for PET, PS and PE, respectively.

The ToF-SIMS spectra measurements were carried out in positive mode using a ToF-SIMS V instrument (ION-TOF GmbH Germany) at UCCS, University Lille 1. This instrument is equipped with a Bi liquid metal ion gun. Pulsed Bi$^+$ primary ions were used for the analysis (25 keV, 1 pA). Surface spectra were taken from an area of 500×500 μm^2 (256×256 pixels—20 scans) and then reconstructed over 100×100 μm^2 areas corresponding to irradiated and non-irradiated zones. Charging effects, due to the primary ion beam, were compensated for using pulsed low-energy electrons (20 eV). The C/H ratios for PS, PE and PET could not be directly determined from ToF-SIMS measurements. Instead, we quantitatively

compared the intensity ratio H$^+$/C$_2$H$_4^+$ for each irradiated spot with the surrounding unirradiated material. This choice of using a molecular fragment instead of a simple ion (C$^+$) is related to the analytical mode used for ToF-SIMS analysis. This setup has two advantages. First, it provides information on the evolution of the carbon skeleton through irradiation (from H$^+$ to C$_{15}$H$_{13}^+$ masses). After irradiation, the light masses (corresponding to short fragments of irradiated matter) are more abundant than heavier molecular fragments (Supplementary Fig. 3). This is the direct result of the chain scission caused by irradiation. Second, this setup reduces the carbon-based (C$^+$) contamination during surface analysis. For these reasons, C$_2$H$_4^+$ moieties were chosen as the proxy for documenting the change in the C/H ratio during irradiation. Only normalized quantities (relative to unirradiated samples) are provided here because the measured intensity ratios are not directly commensurable to the true atomic C/H ratio. Finally, as for isotopic measurements, a typical analysis consisted of bracketing measurements of irradiated spots and unirradiated zones. Error bars reported in this study are 1σ s.d.s of the replicates on non-irradiated areas (three to four analyses), and classical error propagation was applied to normalized values.

Concentrations of organic radicals were derived from electronic paramagnetic resonance (EPR). The EPR spectra were recorded with a Bruker Elexsys E500 operating at 9 GHz at LASIR, University Lille 1. Amplitude modulation and microwave power were set to 2 G and 5 mW, respectively. For ultraviolet irradiation, the film of PET (6×6 mm^2 size) was directly irradiated inside the transmission mode high sensitivity (TMHS) cavity through optical windows. The spin concentration was determined using weak pitch with a double integration procedure. Typically, in this analytical configuration, the absolute error on the spin concentration is better than 5%.

Kinetic analysis of structural and isotopic changes. All of the structural, chemical and isotopic changes presented in this study share a common kinetics during irradiation. A first-order rate law successfully describes such changes. Typically, a first-order rate law relates the time derivative of the concentration of a given species to its instantaneous concentration through a time constant. As the electron dose E is a function of both the stopping power and the electron fluence equation (1), the evolution of a quantity A is described as

$$\frac{d[A]}{dE} = -k_E \cdot [A]. \qquad (2)$$

where k_E is the dose constant. The integration of this relation gives access to the evolution of the quantity A as a function of the electron dose:

$$[A] = [A_0]e^{-k_E \cdot E}, \qquad (3)$$

where $[A_0]$ is the initial intensity of a given feature of the polymer over an increasing dose E. The different patterns of change can be compared because the dose constant is a direct proxy for the rate of molecular change through irradiation. For this purpose, we used normalized quantities (peak areas and ratios).

Based on this fitting procedure, we determined the electron dose required for each sample to reach 90% of the plateau value (Supplementary Table 2). These doses were, in turn, used as threshold values to determine the isotopic signature of each sample at the plateau. All experiments performed at higher doses were taken into account to calculate the characteristic isotopic signature on the plateau. Typically, two data points for each polymer were found to sit on the plateau. The isotopic plateau values for each polymer were determined as the mean of the isotopic data situated on the plateau. The error was taken as the 1σ standard deviation of such data set.

Determination of intramolecular fractionation factors. As for the structural evolution, the isotopic data set shows isotopic plateaus (Fig. 3). The isotopic signatures of irradiated materials at their respective plateaus (given relative to SMOW) are $463 \pm 113\%$, $315 \pm 30\%$ and $272 \pm 60\%$ for PE, PS and PET, respectively. The isotopic signature of each C–H bond was then determined assuming local equilibrium under the electron beam. A simple mass balance was considered because the initial relative proportions of the C–H bonds are known, and all C–H groups are lost at a comparable rate during irradiation, as discussed above. Hence:

$$\left(\frac{D}{H}\right)_{PE} = \left(\frac{D}{H}\right)_{Aliph} \qquad (4)$$

$$\delta D_{Aliph} = 463 \pm 113\ \%o \qquad (5)$$

$$\left(\frac{D}{H}\right)_{PET} = \frac{1}{2} \cdot \left(\frac{D}{H}\right)_{Aliph} + \frac{1}{2} \cdot \left(\frac{D}{H}\right)_{Arom} \qquad (6)$$

$$\delta D_{Arom} = 81 \pm 128\ \%o \qquad (7)$$

$$\left(\frac{D}{H}\right)_{PS} = \frac{2}{8} \cdot \left(\frac{D}{H}\right)_{Aliph} + \frac{5}{8} \cdot \left(\frac{D}{H}\right)_{Arom} + \frac{1}{8} \cdot \left(\frac{D}{H}\right)_{Benz} \qquad (8)$$

$$\delta D_{Benz} = 1192 \pm 173\ \%o \qquad (9)$$

From these molecular δD (relative to SMOW), the intramolecular fractionation factors were easily determined:

$$\alpha_{BenzylicAromatic} = \frac{\left(D/H\right)_{Benz}}{\left(D/H\right)_{Arom}} = 2.03 \pm 0.29 \quad (10)$$

$$\alpha_{AliphaticAromatic} = \frac{\left(D/H\right)_{Aliph}}{\left(D/H\right)_{Arom}} = 1.35 \pm 0.19 \quad (11)$$

The errors of the ratios were propagated to the α values (quadratic sums), and the relative errors of the factors were found to be lower than 14%.

Timescale for a similar irradiation in the disk. In a protoplanetary context, the timescale required for the irradiation to reach a dose comparable to those studied here was determined from the measured present-day energy distribution of the quiescent solar flux[18]. This distribution follows a power law where low-energy electrons (from 1 to 10 keV) are much more abundant than high-energy ones (Supplementary Fig. 5). For the low-energy range (1–10 keV), the integrated flux and the associated electron dose rate were calculated (blue and red areas in Supplementary Fig. 5). The stellar activity of the protostars has been shown to be at least three orders of magnitude higher than today[19,28]. Thus, we considered an electron flux higher than the quiescent present-day solar flux by three orders of magnitude. In this context, only few hundred years are needed to produce the same deposited energy that was experienced in our experimental irradiations (Supplementary Table 3).

Finally, protoplanetary environments are optically thick and cold. Therefore, direct electron (photon or ion) irradiation is essentially effective on a thin and warmer layer of the disk surface. Conversely, close to the midplane, the dust is not exposed to solar particles. This shield effect is, however, poorly efficient because of the vertical motions predicted to occur in a turbulent disk[23]. Current models show that micrometre-sized dust from the disk might have spent up to 5% of its lifetime at altitudes where irradiation is efficient[23]. Therefore, the doses required to reach the plateau values are completely compatible with the typical lifetime of the disk.

References
1. Sephton, M. A. Organic matter in carbonaceous meteorites: past, present and future research. *Phil. Trans. R. Soc. A* **363**, 2729–2742 (2005).
2. Le Guillou, C., Bernard, S., Brearley, A. J. & Remusat, L. Evolution of organic matter in Orgueil, Murchison and Renazzo during parent body aqueous alteration: Insitu investigations. *Geochim. Cosmochim. Acta* **131**, 368–392 (2014).
3. Gardinier, A. *et al.* Solid state CP/MAS 13C NMR of the insoluble organic matter of the Orgueil and Murchison mete- orites: quantitative study. *Earth Planet. Sci. Lett.* **184**, 9–21 (2000).
4. Remusat, L., Derenne, S. & Robert, F. New insight on aliphatic linkages in the macromolecular organic fraction of Orgueil and Murchison meteorites through ruthenium tetroxide oxidation. *Geochim. Cosmochim. Acta.* **69**, 4377–4386 (2005).
5. Binet, L., Gourier, D., Derenne, S., Robert, F. & Ciofini, I. Occurence of abundant diradicaloid moieties in the insoluble organic matter from the Orgueil and Murchison meteorites: a fingerprint of its extraterrestrial origin? *Geochim. Cosmochim. Acta* **68**, 881–891 (2004).
6. Robert, F. & Epstein, S. The concentration and isotopic composition of hydrogen, carbon and nitrogen in carbonaceous meteorites. *Geochim. Cosmochim. Acta.* **46**, 81–95 (1982).
7. Geiss, J. & Gloecker, G. Abundances of deuterium and helium in the protosolar cloud. *Space Science Rev* **84**, 239–250 (1998).
8. Alexander, C. M. O. D. *et al.* The origin and evolution of chondrites recorded in the elemental and isotopic compositions of their macromolecular organic matter. *Geochim. Cosmochim. Acta.* **71**, 4380–4403 (2007).
9. Remusat, L., Palhol, F., Robert, F., Derenne, S. & France-Lanord, C. Enrichment of deuterium in insoluble organic matter from primitive meteorites: a solar system origin? *Earth Planet. Sci. Lett* **243**, 15–25 (2006).
10. Yang, J. & Epstein, S. Interstellar organic matter in meteorites. *Geochim. Cosmochim. Acta* **47**, 2199–2216 (1983).
11. Sandford, S. A., Bernstein, M. P. & Dworkin, J. P. Assessment of the interstellar processes leading to deuterium enrichment in meteoritic organics. *Meteorit. Planet. Sci* **36**, 1117–1133 (2001).
12. Busemann, H. *et al.* Interstellar chemistry recorded in organic matter from primitive meteorites. *Science* **312**, 727–730 (2006).
13. Willacy, K. & Woods, P. M. Deuterium chemistry in protoplanetary disks. II. The inner 30 AU. *Astrophys. J* **703**, 479–499 (2009).
14. Eiler, M. E. & Kitchen, N. Hydrogen isotope evidence for the origin and evolution of the carbonaceous chondrites. *Geochim. Cosmochim. Acta* **68**, 1395–1411 (2004).
15. Geiss, J. & Reeves, H. Deuterium in the solar system. *Astron. Astrophys* **93**, 189–199 (1981).
16. Ceccarelli, C. *et al.* The puzzling detection of D_2CO in the molecular cloud L1689N. *Astron. Astrophys* **381**, 17–20 (2002).
17. Ciesla, F. J. & Sandford, S. A. Organic synthesis via irradiation and warming of ice grains in the solar Nebula. *Science* **336**, 452–454 (2012).
18. Lin, R. P. *et al.* A three-dimensional plasma and energetic particle investigation for the WIND spacecraft. *Space Sci. Rev.* **71**, 125–153 (1995).
19. Feigelson, E. D., Garmire, G. P. & Pravdo, S.H. Magnetic flaring in the pre-main sequence Sun and implications for the early Solar System. *Astrophys. J* **572**, 335–349 (2002).
20. De Gregorio, B. T. *et al.* Isotopic anomalies in organic nanoglobules from Comet 81P/ Wild 2: Comparison to Murchison nanoglobules and isotopic anomalies induced in terrestrial organics by electron irradiation. *Geochim. Cosmochim. Acta* **74**, 4454–4470 (2010).
21. Le Guillou, C., Remusat, L., Bernard, S., Brearley, A. J. & Leroux, H. Amorphization and D/H fractionation of kerogens during experimental electron irradiation: comparison with chondritic organic matter. *Icarus* **226**, 101–110 (2013).
22. Laurent, B. *et al.* Isotopic and structural signature of experimentally irradiated organic matter. *Geochim. Cosmochim. Acta.* **142**, 522–534 (2014).
23. Ciesla, F. J. Residence times of particles in diffusive protoplanetary disk environments. I. Vertical motions. *Astrophys. J* **723**, 514–529 (2010).
24. Robert, F., Derenne, S., Thomen, A., Anquetil, C. & Hassouni, K. Deuterium exchange rate between D_3^+ and organic CH bonds: Implications for D enrichment in meteoritic IOM. *Geochim. Cosmochim. Acta.* **75**, 7522–7532 (2011).
25. Cleeves, I. L. *et al.* The ancient heritage of water ice in the solar system. *Science* **345**, 1590–1593 (2014).
26. Ouchi, I., Nakai, I. & Kamada, M. Anisotropic absorption spectra of polyester films in the ultraviolet and vacuum ultraviolet regions. *Nucl. Instr. Meth. Phys. Res* **199**, 270–274 (2003).
27. Berger, M. J., Coursey, J. S., Zucker, M. A. & Chang, J. *ESTAR, PSTAR, and ASTAR: Computer Programs for Calculating Stopping-Power and Range Tables for Electrons, Protons, and Helium Ions* (version 1.2.3). [Online] Available at http://physics.nist.gov/Star[2012, 06 01]. National Institute of Standards and Technology, Gaithersburg, MD (2005).
28. Benz, A. O. & Güdel, M. Physical processes in magnetically driven flares on the sun, stars, and young stellar objects. *Annu. Rev. Astron. Astrophys.* **48**, 241–287 (2010).

Acknowledgements
We thank C. France-Lanord (CRPG, France) for the bulk D/H analysis of the starting materials. The thorough and constructive comments from M. Chaussidon and N. Dauphas greatly improved the original manuscript. This work was supported by the French ANR Program (2011JS56 004 01, FrIHIDDA), and support for M.R. was provided by the Region Nord-Pas de Calais. The SEM, FTIR and TOF-SIMS apparatus in Lille (France) are hosted by the Chevreul Institute and are supported by the Conseil Regional du Nord-Pas de Calais and the European Regional Development Fund (ERDF). The National NanoSIMS facility at the MNHN was established by funds from the CNRS, Région Ile de France, Ministère délégué à l'Enseignement Supérieur et à la Recherche and the MNHN.

Author contributions
B.L. performed experiments and data analysis with feedback from M.R., L.R. and H.L. and benefitting from an original idea by M.R. and L.R. NanoSIMS, Infrared spectroscopy, EPR spectroscopy and ToF-SIMS analyses were performed by L.R., C.D., H.V. and N.N., respectively. H.L. and J.-M.L. provided expertise on the physics of electron irradiation and polymers, respectively. All co-authors discussed the results. M.R. wrote the manuscript, with important contributions from L.R., F.R, H.L. and B.L.

Additional information

6

Nanoscale infrared spectroscopy as a non-destructive probe of extraterrestrial samples

Gerardo Dominguez[1,2,*], A.S. Mcleod[3,*], Zack Gainsforth[4], P. Kelly[3], Hans A. Bechtel[5], Fritz Keilmann[6], Andrew Westphal[4], Mark Thiemens[2] & D.N. Basov[3]

Advances in the spatial resolution of modern analytical techniques have tremendously augmented the scientific insight gained from the analysis of natural samples. Yet, while techniques for the elemental and structural characterization of samples have achieved sub-nanometre spatial resolution, infrared spectral mapping of geochemical samples at vibrational 'fingerprint' wavelengths has remained restricted to spatial scales $> 10\,\mu m$. Nevertheless, infrared spectroscopy remains an invaluable contactless probe of chemical structure, details of which offer clues to the formation history of minerals. Here we report on the successful implementation of infrared near-field imaging, spectroscopy and analysis techniques capable of sub-micron scale mineral identification within natural samples, including a chondrule from the Murchison meteorite and a cometary dust grain (Iris) from NASA's Stardust mission. Complementary to scanning electron microscopy, energy-dispersive X-ray spectroscopy and transmission electron microscopy probes, this work evidences a similarity between chondritic and cometary materials, and inaugurates a new era of infrared nano-spectroscopy applied to small and invaluable extraterrestrial samples.

[1]Department of Physics, California State University, San Marcos, San Marcos, California 92096-0001, USA. [2]Department of Chemistry and Biochemistry, University of California, San Diego, La Jolla, California 92093, USA. [3]Department of Physics, University of California, San Diego, La Jolla, California 92093, USA. [4]Space Sciences Laboratory, University of California, Berkeley, Berkeley, California 94720, USA. [5]Advanced Light Source, Lawrence Berkeley National Laboratory, Berkeley, California 94720, USA. [6]Ludwig-Maximilians-Universität and Center for Nanoscience, 80539 München, Germany. *These authors contributed equally to this work. Correspondence and requests for materials should be addressed to G.D. (email: gdominguez@csusm.edu).

Infrared (IR) spectroscopy in the frequency range of 400–4,000 cm^{-1} ($\lambda = 25$–2.5 µm) is widely used for the identification of chemical compositions both in the laboratory and in remote observations. However, use of this technique for studying sub-micron scale heterogeneity characteristic of natural samples or individual micron-sized samples such as the cometary dust grains retrieved by the National Aeronautics and Space Administration (NASA)'s Stardust mission[1–6] is severely hampered by the spatial resolution afforded by the diffraction limit and the wavelength of IR light[7,8]. This limitation rules out direct investigations of both sub-micron samples and sub-micron inclusions in larger heterogeneous samples by conventional Fourier transform infrared spectroscopy (FTIR)[9–16]. Yet, as acknowledged among the first-reported findings from the Stardust mission, advances in analytic techniques promise to expose new secrets about the origin and evolution of the solar system contained in the recovered microscopic cometary grains[1]. With regard to IR spectroscopy, this claim cannot be overstated given its reputation as the chemical 'fingerprinting' technique of choice. Moreover, the return of invaluable micron-scale samples from spacecraft missions such as Stardust poses an urgent need for new non-destructive, even non-perturbative, nanoscale probes of chemical structure.

In recent years, developments in scattering-type near-field scanning optical microscopy (s-SNOM) have enabled acquisition of IR spectral information at sub-micron length scales[17,18]. Conceptually, this imaging technique relies on an atomic force microscope (AFM) tip acting as an optical antenna (or 'near-field probe') to focus incident electromagnetic radiation to areas smaller than the wavelength of incident light. Recent applications of the technique include studies of plasmons in graphene and spatial mapping of electronic transitions in correlated oxides[19–25]. By incorporating a broadband mid-IR (MIR) laser and spectroscopic interferometer into an s-SNOM system, investigators have now demonstrated 20-nm-resolved IR nanospectroscopy (nanoFTIR)[26–31]. Yet, to the best of our knowledge, heretofore the only attempt at near-field imaging and spectroscopy of natural geochemical samples has utilized a conventional (thermal) IR source to produce maps of spectral features limited to ~1 µm resolution in a primitive meteorite[32]. However, the absence of background suppression schemes requisite for near-field microscopy places much doubt on their study's efficacy as a *bona fide* sub-diffraction optical probe.

Complemented by scanning electron microscopy (SEM), energy-dispersive spectroscopy (EDS) and transmission electron microscopy (TEM) probes, here we report on the application of IR s-SNOM and nanoFTIR with a metallic near-field probe to identify and distinguish silicate minerals in a sample of Comet 81P/Wild 2 and a portion of the Murchison meteorite, a CM2 chondrite comprised of organics and stony grains (chondrules), with sub-micron (~20 nm) lateral resolution. We also demonstrate that nanoscale IR spectroscopy can sense elemental concentration gradients at the level of a few percent caused by rapid cooling, identify sub-micron intergrowths at the confluence of distinct lattice structures, providing evidence for shock within a chondrule, and directly reveal juxtaposed crystalline and amorphous silicates in the Iris cometary grain. NanoFTIR measurements of Iris are consistent with mounting evidence[3,33,34] (Z. Gainsforth *et al.*, manuscript in preparation) that igneous materials formed at high temperatures deep in the solar nebula were incorporated into cometary bodies, advancing a surprising scenario whereby planetary materials must have been mixed over the grandest scales into the outer solar system[1].

Results

Material identification with nanoFTIR.

Material identification in spatial volumes smaller than the wavelength of light through s-SNOM requires establishing a physical relationship between an incident electromagnetic field, the dielectric properties of the material directly underneath the probe tip, and the consequently scattered electromagnetic field recorded by the microscope (for our experimental set-up see Methods). In the simplest model of s-SNOM, based on the combination of an AFM with IR optics, the incident electromagnetic field irradiates the AFM probe, inducing a dipole of electric charge in its tip, as shown schematically in Fig. 1. In accordance with the boundary conditions of Maxwell's equations at the surface of the dielectric material directly underneath, an interacting mirror dipole is induced within the sample. An electric near-field interaction between the tip and charges from its mirror image modifies the amplitude and phase of backscattered radiation from the probe, which functions therefore as an optical antenna highly sensitive to the dielectric environment near its tip. Using suitable background suppression and techniques, interferometric detection of this backscattered radiation while scanning the sample beneath the tip provides sensitivity to the sample's local permittivity at a resolution comparable to the radius of curvature at the tip apex (≈ 20 nm).

A rudimentary connection between the amplitude and phase of this detected radiation ('near-field signal') and the optical

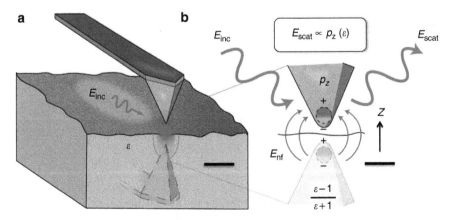

Figure 1 | Principle of near-field microscopy for material identification. (a) The probe tip interacts with at a sample region characterized locally by permittivity ε, a function of wavelength. Scale bar: 10 µm. (b) The incident electromagnetic field E_{inc} induces a strong dipole moment in the tip, which interacts with a mirror dipole in the sample whose strength is determined by the local permittivity. This interaction sensitively affects the tip's vertical polarization p_z, therefore locally modulating the amplitude and phase of the tip-scattered field E_{scat}. Scale bar: 100 nm.

properties of the sample can be achieved through a simple or extended dipole approximation[35,36]. Recently, we have developed more sophisticated models of the optical near-field interaction between the probe and the sample and have applied them with excellent results to the study of graphene–substrate optical coupling[20] and oxide thin films[25]. Our latest and most accurate model[37] interrelates the scattered signal with the dielectric function and optical reflection coefficient of the sample (Supplementary Fig. 1), the same reflectivity accessed by conventional far-field FTIR spectroscopy. Likewise, the imaginary component of the sample's refractive index, often called the IR extinction or absorption coefficient, describes its IR absorption efficiency. Maxima in the frequency-dependent IR absorption coefficient correspond directly to the absorption peaks of 'fingerprint' molecular vibrations measured by conventional transmission-mode FTIR or total internal reflectance spectroscopies.

However, although related, the frequency-dependent scattered field measured by nanoFTIR is not equivalent to a conventional FTIR absorption spectrum. Surface phonons couple particularly strongly to a near-field probe as their electric dipole vibrates resonantly with the incident probe field, shifting peaks in the scattering spectrum to lower energy by an amount commensurate with their strength[38]. Quantitative comparisons between a sample's near-field spectrum (for example, Fig. 2) and its associated conventional FTIR absorption spectrum require that the geometry of the near-field interaction be explicitly taken into account by a realistic model of the tip, such as the one used in this work to 'invert' the scattering amplitude and phase spectra to extract nano-resolved dielectric properties of the material underneath the tip apex[28,37]. Elementary relations connecting the reflection coefficient to the IR absorption coefficient have been presented elsewhere[29] and, for characterization of the cometary particle Iris, we apply these here for the first time to the direct identification of silicate minerals using nanoFTIR (see Methods).

NanoFTIR characterization of a chondrule. Figure 2 schematically depicts our s-SNOM and nanoFTIR system (see Methods) and presents characteristic scattering spectra for a standard crystal of forsterite, one of the most abundant minerals in the solar system and common chondrule constituent. We applied this system first to characterize a sub-millimetre chondrule within the Murchison meteorite, a CM2 chondrite notable for its abundant organics and aqueously altered minerals (preparation in Methods). The chondrule was first coated with ≈ 15 nm of conductive carbon and imaged by SEM; Fig. 3a,b displays backscattered electron images. Elemental compositions (displayed in Fig. 3d) were measured at < 500 nm resolution using EDS. The chondrule

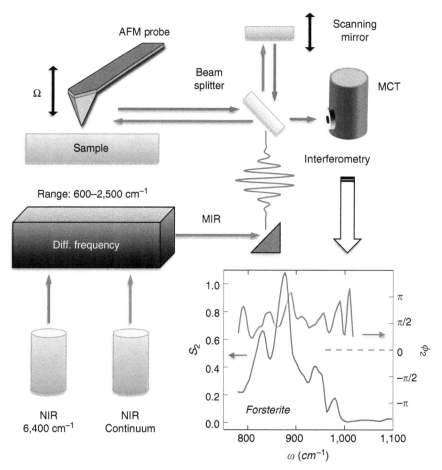

Figure 2 | Implementation of nanoFTIR with a coherent infrared source. A nonlinear optical crystal (GaSe) is used to combine two near-infrared sources in a difference frequency generation scheme, producing coherent MIR radiation that is frequency tunable with a 300-cm^{-1} bandwidth. This beam is focused onto the AFM tip, resulting in strong near-field interactions with the sample and a backscattered field (characterized by amplitude and phase) modulated at the probe-tapping frequency and detected by the HgCdTe (MCT) detector. Interfering this light with a continuously delayed reference beam (scanning mirror) and demodulating at harmonics of the tapping frequency produces an interferogram whose Fourier transform constitutes the desired near-field scattering spectrum. The displayed scattering amplitude and phase spectra are obtained for a forsterite crystal standard (San Carlos olivine).

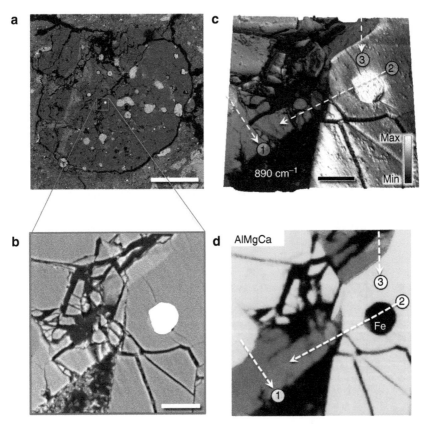

Figure 3 | Nanoscale imaging of a chondrule. (**a**) Backscattered secondary electron (BSE) image of the polished Murchison sample region, with red indicating the field of view for the nanoFTIR images. Scale bar: 100 μm. (**b**) BSE image of this boxed region in **a** where an iron metal grain can be clearly seen. Scale bar: 5 μm. (**c**) Infrared near-field amplitude image acquired at $w = 890\,\mathrm{cm}^{-1}$ superimposed on simultaneously acquired AFM topography (range: 500 nm). Positions of the three (1–3) nanoFTIR line scans presented in Fig. 4 are shown by dashed lines. Scale bar: 5 μm. (**d**) Composite EDS map of the same region again showing the positions of the three (1–3) nanoFTIR line scans presented in Fig. 4. (Green = forsterite, purple = augite, orange = aluminous orthopyroxene.)

was then sputtered with argon to ablate the carbon coat and subsequently imaged with s-SNOM in monochromatic mode (pseudo-heterodyne detection technique[39]) taking illumination from a CO_2 laser at $890\,\mathrm{cm}^{-1}$ to quickly reveal an IR contrast map of the region (Fig. 3c). Optically reflective or vibrationally resonant materials at this frequency display positive imaging contrast. IR scattering contrast is evident among microscopic crystallites, a bright metallic inclusion (iron), and a dark non-resonant background of carbonaceous material (including preparation epoxy).

The chondrule contains Al-, Mg- and Ca-rich silicate phases, and EDS identifies Mg-rich regions (green in Fig. 3d) as forsterite (Mg_2SiO_4). To distinguish extant mineral phases through their vibrational signatures, we collected spectroscopic nanoFTIR line scans 8–10 μm in length across three distinct microcrystalline subregions, acquiring scattered amplitude and phase spectra every 50 nm over the frequency range relevant to the strongly identifying vibrational modes of the silicate minerals (800–1150 cm^{-1}). Scattering amplitude spectra from these line scans are displayed in Fig. 4, exhibiting several spectroscopically distinct subregions. These line scans verify the ubiquitous presence of forsterite (green in Fig. 4) by comparing with the reference spectra (Fig. 2). Compositional boundaries sharper than 100 nm and distinct silicates as small as 200 nm are resolved spectroscopically. Metallic composition of the large iron inclusion (red spectrum in Fig. 4, line scan 2) was confirmed by its exceptionally strong and spectrally uniform induced scattering amplitude, although several superimposed 'void bands' provide evidence for a surface oxide layer of probable terrestrial origin, as

from 'gamma'-FeOOH (lepidocrocite) that exhibits identifying MIR vibrational modes[40]. These resonances are optically 'sensed' through near-field coupling with their metallic host, as leveraged previously for the detection of weak molecular vibrations using metal nanoantennas[41].

Stoichiometries for Al- and Ca-rich regions measured by EDS are consistent with several minerals formed from linear chains of silicate tetrahedra (inosilicates) but were insufficient to unambiguously identify the associated mineral species since EDS is insensitive to the presence of –OH. In such cases, concrete identification of mineral phases has conventionally relied on complementary analytic techniques including X-ray diffraction (XRD), TEM and various transmission-mode X-ray spectroscopies, which although powerful, demand either sample volumes of many cubic microns or thin slices excised from the bulk sample at the cost of man-hours and destroying the nearby sample. Even the electron backscatter diffraction technique, a powerful high-resolution probe of crystal structure (insensitive to glassy structure), requires careful deposition of a conductor over insulating minerals[42]. In stark contrast, nanoFTIR measurements are fundamentally non-destructive, requiring no special sample preparation beyond a surface polish with micron-scale smoothness, which even then might be omitted for smooth or terraced sample regions. Furthermore, since mid-IR energies are insufficient to break chemical bonds or cause ionization, risk to the sample from beam damage during the nanoFTIR measurement is completely implausible when using illuminating radiation intensities $<5\,\mathrm{mW}$ (0.5 mW were used in this work)[43]. Nevertheless, a lamella along line scan 3 (indicated in Fig. 3c,d)

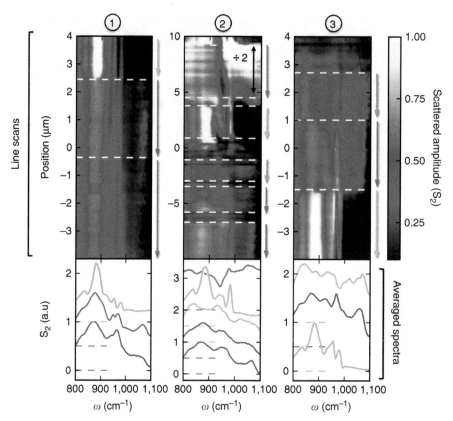

Figure 4 | NanoFTIR characterization of a chondrule. (Line scans) Three nanoFTIR line scans (scattering amplitude versus position) through a chondrule of the Murchison meteorite identify distinct crystalline phases according to their unique near-field spectra; positions of line scan acquisitions are shown in Fig. 3c,d. Line scan 1: a sequence of forsterite (green arrow), Al-poor (blue) and Al-rich (purple) augite. Line scan 2: an iron inclusion with suspected phonon-resonant oxide layer (red, amplitude reduced by 2 ×), followed by forsterite (green), aluminous orthopyroxene (orange), Al-rich and Al-poor augites. Comparatively broad phonon peaks in the final three regions suggest structural disorder. Line scan 3: A progression of aluminous orthopyroxene, augite and forsterite. (Averaged spectra) Position-averaged spectra of these distinct regions are similarly colour coded (offset and re-scaled for clarity, dashes indicate zero lines).

was extracted from our Murchison meteorite sample by focused ion beam and examined by TEM (Supplementary Methods and Supplementary Fig. 4). In this case, we undertook this measure to unambiguously establish a correlative TEM and nanoFTIR spectroscopy study of the same region[44]. Whereas the complete focused ion milling and TEM study required three days of instrument time, nanoFTIR measurements were completed within 3 h.

Analysis of nanoFTIR data inherits from the rich existing literature of vibrational spectroscopy obtained from bulk terrestrial minerals by conventional FTIR. Pyroxenes constitute a solid solution series of single-chain silicates with chemical formula $M^1M^2T_2O_6$ (M^1, M^2 denoting cation sites and T the tetrahedral site, typically Si or Al) including Mg-rich (enstatite) and Ca-rich (augite) members common among chondritic meteorites. Established IR absorption spectra for enstatite[45,46] match vibrational bands of the first several nanoFTIR spectra (Supplementary Fig. 3) of line scan 3 (Fig. 4 in orange), consistent with Mg-dominated occupancy in M^1 and M^2 sites as confirmed by TEM and EDS (Fig. 5b). However, EDS also indicates the Murchison Mg-rich pyroxene contains 5.3 wt% aluminium, making this phase aluminous orthopyroxene (OPX). A discrete change in vibrational character is observed by nanoFTIR upon entry to the Ca-rich phase, which compares favourably to attenuated total internal reflectance spectra of augite[47,48]. Comparatively, such a concrete identification for this compound was otherwise impossible via EDS, since the measured elemental stoichiometry admitted other mineral

assignments (Supplementary Fig. 3). The abrupt disappearance and replacement of several vibrational modes displayed in Fig. 5a (orange to purple) matches an abrupt compositional change observed by EDS with a reduction in Mg by 10 wt% and an increase in Ca by 14 wt%. Accordingly, a phase transition from OPX to CPX (clinopyroxene) was measured by selected area diffraction in TEM (Supplementary Methods). While existing IR studies fail to observe marked differences in vibrational character between ortho- and clinoenstatites[46,49], the monoclinic structure of augite presents yet a distinct space group, C2/c versus the $P2_1/c$ of clinoenstatite. We therefore attribute the resolved boundary in nanoFTIR to differential cation occupations in M^1 and M^2 (ref. 50), establishing nanoFTIR as a sensitive indicator of subtle shifts in elemental chemistry at the nanoscale. We infer these cation concentrations resulted from strong temperature gradients as the host pyroxene rapidly crystallized.

Our correlative TEM subsequently resolved the transition zone between OPX and CPX phases (Fig. 5d) and electron diffraction verified our identification of phases (Fig. 5c and Supplementary Fig. 4). Moreover, a 300-nm-wide superlattice of CPX and OPX lamella (marked OCPX) is revealed through electron diffraction as a continuum of reflections along the a^* axis and shows contrast in high-resolution and darkfield TEM imaging. Although only a few hundred unit cells across, this intergrowth region was first resolved by nanoFTIR, manifesting an anomalous IR vibrational signature at several spectroscopy positions across the OPX/CPX interface in Fig. 5a, particularly through the sharp 'kink' in several vibrational modes suspiciously characteristic of inter-

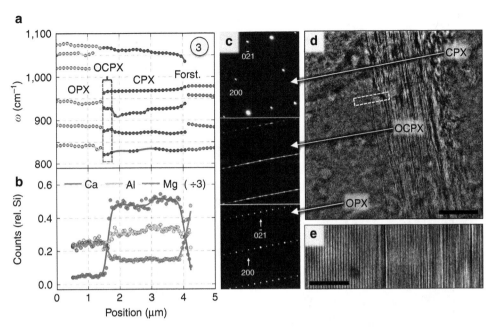

Figure 5 | Correlative nanoFTIR and TEM of sub-micron cation concentration gradients. (**a**) Modal frequencies versus position resolved by nanoFTIR through line scan 3 of the Murchison meteorite. Forsterite (green) gives way to pyroxene phases: calcium-rich clinopyroxene (CPX, purple) and an aluminous orthopyroxene (OPX, orange). Modulations in modal frequencies correlate with increases in cation concentrations at the few percent level as determined by transmission electron EDS (**b**). The spectroscopically distinct 300-nm wide transition band between CPX and OPX observed by nanoFTIR is attributed to an OPX/CPX superlattice (OCPX, dashed region). (**c**) Transmission electron diffraction collected from the respective pyroxene phases confirms the aperiodic lattice superstructure of this band, indicative of crystal shock, as resolved also by (**d**) TEM imaging. Scale bar: 200 nm. (**e**) High-resolution TEM zoom in view onto the boxed region of (**d**) resolves the OPX/OCPX interface. Scale bar: 25 nm.

layer-coupled mode repulsion. The lack of such effect for the $1{,}050\text{-cm}^{-1}$ mode likely associates with its short-range vibrational nature—a 'breathing' mode among isolated silicate tetrahedra (Supplementary Methods)—unaffected by broken long-range crystal symmetry. Such interphase ortho/clinopyroxene regions are a common indicator of crystal shock and rapid cooling[51]. Here we interpret the structure as a consequence of collisional events in the history of Murchison's parent body, in particular the event that may have caused the host chondrule to shatter (Fig. 3a and Supplementary methods). Sensitivity to these shock phenomena establishes nanoFTIR as a promising non-destructive probe of collisional histories among invaluable extraterrestrial samples.

Figure 5a highlights a notable shift in modal frequencies (especially near $950\,\text{cm}^{-1}$ and $1{,}050\,\text{cm}^{-1}$) resolved by nanoFTIR at sub-micron scales in line scan 3, coincident with further lattice displacement of Mg by heavier Ca and Al in CPX at the few-percent level on approach to the adjacent forsterite (Fig. 5b). These vibrational shifts associate with perturbations in the crystal's silicate tetrahedral bond lengths, mediated by commensurate changes in local cation concentration (Supplementary Methods). Vibrational spectroscopy via nanoFTIR can evidently provide a sensitive sub-micron indicator of cation lattice coordination and concentration variations at the few-percent level. We propose that nanoFTIR might be deployed on extraterrestrial pyroxenes to detect T-site occupation by Al and Fe^{3+} that, when compared with other petrographic characteristics, can provide evidence for aqueous alteration or for high fugacity during formation in the solar nebula.

NanoFTIR characterization of a cometary grain. We next applied nanoFTIR to a cometary particle returned from Comet 81P/Wild 2 by the NASA's Stardust recovery mission, a sample for which the non-destructive nature of our probe is of paramount importance. The cometary grain C2052,12,74 ('Iris') was first characterized by synchrotron X-ray, electron, and ion microscopies, as well as secondary ion mass spectroscopy[33]. On a surface of Iris exposed by ultramicrotomy, we first utilized our nanoFTIR system to acquire a broadband integrated image of the particle (Fig. 6a,b). This mode of imaging permits rapid acquisition of relatively large IR maps of the sample in which local contrast indicates the optical response amplitude integrated over the bandwidth of illumination ($800\text{--}1{,}150\,\text{cm}^{-1}$). For example, chunks of Pb–In alloy introduced during sample preparation for secondary ion mass spectroscopy are brightly resolved at the particle periphery. This same region of Iris was also imaged with EDS as shown in Fig. 6c, indicating clear correspondence between the mesostasis region identified previously[33] (Z. Gainsforth et al., manuscript in preparation) and the elemental EDS map. We acquired a nanoFTIR line scan 7 μm in length along the dotted line indicated in Fig. 6b, measuring local scattering amplitude and phase spectra every 100 nm across the mesostasis between two forsterite crystals, and utilized the aforementioned inversion procedure (detailed in Methods and Supplementary Figs 1 and 2) to convert our scattering spectra to a more appropriate series of IR absorption spectra (Fig. 6d) suitable for comparison against conventional IR absorption spectra for candidate silicate minerals[52–54].

Iris is known to consist of olivines set in a groundmass of plagioclase (a solid-solution series of feldspar minerals) with composition 6–26 mol% in Ca-rich anorthite (An) and Na-rich albitic (Ab) glass that formed from quenched igneous cooling. Since the groundmass grew under non-equilibrium conditions, the crystalline plagioclase presents a continuum of chemical compositions including albite and Ca-rich oligoclase

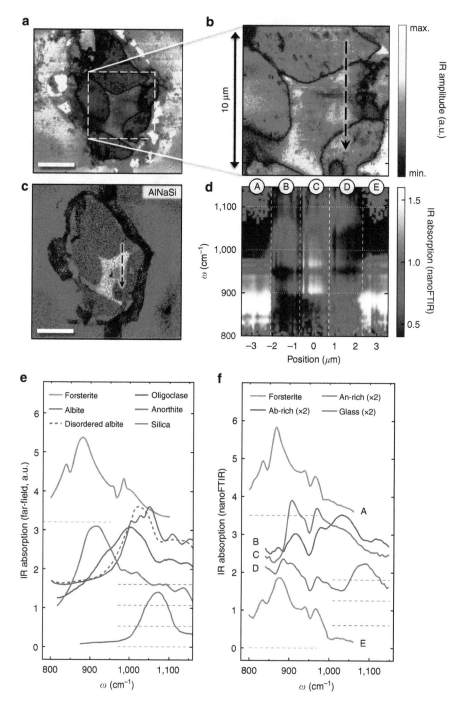

Figure 6 | NanoFTIR characterization of the cometary dust grain Iris. (**a**) Broadband spectrally integrated (800–1,150 cm⁻¹) IR s-SNOM image of Iris. Scale bar: 5 microns. (**b**) Zoom onto the central region; colour-scale for spectrally integrated scattering amplitude at right, with nanoFTIR line scan position indicated by the arrow. (**c**) Composite EDS map displaying relative atomic concentrations of aluminium (red), sodium (green), and silicon (blue). Silicon-rich forsterite crystals as well as the surrounding SiO₂ aerogel appear blue, whereas the groundmass reveals feldspar (white) and glass (red and pink). Scale bar: 5 microns. (**d**) Using a realistic model of the near-field probe-sample interaction, position-resolved IR absorption spectra were extracted from the nanoFTIR line scan across two forsterite crystals (regions labelled A, E) and three distinct glassy phases of plagioclase and mesostasis (B, C, D). (**e**) Literature IR absorption spectra from forsterite, various plagioclase feldspars and amorphous silica. Oligoclase and anorthite spectra are from powdered samples[51], silica from a thin film[53] and all others from bulk (crystalline) samples[52]. (**f**) Region-averaged IR absorption spectra extracted from nanoFTIR line scan positions indicated in **a** (labelled by region). Extracted forsterite spectra agree well with reported absorption spectra[52], whereas plagioclase regions exhibit albite (Ab)-rich, anorthite (An)-rich and glassy signatures according to their phonon absorption frequencies. Sharp anorthite and oligoclase nanoFTIR peaks imply crystalline material, whereas reported spectra for powders (**e**) display added broadening[57,59]. The surprisingly broad albite peak resolved by nanoFTIR is consistent with strong Al-Si disorder[54].

(compositionally intermediate between Ab and An). The remaining liquid was unable to crystallize at all because of rapid cooling. The transition from crystalline to amorphous material is

observable in our line scan (Fig. 6d), wherein three distinct spectral signatures are resolved. Representative spatially averaged nanoFTIR absorption spectra derived from regions bounded by dashed lines in Fig. 6d are shown in Fig. 6f for clarity. Figure 6e displays literature-derived IR absorption spectra for mineral standards that present good matches to these regions based on a comparison of prominent IR absorption band frequencies. Spectra for the purple and red phases (Fig. 6f) exhibit vibrational bands at nearly identical frequencies but at varying intensity, consistent with An-rich and Ab-rich regions, respectively. Accordingly, EDS mapping ascribes elemental abundances from these two regions to oligoclase with a slowly varying composition of 20–26 An mol%.

Whereas the anorthite and oligoclase phonon bands resolved by nanoFTIR at 910 and 970 cm^{-1} appear expectedly sharper than those observed in far-field absorption spectra for powders[55–57] at ~915 and ~990 cm^{-1} (Fig. 6e), the measured albite peak is notably broad (70 cm^{-1} full-width at half-maximum; see Supplementary Fig. 2), comparing most favourably to spectra for crystals with reduced Al–Si order in the tetrahedral site such as effected by high-temperature exposure (>1,000 K) (ref. 54). On the basis of its coincident presence among far-field IR spectra of Iris (Supplementary Fig. 5), the dip-like feature observed throughout our line scan near 950 cm^{-1} appears to be intrinsic in nature, whether associated with a gap in phonon modes of forsterite or with the absence of absorption between sharp peaks associated with An or oligoclase. Whereas EDS resolution was limited to about 750 nm, nanoFTIR resolves variations in spectral absorption at the scale of 100 nm. Lacking gradual shifts in absorption frequencies, the most likely explanation for such sharp well-defined variations in nanoFTIR is the detection of two centrally embedded crystallites of anorthite and oligoclase resolved at the sub-micron scale. Comparatively, TEM was also able to detect one such crystallite (Z. Gainsforth et al., manuscript in preparation) although its spatial relations were unfortunately scrambled (not atypically) from the requisite extraction of a thin microtomed slice of sample.

Moreover, Fig. 6d reveals a third phase (blue in Fig. 6f) where peak positions and amplitudes vary gradually as a function of position. EDS excludes the possibility that this region is plagioclase on the basis of stoichiometry, suggesting rather an albitic glass, and TEM diffraction and brightfield measurements verify the presence of glass in contact with oligoclase elsewhere within this particle (Z. Gainsforth et al., manuscript in preparation). NanoFTIR spectra in this region (Fig. 6d, f—blue) show the growth of a peak near 1,100 cm^{-1} similar to that seen in amorphous SiO_2 standards[53], while retaining some but not all of the peaks expected from plagioclase. This suggests an albitic glass populated by under-resolved nanocrystals of oligoclase, again indicative of rapid cooling from a melt. Demonstrably, nanoFTIR presents a novel means of optical identification and potential compositional characterization[58,59] for bona fide amorphous silicate materials in returned microscale cometary dust grains, as well as in similarly sized interplanetary dust particles or aerosols. The composition of Iris shown here is decidedly inconsistent with the long-held view that comets formed in the outer solar system from agglomerated circum- or interstellar grains[60], comprising—for example, dust, ice and silicates amorphized by radiation exposure. Instead, the direct IR spectroscopic evidence presented here for igneous crystalline and amorphous materials coexisting at the micron-scale within a cometary dust grain adds to mounting evidence for a common history of high-temperature formation shared by cometary and chondritic materials deep within the solar nebula[33,34] (Z. Gainsforth et al., manuscript in preparation). This scenario awaits a suitable explanation for the requisite mixing of planetary materials into the outer solar system that must have enabled their incorporation among remote cometary bodies such as 81P/Wild 2 (ref. 1).

Discussion

We have shown that IR s-SNOM is capable of providing IR spectral identification of natural samples at deeply sub-micron spatial resolutions, probing details of material composition not easily accessed by conventional means. Here we demonstrate this capability within a macroscopic meteorite sample (Murchison) as well as a cometary dust grain (Iris) from NASA's sample return mission, Stardust. Through application of a sophisticated model of the probe–sample near-field interaction, we have shown for the first time that nanoFTIR-extracted IR absorption spectra can be used to quantitatively probe and identify the underlying geochemistry of natural samples at sub-micron resolutions, enabling identification of resonance frequencies and relative levels of structural disorder in a fashion that removes the influence of complex probe–sample interactions. These samples span the size range encountered in the analysis of natural samples.

Together with corresponding elemental maps from EDS, our nanoFTIR line scans through a chondrule in Murchison clearly capture variations in minerology at sub-micron resolutions and are verified by TEM. Furthermore, our nanoFTIR line scans are capable of resolving subtle shifts in vibrational frequencies associated with small variations in silicate cation concentrations at the sub-micron scale. Future work may explore these shifts in a quantitative manner via first-principles calculations. Our nanoFTIR survey of Iris demonstrates that s-SNOM is also sensitive to contrasts in crystallinity, as indicated by the prominence and width of phonon absorption bands within the plagioclase, and future work may establish quantitative measures of disorder. Whereas the physical mechanism of nanoFTIR spectroscopy is relatively settled, and practical quantitative 'inversion' for the interpretation of mineral spectra is now well demonstrated here and elsewhere[37], fully leveraging its sensitivity will call for a detailed empirical or ab anitio approach connecting nanoscale geochemistry to the local phonon response and associated IR permittivity.

These results highlight the high potential of nanoFTIR for geochemical studies of crystallization dynamics or other processes producing variations in the chemical environment over sub-micron length scales, inaccessible to conventional diffraction-limited FTIR. Having demonstrated the application of SNOM-based nanoFTIR for characterizing and imaging silicate minerals, it is hoped that this technique can be applied for studying the distribution of more volatile chemical components found within extraterrestrial samples. This sub-micron mapping could in turn, through strong correlations between minerals such as SiC (well-known presolar astromaterial) and organics, for example, provide new insights into the catalytic role that mineral surfaces may play in the production of prebiotic molecules in the early solar system and planetary systems at large. Finally, the ability to non-destructively characterize volatile species before more destructive/altering techniques are applied to precious and unique returned samples is a significant advance. Future applications of IR s-SNOM and nanoFTIR may also include distribution studies of individual terrestrial aerosol mixing states, examination of synthetic and natural non-equilibrium solids, and sub-micron identification of hydrated and organic chondrite phases. Furthermore, this technique should be suitable for the analysis of additional returned samples from missions including NASA's Stardust spacecraft[1,2,6] and Hayabusa[61], as well as interplanetary dust particles and other primitive solar system materials possessing sub-micron heterogeneity[62].

Methods

Broadband IR source and nanoFTIR implementation. NanoFTIR measurements were performed using instrumentation at the Department of Physics at the

University of California, San Diego. We achieved broadband illumination of our samples using a broadband MIR-coherent source (mod. BB from www.lasnix.com), combining two near-infrared laser beams in a GaSe crystal for difference frequency generation of coherent MIR radiation (100 fs pulses, 300 cm^{-1} bandwidth, tunable across 600–2,500 cm^{-1}, ~3 μW per cm^{-1}). This beam is coupled to an asymmetric Michelson interferometer, allowing measurement of the amplitude and phase of the backscattered radiation with 3 cm^{-1} spectral resolution[28,63]. Our nanoFTIR implementation is depicted schematically in Fig. 2.

Because the scattered signal is a strong function of the distance between the near-field probe and the sample surface, we make use of the AFM probe's tapping capability (tapping mode at a frequency $\Omega \approx 250$ kHz) to modulate the scattered signal and enable distinction from background scattering. The near-exponential dependence of the near-field interaction on the tip–sample distance d implies that scattered field components modulated at harmonics $n\Omega$ with $n \geq 2$, are directly attributable to the near-field polarization of the tip. Sub-diffractional contrasts in the scattered field at these harmonics therefore correspond with variations in the local chemical composition of the sample[38]. NanoFTIR data presented in this work were acquired by lock-in detection of the scattered signal at the second harmonic (2Ω) of the probe's tapping frequency, resulting in second harmonic amplitude S_2 and phase φ_2 signals, whereas imaging data utilize the third harmonic. Finally, quantitative spectral information was obtained by normalizing the frequency-dependent second harmonic signals acquired from sample regions to those spectra acquired from a reference material using the same near-field probe. A silicon wafer with a 100-nm surface layer of evaporated gold (Au) was used for normalization.

Extraction of dielectric properties from nanoFTIR spectra. The lightning rod model of probe–sample near-field interaction realistically relates the experimental measurables of s-SNOM to the sample's dielectric properties near the probe apex. While this sophisticated model does not yield a closed-form relation, it can still be applied in the inverse sense to extract the self-consistent surface reflection coefficients $\beta(\omega) \approx \frac{\varepsilon-1}{\varepsilon+1}$ from sub-50-nm-resolved nanoFTIR spectra[37], provided that sample material is bulk-like (>100 nm in thickness), the spectra are referenced to a known sample material such as gold, and that the probe geometry approximates a metal cone with rounded tip apex. The inversions performed in this work assume a tip radius of 40 nm (the sharpest feature size resolvable by concurrent AFM), a cone half-angle of 20° and probe-tapping amplitude of 60 nm, consistent with probe specifications (Arrow NCPt from NanoWorld) and measurement parameters. The 'effective' (indicated by brackets) dielectric constant $\varepsilon(\omega)$ and absorption coefficient $\kappa(\omega)$ of the material are then given by[29]

$$\langle \varepsilon \rangle = \frac{1-\beta}{1+\beta} \quad \text{and} \quad \langle \kappa \rangle = \text{Im}\left[\sqrt{\varepsilon}\right] = \sqrt{\frac{\sqrt{\langle\varepsilon_1\rangle^2 + \langle\varepsilon_2\rangle^2} - \langle\varepsilon_1\rangle}{2}}.$$

In the case of optically uni- or biaxial crystals, the dielectric properties thus extracted are effective in the sense that they convey optical activity only along crystal axes probed by the incident electric field. However, because near-fields from the probe in the vicinity of the sample are predominantly surface-normal polarized, crystal axes are probed to varying degree, resulting in near-field spectra that reveal all infrared-active vibrational modes possessing a dipole moment perpendicular to the surface[64]. This raises the possibility that 'missing' absorption peaks in $\kappa(\omega)$ could potentially hinder direct comparison to conventional absorption spectra, although this difficulty was not encountered throughout this study.

Sample preparation. We prepared a freshly exposed sample of the Murchison meteorite, a primitive CM2 chondrite, by cleaving an ~5-mm-sized chunk from a much larger piece of Murchison. This fresh chunk was placed at the bottom of a cylindrical mould and embedded in resin (Epo-Fix, Electron Microscopy Sciences). After curing for 24 h, the epoxy cylinder with the Murchison sample was polished using an optical polisher until a region of the sample was exposed. A qualitative determination of optical flatness was achieved by polishing with poly-diamond abrasives, with the final sub-micron polish step performed by hand on a velvet cloth. The sample's flatness was confirmed by briefly imaging in a Tescan Vega 3 SEM and through AFM in the s-SNOM apparatus.

Samples of cometary dust grains from Comet 81P/Wild 2 (Wild 2), captured in a nano-porous silica aerogel by the Stardust mission (1), were extracted and prepared for laboratory analysis using standard techniques[65]. A notable Stardust particle named Iris (from track C2052,12,74) has been analyzed with a variety of techniques including Scanning Transmission X-ray Microscopy and TEM[33]. Using the nanoFTIR set-up that was described previously, we present the first sub-micron IR spectral maps of a cometary sample.

TEM analysis. Murchison line scan 3 (Figs 3d and 5) was verified by S/TEM. After nanoFTIR measurements, a lamella was removed from the meteorite using focused ion beam milling and examined by TEM at the National Center for Electron Microscopy. The lamella exactly traced the nanoFTIR path so that the IR spectrum at each point might be directly compared with TEM imaging, diffraction and EDS composition. Diffraction and imaging in brightfield and darkfield were carried out using a Zeiss Libra 200 MC at 200 keV with an in-column Omega energy filter. With the energy filter, ΔE was ~10 eV centred around the zero-loss peak.

STEM/EDS was carried out using a Phillips CM200 with an Oxford Inca Si(Li) energy-dispersive detector.

References

1. Brownlee, D. *et al.* Comet 81P/Wild 2 under a microscope. *Science* **314**, 1711–1716 (2006).
2. Sandford, S. A. *et al.* Organics captured from Comet 81P/Wild 2 by the Stardust Spacecraft. *Science* **314**, 1720–1724 (2006).
3. McKeegan, K. D. *et al.* Isotopic compositions of cometary matter returned by Stardust. *Science* **314**, 1724–1728 (2006).
4. Keller, L. P. *et al.* Infrared spectroscopy of Comet 81P/Wild 2 samples returned by Stardust. *Science* **314**, 1728–1731 (2006).
5. Flynn, G. J. *et al.* Elemental compositions of Comet 81P/Wild 2 samples collected by Stardust. *Science* **314**, 1731–1735 (2006).
6. Zolensky, M. E. *et al.* Mineralogy and petrology of Comet 81P/Wild 2 nucleus samples. *Science* **314**, 1735–1739 (2006).
7. Maras, A. *et al.* In situ synchrotron infrared reflectance study of olivine microcrystals in meteorite matrices. *Meteorit. Planet. Sci.* **36**, A121 (2001).
8. Cooney, T. F., Scott, E. R. D., Krot, A. N., Sharma, S. K. & Yamaguchi, A. Vibrational spectroscopic study of minerals in the Martian meteorite ALH84001. *Am. Mineral.* **84**, 1569–1576 (1999).
9. Keller, L. P. *et al.* Sulfides in interplanetary dust particles: a possible match to the 23 μm feature detected by the Infrared Space Observatory. *Lun. Plan. Sci.* **31**, 1860 (2000).
10. Raynal, P. I. *et al.* Synchrotron infrared microscopy of micron-sized extraterrestrial grains. *Planet. Space Sci.* **48**, 1329–1339 (2000).
11. Keller, L. P. *et al.* Identification of iron sulphide grains in protoplanetary disks. *Nature* **417**, 148–150 (2002).
12. Keller, L. P. *et al.* The nature of molecular cloud material in interplanetary dust. *Geochim. Cosmochim. Acta* **68**, 2577–2589 (2004).
13. Morlok, A., Jones, G. C., Grady, M. M. & Stansbery, E. FT-IR micro-spectroscopy of fine-grained planetary materials: further results. *Lun. Plan. Sci.* **35**, 1622 (2004).
14. Matrajt, G. *et al.* FTIR and Raman analyses of the Tagish Lake meteorite: relationship with the aliphatic hydrocarbons observed in the diffuse interstellar medium. *Astron. Astrophys.* **416**, 983–990 (2004).
15. Matrajt, G. *et al.* FTIR analysis of the organics in IDPs: comparison with the IR spectra of the diffuse interstellar medium. *Astron. Astrophys.* **433**, 979–995 (2005).
16. Flynn, G. J., Keller, L. P., Feser, M., Wirick, S. & Jacobsen, C. The origin of organic matter in the solar system: evidence from the interplanetary dust particles. *Geochim. Cosmochim. Acta* **67**, 4791–4806 (2003).
17. Knoll, B. & Keilmann, F. Near-field probing of vibrational absorption for chemical microscopy. *Nature* **399**, 134–137 (1999).
18. Taubner, T., Hillenbrand, R. & Keilmann, F. Nanoscale polymer recognition by spectral signature in scattering infrared near-field microscopy. *Appl. Phys. Lett.* **85**, 5064–5066 (2004).
19. Fei, Z. *et al.* Infrared nanoscopy of dirac plasmons at the graphene–SiO$_2$ interface. *Nano Lett.* **11**, 4701–4705 (2011).
20. Fei, Z. *et al.* Gate-tuning of graphene plasmons revealed by infrared nano-imaging. *Nature* **487**, 82–85 (2012).
21. Qazilbash, M. M. *et al.* Mott transition in VO$_2$ revealed by infrared spectroscopy and nano-imaging. *Science* **318**, 1750–1753 (2007).
22. Jones, A. C., Berweger, S., Wei, J., Cobden, D. & Raschke, M. B. Nano-optical investigations of the metal-insulator phase behavior of individual VO$_2$ microcrystals. *Nano Lett.* **10**, 1574–1581 (2010).
23. Atkin, J. M., Berweger, S., Jones, A. C. & Raschke, M. B. Nano-optical imaging and spectroscopy of order, phases, and domains in complex solids. *Adv. Phys.* **61**, 745–842 (2012).
24. Chen, J. *et al.* Optical nano-imaging of gate-tunable graphene plasmons. *Nature* **487**, 77–81 (2012).
25. Zhang, L. M. *et al.* Near-field spectroscopy of silicon dioxide thin films. *Phys. Rev. B* **85**, 075419 (2012).
26. Amarie, S., Ganz, T. & Keilmann, F. Mid-infrared near-field spectroscopy. *Opt. Express* **17**, 21794–21801 (2009).
27. Huth, F. *et al.* Nano-FTIR absorption spectroscopy of molecular fingerprints at 20 nm spatial resolution. *Nano Lett.* **12**, 3973–3978 (2012).
28. Amarie, S. & Keilmann, F. Broadband-infrared assessment of phonon resonance in scattering-type near-field microscopy. *Phys. Rev. B* **83**, 045404 (2011).
29. Govyadinov, A. A., Amenabar, I., Huth, F., Carney, P. S. & Hillenbrand, R. Quantitative measurement of local infrared absorption and dielectric function with tip-enhanced near-field microscopy. *J. Phys. Chem. Lett.* **4**, 1526–1531 (2013).
30. Amarie, S. *et al.* Nano-FTIR chemical mapping of minerals in biological materials. *Beilstein J. Nanotechnol.* **3**, 312–323 (2012).
31. Dai, S. *et al.* Tunable phonon polaritons in atomically thin van der Waals crystals of boron nitride. *Science* **343**, 1125–1129 (2014).

32. Kebukawa, Y. *et al.* Spatial distribution of organic matter in the Bells CM2 chondrite using near-field infrared microspectroscopy. *Meteoritics* **45**, 394–405 (2010).

33. Ogliore, R. C. *et al.* Incorporation of a late-forming chondrule into Comet Wild 2. *Astrophys Lett.* **745**, L19 (2012).

34. Nakamura, T. *et al.* Chondrulelike objects in short-period Comet 81P/Wild 2. *Science* **321**, 1664–1667 (2008).

35. Keilmann, F. & Hillenbrand, R. in *Nano-Optics and Near-Field Optical Microscopy* (eds Zayats, A. & Richards, D.) 235–265 (Artech House, 2009).

36. Cvitkovic, A., Ocelic, N. & Hillenbrand, R. Analytical Model for Quantitative Prediction of Material Contrasts in Scattering-Type Near-Field Optical Microscopy. *Optics Express* **15**, 8550–8565 (2007).

37. McLeod, A. S. *et al.* Model for quantitative near-field spectroscopy and the extraction of nanoscale-resolved optical constants. *Phys. Rev. B* **90**, 085136 (2014).

38. Hillenbrand, R., Taubner, T. & Keilmann, F. Phonon-enhanced light-matter interaction at the nanometre scale. *Nature* **418**, 159–162 (2002).

39. Ocelic, N., Huber, A. & Hillenbrand, R. Pseudoheterodyne detection for background-free near-field spectroscopy. *Appl. Phys. Lett.* **89**, 101124 (2006).

40. Glotch, T. D. & Rossman, G. R. Mid-infrared reflectance spectra and optical constants of six iron oxide/oxyhydroxide phases. *Icarus* **204**, 663–671 (2009).

41. Alonso-González, P. *et al.* Experimental verification of the spectral shift between near- and far-field peak intensities of plasmonic infrared nanoantennas. *Phys. Rev. Lett.* **110**, 203902 (2013).

42. Prior, D., Mariani, E. & Wheeler, J. in *Electron Backscatter Diffraction in Materials Science* (eds Schwartz, A. J., Kumar, M., Adams, B. L. & Field, D. P.) 345–360 (Springer, 2009).

43. Holman, H. Y. N., Martin, M. C. & McKinney, W. R. Synchrotron-based FTIR spectromicroscopy: cytotoxicity and heating considerations. *J. Biol. Phys.* **29**, 275–286 (2003).

44. Stiegler, J. M., Tena-Zaera, R., Idigoras, O., Chuvilin, A. & Hillenbrand, R. Correlative infrared–electron nanoscopy reveals the local structure-conductivity relationship in zinc oxide nanowires. *Nat. Commun.* **3**, 1131 (2012).

45. Demichelis, R. *et al.* The infrared spectrum of ortho-enstatite from reflectance experiments and first-principle simulations. *Mon. Not. R. Astron. Soc.* **420**, 147–154 (2012).

46. Chihara, H., Koike, C., Tsuchiyama, A., Tachibana, S. & Sakamoto, D. Compositional dependence of infrared absorption spectra of crystalline silicates. I. Mg-Fe pyroxenes. *Astron. Astrophys.* **391**, 267–273 (2002).

47. Dowty, E. & Clark, J. R. Crystal structure refinement and optical properties of a Ti^{3+} fassaite from the Allende meteorite. *Am. Mineral.* **58**, 230–242 (1973).

48. Simon, S. B. & Grossman, L. A comparative study of melilite and fassaite in types B1 and B2 refractory inclusions. *Geochim. Cosmochim. Acta* **70**, 780–798 (2006).

49. Koike, C. *et al.* Absorption spectra of Mg-rich Mg-Fe and Ca pyroxenes in the mid- and far-infrared regions. *Astron. Astrophys.* **363**, 1115–1122 (2000).

50. Morimoto, N. *et al.* Nomenclature of pyroxenes. *Am. Mineral.* **73**, 1123–1133 (1988).

51. Leroux, H. Microstructural shock signatures of major minerals in meteorites. *Eur. J. Mineral.* **13**, 253–272 (2001).

52. Downs, R. T. *Program and Astracts of the 19th General Meeting of the International Mineralogical Association*, O03–O13 (Kobe, Japan, 2006).

53. Nakamura, M., Kanzawa, R. & Sakai, K. Stress and density effects on infrared absorption spectra of silicate glass films. *J. Electrochem. Soc.* **133**, 1167–1171 (1986).

54. Zhang, M. *et al.* Exsolution and Al-Si disorder in alkali feldspars; their analysis by infrared spectroscopy. *Am. Mineral.* **82**, 849–857 (1997).

55. Balan, E. *et al.* Line-broadening effects in the powder infrared spectrum of apatite. *Phys. Chem. Minerals* **38**, 111–122 (2011).

56. Ruppin, R. & Englman, R. Optical phonons of small crystals. *Rep. Prog. Phys.* **33**, 149 (1970).

57. Genzel, L. & Martin, T. P. Infrared absorption by surface phonons and surface plasmons in small crystals. *Surf. Sci.* **34**, 33–49 (1973).

58. Day, D. E. & Rindone, G. E. Properties of soda aluminosilicate glasses: I, refractive index, density, molar refractivity, and infrared absorption spectra. *J. Am. Chem. Soc.* **45**, 489–496 (1962).

59. Roy, B. N. Infrared spectroscopy of lead and alkaline-earth aluminosilicate glasses. *J. Am. Chem. Soc.* **73**, 846–855 (1990).

60. Fernandez, J. A. & Jockers, K. Nature and origin of comets. *Rep. Prog. Phys.* **46**, 665 (1983).

61. Nakamura, T. *et al.* Itokawa dust particles: a direct link between S-type asteroids and ordinary chondrites. *Science* **333**, 1113–1116 (2011).

62. Rietmeijer, F. J. M. Interplanetary dust particles. *Rev. Mineral Geochem.* **36**, 2.1–2.95 (1998).

63. Keilmann, F. & Amarie, S. Mid-infrared frequency comb spanning an octave based on an Er fiber laser and difference-frequency generation. *J. Infrared Milli. Terahz Waves* **33**, 479–484 (2012).

64. Amenabar, I. *et al.* Structural analysis and mapping of individual protein complexes by infrared nanospectroscopy. *Nat. Commun.* **4**, 2890 (2013).

65. Westphal, A. J. *et al.* Aerogel keystones: Extraction of complete hypervelocity impact events from aerogel collectors. *Meteoritics* **39**, 1375–1386 (2004).

Acknowledgements

This work and G.D. were supported by NASA's Laboratory Analysis of Returned Samples program (# NNX11AF24G). The Advanced Light Source and National Center for Electron Microscopy is supported by the Director, Office of Science, Office of Basic Energy Sciences of the US Department of Energy under Contract No. DE-AC02-05CH11231. A.S.M acknowledges the support from a US Department of Energy Office of Science graduate research fellowship.

Author contributions

G.D developed the concept of nano-scale IR s-SNOM characterization of extraterrestrial samples and G.D., A.S.M., Z.G., M.T., and D.N.B. conceived and designed the experiments. A.S.M. performed the s-SNOM imaging and nano-spectroscopy experiments and together with P.K. developed and applied a framework for extraction of nano-scale IR extinction spectra. G.D. and Z.G. polished and mounted the Murchison meteorite sample. Z.G. prepared the Iris cometary grain, prepared the FIB lamella, and characterized both samples by SEM, EDS, and TEM. H.A.B. performed synchrotron-based far-field infrared spectroscopy on the Iris cometary grain. All authors discussed the data. A.S.M. and G.D. wrote the manuscript with support from Z.G. and D.N.B.

Additional information

7

Aerosol influence on energy balance of the middle atmosphere of Jupiter

Xi Zhang[1], Robert A. West[2], Patrick G.J. Irwin[3], Conor A. Nixon[4] & Yuk L. Yung[5]

Aerosols are ubiquitous in planetary atmospheres in the Solar System. However, radiative forcing on Jupiter has traditionally been attributed to solar heating and infrared cooling of gaseous constituents only, while the significance of aerosol radiative effects has been a long-standing controversy. Here we show, based on observations from the NASA spacecraft Voyager and Cassini, that gases alone cannot maintain the global energy balance in the middle atmosphere of Jupiter. Instead, a thick aerosol layer consisting of fluffy, fractal aggregate particles produced by photochemistry and auroral chemistry dominates the stratospheric radiative heating at middle and high latitudes, exceeding the local gas heating rate by a factor of 5–10. On a global average, aerosol heating is comparable to the gas contribution and aerosol cooling is more important than previously thought. We argue that fractal aggregate particles may also have a significant role in controlling the atmospheric radiative energy balance on other planets, as on Jupiter.

[1] Department of Earth and Planetary Sciences, University of California Santa Cruz, Santa Cruz, California 95064, USA. [2] Jet Propulsion Laboratory, California Institute of Technology, 4800 Oak Grove Drive, Pasadena, California 91109, USA. [3] Atmospheric, Oceanic and Planetary Physics, University of Oxford, Clarendon Laboratory, Parks Road, Oxford OX1 3PU, UK. [4] NASA Goddard Space Flight Center, Greenbelt, Maryland 20771, USA. [5] Division of Geological and Planetary Sciences, California Institute of Technology, Pasadena, California 91125, USA. Correspondence and requests for materials should be addressed to X.Z. (email: xiz@ucsc.edu).

As on Earth, Jupiter's atmospheric temperature profile exhibits a strong inversion above the tropopause[1], implying that its middle atmosphere, or the 'stratosphere', is convectively inhibited. Therefore, the energy budget should be dominated by radiation and the stratified middle atmosphere is in global radiative equilibrium. A first-order question is: which constituents in the atmosphere control this energy balance? About half of the incoming solar radiation on Jupiter penetrates deep into the troposphere and one third is reflected back to space (Fig. 1)[2]. The bulk constituents, hydrogen and helium, are not radiatively active except via H_2–H_2 and H_2–He collisional-induced absorption (CIA) at pressures >10 hPa (refs 3,4). The next most abundant gas, methane (CH_4), diffuses upward from the deep atmosphere and heats the stratosphere by absorbing the near-infrared solar flux[3–8]. The methane photochemical products acetylene (C_2H_2) and ethane (C_2H_6), together with H_2–H_2 and H_2–He CIA, absorb the upward mid-infrared radiation from the troposphere and re-radiate it to space, resulting in an efficient net cooling of the middle atmosphere[3–9] to compensate the solar heating.

The global maps of temperature and C_2 hydrocarbons were recently retrieved from the Jupiter flyby data from Cassini and Voyager-1 spacecraft in 2000 (refs 4,10–12) and 1979 (refs 4,12), respectively. On the basis of a state-of-the-art radiative transfer model (see Methods section), we investigate the global energy balance of Jupiter[4]. Surprisingly, the global average cooling flux by gaseous constituents in the middle atmosphere is estimated to be ~ 1.4 W m^{-2}, about 1.5 times larger than solar flux absorbed by the stratospheric CH_4 (~ 0.9 W m^{-2}; Fig. 1). Vertically, the gas solar heating rate is substantially smaller than the gas thermal cooling at pressures >10 hPa (ref. 4). The energy imbalance consistently revealed by the Voyager and Cassini data is not a seasonal effect because Jupiter has nearly zero obliquity. The Jupiter–Sun distance was different for the two flybys, varying

from northern fall equinox (Voyager) to the northern summer solstice (Cassini), but the global average heating is not altered significantly. Long-term ground-based observations from 1980 to 2000 also show that the global average temperature at 20 hPa does not substantially vary with time[10], and thus neither does the thermal radiative cooling. The violation of the radiative energy equilibrium thereby suggests the presence of an additional strong heat source other than CH_4 in the middle atmosphere of Jupiter, which absorbs the missing ~ 0.5 W m^{-2}.

Here we show that the missing heat source is aerosols, the end product of atmospheric chemistry on Jupiter. As a result of photochemistry, with the help of auroral chemistry, especially at high latitudes (or the 'auroral zone') where high-energy particles penetrate into the atmosphere, complex hydrocarbon compounds can form and eventually coagulate and condense as aerosols, or haze particles[13,14]. On the basis of Cassini imaging science subsystem (ISS) observations[15], here we derive the globally averaged solar flux absorbed by the stratospheric aerosols of ~ 0.5 to 0.7 W m^{-2}, more than half of the amount due to CH_4 (Fig. 1). The aerosol heating is predominant at middle and high latitudes, exceeding the local gas heating rate by a factor of 5–10. For the first time, we estimate the possible aerosol cooling effect, which might be as important as the cooling via hydrocarbons at high latitudes. We conclude that the aerosols maintain the atmospheric energy balance and must be partially responsible for the stratospheric temperature inversion. That the photochemistry and auroral chemistry control the atmospheric energetics via the production of aerosols suggests that Jupiter exhibits a new regime of atmospheric energy balance that is different from that of the Earth.

Results

Aerosol heating effect. The global aerosol map has been revealed from images acquired by the ISS during its Jupiter flyby[15]. At low latitudes (40° S to 25° N), an optically thin layer composed of compacted particles with radii ~ 0.2–0.5 μm is found to be concentrated at ~ 50 hPa. The rest of the atmosphere is covered by an optically thick aerosol layer at 10–20 hPa composed of fluffy, fractal particles aggregated from hundreds to thousands of ten-nanometre size monomers, similar to the haze particles on Titan[16,17]. The fractal dimension of these aggregates is assumed as 2, meaning that their geometric structure of the aggregate particles lies between a long linear chain (fractal dimension of 1) and a fully compacted cluster (fractal dimension of 3). This type of fractal aggregates is consistent with the ISS observations and the polarization observations of Jupiter, whereas spherical particles are not[16,17]. Fractal aggregates are known to be much more absorbing than spherical particles in the ultraviolet and visible wavelengths[18]. For instance, from 0.2 to 1 μm, an aggregate particle composed of a thousand ten-nanometre monomers can absorb twice as much of the solar flux as an ensemble of 0.07 μm individual spherical particles (assumed in ref. 19) with the same extinction optical depth, because the former is less scattering than the latter. Figure 2b shows that, at high latitudes, the opacity of the fractal aggregates on Jupiter is considerably larger than the CH_4 opacity in the ultraviolet and visible ranges, indicating that these particles can absorb a significant fraction of solar energy at wavelengths where the solar blackbody radiation peaks (Fig. 2a). In addition, long-term observations suggest that the seasonal variation of the Jovian north–south polarization asymmetry is only $\sim 0.5\%$ (ref. 20). The lack of strong temporal variation implies that the fractal aggregates constitute a steady heat source.

The radiative heating calculations demonstrate that the middle atmosphere of Jupiter is heated by two components: aerosols in

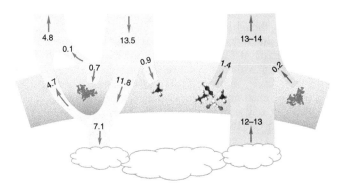

Figure 1 | Globally averaged heating and cooling fluxes on Jupiter. The heating (yellow branch) and cooling (cyan branch) fluxes are in units of W m^{-2}. The stratosphere is shaded. The heating flux is associated with the incoming solar radiation and the cooling flux is related to the outgoing thermal radiation. Of the 13.5 W m^{-2} of solar radiation incident to Jupiter's atmosphere, 0.1 W m^{-2} is reflected back to space and 11.8 W m^{-2} is transmitted to the troposphere. Tropospheric hazes and clouds absorbed 7.1 W m^{-2} and 4.7 W m^{-2} is reflected back to space[2]. The remainder of the solar energy is absorbed in the middle atmosphere by fractal haze particles (0.7 W m^{-2}) and CH_4 gas molecules (0.9 W m^{-2}). The total outgoing thermal radiation from our radiative calculation is ~ 13–14 W m^{-2}, consistent with that from Cassini and Voyager observations[9]. The thermal cooling flux is mainly emitted from the troposphere (12–13 W m^{-2}). In the middle atmosphere, the net cooling flux is 1.4 W m^{-2} emitted by gas molecules H_2, CH_4, C_2H_2, and C_2H_6 (black and white molecule diagrams). The upper limit of the outgoing thermal flux from the fractal aggregates (blue diagrams) is ~ 0.2 W m^{-2} as determined in this study.

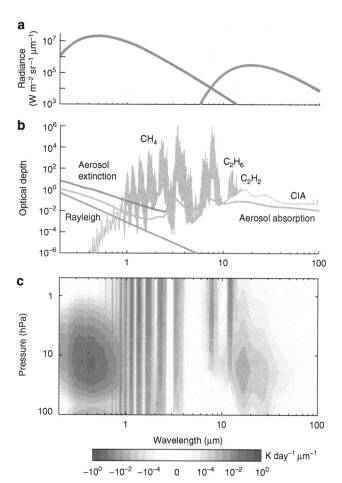

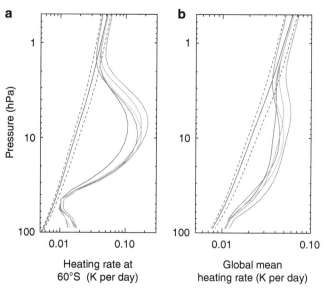

Figure 3 | Vertical heating rate profiles. (**a**) Zonally averaged heating rates at 60° S; (**b**) Globally averaged heating rates. The gas-only calculations are shown in black. The dashed black lines show the possible gas heating rates due to the uncertainty of CH_4 profiles. The coloured lines represent different aerosol retrieval solutions. Cases H1-H5 correspond to the green, red, purple, blue and orange curves, respectively. See Table 1 for detailed input information of the cases.

Figure 2 | Spectrally resolved heating and cooling rates and corresponding energy fluxes and opacity. (**a**) Globally averaged solar radiation received by Jupiter, approximated by a blackbody of 5,778 K (red) and Jupiter thermal radiation in the stratosphere approximated by a blackbody of 150 K (blue). (**b**) Total optical depth from the top of the atmosphere to 100 hPa as a function of wavelength at 60° S. The gas optical depth (grey) includes H_2-H_2 and H_2-He CIA and CH_4, C_2H_2 and C_2H_6 absorption. The non-gas components include Rayleigh scattering (blue), fractal aggregate aerosol extinction (red) and the aerosol absorption (orange). (**c**) Spectrally resolved zonally averaged solar heating (0.2–5 μm) and cooling (5–100 μm) map at 60° S. Absolute values of the heating/cooling rates that are $<10^{-6}$ K per day per μm are not shown. Solar heating dominates shortwards of 5 μm while Jupiter thermal cooling (shown in negative values here) dominates longwards of 5 μm. Contributions from the H_2-H_2 and H_2-He CIA and gas vibrational-rotational bands are shown. Aerosol heating is important in the ultraviolet and visible regions and aerosol cooling is important in the mid-infrared region beyond 11 μm.

the ultraviolet and visible wavelengths and CH_4 in the near-infrared. Figure 2c shows the spectrally resolved zonally averaged heating rate as a function of wavelength and pressure at 60° S. At near-infrared wavelengths longer than 0.9 μm, strong CH_4 bands completely dominate the heating with minor contributions from H_2-H_2 and H_2-He CIA (Fig. 2b). At shorter wavelengths, heating by the fractal aggregates is predominant. The maximum aerosol heating occurs near the wavelength of the solar spectrum peak (~0.5 μm), but slightly shortwards owing to the increasing absorption of aerosols towards the ultraviolet. At 60° S, the integrated aerosol heating rate over all wavelengths can reach ~0.2 K per day at 10 hPa, where it exceeds the CH_4 heating rate by a factor of 10 (Fig. 3a). The aerosol heating rate appears to

decrease rapidly above 5 hPa and does not contribute significantly to the total heating rate in the upper stratosphere. However, this conclusion is not certain since the resolution of the near-infrared spectra[21] is not sufficient to fully characterize the upper stratosphere. On the basis of higher-resolution observations[22], another aerosol layer was found above 5 hPa at the poles that might contribute to the local heating rate in the upper stratosphere, though not to the total energy budget of the middle atmosphere due to its lower density at those levels.

On the basis of Cassini observations, the globally averaged heat flux absorbed by the stratospheric aerosols is ~0.5–0.7 W m^{-2}, more than half of the amount due to CH_4 (Fig. 1). The maximum globally averaged aerosol heating rate of ~0.03 K per day occurs at ~10 hPa, comparable to the total near-infrared CH_4 heating rate at the same pressure level. But the globally averaged heating rate including aerosols is about twice that of the gas-only heating rate at pressures >20 hPa (Fig. 3b). Spatially, the aerosol heating is predominant at middle and high latitudes, which is attributable to the optically thick fractal aggregates layer (Fig. 4b). Therefore, the aerosol heating naturally, if not coincidentally, compensates for the energy deficit due to gas heating and cooling at the appropriate pressure levels (Fig. 4e).

Owing to the existence of degenerate solutions in the interpretation of ISS observations[15], we performed sensitivity tests to estimate the uncertainty ranges of the aerosol heating rate. The tropospheric haze and cloud are treated as an effective cloud layer in the troposphere[15]. Our tests show that the effective cloud albedo and phase function within the retrieved uncertainties has insignificant effect on the stratospheric heating rate. The heating rate is more sensitive to the total optical depth and single scattering albedo of the stratospheric aerosols. Given the constraints from the Cassini ISS observations, testing the sensitivity of the heating rate to each individual aerosol parameter is inappropriate. However, multiple solutions still exist in the aerosol retrieval[15]. Therefore, we adopted five typical retrieval solutions for the middle and high latitudes, namely cases

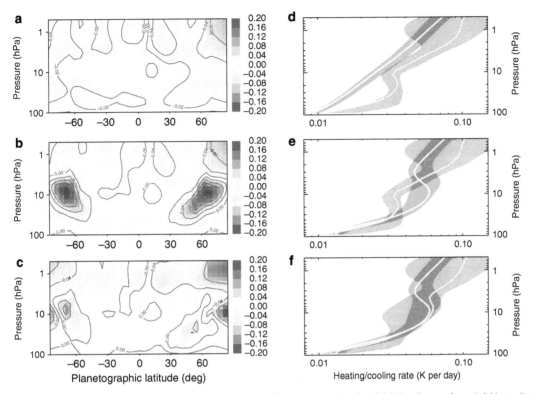

Figure 4 | Radiative balance calculation results in the middle atmosphere of Jupiter based on Cassini flyby observations. (**a**) Net radiative heating rate map (in units of K per Earth day) without aerosols; (**b**) net radiative heating rate map with aerosol heating; (**c**) net radiative heating rate map with aerosol heating and cooling; (**d**) globally averaged heating (yellow with pink shading) and cooling (cyan with blue shading) profiles without aerosols; (**e**) globally averaged heating and cooling profiles with aerosol heating; and (**f**) globally averaged heating and cooling profiles with aerosol heating and cooling. The uncertainty ranges are shaded.

Table 1 | Sensitivity cases for ISS retrieval and heating rate calculation.

Case	CH$_4$ mixing ratio	k (UV1)	k (CB3/MT3)	Colour
H1 (Nominal)	1.8×10^{-3}	2×10^{-2}	1×10^{-3}	Green
H2	1.8×10^{-3}	6×10^{-3}	1×10^{-4}	Red
H3	1.8×10^{-3}	8×10^{-2}	4×10^{-3}	Purple
H4	1.5×10^{-3}	2×10^{-2}	1×10^{-3}	Blue
H5	2.5×10^{-3}	2×10^{-2}	1×10^{-3}	Orange

H1–H5 in Table 1. Those cases were designed to explore the parameter space within the uncertainties of the CH$_4$ mixing ratio[4,23] and the imaginary part of the refractive indices (hereafter k values) of the UV1 and CB3/MIT3 channels of Cassini ISS[15]. All cases provide good fits to the limb-darkening observations from Cassini ISS (Table 1).

The heating rate calculation at 60° S (Fig. 3a) shows that the maximum heating rate including aerosols is about a factor of 2 larger than the minimum at ~10 hPa. The maximum (H3) and minimum (H2) occurs when the k value of the UV1 channel reaches the upper and lower bound, respectively, outside which the UV1 limb-darkening profiles cannot be explained[15]. On a global average (Fig. 3b), the maximum heating rate including aerosols is about a factor of 1.5 larger than the minimum (shaded in red in Fig. 4e,f).

Aerosol cooling effect. Aerosols could also cool the middle atmosphere of Jupiter but this effect has not been explored in previous studies. The mid-infrared optical properties of aerosols produced in a hydrogen-dominated environment have not been

measured experimentally. But at wavelengths shorter than 2.5 µm, previous laboratory experiments found that the k values of aerosols produced in a CH$_4$/H$_2$ gas mixture could be either larger or smaller than their counterpart in the CH$_4$/N$_2$ mixture, depending on the chemical composition and environmental pressure[24,25]. Therefore, the fractal aggregates on Jupiter might have non-negligible opacity in the mid-infrared compared with H$_2$–H$_2$ and H$_2$–He CIA if their optical constants behave like the aerosols on Titan, which are strongly absorbing with almost no scattering beyond 5 µm (Fig. 2b)[26,27]. To estimate the possible cooling effect from aerosols, we included aerosol absorption in a non-linear inversion model to fit the spectra from Cassini CIRS[28]. However, owing to insufficient sensitivity of CIRS observations to the Jovian aerosol opacity, a pure-gas (that is, non-detection of aerosols) model is also able to fit the spectra[4,10–12]. Future analysis on the possible C–H bending vibrational features of aerosols at 1,380 and 1,460 cm^{-1} that have been detected on Titan[27] might provide more constraints on the infrared opacity and chemical structure of aerosol particles on Jupiter. Here we aim to derive the upper limit of aerosol opacity from the CIRS spectra and estimate the upper bound of the aerosol thermal infrared cooling.

For each latitude, we included aerosols in the Non-linear optimal Estimator for MultivariatE spectral analySIS (NEMESIS) model[28] and retrieved the temperature profile and the mixing ratios of C$_2$H$_2$ and C$_2$H$_6$ following the procedure detailed in ref. 4. Owing to the lack of Jupiter-analogue aerosol measurements in mid-infrared, we tested several k values based on the laboratory tholin results[26] and recently derived k values from Cassini observations on Titan[27]. The latter shows cooling 2–3 times smaller than the former and exhibits different wavelength dependence. We gradually increased the k values

Table 2 | Sensitivity cases for CIRS retrieval and cooling rate calculation.

Case	k values in the mid-infrared	χ^2/N (600–850 cm^{-1})	χ^2/N (1,225–1,325 cm^{-1})	Colour
C1	Pure gas, no aerosol	0.544	1.094	Red
C2	Titan tholin experiment[26]	0.543	1.042	Blue
C3	Titan CIRS observations[27]	0.541	1.090	Orange
C4 (Nominal)	C2 values divided by 2	0.543	1.051	Green
C5	Five times of C2 values	0.607	1.964	Brown

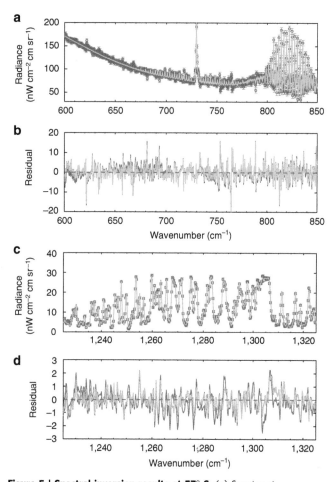

Figure 5 | Spectral inversion results at 57° S. (a) Spectra at 600–850 cm^{-1} region (H$_2$–H$_2$ and H$_2$–He CIA, C$_2$H$_2$ and C$_2$H$_6$ bands); (b) fitting residual at 600–850 cm^{-1} region; (c) spectra at 1,225–1,325 cm^{-1} region (CH$_4$ bands); (d) fitting residual at 1,225–1,325 cm^{-1} region. CIRS observations are shown as black circles. The red, blue, orange, green and brown colours represent NEMESIS retrieval cases C1–C5, respectively. The goodness of fit (χ^2/N where N is the number of measurements) in the 600–850 cm^{-1} region is $\sim$0.5–0.6 for each case. In the CH$_4$ band, the goodness of fit is around unity for each case except for the brown case ($\chi^2/N = 1.96$), which does not fit the CIRS spectra. See Table 2 for detailed information of the cases.

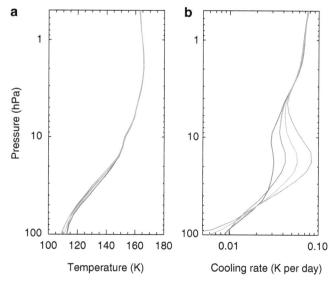

Figure 6 | Vertical temperature and cooling rate profiles at 57° S. (a) Retrieved temperature profiles; (b) corresponding zonally averaged cooling rates. The red, blue, orange and green colours represent cases C1–C4, respectively. The C5 case is not used because it cannot explain the CIRS observations.

until the CIRS spectra cannot be fitted within the measurement uncertainties. Table 2 summarized five typical tests. For each choice of optical constants, we performed an atmospheric retrieval of the Cassini CIRS spectra using NEMESIS model to constrain the aerosol opacity. Figure 5 illustrates the retrieval fitting results and residual values at 57° S and the values of goodness of fit are shown in Table 2. The retrieved temperature profiles are shown in Fig. 6a. If we enhance the optical constants

from ref. 25 by a factor of 5 (case C5), we are not able to fit the CH$_4$ emission spectra (Fig. 5). Overall, we estimated the upper limit of the aerosol optical depth in the mid-infrared wavelengths to be $\sim$0.1 at 100 hPa at high latitudes. This places an upper limit on the aerosol contribution to the globally averaged cooling flux of $\sim$0.2 W m^{-2} in the stratosphere (Fig. 1).

The aerosol cooling effect could be significant at middle and high latitudes where the particles are abundant but negligible at low latitudes. When the aerosol opacity is included, the cooling rate increases in the aerosol layer (Fig. 6b). At 57° S, with a moderate choice of k (case C4), the zonally averaged cooling rate including aerosols is $\sim$0.06 K per day. This is about two times larger than the gas-only cooling rate (case C1) at the pressures where the aerosol mixing ratio peaks ($\sim$10–20 hPa, Fig. 6b). The globally averaged aerosol cooling rate could be comparable to the gas cooling rate at 20 hPa and partially compensate for the aerosol heating effect (Fig. 4f). On the other hand, the cooling rates from the aerosol cases are smaller than the pure-gas case at pressure levels below the aerosol layer (Fig. 6b), a result of energy conservation as constrained by the total emission observed by CIRS. Indeed, stronger aerosol absorption leads to a colder retrieved temperature profile (Fig. 6a). For instance, the temperature profile at 57° S from the C2 case is $\sim$3 K colder than the case without aerosols (C1 case) below $\sim$40 hPa. This colder temperature leads to a smaller cooling effect than the pure-gas case (C1) at pressures $> \sim$40 hPa where the cooling is dominated by H$_2$–H$_2$ and H$_2$–He CIA (Fig. 6b).

We estimate the possible aerosol cooling rate for each latitude and their influences on the globally averaged cooling rate by combining the 'ensemble uncertainty'[4] of the temperature and gas abundances and the aerosol cooling rate tests. The uncertainty range including the aerosol contribution (Fig. 4f) is larger than that without aerosols (Fig. 4d) because the aerosols remain elusive from the CIRS spectra. Global radiative equilibrium is achievable when both aerosol heating and cooling are included (Fig. 4f).

Discussion

Owing to the lack of sufficient observational evidence before, the importance of aerosol heating in the middle atmosphere of Jupiter has long been a controversial question. It was suggested since the 1970s that the Jovian aerosols might absorb solar radiation and heat the atmosphere[6–8]. In the 1990s, based on the International Ultraviolet Explorer and Voyager-2 data, aerosol heating was shown to have a large impact on the atmospheric circulation[29]. However, a later analysis using Hubble Space Telescope (HST) images found that the aerosols have relatively insignificant effects[19]. With better constraints on the spatial distribution and optical properties of aerosols, Cassini observations confirm a pronounced aerosol heating at high latitudes on Jupiter. Aside from the aerosol cooling effect, our conclusion is qualitatively consistent with previous estimate from the International Ultraviolet Explorer and Voyager-2 data[29] but with a much better spatial coverage. However, this work disagrees with heating rates derived solely on the basis of HST data[3,19].

The major difference between the current work and previous studies is probably attributable to the fractal nature of the aerosol and aerosol spatial distribution. Through the multi-channel–multi-phase retrieval on Cassini images, we can characterize the fractal aggregates in great detail, including the optical depth, single scattering albedo and phase function of the particles. On the basis of low-phase-angle images alone[19], the polar aerosols were assumed as tiny spherical particles of $\sim 0.07\,\mu m$ in radius instead of the submicron size aggregate particles. The latter have lower single scattering albedo than the spherical particles. Furthermore, with a larger particle size, the fractal aggregates have less backscattering than the spherical particles. Owing to the above two reasons, a larger total optical depth of the fractal aggregates is required to explain the low-phase-angle I/F observations, leading to a larger heating rate in this study, than that in ref. 19, at the south pole. Note that the k values used in ref. 19 are lower than our nominal model but still within the sensitivity test range in our study.

The vertical distributions of aerosols in previous studies[19,29] are based on microphysical simulations that are inconsistent with the near-infrared observations[21]. Previous studies adopted a haze layer located above 10 hPa at polar regions, while the near-infrared spectra reveal a main haze layer at 10–20 hPa (ref. 20). West et al.[29] only sampled two latitudes and estimated the other latitude information by scaling. The HST images[19] have a good latitudinal coverage, but the aerosol heating appears influential only at the south pole, not at middle latitudes or at the north pole. Moreno et al.[19] reported the aerosol optical depth at the north pole one order of magnitude less than that at the south pole. This result is inconsistent with other observations. For example, recent high-resolution ground-based near-infrared spectra[22] concluded that the near-infrared haze optical depth at the northern pole is comparable to that at the southern pole. The Cassini images in low and high phase angles also revealed that the haze ultraviolet optical depth at north high latitudes is not significantly less than its south counterpart[15]. Furthermore, Cassini images show that the haze optical depth in the ultraviolet channel can approach unity down to 100 hPa at middle latitudes[15]. Including the haze

contribution at those latitudes would greatly enhance the aerosol heating. This will not only influence the local heat balance but also on the global energy equilibrium, especially at 10–20 hPa, as shown in our study.

Several other factors might also attribute to the difference between the current work and previous studies. For instance, we have a much better global coverage of the temperature, hydrocarbons and aerosols based on Cassini observations. The cooling rate in ref. 3 was likely to be underestimated because the temperature profile from Galileo entry probe is shown to be colder than the globally averaged temperature profile from Voyager and Cassini observations[4], albeit the cooling rate in ref. 3 is still slightly larger than the gas heating rate at pressures $>10\,hPa$. The spectroscopic data of CH_4, C_2H_2 and C_2H_6 have been greatly improved in the last decade (see Methods section for gas opacity). The state-of-the-art line data allow us to adopt the line-by-line approach to resolve the vibrational–rotational line shape of hydrocarbons[4], the most accurate radiative transfer method to estimate the gas heating and cooling rate.

Another possible heating mechanism in the middle atmosphere is energy dissipation of upward propagating gravity waves from the troposphere. However, as per previous studies[30,31], this hypothesis has several defects. First, there is little evidence of stratospheric gravity waves at middle and high latitudes. Second, there is no direct evidence of wave breaking in the lower stratosphere of Jupiter. Third, gravity wave breaking could either heat or cool the middle atmosphere but the net effect is difficult to quantify[30].

Unlike the Earth, on which the photochemical product (ozone) only dominates the atmospheric radiative heating, Jupiter might exhibit a different regime of atmospheric energy balance where both the heating and cooling are significantly controlled by the photochemistry and auroral chemistry via the production of aerosols and C_2 hydrocarbons. Aside from the first-order global energy balance, aerosol heating and cooling on Jupiter also influence the spatial distributions of radiative forcing, which has a significant impact on the large-scale dynamical circulation in the middle atmosphere[5,19,29]. The NASA JUNO spacecraft, arriving at Jupiter in 2016, will provide more insights on the aurora processes and aerosol formation in the polar region.

Jupiter is the second planetary body and the first hydrogen-dominated planet that shows evidence of hydrocarbon aerosols playing a significant role in regulating the radiation flux and most probably the circulation of its middle atmosphere. The other one is Titan[32]. Although Jupiter's atmosphere is primarily dominated by hydrogen, Titan's is dominated by nitrogen, both of these atmospheres produce fluffy, fractal aggregate particles, suggesting that fractal aggregates might be a ubiquitous result of hydrocarbon chemistry. In view of the existence of hydrocarbon aerosols in many other atmospheres dominated by hydrogen or nitrogen, such as those of Saturn[33], Uranus[34] and Neptune[35], the early Earth[18], and possibly some exoplanets[36–40], we hypothesize that the heating and cooling from fractal aggregates could also be important for determining the radiative energy distribution and climate evolution on these planets. Owing to their strong heating effects, fractal aggregates play a significant role in creating the temperature inversion in the lower stratosphere of Jupiter. They might also be partially responsible for the temperature inversions observed on the other giant planets in the Solar System, but neglected in previous studies[41]. A typical feature of the fluffy aggregate particle is its extremely low density, which might help to explain the existence of very high and thick haze layers at pressures $<0.1\,hPa$, as seen on the Neptune and sub-Neptune size planets GJ436b (ref. 39) and GJ1214b (ref. 40). A thorough study of fractal aggregates will shed light on how to characterize these particles in future photometry and polarization observations.

Methods

Radiative heating and cooling model. For heating rate calculations between 0.20 and 0.94 μm where the aerosol contribution is significant, we use a multiple scattering model based on the C version of the discrete ordinates radiative transfer code (DISORT Program for a multi-layered plane-parallel medium)[42]. The phase function and cross sections of low-latitude particles were calculated based on Mie theory, while that of the fractal aggregates at middle and high latitudes were computed using a parameterization method for the aggregates with a fractal dimension of two[17]. The parameterization is based on electromagnetic scattering computations using the multi-sphere method[43]. In the heating rate calculations, we use 32 streams to characterize the intensity angular distribution, which displays almost no difference from the 64-stream case. A Gaussian quadrature method with 10 zenith angles is used to average the heating rates longitudinally. The spectral resolution is 0.001 μm. The effective cloud albedo in the troposphere is interpolated between 0.20 and 0.94 μm based on the retrieved albedo from the UV1 channel and CB3/MT3 channels of Cassini ISS[15].

At longer wavelengths (0.94–200 μm), our calculations adopt the line-by-line approach, based on a state-of-the-art high-resolution radiative heating and cooling model for the stratosphere of Jupiter, which has been rigorously validated against simple but realistic analytical solutions[4]. The CH$_4$ heating rate calculation from 0.94 to 10 μm is performed with a spectral resolution of 0.005 cm^{-1} to resolve the CH$_4$ spectral line shape using the most current CH$_4$ line lists (see discussion for gas opacity below). The thermal cooling rate from 50 to 2,500 cm^{-1} (4–200 μm) is calculated with a spectral resolution of 0.001 cm^{-1}.

We calculate heating and cooling rates for every latitude to produce the latitude-pressure two-dimensional maps. The specific heat of Jupiter's atmosphere is taken as 1.0998×10^4 J kg^{-1} K^{-1} (ref. 44). The globally averaged profiles are obtained via an area-weighted mean from 90° N to 90° S. Since the data quality at latitudes north of 70° N and south of 70° S is not sufficiently good for rigorous atmospheric retrieval, we do not derive the atmospheric characteristics from the ISS images and CIRS spectra. Instead, we assume that the vertical profiles of aerosol, temperature and gases are identical to their values at 70° north and south, respectively.

This assumption might introduce some uncertainty in our estimate of the global energy balance because the heating and cooling in the polar region, especially in the infrared aurora region, are not negligible. The polar aurora is known to be highly variable both temporally and spatially[1]. According to our limited data on the polar regions from Galileo[45] and Cassini[46] spacecraft as well as ground-based observations[47,48], both regions poleward of 70° N and 70° S could be different from that ~65°–70°. For example, the Cassini CIRS instrument[46] detected a variation of thermal emission over the C$_2$H$_2$ and C$_2$H$_6$ bands at regions poleward of 65°, within about a factor of 2. However, because the surface area poleward of 70° amounts to merely 6% of the total surface area of Jupiter, increasing the polar cooling rate by a factor of 2 will probably introduce an uncertainty of only ~6% of the total cooling rate, which is still well-located within our estimated uncertainty range (Fig. 4f). On the other hand, the pressure level of infrared aurora source has not been precisely determined. If the emission originates from the upper stratosphere (for example, above 0.1 mbar level), the aurora and its variability might have little impact on the thermal cooling rate at pressures we focus here. We should also point out that the aurora might also be associated with some heating mechanisms in the polar region that have not been considered in this study. Future analysis of the Jovian polar region will provide more information on the local energetics.

The aerosol and gas opacities used in the radiative calculations are discussed below.

Aerosol opacity sources. In the visible and near-infrared wavelengths, the early laboratory studies[24–26] measured the optical properties of Titan-analogue aerosols (that is, in the nitrogen environment) and Jupiter-analogue aerosols (that is, in the hydrogen environment), respectively. Using the radio frequency glow discharge in a CH$_4$/H$_2$ gas mixture, it was found that the refractive indices of these aerosols are consistent with high-phase-angle photometry data of Uranus by Voyager-2 at 0.55 μm (ref. 24). Compared with aerosols produced in the CH$_4$/N$_2$ mixture[26], the imaginary part of the refractive index of the Jupiter-analogue aerosols could be either larger or smaller, depending on the chemical composition. Unfortunately, ref. 24 has been the only laboratory experiment of the CH$_4$/H$_2$ gas mixture to date. It should be noted that the subsequent Titan-analogue laboratory measurements show that the aerosol properties are significantly influenced by the gas composition and environmental pressure[25].

Zhang et al.[15] combined the ground-based infrared spectra with the Cassini ISS observations at both low and high phase angles, and determined the k values for the UV1 channel (0.258 μm) and the infrared channels (CB3 at 0.938 μm and MT3 at 0.889 μm). Owing to the existence of degeneracy, the k value at UV1 channel varies from 0.008 to 0.02 and that at 0.9 μm varies from 0.0001 to 0.004. Different choices of k would imply different solutions to fit the ISS data, such as the radius of the monomers from 10 to 40 nm, the number of monomers per aggregate particle from 100 to 1,000, and the abundance of particles. In all possible solutions, the total aerosol opacity only changes by ~30% among all solutions. The corresponding aerosol heating rate does not change significantly.

For the heating rate calculations, we choose the model parameters for the k values from ref. 15 as our nominal case (H1 in Table 1), with a careful sensitivity study in the radiatiave heating calculation. The k value is ~0.02 at 0.258 μm and

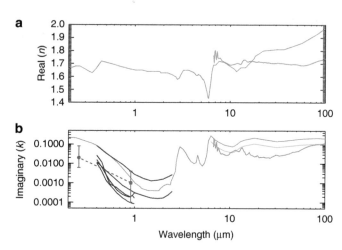

Figure 7 | Refractive indices of aerosols from 0.2 to 100 μm. (a) Real part of the refractive index; (**b**) imaginary part. The blue one is from the Titan tholin experiment[26]; orange is from Titan CIRS observations[27]; green is same as the blue one but reduced by a factor of 2 (case C4); black is the Jupiter-analogue aerosol experiment[24]; red is the derived from Cassini ISS observations[15], with an interpolation in the coordinate of linear wavelength and logarithmic k value. The red dashed line is used in the nominal case for the heating rate calculation (case H1) in this study.

0.001 at ~0.9 μm. We performed an interpolation for the wavelengths between 0.20 and 0.94 μm in the coordinate of linear wavelength and logarithmic k value. The interpolation can be justified by the approximate same trend shown in the laboratory measurements (Fig. 7). As in ref. 15, we adopt the real part of refractive index (n) from ref. 26.

Gas opacity sources. From 0.20 to 0.94 μm, the CH$_4$ opacity is based on ref. 49 and Rayleigh scattering optical depth is taken from ref. 14. From 0.94 to 10 μm, a line-by-line calculation is performed. We compared several CH$_4$ line databases, including the HITRAN2012 (refs 50,51), the database from ref. 22, and the M5 database[52]. For the CH$_4$ broadening width, it has been suggested that line widths in the Jovian atmosphere (H$_2$–He mixture) are similar to those in the Earth's atmosphere (N$_2$–O$_2$ mixture)[22]. All the above three CH$_4$ opacity sources result in a consistent heating rate in our calculation. However, none of the above opacity sources covers the CH$_4$ band between 0.94 and 1.1 μm. Recently, a '10 to 10' CH$_4$ line database is computed from first principles[53]. This database is shown to be roughly consistent with the HITRAN2012 CH$_4$ data in their overlapping near-infrared wavelengths (personal communication with J. Bailey) and therefore is helpful to fill the gap between 0.94 and 1.1 μm in our calculation. For this CH$_4$ band, due to the lack of laboratory measurements of line shape parameters, we adopted an average pressure-broadened half-width of 0.06 cm^{-1} for foreign-broadening and 0.077 cm^{-1} for self-broadening at 1 bar pressure, and the temperature dependence exponent is ~0.85 (refs 52,54). The contribution from this band to the total heating rate is negligible (Fig. 2c). For the heating rate calculation, we obtain the near-infrared H$_2$–H$_2$ CIA absorption from refs 55,56 and H$_2$–He CIA absorption from refs 57–60.

The thermal cooling rate is calculated from 50 to 2,500 cm^{-1} (4–200 μm). The opacity sources of CH$_4$, C$_2$H$_2$ and C$_2$H$_6$ are obtained from HITRAN2012 with hydrogen-broadening widths[4,61]. Fractal aggregates are treated as pure absorbers in the thermal wavelengths due to their negligible single scattering albedo in the mid-infrared. The mid-infrared H$_2$–H$_2$ and H$_2$–He CIAs are obtained from ref. 62. Figure 2c shows the spectrally resolved heating/cooling rate as a function of wavelength and pressure, in which one can see the contributions from the CIAs and different vibrational–rotational bands from gases in Fig. 2b.

Spectral inversion model. The temperature and hydrocarbon distributions are simultaneously retrieved from the Cassini CIRS spectra[4,10–12] using the NEMESIS algorithm[28]. This inversion model has been used in previous studies involving CIRS data retrieval[4,11,12]. In this study we extended the previous retrieval framework to include aerosol absorption in the mid-infrared wavelengths. No scattering calculations were needed as the fractal aggregate particles are significant absorbers and have negligible single scattering albedos at these wavelengths.

References

1. Moses, J. I. et al. in *Jupiter-The Planet, Satellites and Magnetosphere* (eds Bagenal, F., Dowling, T. & McKinnon, W.) 129–157 (Cambridge University Press, 2004).

2. Hanel, R., Conrath, B., Herath, L., Kunde, V. & Pirraglia, J. Albedo, internal heat, and energy balance of Jupiter: Preliminary results of the Voyager infrared investigation. *J. Geophys. Res.* **86**, 8705–8712 (1981).

3. Yelle, R. V., Griffith, C. A. & Young, L. A. Structure of the Jovian stratosphere at the Galileo probe entry site. *Icarus* **152**, 331–346 (2001).

4. Zhang, X. *et al.* Radiative forcing of the stratosphere of Jupiter, part I: atmospheric cooling rates from Voyager to Cassini. *Planet. Space Sci.* **88**, 3–25 (2013).

5. Conrath, B. J., Gierasch, P. J. & Leroy, S. S. Temperature and circulation in the stratosphere of the outer planets. *Icarus* **83**, 255–281 (1990).

6. Wallace, L., Prather, M. & Belton, M. J. S. The thermal structure of the atmosphere of Jupiter. *Astrophys. J.* **193**, 481–493 (1974).

7. Cess, R. & Chen, S. The influence of ethane and acetylene upon the thermal structure of the Jovian atmosphere. *Icarus* **26**, 444–450 (1975).

8. Appleby, J. F. & Joseph, S. H. Radiative-convective equilibrium models of Jupiter and Saturn. *Icarus* **59**, 336–366 (1984).

9. Li, L. *et al.* Emitted power of Jupiter based on Cassini CIRS and VIMS observations. *J. Geophys. Res. Planets* **117**, E11 (2012).

10. Simon-Miller, A. A. *et al.* Jupiter's atmospheric temperatures: from Voyager IRIS to Cassini CIRS. *Icarus* **180**, 98–112 (2006).

11. Nixon, C. A. *et al.* Meridional variations of C_2H_2 and C_2H_6 in Jupiter's atmosphere from Cassini CIRS infrared spectra. *Icarus* **188**, 47–71 (2007).

12. Nixon, C. A. *et al.* Abundances of Jupiter's trace hydrocarbons from Voyager and Cassini. *Planet. Space Sci.* **58**, 1667–1680 (2010).

13. Wong, A. S., Yung, Y. L. & Friedson, A. J. Benzene and haze formation in the polar atmosphere of Jupiter. *Geophys. Res. Lett.* **30**, 1–4 (2003).

14. West, R. A. *et al.* in *Jupiter-The Planet, Satellites and Magnetosphere* (eds Bagenal, F., Dowling, T. & McKinnon, W.) 79–104 (Cambridge University Press, 2004).

15. Zhang, X., West, R. A., Banfield, D. & Yung, Y. L. Stratospheric aerosols on Jupiter from Cassini observations. *Icarus* **226**, 159–171 (2013).

16. West, R. A. & Smith, P. H. Evidence for aggregate particles in the atmospheres of Titan and Jupiter. *Icarus* **90**, 330–333 (1991).

17. Tomasko, M. G. *et al.* A model of Titan's aerosols based on measurements made inside the atmosphere. *Planet. Space Sci.* **56**, 669–707 (2008).

18. Wolf, E. T. & Toon, O. B. Fractal organic Hazes provided an ultraviolet shield for early Earth. *Science* **328**, 1266–1268 (2010).

19. Moreno, F. & Sedano, J. Radiative balance and dynamics in the stratosphere of Jupiter: results from a latitude-dependent aerosol heating model. *Icarus* **130**, 36–48 (1997).

20. Starodubtseva, O. M., Akimov, L. A. & Korokhin, V. V. Seasonal variations in the north–south asymmetry of polarized light of Jupiter. *Icarus* **157**, 419–425 (2002).

21. Banfield, D., Conrath, B. J., Gierasch, P. J. & Nicholson, P. D. Near-IR spectrophotometry of Jovian aerosols—meridional and vertical distributions. *Icarus* **134**, 11–23 (1998).

22. Kedziora-Chudczer, L. & Bailey, J. Modelling the near-IR spectra of Jupiter using line-by-line methods. *Mon. Not. R. Astron. Soc.* **414**, 1483–1492 (2011).

23. Wong, M. H., Mahaffy, P. R., Atreya, S. K., Niemann, H. B. & Owen, T. C. Updated Galileo probe mass spectrometer measurements of carbon, oxygen, nitrogen, and sulfur on Jupiter. *Icarus* **171**, 153–170 (2004).

24. Khare, B. N., Sagan, C., Thompson, W., Arakawa, E. & Votaw, P. Solid hydrocarbon aerosols produced in simulated Uranian and Neptunian stratospheres. *J. Geophys. Res.* **92**, 15067–15082 (1987).

25. Imanaka, H. *et al.* Laboratory experiments of Titan tholin formed in cold plasma at various pressures: implications for nitrogen-containing polycyclic aromatic compounds in Titan haze. *Icarus* **168**, 344–366 (2004).

26. Khare, B. N. *et al.* Optical constants of organic tholins produced in a simulated Titanian atmosphere: from soft X-ray to microwave frequencies. *Icarus* **60**, 127–137 (1984).

27. Vinatier, S. *et al.* Optical constants of Titan's stratospheric aerosols in the 70–1,500 cm^{-1} range constrained by Cassini/CIRS observations. *Icarus* **219**, 5–12 (2012).

28. Irwin, P. G. J. *et al.* The NEMESIS planetary atmosphere radiative transfer and retrieval tool. *J. Quant. Spectrosc. Radiat. Transfer* **109**, 1136–1150 (2008).

29. West, R. A., Friedson, A. J. & Appleby, J. Jovian large-scale stratospheric circulation. *Icarus* **100**, 245–259 (1992).

30. Young, L. A., Yelle, R. V., Young, R., Seiff, A. & Kirk, D. B. Gravity waves in Jupiter's stratosphere, as measured by the Galileo ASI experiment. *Icarus* **173**, 185–199 (2005).

31. Watkins, C. & Cho, J. Y. K. The vertical structure of Jupiter's equatorial zonal wind above the cloud deck, derived using mesoscale gravity waves. *Geophys. Res. Lett.* **40**, 472–476 (2013).

32. Tomasko, M. G. *et al.* A model of Titan's aerosols based on measurements made inside the atmosphere. *Planet. Space Sci.* **56**, 669–707 (2008).

33. West, R. A., Baines, K. H., Karkoschka, E. & Sánchez-Lavega, A. in *Saturn from Cassini–Huygens* (eds Dougherty, M. K., Esposito, L. W. & Krimigis, S. S. M.) 161–179 (Springer, 2009).

34. Pollack, J. B. *et al.* Nature of stratospheric haze on Uranus: evidence for condensed hydrocarbons. *J. Geophys. Res.* **92**, 15037–15065 (1987).

35. Baines, K. H. & Hammel, H. B. Clouds, hazes and stratospheric methane abundance ratio in Neptune. *Icarus* **109**, 20–39 (1994).

36. Sing, D. K. *et al.* Hubble Space Telescope transmission spectroscopy of the exoplanet HD 189733b: high-altitude atmospheric haze in the optical and near-ultraviolet with STIS. *Mon. Not. R. Astron. Soc.* **416**, 1443–1455 (2011).

37. Sing, D. K. *et al.* HST hot-Jupiter transmission spectral survey: evidence for aerosols and lack of TiO in the atmosphere of WASP-12b. *Mon. Not. R. Astron. Soc* **436**, 2956–2973 (2013).

38. Deming, D. *et al.* Infrared transmission spectroscopy of the exoplanets HD 209458b and XO-1b using the wide field camera-3 on the Hubble Space Telescope. *Astrophys. J.* **774**, 95 (2013).

39. Knutson, H. A., Benneke, B., Deming, D. & Homeier, D. A featureless transmission spectrum for the Neptune-mass exoplanet GJ436b. *Nature* **505**, 66–68 (2014).

40. Kreidberg, L. *et al.* Clouds in the atmosphere of the super-Earth exoplanet GJ1214b. *Nature* **505**, 69–72 (2014).

41. Robinson, T. D. & Catling, D. C. Common 0.1-bar tropopause in thick atmospheres set by pressure-dependent infrared transparency. *Nat. Geosci.* **7**, 12–15 (2014).

42. Buras, R., Dowling, T. & Emde, C. New secondary-scattering correction in DISORT with increased efficiency for forward scattering. *J. Quant. Spectrosc. Radiat. Transfer* **112**, 2028–2034 (2011).

43. Mishchenko, M. I., Travis, L. D. & Mackowski, D. W. T-matrix computations of light scattering by nonspherical particles: A review. *J. Quant. Spectrosc. Radiat. Transfer* **55**, 535–575 (1996).

44. Irwin, P. G. J. *Giant Planets of Our Solar System: Atmospheres, Composition, and Structure* (Springer, 2009).

45. Rages, K., Beebe, R. & Senske, D. Jovian stratospheric hazes: the high phase angle view from Galileo. *Icarus* **139**, 211–226 (1999).

46. Kunde, V. G. *et al.* Jupiter's atmospheric composition from the Cassini thermal infrared spectroscopy experiment. *Science* **305**, 1582–1586 (2004).

47. Kostiuk, T., Romani, P. N., Espenak, F., Livengood, T. A. & Goldstein, J. J. Temperature and abundances in the Jovian auroral stratosphere, 2. Ethylene as a probe of the microbar region. *J. Geophys. Res. Planets* **98**, 18823–18830 (1993).

48. Livengood, T. A., Kostiuk, T., Espenak, F. & Goldstein, J. J. Temperature and abundances in the Jovian auroral stratosphere, 1. Ethane as a probe of the millibar region. *J. Geophys. Res. Planets* **98**, 18813–18822 (1993).

49. Karkoschka, E. & Tomasko, M. G. Methane absorption coefficients for the Jovian planets from laboratory, Huygens, and HST data. *Icarus* **205**, 674–694 (2010).

50. Brown, L. R. *et al.* Methane line parameters in the HITRAN2012 database. *J. Quant. Spectrosc. Radiat. Transfer* **130**, 201–219 (2013).

51. Rothman, L. S. *et al.* The HITRAN2012 molecular spectroscopic database. *J. Quant. Spectrosc. Radiat. Transfer* **130**, 4–50 (2013).

52. Sromovsky, L. A., Fry, P. M., Boudon, V., Campargue, A. & Nikitin, A. Comparison of line-by-line and band models of near-IR methane absorption applied to outer planet atmospheres. *Icarus* **218**, 1–23 (2012).

53. Yurchenko, S. N., Tennyson, J., Bailey, J., Hollis, M. D. & Tinetti, G. Spectrum of hot methane in astronomical objects using a comprehensive computed line list. *Proc. Natl Acad. Sci. USA* **111**, 9379–9383 (2014).

54. Nikitin, A. V. *et al.* GOSAT-2009 methane spectral line list in the 5,550–6,236 cm^{-1} range. *J. Quant. Spectrosc. Radiat. Transfer* **111**, 2211–2224 (2010).

55. Borysow, A. Collision-induced absorption coefficients of H_2 pairs at temperatures from 60 K to 1,000 K. *Astron. Astrophys.* **390**, 779–782 (2002).

56. Borysow, A., Trafton, L., Frommhold, L. & Birnbaum, G. Modeling of pressure-induced far-infrared absorption spectra: Molecular hydrogen pairs. *Astrophys. J.* **296**, 644–654 (1985).

57. Borysow, J., Frommhold, L. & Birnbaum, G. Collison-induced rototranslational absorption spectra of H_2-He pairs at temperatures from 40 to 3000 K. *Astrophys. J.* **326**, 509–515 (1988).

58. Borysow, A. & Frommhold, L. Collision-induced infrared spectra of H_2-He pairs at temperatures from 18 to 7,000 K. II-Overtone and hot bands. *Astrophys. J.* **341**, 549–555 (1989).

59. Borysow, A., Frommhold, L. & Moraldi, M. Collision-induced infrared spectra of H_2-He pairs involving 0-1 vibrational transitions and temperatures from 18 to 7,000 K. *Astrophys. J.* **336**, 495–503 (1989).

60. Borysow, A. New model of collision-induced infrared absorption spectra of H_2-He pairs in the 2-2.5 μm range at temperatures from 20 to 300 K: An update. *Icarus* **96**, 169–175 (1992).

61. Orton, G. S. *et al.* Semi-annual oscillations in Saturn's low-latitude stratospheric temperatures. *Nature* **453**, 196–199 (2008).

62. Orton, G. S., Gustafsson, M., Burgdorf, M. & Meadows, V. Revised ab initio models for H_2-H_2 collision-induced absorption at low temperatures. *Icarus* **189**, 544–549 (2007).

Acknowledgements

We thank E. Karkoschka, L. Brown, G. Orton, J. Bailey, T. Kostiuk, A. Showman and L. Li for useful discussions and comments. Special thanks to M. Gerstell, P. Gao, R. Hu, P. Kopparla, C. Li, M.C. Liang, S. Newman, R.L. Shia, M. Wong, X. Xi and Q. Zhang for proofreading the manuscript. The early phase of this research was supported by the Outer Planets Research program via NASA Grant JPL 1452240 to the California Institute of Technology. R.A.W. and C.A.N. are supported by the NASA Cassini project. P.G.J.I. acknowledges the support of the UK Science and Technology Facilities Council.

Author contributions

X.Z. carried out the radiative modelling and CIRS spectral retrieval; R.A.W. provided the ISS data; C.A.N. provided the CIRS data; R.A.W. and Y.L.Y. helped with radiative modelling; P.G.J.I. and C.A.N. helped with spectral inversion modelling; all authors contributed to the paper writing.

Additional information

Competing financial interests: The authors declare no competing financial interests.

8

Reconstructing the transport history of pebbles on Mars

Tímea Szabó[1,2], Gábor Domokos[2], John P. Grotzinger[3] & Douglas J. Jerolmack[1]

The discovery of remarkably rounded pebbles by the rover Curiosity, within an exhumed alluvial fan complex in Gale Crater, presents some of the most compelling evidence yet for sustained fluvial activity on Mars. While rounding is known to result from abrasion by inter-particle collisions, geologic interpretations of sediment shape have been qualitative. Here we show how quantitative information on the transport distance of river pebbles can be extracted from their shape alone, using a combination of theory, laboratory experiments and terrestrial field data. We determine that the Martian basalt pebbles have been carried tens of kilometres from their source, by bed-load transport on an alluvial fan. In contrast, angular clasts strewn about the surface of the Curiosity traverse are indicative of later emplacement by rock fragmentation processes. The proposed method for decoding transport history from particle shape provides a new tool for terrestrial and planetary sedimentology.

[1] Department of Earth and Environmental Science, University of Pennsylvania, 251 Hayden Hall, 240 South 33rd Street, Philadelphia, Pennsylvania 19104, USA.
[2] Department of Mechanics, Materials and Structures, Budapest University of Technology and Economics, Műegyetem rkp. 1-3. K261, Budapest 1111, Hungary.
[3] Division of Geological and Planetary Sciences, California Institute of Technology, 1200 East California Boulevard, Pasadena, California 91125, USA.
Correspondence and requests for materials should be addressed to D.J.J. (email: sediment@sas.upenn.edu).

Gale Crater (Fig. 1), the landing site for the Mars Curiosity rover, is estimated to have formed ~3.6 billion years ago[1]. Numerous erosional drainage networks debouch into the crater, which have built a series of merged alluvial fans that fringe the interior of the crater rim. Curiosity landed on top of an exhumed alluvial fan complex, and only several hundred metres from the distal end of a younger, better preserved alluvial fan, the Peace Vallis fan (Fig. 1). This fan exhibits a steep and channelized upper portion (slope, $S \approx 3\%$, length ≈ 10 km) that transitions into a less steep, unchannelized lower region ($S \approx 1\%$, length ≈ 3–4 km)[1]. The discovery of rounded pebbles, near the landing site at Bradbury Rise, provided on-the-ground confirmation of a fluvial depositional environment for the exhumed Gale crater sedimentary rocks that were of uncertain origin prior to landing[2]. Deposits from several sites contained rounded to sub-rounded particles, millimetres to centimetres in diameter, that were mixed with sand to form conglomerates[2,3]. A paleohydraulic reconstruction indicates that the gravel was transported as bed load—that is, by rolling, sliding and hopping along the river bed—and this interpretation is strongly supported by the observed (imbricated) fabric of the pebbles preserved in outcrop[2,4]. Fluvial deposits with interstratified conglomerate facies extend across a distance of at least 9 km, and define

a stratigraphic succession that is at least many tens of metres thick[5-7]. These outcrops are exhumed alluvial fan deposits that predate the Peace Vallis fan, revealing a complex depositional history. Based on terrestrial studies, it was suggested that a transport distance of at least 'several kilometres' was required for fluvial abrasion to produce the observed rounding, and, therefore, that the associated climate at Gale Crater, Mars was very different from the hyperarid and cold conditions of today[2]. Determining how much abrasion has occurred for these pebbles, and how far they have travelled, could significantly improve reconstructions of paleoenvironment and provenance[1]. To interpret these data further, we seek a generic and quantitative pattern for the shape evolution of pebbles under collisional abrasion by bed-load transport.

For the idealized case of a single particle striking a wall, it has been demonstrated that abrasion is a diffusive process; that is, the erosion rate at any point on the surface of a pebble is proportional to the local curvature[8,9]. Collision among like-sized particles is different in detail, but remains predominantly diffusive[10,11]. As a consequence, initially blocky particles first rapidly round as high-curvature regions are worn off, and then this rounding slows as the particle becomes rounder. Abrasion rate depends on collision energy, frequency of impacts and material properties[12,13], factors not considered in the idealized geometric model (ref. 14 for a first attempt to include collision energy). However, by casting shape as a function of mass loss—rather than time or distance—this model was shown to accurately predict the evolution of an initially cuboid particle colliding with the wall of a drum[9]. Of course the reality of pebble abrasion in a natural river is far from this simple picture. Some important differences are: collisions are typically among numerous particles having a variety of sizes; particles move as bed load driven by a turbulent fluid; and initial particle shapes are varied, and not cuboid. Despite these differences, a recent field study demonstrated that the downstream evolution of pebble shape in a natural river exhibited patterns entirely consistent with the idealized model[15].

In this paper, we present new experimental results and analysis of terrestrial field data, that suggest that the shape of river-transported pebbles is a unique function of the fractional mass lost due to abrasion. We use this result to show how the distance a pebble has travelled from its source may be estimated using shape alone. This new tool is validated on an alluvial fan on Earth, and then used to interpret the Martian conglomerates. Our findings indicate that the rounded Martian pebbles have been transported tens of kilometres, and point to the northern rim of Gale Crater as a likely source.

Results

New experiments and field data. We seek quantitative relations between pebble shape and mass loss for particles transported as bed load. For comparison of Martian and terrestrial data, shape parameters must be estimated from two-dimensional image data (Fig. 2) and be sensitive to abrasion. Based on previous work[9,15], we select the following: isoperimetric ratio (or circularity), IR; convexity, C; and the ratio of short and long axis lengths (axis ratio), b/a (Fig. 3). Although the qualitative evolution of pebble shapes under collisional abrasion is general (Fig. 2a), quantitative trends depend on the initial shapes of particles[9]. In the headwaters of rivers, the initial particles are typically rock fragments produced from weathering[16]. It has recently been discovered that rock fragments generated from a variety of processes—from slow weathering to gentle breakage to explosion—exhibit similar shape characteristics, as a consequence of brittle fracture[17]. This fortunate convergence implies that pebble shape evolution trends may also be quantitatively similar.

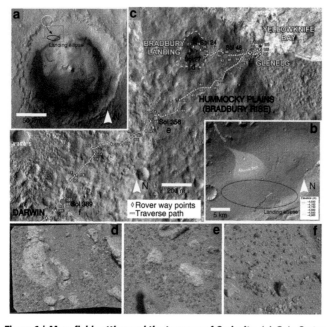

Figure 1 | Mars field setting and the traverse of Curiosity. (a) Gale Crater, with location of the Curiosity landing ellipse. White circle highlights eroded channel feeding the northern crater rim, which has been proposed to be the sediment source for Bradbury Rise conglomerates. Arrow indicates sediment transport direction, red box shows area in **b**. **(b)** The landing site (within the red box) and Curiosity's traverse for Sols (Martian days) 0–403 (yellow line) and Sols 403–817 (black dashed line). The Peace Vallis alluvial fan extends down dip and into Curiosity's landing ellipse. The landing ellipse also contains exhumed alluvial fan deposits that predate Peace Vallis, and define a bajada, which once depositionally infilled the crater margin[5-7]. Red box shows area in **c**. **(c)** Expanded view of Curiosity's traverse for Sols 0–403, with the locations of images studied in this paper (red dots). **(d)** Rounded pebbles at Sol 27, Link outcrop. Image number: CX00027MR0030530F399886415VA. **(e)** Rounded pebbles at Sol 356. Image number: 0356MR1452001000E1_DXXX. **(f)** Angular clasts at Sol 389. Image number: 0389ML1600090000E1_DXXX. On **d-f**, analysed grains are enhanced with purple colour. Image credits: **(a)** NASA/JPL-Caltech/ESA/DLR/FU Berlin/MSSS; **(b,c)** NASA/JPL-Caltech/Univ. of Arizona; **(d-f)** NASA/JPL-Caltech/MSSS.

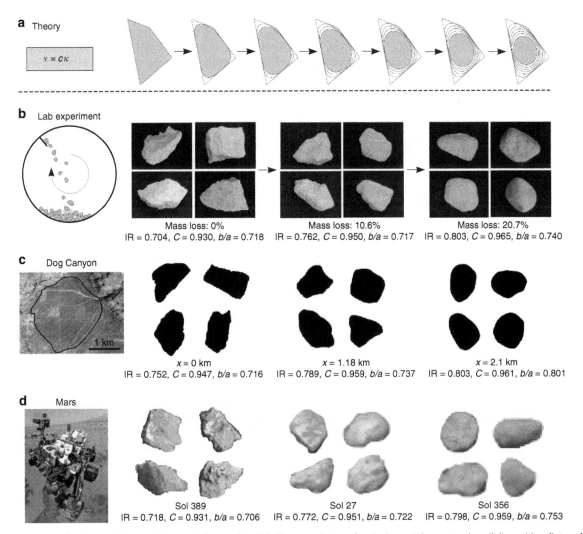

Figure 2 | Qualitative shape trends from theory and observation. (a) Shape evolution of a single particle constantly colliding with a flat surface is described by Firey's equation[8] $v = c\kappa$, where v is the speed of abrasion in the inward normal direction, c is a constant and κ is the local curvature. This is illustrated on a quadrangle. **b–d** show example pebbles from each system studied, with comparable shape parameters as indicated beneath each image (IR: circularity, C: convexity, b/a: axis ratio, Fig. 3). **(b)** Sketch of the rotating drum experiment, limestone pebble samples ($a \approx 15$–35 mm) and mean shape parameter values after 0, 10.6 and 20.7% mass loss. **(c)** Aerial image of Dog Canyon fan, example limestone pebble contours ($a \approx 20$–40 mm) and mean shape parameter values at $x = 0$, $x = 1.18$ and $x = 2.10$ km. Grains were collected from the active channel denoted by the blue line. **(d)** A few Martian grain contours ($b \approx 2$–32 mm; ref. 2) and mean shape parameter values at Sols 389, 27 and 356. Image credits: **(c)** Google Earth; **(d)** NASA/JPL-Caltech/MSSS.

To examine these ideas, we conducted a new set of experiments that simulated abrasion in a more natural manner than our previous work[9]. Eighty limestone fragments with a size range of $a = 15$–35 mm were placed in a small rotating drum (diameter 20 cm, rotation rate 50 r.p.m.) with a paddle, so that grains were lifted and dropped causing inter-particle collisions (Fig. 2b). The pebbles were removed from the drum after a certain number of rotations (Methods), and their shapes and mass were recorded (Fig. 3). Shape evolution follows the same trends as previous single-particle results[9], but the curves are shifted in space due to the difference in initial particle shapes.

We compare field data from a steep, mountain river in Puerto Rico[15] to these new experiments, by re-casting downstream shape changes as a function of mass loss (Methods). We note first that the initial shapes of the Puerto Rico volcaniclastic pebbles are, within error, identical to the crushed limestone particles used in the experiments (Fig. 3). Shape evolution trends are also in reasonable agreement with experiments, given the vast differences in transport conditions between the drum and the natural river (Fig. 3). This agreement supports the possibility of a generic,

quantitative relation between pebble shape and mass loss for collisional abrasion.

To explore the consequences of this finding in a depositional environment more comparable to the Martian deposits, we collected downstream pebble shape data on the Dog Canyon alluvial fan in New Mexico, USA (Fig. 2c). Particle shapes were determined from images while mass was not measured, so the data are comparable to available measurements on Mars. The profile of Dog Canyon fan is similar to Peace Vallis—which may or may not be representative of the older exhumed alluvial fan deposits—although shorter in length. The upper fan is channelized and steeper ($S \approx 4\%$, length ≈ 2 km), with limestone gravel that decreases from ~ 40 to 20 mm (similar to experiments). Channels disappear at the gravel-sand transition, beyond which lies a mixed sand-gravel region with a lower slope ($S \approx 1\%$)[18]. The latter is indicative of an environment that would produce conglomerates similar to those seen on Mars. Initial pebble shapes at the apex of the fan are slightly more rounded than fragments; this is to be expected, as some abrasion is likely to occur within the upstream canyon (Fig. 3). Downstream shape evolution appears similar to the other data; however, it cannot be directly compared since mass loss is unknown.

It has been demonstrated that the mass of pebbles (M) decreases exponentially with downstream distance (x) in alluvial rivers,

$$M = M_0 e^{-kx}, \qquad (1)$$

where k is an empirically determined 'diminution coefficient'[9,19] and M_0 is initial mass. This decrease is caused both by abrasion, and by size-selective sorting in which less massive particles travel farther downstream[20,21]. Both processes contribute, in unknown proportions, to the observed value for the diminution coefficient, that is, $k = k_a + k_s$. It is generally accepted that sorting is dominant over abrasion in many rivers ($k_s \gg k_a$); but abrasion is still significant, and is likely dominant in settings where sediment storage is limited. It has recently been shown that the effects of abrasion (k_a) may be isolated by examining pebble shape[9,15]. If we assume that the derived shape/mass-loss curve is

general, the Dog Canyon shape data indicate a mass loss due to abrasion of 15% ($M/M_0 = 0.85$) over $x = 2$ km (Fig. 3). (Volume estimates from measured size indicate an overall mass reduction of close to 90%, consistent with the dominance of size-selective sorting in this strongly depositional alluvial fan setting[18].) The resulting estimate for the abrasion diminution coefficient from equation (1), $k_a \sim 10^{-1}$ km^{-1}, is consistent with previous experiments that simulated collisional abrasion of similar-sized limestone pebbles driven by a water current[12]. This agreement indicates that pebble shape alone may be used to provide an estimate of travel distance.

Analysis of Martian rocks. We measured the shape of 261 and 304 particles at 2 distinct locations across Bradbury Rise (Fig. 1), where rounded pebbles associated with conglomerate deposits were identified[2,3,22,23]. We did not attempt to measure particle size from these oblique images, but previous studies indicate a range of 2–32 mm with a median diameter $b \approx 10$ mm. Contours of unobscured pebbles were traced with a resolution of ~ 70 contour points per particle (Methods). We used the same methods also to examine the shape of angular clasts observed during Curiosity's traverse, at three selected locations (Fig. 1). These angular clasts are strewn about the Martian surface, are not related to the ancient lithified conglomerates, and have been interpreted as more recently emplaced impact breccia clasts[3]. They are readily distinguished from the ancient stream-transported rounded pebbles by their shape and lack of association with any outcrop (Figs 2d and 1f). We remark that the hypothesized flow direction on the alluvial fan complex differs from the rover's transect[1], therefore we did not attempt to find any trend in the shape data along the rover's transect; the main consideration in selecting the sites was to obtain a sufficient sample size of particles within a single image.

Measured shape parameters for the angular clasts are nearly identical to the terrestrial fragments (Fig. 3, Supplementary Fig. 1), indicating these particles were formed by fragmentation processes[17] and have experienced no fluvial transport. Although we do not have shape measurements for particles at the headwaters of the streams feeding into Gale Crater, we infer that observed clast shapes are likely representative of the initial (pre-abrasion) conditions for the rounded pebbles. The rounded pebbles are distinct from the clasts; all shape parameters indicate

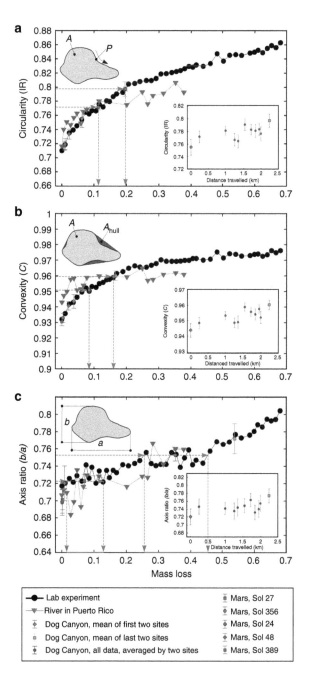

Figure 3 | Quantitative shape evolution as a function of mass loss. Upper left insets: definition of shape parameters. (**a**) Circularity (or isoperimetric ratio), defined as $IR = 4\pi A/P^2$, where A is the area and P is the perimeter of the pebble's projection in the $a - b$ plane[42]. (**b**) Convexity, $C = A/A_{hull}$, where A_{hull} is the area of the convex hull[15]. (**c**) Axis ratio, the ratio of the short (b) and long (a) axis lengths. Lower right insets: evolution of shape parameters against transport distance from the apex of Dog Canyon alluvial fan. Neighbouring sites were paired and averaged to form 11 data points from the 22 sites sampled (Methods). Main diagrams show evolution of shape parameters against mass loss in the experiment (black dots), and in the river from Puerto Rico (grey triangles)[15]. Coloured markers represent mean shape parameter values, with error bars showing the s.e. Rounded Mars pebble values (red markers, Supplementary Table 1) were projected onto the experimental curves (red horizontal arrows) to estimate mass loss (red vertical arrows); the difference in shape values between the two populations is interpreted as inter-site variability rather than a reflection of any trend. Blue markers represent angular clasts from Mars. Dog Canyon results (green and magenta markers) were also projected onto the experimental curves; data suggest particles at fan apex ($x = 0$ km) are slightly abraded due to transport in the upstream canyon ($\approx 5\%$ mass loss), and experience $\approx 15\%$ mass loss due to bed-load transport over a 2-km distance down the alluvial fan.

that significant abrasion has occurred (Fig. 3, Supplementary Table 1). According to the terrestrial shape evolution curves, the changes in IR and C associated with the difference between angular clasts and rounded pebbles correspond to $\sim$10 and 20% mass loss, respectively, for the two sites. Axis ratio measurements allow for up to 45% mass loss, although there is much greater uncertainty (Fig. 3).

Discussion

Peace Vallis—and, presumably, the exhumed underlying fan associated with the Bradbury rise conglomerates—is similar to the Dog Canyon alluvial fan in many respects. One important difference, however, is that the Gale pebbles are basaltic in composition rather than limestone. Experiments by Attal and Lavé[12] indicate that the abrasion rates for (igneous) volcanic rocks, although highly variable, are an order of magnitude smaller than for limestone under identical transport conditions. Accordingly, we expect $k_a \sim 10^{-2}\,\mathrm{km}^{-1}$; this value is also consistent with compiled field data for rivers with negligible sorting (that is, $k \approx k_a$) (ref. 20). If we adopt this value for k_a, and take a representative value of 20% mass loss, equation (1) would produce an estimate that the pebbles exposed at Bradbury Rise have been transported a distance of $x \approx 20\,\mathrm{km}$ from their source.

The above calculation does not take into account the reduced gravity (g) on Mars (for example, ref. 24). The relation between pebble mass loss and downstream distance should be a function of: the energy of individual collisions; and the hop length of an individual pebble, which determines the number of collisions per unit distance downstream[9,13]. Abrasion rate is proportional to kinetic energy $= 1/2 m v_s^2$, where m and v_s are pebble mass and collision velocity, respectively[12,13,25]. We make the simplifying assumption that v_s is proportional to pebble settling velocity, which may be approximated in the large-particle limit ($b \gg 10^0\,\mathrm{mm}$) as $w_s \sim \sqrt{Rgb}$ where R is relative submerged density[26]. From this, we might naively expect that $k_{a,\mathrm{Mars}}/k_{a,\mathrm{Earth}} = w_{s,\mathrm{Mars}}^2/w_{s,\mathrm{Earth}}^2 = g_{\mathrm{Mars}}/g_{\mathrm{Earth}} \approx 0.38$. Considering pebble hop length, it has been shown experimentally that a characteristic hop-length scale (l_d) of particles in bed load is $l_d = u_s t_d \sim b u_s/w_s$, where u_s and t_d are the particle horizontal velocity and settling time, respectively[27]. Assuming gravity-driven (normal) flow conditions, $u_s \sim \sqrt{g}(\sqrt{hS} - \sqrt{\tau_* Rb})$, where h is river flow depth and τ_* is the threshold dimensionless stress for initiation of motion[27]. If, following others[28,29], we assume that τ_* is the same for Earth and Mars, we see that $u_s \sim \sqrt{g}$. From these arguments it appears that the smaller g of Mars reduces both the settling velocity and horizontal velocity of sediment grains, the latter because the smaller gravitational force results in slower Martian river-current velocities. Considering the hop length of pebbles, both particle velocity terms scale as $\sqrt{g}$ and thus gravity cancels out of the problem. This stands in contrast to the case of aeolian (wind-blown) bed-load transport on Mars, where it has been predicted that particle hops are farther and faster than on Earth[30]; however, this is due mostly to the large differences in the density and viscosity of the atmosphere between the two planets. Fluvial transport is (presumably) driven by water on both planets, and its density and viscosity vary only modestly with temperature. It is worth noting that, in the small particle limit ($b < < 10^0\,\mathrm{mm}$), the settling velocity scales linearly with g (ref. 26), and thus the hop length of sand-size and smaller particles would be expected to depend on gravity as $l_d \sim 1/\sqrt{g}$.

We deduce from this analysis that, to first order, reduced collision energy is likely the dominant effect of gravity on pebble abrasion in bed load. The same may also be true for the case of erosion of bedrock channels by bed-load abrasion. From the

calculations above, incorporating this effect changes the estimated transport distance of the Martian pebbles by a factor of $1/0.38$ to $x \approx 50\,\mathrm{km}$. Previous studies used compositional information to suggest that the source area for the conglomerates was the northern rim of Gale Crater[5,31]. Our quantitative estimates for transport distance support this view. In all likelihood, fluvial transport has carried the pebbles tens of kilometres from their source. The distance to the northern rim of Gale Crater, and the outlet of an erosional drainage basin, is $\sim$30 km (Fig. 1). We conclude that the rounded pebbles were sourced from fluvial erosion of the northern rim, and were deposited along the lower reaches of an alluvial fan complex. Subsequent erosion, likely by wind, has exhumed the fan to produce intermittent exposures across Bradbury Rise.

It is difficult to assign uncertainty estimates to the calculations presented here, and we emphasize that our results should be interpreted in terms of order of magnitude. Transport equations account for the influence of gravity on dimensional grounds, but include empirical coefficients determined from terrestrial data that might be influenced by gravity in unknown ways. From a measurement perspective, it appears that the IR and convexity provide more reliable estimates for mass loss than does the axis ratio. One reason for this is that—as pointed out previously[9]—pebbles may lose close to half of their mass without a significant change in the axis ratio. The parameters IR and C are most sensitive to the initial phase of abrasion, while b/a is least sensitive. Another possible effect worthy of examination is the influence of grain fabric on the axis ratio. While Domokos et al.[17] found a universal distribution of axis ratios for fragmented rocks of many lithologies, the samples and simulations examined homogeneous materials. The range of axis ratios for natural fragments formed from rocks with significant heterogeneity (for example, fractures and foliations) may be more varied.

This study takes advantage of two general principles of particle shape, to provide a new tool for estimating the transport distance of fluvial pebbles. The first is that particles formed by fragmentation, regardless of the particular process, have similar shape[17]. This recently established result from terrestrial studies[17] is now extended to Mars, and is consistent with the hypothesis that the angular clasts observed along Curiosity's traverse are impact breccia[3] (although aeolian abrasion may also form angular ventifacts[32]). The second principle is that the shape evolution of these initially fragmented particles under collisional bed-load abrasion follows a single curve, when cast as a function of mass loss. This is supported by the similarity of results from a natural river and a simple drum experiment, and their consistency with the geometric theory of abrasion[9,11]. Together these two ideas provide a means to estimate mass loss due to abrasion from bed-load transport, using shape alone. Transport distance may then be estimated using equation (1), if a value for k_a can be reasonably constrained. From field and experimental data, we determine that k_a is of order $10^{-2}\,\mathrm{km}^{-1}$ for common quartzite and igneous gravel in rivers on Earth, and propose that this parameter may scale linearly with gravity resulting in reduced abrasion rates on Mars.

The technique applied here to ancient Martian conglomerates could just as well be used for ancient and modern deposits on Earth and river pebbles on other planetary bodies such as Titan[33]. Determining mass loss from pebble shape could help to determine the contribution of pebble abrasion to sand and silt production in rivers[34]. Estimating transport distance from shape provides a new means for assessing sediment provenance. The theory underlying shape evolution is purely geometric[11], and therefore should apply to all scales so long as the basic assumptions are fulfilled. It has already been shown that the geometric model captures salient features of the shapes of asteroids abraded by collisions with meteorites[35]. We propose that our findings on pebble shape evolution may be extended to aeolian settings, which could find similar applications in sediment provenance studies and for quantifying dust production resulting from sand abrasion[34,36]. This would also allow grain-scale rover measurements to inform our evolving understanding of the frequency and magnitude of dune activity on present-day Mars[37–39]. A quantitative comparison between fluvial and aeolian environments is not yet possible, as data regarding the latter are insufficient at present. A recent study of sand shape in a terrestrial gypsum dune field[40] is encouraging, as reported trends are qualitatively consistent with our findings.

Methods

Data collection. Laboratory particles were created from soft limestone blocks with initial sizes in the range 50–70 mm, sourced from Sóskút, Hungary. These rocks were chosen because: they are easily crushed, allowing creation of a desired initial particle size range; and they erode quickly by abrasion—but do not fragment—in the drum, so experiments may be conducted efficiently. The blocks were crushed with a hammer to produce naturally shaped fragments in the size range of $a = 15$–35 mm, similar to Mars pebbles and also Dog Canyon. We checked that their shape distribution matched that of natural rock fragments, which are known to follow a universal distribution regardless of the rock type[17] (Supplementary Fig. 1). The crushed limestone grains were abraded at 50 r.p.m. in the rotating drum. Using a drop height of $h = 20$ cm equal to the drum diameter, collision velocity may be approximated as $\sqrt{2gh} = 2\,\mathrm{m\,s^{-1}}$; this produces collision energies comparable to those expected for fluvial transport of similar-sized grains[34]. The experiment was stopped every n rotations, at which point: dust was removed from the drum to prevent frictional abrasion; the total weight of the grains was measured; and all grains were imaged on a large black board, which provided high contrast (Fig. 2). The interval n was increased approximately logarithmically as the experiment progressed, to sample at intervals of roughly equal mass-loss fraction (equation (1)); $n = 5$ from 0 to 10 rotations, $n = 10$ from 10 to 80 rotations, $n = 20$ from 80 to 200 rotations, $n = 50$ from 200 to 1,000 rotations, and $n = 100$ from 1,000 to 2,500 rotations.

Grain size data from the Dog Canyon alluvial fan were reported previously[18], including details on the setting and sampling locations. Here we report new shape data measured at these same locations, which represent 22 cross-sections of the active channel (Fig. 2). At each site, 20 pebbles were collected from the channel bottom following the Wolman pebble count method[41]. Each pebble was placed on a rigid, high-contrast board and imaged. Since sample size at each site was small, spatially neighbouring data were paired to form 11 data points on Fig. 3, thus each of them averaging the data of 40 pebbles. Pebble shape data for the Rio Mameyes—the mountain stream in northeastern Puerto Rico referenced in this paper—were collected following a similar procedure. Main results and details on the sampling and imaging methods were reported previously[15]. Shape data from that study were re-plotted here against fraction mass loss resulting from abrasion (instead of transport distance), where the mass-loss fraction was estimated from $M/M_0 = e^{-0.053x}$ based on a numerical model fit to the data[15] (Discussion in the cited paper). This translates to a value $k_a = 0.053\,\mathrm{km^{-1}}$ in equation (1).

Image analysis. Grain contours were manually traced in Adobe Photoshop for the Martian pebbles, since image contrast was too low for automated methods. For the Dog Canyon fan pebbles, contours were semi-automatically traced by the Quick Selection Tool in Adobe Photoshop, which is able to detect edges of objects based on contrast and colour changes between the object and its background. This same procedure was used to process the Puerto Rico river pebbles[15] and laboratory experiments. After determining the contours, all images were converted to binary images (Supplementary Fig. 2) and imported to Matlab.

Measured shape parameters are sensitive to the resolution of the contours, a factor that has not been quantitatively assessed up to now. We examined this scale dependence by downsampling experimental images at different resolutions (Supplementary Fig. 3), which showed a significant effect. To circumvent this issue and allow comparison of data from different settings, pebble contours should be determined from approximately the same resolution for all images. The resolution for individual Martian pebbles is very low, on average 70 pixels per grain contour. Accordingly, all other images were resized so that the mean resolution was ~70 pixels per grain contour for each population (Supplementary Table 2). Resizing was performed with Matlab's bicubic interpolation method, where the output pixel value is a weighted average of pixels in the nearest four-by-four neighbourhood.

Martian images were checked for statistical convergence of shape parameters, and results verify that sample numbers for each image were sufficiently large (Supplementary Fig. 4). This is reflected also by the very small errors of the mean values (Supplementary Table 1).

References

1. Palucis, M. C. *et al.* The origin and evolution of the Peace Vallis fan system that drains to the Curiosity landing area, Gale Crater, Mars. *J. Geophys. Res. Planets* **119**, 705–728 (2014).

2. Williams, R. M. E. *et al.* Martian fluvial conglomerates at Gale crater. *Science* **340**, 1068–1072 (2013).

3. Yingst, R. A. *et al.* Characteristics of pebble- and cobble-sized clasts along the Curiosity rover traverse from Bradbury Landing to Rocknest. *J. Geophys. Res. Planets* **118**, 2361–2380 (2013).

4. Jerolmack, D. J. Pebbles on mars. *Science* **340**, 1055–1056 (2013).

5. Grotzinger, J. P. *et al.* A habitable fluvio-lacustrine environment at Yellowknife Bay, Gale Crater, Mars. *Science* **343**, 1242777 (2014).

6. Grotzinger, J. P. *et al.* Geologic framework for Aeolis Palus bedrock, and its relationship to Mt. Sharp, Mars. in *AGU 2014 Fall Meeting, Abstract P42C-01* (San Francisco, CA, USA, 2014).

7. Grotzinger, J. P. *et al.* Fluvial-lacustrine sedimentology and stratigraphy, Gale Crater, Mars. in *GSA Annual Meeting, Abstract 85-6* (Vancouver, 2014).

8. Firey, W. J. Shapes of worn stones. *Mathematika* **21**, 1–11 (1974).

9. Domokos, G., Jerolmack, D. J., Sipos, A. Á. & Török, Á. How river rocks become round: resolving the shape-size paradox. *PLoS ONE* **9**, e88657 (2014).

10. Bloore, F. J. The shape of pebbles. *Math. Geol.* **9**, 113–122 (1977).

11. Domokos, G. & Gibbons, G. W. The evolution of pebble size and shape in space and time. *Proc. R. Soc. Lond. A* **468**, 3059–3079 (2012).

12. Attal, M. & Lavé, J. Pebble abrasion during fluvial transport: Experimental results and implications for the evolution of the sediment load along rivers. *J. Geophys. Res. Earth Surf.* **114**, F04023 (2009).

13. Sklar, L. S. & Dietrich, W. E. A mechanistic model for river incision into bedrock by saltating bed load. *Water Resour. Res.* **40**, W06301 (2004).

14. Domokos, G. & Gibbons, G. W. Geometric and physical models of abrasion. 23 (2013) Preprint at *arXiv:1307.5633*.

15. Miller, K. L., Szabó, T., Jerolmack, D. J. & Domokos, G. Quantifying the significance of abrasion and selective transport for downstream fluvial grain size evolution. *J. Geophys. Res. Earth Surf.* **119**, 2412–2429 (2014).

16. Adams, J. Wear of unsound pebbles in river headwaters. *Science* **203**, 171–172 (1979).

17. Domokos, G., Kun, F., Sipos, A. Á. & Szabó, T. Universality of fragment shapes. *Sci. Rep.* **5**, 9147 (2015).

18. Miller, K. L., Reitz, M. D. & Jerolmack, D. J. Generalized sorting profile of alluvial fans. *Geophys. Res. Lett.* **41**, 7191–7199 (2014).

19. Sternberg, H. Untersuchungen Über Längen-und Querprofil geschiebeführender Flüsse. *Z. Naturforsch.* **XXV**, 483–506 (1875).

20. Morris, P. & Williams, D. A worldwide correlation for exponential bed particle size variation in subaerial aqueous flows. *Earth Surf. Proc. Land.* **24**, 835–847 (1999).

21. Fedele, J. J. & Paola, C. Similarity solutions for fluvial sediment fining by selective deposition. *J. Geophys. Res. Earth Surf.* **112**, F02038 (2007).

22. Newsom, H. E. *et al.* Gale crater and impact processes—Curiosity's first 364 Sols on Mars. *Icarus* **249**, 108–128 (2015).

23. Vasavada, A. R. *et al.* Overview of the Mars Science Laboratory mission: Bradbury Landing to Yellowknife Bay and beyond. *J. Geophys. Res. Planets* **119**, 1134–1161 (2014).

24. Komar, P. D. Modes of sediment transport in channelized water flows with ramifications to the erosion of the Martian outflow channels. *Icarus* **42**, 317–329 (1980).

25. Anderson, R. S. Erosion profiles due to particles entrained by wind: application of an eolian sediment-transport model. *Geol. Soc. Am. Bull.* **97**, 1270–1278 (1986).

26. Ferguson, R. I. & Church, M. A simple universal equation for grain settling velocity. *J. Sediment. Res.* **74**, 933–937 (2004).

27. Lajeunesse, E., Malverti, L. & Charru, F. Bed load transport in turbulent flow at the grain scale: Experiments and modeling. *J. Geophys. Res. Earth Surf.* **115**, F04001 (2010).

28. Sagan, C. & Bagnold, R. A. Fluid transport on Earth and aeolian transport on Mars. *Icarus* **26**, 209–218 (1975).

29. Kleinhans, M. G. Flow discharge and sediment transport models for estimating a minimum timescale of hydrological activity and channel and delta formation on Mars. *J. Geophys. Res. Planets* **110**, E12003 (2005).

30. Almeida, M. P., Parteli, E. J., Andrade, J. S. & Herrmann, H. J. Giant saltation on Mars. *Proc. Natl Acad. Sci. USA* **105**, 6222–6226 (2008).

31. McLennan, S. M. *et al.* Elemental geochemistry of sedimentary rocks at Yellowknife Bay, Gale crater, Mars. *Science* **343**, 1244734 (2014).

32. Bridges, N. T. *et al.* The rock abrasion record at Gale crater: Mars Science Laboratory results from Bradbury landing to Rocknest. *J. Geophys. Res. Planets* **119**, 1374–1389 (2014).

33. Burr, D. M., Emery, J. P., Lorenz, R. D., Collins, G. C. & Carling, P. A. Sediment transport by liquid surficial flow: application to Titan. *Icarus* **181**, 235–242 (2006).

34. Jerolmack, D. J. & Brzinski, T. A. Equivalence of abrupt grain-size transitions in alluvial rivers and eolian sand seas: a hypothesis. *Geology* **38**, 719–722 (2010).

35. Domokos, G., Sipos, A. Á., Szabó, G. y. M. & Várkonyi, P. L. Formation of sharp edges and planar areas of asteroids by polyhedral abrasion. *Astrophys. J. Lett.* **699**, L13 (2009).

36. Bullard, J. E., McTainsh, G. H. & Pudmenzky, C. Aeolian abrasion and modes of fine particle production from natural red dune sands: an experimental study. *Sedimentology* **51**, 1103–1125 (2004).

37. Bridges, N. T. *et al.* Earth-like sand fluxes on Mars. *Nature* **485**, 339–342 (2012).

38. Hansen, C. J. *et al.* Seasonal erosion and restoration of Mars' northern polar dunes. *Science* **331**, 575–578 (2011).

39. Silvestro, S., Fenton, L. K., Vaz, D. A., Bridges, N. T. & Ori, G. G. Ripple migration and dune activity on Mars: evidence for dynamic wind processes. *Geophys. Res. Lett.* **37**, L20203 (2010).

40. Jerolmack, D. J., Reitz, M. D. & Martin, R. L. Sorting out abrasion in a gypsum dune field. *J. Geophys. Res. Earth Surf.* **116**, F02003 (2011).

41. Wolman, M. G. A method of sampling coarse river-bed material. *Eos Trans AGU* **35**, 951–956 (1954).

42. Blott, S. J. & Pye, K. Particle shape: a review and new methods of characterization and classification. *Sedimentology* **55**, 31–63 (2008).

Acknowledgements

We thank Rebecca Williams and Nicolas Mangold for their help in locating the Mars conglomerates and their guidance in Curiosity's raw data; Kim Miller for providing data from Dog Canyon; and Sarolta Bodor for help in the drum experiment. Research was supported by the US National Science Foundation Luquillo Critical Zone Observatory (EAR-1331841) to D.J.J. a Korányi Fellowship to T.S.Z. Hungarian OTKA grant 104601 to G.D. and T.Z. and the NASA Astrobiology Institute and Mars Science Laboratory Mission to J.P.G.

Author contributions

T.S.Z. performed all of the data analysis; G.D. led the laboratory experiment; J.P.G. developed the Martian geology; and D.J.J. supervised the research. All authors contributed to writing of the document and interpreting the results.

Additional information

9

The puzzling Venusian polar atmospheric structure reproduced by a general circulation model

Hiroki Ando[1], Norihiko Sugimoto[2], Masahiro Takagi[3], Hiroki Kashimura[4], Takeshi Imamura[1] & Yoshihisa Matsuda[5]

Unlike the polar vortices observed in the Earth, Mars and Titan atmospheres, the observed Venus polar vortex is warmer than the midlatitudes at cloud-top levels (~ 65 km). This warm polar vortex is zonally surrounded by a cold latitude band located at $\sim 60°$ latitude, which is a unique feature called 'cold collar' in the Venus atmosphere. Although these structures have been observed in numerous previous observations, the formation mechanism is still unknown. Here we perform numerical simulations of the Venus atmospheric circulation using a general circulation model, and succeed in reproducing these puzzling features in close agreement with the observations. The cold collar and warm polar region are attributed to the residual mean meridional circulation enhanced by the thermal tide. The present results strongly suggest that the thermal tide is crucial for the structure of the Venus upper polar atmosphere at and above cloud levels.

[1]Institute of Space and Astronautical Science, Japan Aerospace Exploration Agency, Sagamihara, Kanagawa 252-0222, Japan. [2]Research and Education Center for Natural Sciences, Department of Physics, Keio University, Yokohama, Kanagawa 223-8521, Japan. [3]Faculty of Science, Kyoto Sangyo University, Kita-ku, Kyoto 603-8555, Japan. [4]Japan Agency for Marine-Earth Science and Technology, Yokohama, Kanagawa 236-0001, Japan. [5]Department of Astronomy and Earth Sciences, Tokyo Gakugei University, Koganei, Tokyo 184-8501, Japan. Correspondence and requests for materials should be addressed to H.A. (email: hando@ac.jaxa.jp).

A lot of infrared and radio occultation measurements in the Venus exploration missions show that the polar vortex is warmer than the midlatitudes at cloud-top levels and surrounded by a zonal cold latitude band called the cold collar located around ~60° latitude at ~65 km (refs 1–4). The morphology seen in the polar vortex changes temporally and sometimes shows a hot dipole or S-shaped structure[3,5–7]. Furthermore, these unique structures have been observed in both north and south polar regions.

While a lot of observational studies about the atmospheric structure in the Venus polar region have been performed, there were no numerical studies that succeeded in reproducing these unique structures in realistic model settings. Lee et al.[8] have reported that the cold collar and the enhanced warm polar region (hot polar spot) are reproduced in their general circulation model (GCM) results and indicated that these structures are related to the reversal of the latitudinal temperature gradient, which is caused by the zonal jet produced by mean meridional circulation (MMC). However, the zonal-mean wind reproduced in their model is not consistent with recent measurements[9]. In addition, the MMC is driven by Newtonian cooling with constant relaxation time of 25 Earth days, in which diurnal variation of the solar heating is excluded. Yamamoto and Takahashi[10,11] have also reproduced the polar vortex in their GCM with an artificial forcing of zonal wavenumber one at the bottom boundary to excite Kelvin and/or Rossby waves. Although they indicated that the polar diurnal tide enhances the cold collar and warm polar region, their results are not consistent with optical and radio occultation measurements for the following reasons: (1) the horizontal temperature distribution has large non-uniformity; (2) the location of the temperature minimum does not move in the longitudinal direction but is fixed at a specific geographical longitude; and (3) the temperature minimum corresponding to the cold collar is not seen in the latitude–height distribution of temperature in their calculation.

The Venus polar region also has interesting features in terms of material transport. Recent optical measurements showed that a significant infrared absorption signature by CO, which is mainly created photochemically above the cloud layer, exists over Venus' south pole, coinciding with bright dipole features seen in the polar vortex and the CO abundance increases with latitude above Venus' nightside cloud top[12,13]. Furthermore, the clouds in the polar region are thicker than those in the cold collar[14]. By using a two-dimensional model including a simple cloud microphysics scheme, Imamura and Hashimoto[15] suggested that the clouds might become thick with latitude by the cloud material transport associated with the MMC. However, the structure of the MMC is simply assumed in their model. It is necessary to consider how its structure is related to the dynamical process.

A thermal tide is the planetary scale wave excited by the solar heating. In the Venus atmosphere, it is strongly excited at the cloud levels because a large part of the solar flux is absorbed there[16–18]. The horizontal wind velocity associated with the thermal tide is estimated to be >40 m s^{-1} at the cloud top. It is also indicated that superrotation at the cloud levels is affected significantly by the thermal tide[19,20]. However, its effect on the atmospheric structure in the polar region has not yet been examined.

In this study, we investigate the structure of the Venus upper polar atmosphere by using a GCM named Atmospheric GCM for the Earth Simulator (AFES) for Venus (see Methods for details). To examine the dynamical effects of the thermal tide, we perform two numerical experiments with observation-based distributions of the solar heating: one with the diurnal components (Case A) and one without them (Case B). The thermal tide is excited only in Case A. Our study shows that the cold collar and warm polar region are

due to the residual mean meridional circulation (RMMC) intensified by the thermal tide, implying that the atmospheric structure at and above cloud levels in the Venus polar region is strongly attributed to the thermal tide. Furthermore, our results also indicate that the observed meridional distributions of cloud thickness and CO abundance above the cloud top are associated with the RMMC enhanced by the thermal tide.

Results

Background zonal wind and temperature distributions. The model atmosphere reached quasi-steady states within approximately four Earth years, and the quasi-steady states were maintained for more than ten Earth years[21]. In the following, we focus on the atmospheric structure in the northern hemisphere, because almost the same structure was observed in the southern hemisphere.

Figure 1 shows meridional cross-sections of the zonally and temporally averaged zonal wind and temperature obtained in Case A. The fast mean zonal wind is maintained at 60–80 km. Above ~80 km, the mean zonal wind decreases very sharply in the vertical direction; the velocity is <10 m s^{-1} at 90 km. In the temperature field, the meridional gradient changes its sign near the cloud-top level (~70 km). That is, the zonal wind and temperature in our results are consistent with the gradient wind balance. As shown later, the warm polar region is related to the positive gradient of the temperature above that level. Significant negative shear of the mean zonal wind is attributed to the thermal tide, since it is theoretically predicted that the thermal tide accelerates the mean zonal wind at 50–70 km and decelerates at 80–100 km (ref. 20). Another notable feature is that the zonal wind velocity is almost unchanged vertically at 40–80 km levels in the polar region. These features of the zonal wind and the temperature are in close agreement with those obtained in recent observations[4,9]. Figure 1 also shows that a midlatitude jet is formed around the cloud-top level (~70 km) at 20°–50° latitudes. The jet axis is located more equatorwards than that obtained in previous works[8,10,11]. This difference may also be explained by the acceleration effect of the thermal tide as noted above. The meridional distribution of the zonal wind is in close agreement with that reported by Machado et al.[22], who retrieved the wind speed in the dayside by the ground-based Doppler velocimetry measurements. In the work of Piccialli et al.[9], in which the meridional-height distributions of the zonal wind are estimated from those of the temperature by assuming the gradient wind balance, the positions of the midlatitude jets are about 15° polewards. It is noted, however, the estimated wind distribution strongly depends on the assumed meridional distribution of the zonal wind at the bottom boundary altitude. Sánchez-Lavega et al.[23] also observed the wind distributions during the local time from 6 to 17 h, and indicated that the zonal wind speed increases from the morning to the afternoon due to the thermal tide at the upper cloud level in a latitude range of 50° S–75° S. This result is quite consistent with our present results. The zonal wind speed at middle and high latitudes is enhanced by the thermal tide by 20–30 m s^{-1} (not shown). It is also noted that, as shown in Sugimoto et al.[24], the strength (or velocity) and the position of the midlatitude jet vary with time in the present model, although the variation width of the latter in the latitudinal direction is not more than ~10°. These fluctuations may contribute to the difference between the present results and the observations. As shown later, the position of this jet is important to the cold collar.

Temporal variation of temperature. Figure 2 shows temporal variations of the zonal-mean temperature in a latitude range of 30°–90° at ~75 km (the pressure level of 1×10^3 Pa), ~68 km

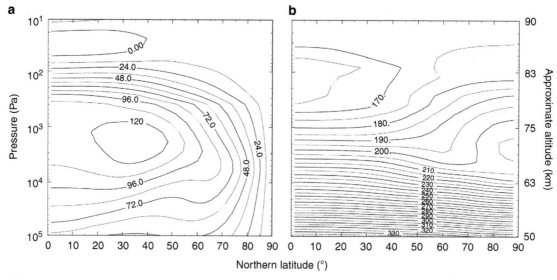

Figure 1 | Meridional cross-sections of the zonally and temporally averaged data in Case A. (a) Zonal wind (m s $^{-1}$) and **(b)** Temperature (K). Averaged period is two Venusian solar days (234 Earth days) after settling into the quasi-steady state.

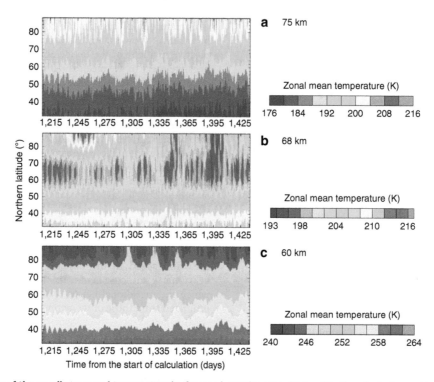

Figure 2 | Time evolutions of the zonally averaged temperature in the quasi-steady state obtained in Case A. Each panel is depicted for two Venusian solar days (234 Earth days) at the altitude of **(a)** ~ 75 km (the pressure level of 1×10^3 Pa), **(b)** ~ 68 km (4×10^3 Pa) and **(c)** ~ 60 km (2×10^4 Pa), respectively. The latitude range is from 30° N to 90° N.

(4×10^3 Pa) and ~ 60 km (2×10^4 Pa) obtained in Case A. At ~ 75 km, the polar region is always warmer than low latitudes with short- and long-period fluctuations (4–5 and ~ 30 Earth days, respectively). The meridional temperature difference is ~ 30 K, which is a little larger than that estimated from the radio occultation measurements[4,9]. At ~ 68 km, the cold collar (local minimum of the temperature) clearly appears at latitudes of 60°–70°. The temperature in the cold collar is colder than the surroundings by ~ 10 K, and qualitatively consistent with the radio occultation measurements[2,4,9]. The short-period fluctuation is dominant at this level, which may be caused by the baroclinic instability waves excited in the midlatitudes at the cloud levels[24]. At ~ 60 km, the temperature monotonically decreases with

latitude. Similarly to ~ 75 km, fluctuations with both short and the long periods appear at this level.

Figure 3 shows time evolution of the horizontal temperature distribution at ~ 68 km obtained in Case A. The cold collar surrounds the warmer polar region at $\sim 60°$ N. The maximum temperature difference between 60° N and the pole is ~ 20 K. The warm polar region rotates around the pole and varies in shape temporally. An S-shaped vortex structure also appears in Fig. 3g. It rotates around the pole, but the centre of rotation is offset from the pole. These features are in close agreement with the cold collar and the polar vortex observed by infrared measurements[1,3,5,6], although the altitude of the cold collar reproduced in the present model is slightly higher than that estimated in such observations.

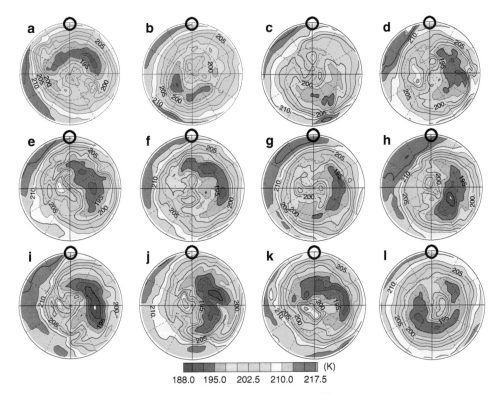

Figure 3 | Time evolution of the temperatures (K) at ∼68 km (the pressure level of 4 × 10³ Pa) in Case A in the polar plot. The latitude range is from 30°N to 90°N. The black circle represents the local solar noon. Each panel of (**a**) to (**l**) shows temperature distribution from 1414 to 1425 days from the start of the calculation. The time interval between the adjacent panels is one Earth day.

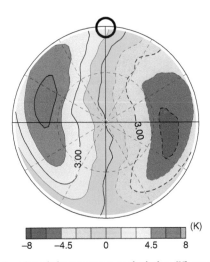

Figure 4 | Polar plot of the temperature deviation (K) associated with the thermal tide at ∼68 km (the pressure level of 4 × 10³ Pa) obtained in Case A. The latitude range is from 30° N to 90° N. The black circle represents the local solar noon. Average period of data is 12 Earth days corresponding to the time interval for **a** to **l** in Fig. 3.

Morphology in the Venus polar region. As shown in Fig. 3, zonal components with wavenumbers of zero, and one or two are predominant in the temperature distribution in the cold collar and the polar vortex. To elucidate the wave contributions, we extract the temperature components associated with the thermal tide and the short-period (transient) waves by using a frequency filter[25], which are shown in Figs 4 and 5, respectively. Figures 3 and 4 indicate that the enhanced cold part in the cold collar is

caused by the cold temperature phase of the thermal tide, which is consistent with Yamamoto and Takahashi[11]. The cold part of the cold collar is nearly always expanded into the morning side, and it sometimes surrounds the polar vortex completely (for example, Fig. 3b,g,k,l). Comparing Fig. 5 with Fig. 3, we found that this expansion is induced by the cold temperature phase of the transient waves. These results show that the cold collar is composed of the superposition of the zonally averaged basic field, the thermal tide and the transient waves.

It is also suggested from Figs 3 and 5 that the small-scale fluctuations seen in the polar vortex are mainly caused by the warm temperature phase of the transient waves. The morphology seen in the polar vortex is strongly affected by the transient waves, implying that the hot dipole or S-shaped structure observed by VIRTIS, the Visible and InfraRed Thermal Imaging Spectrometer on board Venus Express, would reflect the horizontal structure of the transient waves.

Effect of thermal tide. Figure 6a,b show latitude–height distributions of the zonally and temporally averaged zonal wind and temperature over two Venusian solar days (∼234 Earth days) obtained in Cases A and B. The axis of the midlatitude jet in Case A is located at a lower latitude and altitude than in Case B. This is likely due to the momentum transport by the thermal tide as mentioned above. In Case A, the temperature below 70 km altitude decreases with latitude up to 60° N in association with positive vertical shear of the mean zonal wind, whereas it increases with latitude poleward of 70° N. The remarkable cold collar is observed at 67–70 km levels at 60°–70° latitudes along with the polar warm region. In Case B, on the other hand, the temperature monotonically decreases with latitude at levels below 75 km height. Above 75 km, the temperature in the polar region is

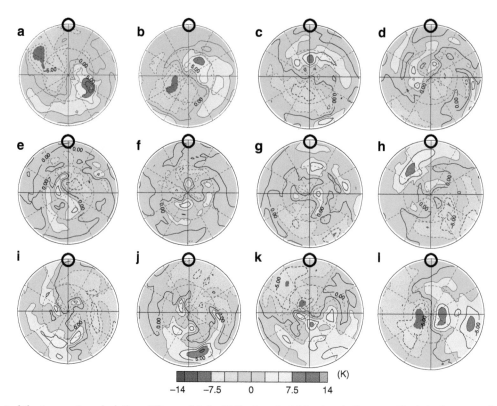

Figure 5 | Polar plot of the temperature deviations (K) associated with the transient (short period) waves. The latitude range is from 30° N to 90° N. The black circle represents the local solar noon. Each panel of (**a**) to (**l**) shows temperature distribution from 1414 to 1425 days from the start of the calculation, corresponding to the time in Fig. 3. The time interval between the adjacent panels is one Earth day.

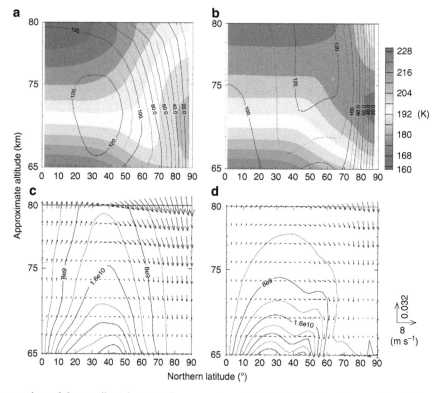

Figure 6 | Meridional cross-sections of the zonally and temporally averaged zonal wind (solid line) and temperature (colour shade) and the horizontal and vertical components of the residual mean meridional circulation (vector) and mass stream fuction (contour). (**a**) Zonal wind and temperature in Case A. (**b**) Those in Case B. (**c**) Residual mean meridonal circulation vector and mass stream function in Case A. (**d**) Those in Case B. Averaged period is two Venusian solar days (234 Earth days) after settling into the quasi-steady state.

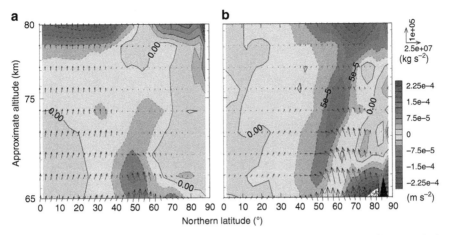

Figure 7 | Meridional cross-sections of the EP flux (vectors) and its divergence (colours). (a,b) Cases A and B, respectively. Average period of data is two Venusian solar days (234 Earth days) after settling into the quasi-steady state.

slightly warmer than in midlatitudes, but the temperature minimum corresponding to the cold collar is not reproduced, which is different from Case A.

RMMC defined by the transformed Eulerian-mean equations[26] (Supplementary Note 1) is shown in Fig. 6c,d for Cases A and B, respectively. The RMMC represents the Lagrangian mean meridional circulation approximately. In Case A, the RMMC above the cloud-top level (~70 km) reaches the polar region and remarkable downward motion exists there, which warms the atmosphere through adiabatic heating and forms the warm polar region. Also in Case B, the downward motion of the RMMC is observed in the polar region. However, it is 2–3 times slower than that in Case A (Supplementary Fig. 1). The adiabatic heating due to the downward motion in the polar region in Case B is much weaker than that in Case A.

These results imply that the thermal structure of Venus upper polar atmosphere (that is, the warm polar region and the cold collar) is attributed to the RMMC enhanced by the thermal tide. To investigate the effect of the thermal tide on the RMMC, we examined the momentum and heat balances in the transformed Eulerian-mean equations. The Eliassen–Palm (EP) flux and its divergence obtained for Cases A and B are shown in Fig. 7. In Case A, strong convergence of the EP flux (shown by blue shades) exists at latitudes equatorward of 55° N-60° N above 75 km, which is balanced mainly with upward advection of the negative vertical shear of the mean zonal wind due to the RMMC. On the other hand, the EP flux is divergent (shown by red shades) at latitudes poleward of 55° N-60° N. This divergence is balanced with the sum of poleward advection of the absolute vorticity and downward advection of the negative vertical shear of the mean zonal wind (Supplementary Fig. 2). Since these vertical motions of the RMMC induce adiabatic cooling and heating at the lower and higher latitudes, respectively, the clear structure of the cold collar has been reproduced in Case A.

In Case B, the EP flux is almost absent at lower latitudes as shown in Fig. 7b. Since the thermal tide is excluded in this case, it is inferred that the EP flux at the low latitudes obtained in Case A is induced by the thermal tide. Convergence of the EP flux is predominant at latitudes poleward of 20° N except at 68–77 km altitudes in the polar region, which is mainly balanced with poleward advection of the absolute vorticity due to the RMMC (Supplementary Fig. 2). The vertical advection can be neglected in Case B because both the vertical motions of the RMMC and the vertical shear of the mean zonal wind are much weaker than those obtained in Case A.

Furthermore, to clarify the contribution of the thermal tide in Case A, we decompose the EP flux into long- and short-period components by using a frequency filter with a period of 10 Earth days, which correspond to the thermal tide and the transient waves, respectively. Figure 8 shows the EP flux divergence due to the thermal tide and the transient waves. Compared with Fig. 7a, a large part of the total EP flux divergence can be ascribed to the thermal tide, and the transient waves may not be neglected in the polar region below 70 km. These results indicate that the different dynamical states have been established in Cases A and B due to the thermal tide. In Case A, the RMMC is enhanced by the thermal tide, which induces the adiabatic polar warming, creating the cold collar (see Supplementary Note 2 and Supplementary Figs 3 and 4 for more details).

Discussion

Our idea that the unique Venusian polar atmospheric structure is attributed to the RMMC enhanced by the thermal tide is different from Yamamoto and Takahashi[11], who suggested that warm polar region is induced by the warm temperature phase of the diurnal tide. Imamura[27] has pointed out that the RMMC above the cloud layer can be induced by upward propagating Rossby and/or gravity waves, but the thermal tide has not been taken into account in his study. Lebonnois et al.[28] also performed GCM simulations, which includes more realistic physical parameterization schemes of the radiative transfer and boundary layer processes. They reported that the cold collar was not reproduced in their results, and inferred that the interaction between temperature and cloud structures, which was not included in their model, might be important for the cold collar formation. However, since the realistic cold collar has been reproduced in our model with the simplified radiative process, it is suggested that the cloud-radiative feedback is not essential for the cold collar. It should be also noted that an equatorial jet appeared in their results. This is different from our results, in which the jet is located at the midlatitudes. Because the temperature distribution is closely related with the mean zonal wind through the gradient wind balance, the realistic mean zonal wind and the RMMC enhanced by the thermal tide are crucial to the unique atmospheric structure seen in the Venus polar region.

In addition, we might be able to explain the recent infrared observational results about the meridional distributions of CO abundance and cloud thickness by taking into account the fact that the meridional and vertical motions of the RMMC shown in

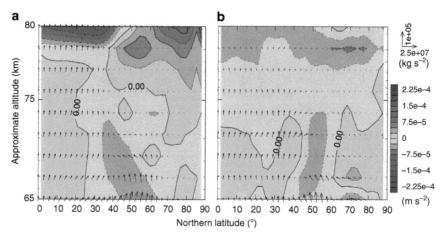

Figure 8 | Meridional cross-sections of the EP flux (vectors) and its divergence (colour) in Case A. (**a**) Thermal tide and (**b**) Transient (short period) waves. Averaged period of data is two Venusian solar days (234 Earth days) after settling into the quasi-steady state.

Fig. 6c approximately represent the material transport. It is inferred that the poleward material transport above the cloud top due to the RMMC leads to high CO abundance in the polar region, because on the way from the equator to the pole, an air parcel keeps gaining CO, which is created photochemically above the cloud layer, until it sinks in the polar region. Furthermore, H_2SO_4 liquid particles as well as CO are transported towards the polar region and grow in the same manner. The temperature at which H_2SO_4 liquid particles evaporate is almost equal to that at the cloud bottom altitude of ~ 50 km (refs 15,29). This temperature is much higher than that observed at altitudes above 65 km, where the RMMC transports the H_2SO_4 particles towards the poles in this study. Therefore, the clouds in the polar region are expected to become thicker than those in the cold collar. This material transport expected from the present result is qualitatively consistent with Imamura and Hashimoto[15].

Our simulations elucidate the importance of the thermal tide for the Venus atmospheric circulation around the cloud-top level (~ 70 km). The cold collar and the warm polar region, which are related to the zonal wind distribution through gradient wind balance, are realistically reproduced in the presence of the diurnal heating and can be explained by the downward motion of the RMMC enhanced by the thermal tide. This is qualitatively analogous to a sudden stratospheric warming in the Earth atmosphere, which is related to the meridional circulation induced by upward propagating Rossby waves[30]. In addition, the observed meridional distributions of cloud thickness and CO abundance above the cloud top are expected to be associated with the RMMC enhanced by the thermal tide. This is a new insight provided by the present work and is helpful for interpreting numerous observations and features of the Venus upper atmosphere.

Although the importance of the thermal tide is revealed in this study, further details of the mechanism how the structure of the RMMC is determined should be elucidated by investigating the wave propagation in our model and analysing observational data in the near future. In this December, the Venus Climate Orbiter 'Akatsuki' will start observations[31,32]. Data provided by Akatsuki will also greatly help us to understand the real Venus atmosphere.

Methods

Set-up of AFES for Venus. AFES for Venus is a full nonlinear GCM on the sphere constructed for the Venus atmosphere[21,24]. The basic equations are primitive equations in sigma coordinates without topography. The physical parameter values applied in this study are as follows: planet radius is 6,050 km; gravity acceleration 8.87 m s^{-2}; planetary rotation period 243 Earth days; and standard surface

pressure 92×10^5 Pa. The specific heat at constant pressure C_p is assumed to be constant at 1,000 J kg^{-1} K^{-1} for simplicity. It is noted that C_p varies approximately from 1.18×10^3 to 7.38×10^2 J kg^{-1} K^{-1} in the Venus atmosphere[33]. The resolution is set to T42L60 (128×64 horizontal grids and 60 vertical levels). The vertical domain extends from the ground to ~ 120 km, with an almost constant grid spacing of 2 km. The model includes horizontal and vertical eddy diffusion. The former is represented by horizontal hyper-viscosity ∇^4 and has a relaxation time for the largest wavenumber of 0.1 Earth days. The vertical diffusion coefficient is assumed to be constant at 0.15 m^2 s^{-1}. Rayleigh friction is employed at the lowest level to mimic surface friction. A sponge layer is used to damp eddy components of the wind above 80 km. However, its effect becomes dominant above 100 km because the damping coefficient is very small at 80–100 km levels.

Radiative processes. Radiative transfer in the infrared region is represented simply by Newtonian cooling with the coefficients based on previous studies[20,34]. Furthermore, the temperature field is relaxed to the prescribed horizontally uniform distribution, whose vertical profile is based on the Venus International Reference Atmosphere. The period of the data analysis in this study is two Venusian solar days (234 Earth days) after settling into the quasi-steady state. The solar heating is based on previous studies[34,35]. In the present study, it is decomposed into a zonal-mean component and deviation from the zonal mean (diurnal component), which excites the mean meridional (Hadley) circulation and the thermal tide, respectively. Case A includes both the components, whereas Case B includes only the zonal mean. See Sugimoto et al.[21,24] for further details of the model settings.

References

1. Taylor, F. W. et al. Structure and meteorology of the middle atmosphere of Venus: infrared remote sensing from the Pioneer Orbiter. J. Geophys. Res. **85**, 7963–8006 (1980).
2. Newman, M., Schubert, G., Kliore, A. J. & Patel, I. R. Zonal winds in the middle atmosphere of Venus from Pioneer Venus radio occultation data. J. Atmos. Sci. **41**, 1901–1913 (1984).
3. Piccioni, G. et al. South-polar features on Venus similar to those near the north pole. Nature **450**, 637–640 (2007).
4. Tellmann, S., Pätzold, M., Häusler, B., Bird, M. K. & Tyler, G. L. Structure of the Venus neutral atmosphere as observed by the radio science experiment VeRa on Venus Express. J. Geophys. Res. **114**, E00B36 (2009).
5. Garate-Lopez, I. et al. A chaotic long-lived vortex at the southern pole of Venus. Nat. Geosci. **6**, 254–257 (2013).
6. Luz, D. et al. Venus's southern polar vortex reveals precessing circulation. Science **332**, 577–580 (2011).
7. Garate-Lopez, I., Garcia Munoz, A., Hueso, R. & Sanchez-Lavega, A. Instantaneous three-dimensional thermal structure of the Southern Polar Vortex of Venus. Icarus **245**, 16–31 (2015).
8. Lee, C., Lewis, S. R. & Read, P. L. A numerical model of the atmosphere of Venus. Adv. Space Res. **36**, 2142–2145 (2005).
9. Piccialli, A. et al. Dynamical properties of the Venus mesosphere from the radio-occultation experiment VeRa onboard Venus Express. Icarus **217**, 669–681 (2012).
10. Yamamoto, M. & Takahashi, M. Venusian middle-atmospheric dynamics in the presence of a strong planetary-scale 5.5-day wave. Icarus **217**, 702–713 (2012).

11. Yamamoto, M. & Takahashi, M. Dynamics of polar vortices at cloud top and base on Venus inferred from a general circulation model: case of a strong diurnal thermal tide. *Planet Space Sci.* **113**, 109–119 (2015).

12. Irwin, P. G. *et al.* Spatial variability of carbon monoxide in Venus' mesosphere from Venus Express/visible and infrared thermal imaging spectrometer measurements. *J. Geophys. Res.* **113**, E00B01 (2008).

13. Grassi, D. *et al.* The Venus nighttime atmosphere as observed by the VIRTIS-M instrument. Average fields from the complete infrared data set. *J. Geophys. Res.* **119**, 837–849 (2014).

14. Lee, Y. J. *et al.* Vertical structure of the Venus cloud top from the VeRa and VIRTIS observations onboard Venus Express. *Icarus* **217**, 599–609 (2012).

15. Imamura, T. & Hashimoto, G. Venus cloud information in the meridional circulation. *J. Geophys. Res.* **103**, 31349–31366 (1998).

16. Pechmann, A. P. & Ingersoll, J. B. Venus lower atmosphere heat balance. *J. Geophys. Res.* **85**, 8219–8222 (1980).

17. Takagi, M. & Matsuda, Y. Sensitivity of thermal tides in the Venus atmosphere to basic zonal flow and Newtonian cooling. *Geophys. Res. Lett.* **32**, L02203 (2005).

18. Takagi, M. & Matsuda, Y. Dynamical effect of thermal tides in the lower Venus atmosphere. *Geophys. Res. Lett.* **33**, L13102 (2006).

19. Newman, M. & Leovy, C. Maintenance of strong rotational winds in Venus' middle atmosphere by thermal tides. *Science* **257**, 647–650 (1992).

20. Takagi, M. & Matsuda, Y. Effects of thermal tides on the Venus atmospheric superrotation. *J. Geophys. Res.* **112**, D09112 (2007).

21. Sugimoto, N., Takagi, M. & Matsuda, Y. Baroclinic instability in the Venus atmosphere simulated by GCM. *J. Geophys. Res.* **119**, 1950–1968 (2014).

22. Machado, P., Luz, D., Widemann, T., Lellouch, E. & Witasse, O. Mapping zonal winds at Venus's cloud tops from ground-based Doppler velocimetry. *Icarus* **221**, 248–261 (2012).

23. Sánchez-Lavega, A. *et al.* Variable winds on Venus mapped in three dimensions. *Geophys. Res. Lett.* **35**, L13204 (2008).

24. Sugimoto, N., Takagi, M. & Matsuda, Y. Waves in a Venus general circulation model. *Geophys. Res. Lett.* **41**, 7461–7467 (2014).

25. Duchon, C. E. Lanczos filtering in one and two dimensions. *J. Atmos. Sci.* **18**, 1016–1022 (1979).

26. Andrews, D. G., Holton, J. R. & Leovy, C. B. *Middle Atmosphere Dynamics* (Academic, 1987).

27. Imamura, T. Momentum balance of the Venusian midlatitude mesosphere. *J. Geophys. Res.* **102**, 6615–6620 (1997).

28. Lebonoiss, S. *et al.* Superrotation of Venus' atmosphere analyzed with a full general circulation model. *J. Geophys. Res.* **115**, E06006 (2010).

29. Ayers, G. P., Gillett, R. W. & Gras, J. L. On the vapor pressure of sulfuric acid. *Geophys. Res. Lett.* **7**, 433–436 (1980).

30. Matsuno, T. A dynamical model of the stratospheric sudden warming. *J. Atmos. Sci.* **28**, 1479–1494 (1971).

31. Nakamura, M. *et al.* Overview of Venus orbiter, Akatsuki. *Earth Planets Space* **63**, 443–457 (2011).

32. Nakamura, M. *et al.* Return to Venus of the Japanese Climate Orbiter AKATSUKI. *Acta Astronaut.* **93**, 384–389 (2014).

33. Seiff, A. *et al.* Models of the structure of the atmosphere of Venus from the surface to 100 kilometers altitude. *Adv. Space Res.* **5**, 3–58 (1985).

34. Crisp, D. Radiative forcing of the Venus mesosphere: I. solar fluxes and heating rates. *Icarus* **67**, 484–514 (1986).

35. Tomasko, M. G., Doose, L. R., Smith, P. H. & Odell, A. P. Measurements of the flux of sunlight in the atmosphere of Venus. *J. Geophys. Res.* **85**, 8167–8186 (1980).

Acknowledgements

We are grateful to Dr Y.J. Lee, Dr S. Murakami and Dr K. Takaya for the useful suggestions. This study was conducted under the joint research project of the Earth Simulator Center entitled 'Simulations of Atmospheric General Circulations of Earth-like Planets by AFES' and partly supported by JSPS KAKENHI Grant Number 25400470.

Author contributions

H.A. and H.K. analysed the simulation data; N.S. performed the numerical simulations; M.T. developed the numerical model; all the authors contributed to the analysis and theoretical interpretations.

Additional information

10

Wave energy budget analysis in the Earth's radiation belts uncovers a missing energy

A.V. Artemyev[1,†], O.V. Agapitov[2,†], D. Mourenas[3], V.V. Krasnoselskikh[1] & F.S. Mozer[2]

Whistler-mode emissions are important electromagnetic waves pervasive in the Earth's magnetosphere, where they continuously remove or energize electrons trapped by the geomagnetic field, controlling radiation hazards to satellites and astronauts and the upper-atmosphere ionization or chemical composition. Here, we report an analysis of 10-year Cluster data, statistically evaluating the full wave energy budget in the Earth's magnetosphere, revealing that a significant fraction of the energy corresponds to hitherto generally neglected very oblique waves. Such waves, with 10 times smaller magnetic power than parallel waves, typically have similar total energy. Moreover, they carry up to 80% of the wave energy involved in wave–particle resonant interactions. It implies that electron heating and precipitation into the atmosphere may have been significantly under/over-valued in past studies considering only conventional quasi-parallel waves. Very oblique waves may turn out to be a crucial agent of energy redistribution in the Earth's radiation belts, controlled by solar activity.

[1] LPC2E/CNRS, 3A, Avenue de la Recherche Scientifique, 45071 Orleans Cedex 2, France. [2] Space Sciences Laboratory, University of California, 7 Gauss Way, Berkeley, California 94720, USA. [3] CEA, DAM, DIF, F-91297 Arpajon Cedex, France. † Present addresses: Space Research Institute (IKI) 117997, 84/32 Profsoyuznaya Street, Moscow, Russia. (A.V.A.); Astronomy and Space Physics Department, National Taras Shevchenko University of Kiev, 2 Glushkova Street, 03222 Kiev, Ukraine (O.V.A.). Correspondence and requests for materials should be addressed to A.V.A. (email: ante0226@gmail.com).

Since whistler-mode waves regulate fluxes of trapped electrons[1,2] and their precipitation rate[3-5] in the upper atmosphere[6-8], accurately determining the wave energy budget in the outer radiation belt has lately become an outstanding challenge for the scientific community[9]. Owing to the sparse wave data obtained by early satellites, their poor coverage of high latitudes and mainly one-component field measurements, and as linear theory was showing much higher parallel wave growth, scientists have commonly relied on the assumption that chorus waves were mainly field-aligned, that is, their propagation was weakly oblique with respect to the geomagnetic field[10,11]. Moreover, crucial theoretical works[12,13] in this area have demonstrated early on that the most important wave field component determining the wave–particle coupling efficiency was generally the magnetic field one, at least over a reasonably large range of wave obliquity. As a result, previous wave statistics focused on the sole magnetic field component of the full wave energy—showing indeed a clear prevalence of parallel waves in the equatorial region sampled by most satellites[14-17].

No global study on the basis of satellite measurements since then has led to a real revision of this conventional picture. Although some studies[16-19] and ray-tracing simulations[20] have recently hinted at both the possible presence and potential importance of very oblique whistler-mode waves, they failed to grasp the full extent of the implications, owing either to their continuing focus on statistics of the sole magnetic field component or to their use of statistical averages over such wide ranges of geomagnetic conditions that the effects of oblique waves have become blurred. Here, we study the full wave energy distribution of whistler waves, including both magnetic and electric field components. Our work suggests that the unexpected presence of a very large electrostatic energy, hitherto missing in past statistics of wave intensity and stored in very oblique waves, may profoundly change the current understanding of both the actual wave generation mechanisms and the processes of wave-induced electron scattering, acceleration and loss in the magnetosphere.

Results

Statistics of wave energy. To compare the impact of oblique and parallel waves in the formation and evolution of keV to MeV electron fluxes in the inner magnetosphere, a reasonable approach consists in first estimating the energy density of both wave populations. Such a global survey is presented in Fig. 1. Here, we make use of 10 years of wave measurements performed by Cluster satellites[16] to evaluate the wave energy distribution throughout much of the Earth's inner magnetosphere as a function of wave obliquity and L-shell (the equatorial distance to the centre of the Earth normalized to Earth's radius). The energy density W of whistler-mode waves is determined by wave electric **E** and magnetic **B** field vectors through a complex relationship involving the tensor of absolute permittivity (see equation (1) in Methods section). Using the cold plasma dispersion relation for whistlers, W depends only on wave characteristics such as magnetic amplitude B, frequency ω, wave-normal angle θ (which defines the wave obliquity with respect to the geomagnetic field) and refractive index $N = kc/\omega$ (with k is the wave vector and c is the velocity of light).

Figure 1 with two-dimensional maps of wave energy W clearly shows that the proportion of very oblique waves, propagating near the resonance cone angle (that is, near 90°), is generally similar to or even larger than the proportion of quasi-parallel waves for L = 3 to 6 during moderate geomagnetic activity (defined by index $K_p < 3$) on the dayside. On the nightside or during more disturbed periods such that $K_p > 3$ (that is, geomagnetic storms or substorms), the amount of very oblique waves is sensibly reduced. The latter reduction stems probably from the presence of higher fluxes of hot (~ 100 eV to 1 keV) plasmasheet electrons injected in the midnight region during disturbed periods[21]. Numerical ray-tracing simulations have shown that such hot electrons can damp very oblique waves propagating near their resonance cone[19,20,22].

The present results therefore challenge the conventional assumption of predominantly quasi-parallel whistler-mode waves in the outer radiation belt. A big, missing slice of the wave energy appears to be stored in very oblique waves—which are mainly made up of electrostatic energy[23]. Although most oblique waves are observed away from the equator, significant amounts moreover exist close to it. It strongly suggests that the widely accepted theory of parallel wave generation near the equator by

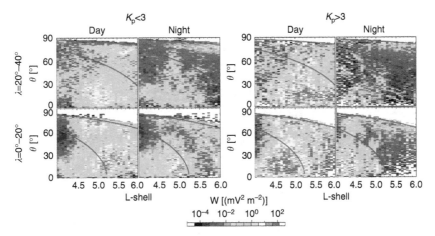

Figure 1 | Distribution of the energy of whistler waves in the Earth radiation belts. The distribution of the density of whistler wave energy W (in mV^2 m^{-2}) is displayed in the (L,θ) space. Data are shown for two ranges of magnetic latitude (the near-equator region with $|\lambda| \in [0°, 20°]$ and the high-latitude region with $|\lambda| \in [20°, 40°]$), for day and night sectors, and for low ($K_p < 3$) and high ($K_p > 3$) geomagnetic activity. Red curves show the position of Gendrin θ_g and resonance cone θ_r angles (where $\cos\theta_g \approx 2\omega/\Omega_c$, $\cos\theta_r \approx \omega/\Omega_c$ and Ω_c is the local electron gyrofrequency). Both angles are calculated with the mean frequency of spacecraft observations, making use of precise plasma density measurements from IMAGE[39]. In the present figure, the wave refractive index has been limited to <100 in agreement with rough but conservative upper bounds due to Landau damping by average levels of hot electrons[19]. Three frequency channels have been taken into account: 2,244.9, 2,828.4 and 3,563.6 Hz, covering almost the full range from 2 to 4 kHz. Each channel is used in the corresponding L-shell range to measure only waves in whistler-mode frequency range.

an unstable electron population exhibiting a temperature anisotropy[11,24,25] might need to be supplemented with some new mechanism allowing the direct generation of very oblique waves there. This could require the presence of additional energetic electron populations differing subtly from the commonly assumed ones.

What are the consequences of the large energy of oblique waves on the dynamics of energetic particles? As their name suggests, wave–particle resonant interactions are controlled not only by wave intensity, but also by the actual efficiency of the resonant interactions. Waves must be in resonance with particles, implying that a certain relationship must be fulfilled between particle energy, pitch angle, wave frequency and obliquity (see equation (2) in Methods section). As a result, only a small portion of the total wave energy density actually corresponds to resonant waves[12]. The wave–particle coupling efficiency Φ, which depends also on cyclotron or Landau resonance harmonic number and on wave field components, provides the exact portion of wave energy interacting resonantly with particles[13,26], finally yielding the resonant wave energy $\Theta^2 = B^2 \Phi^2 / N^2$. Figure 2 shows the wave energy density Θ^2 of resonant waves plotted in the same manner as the wave energy density previously, using measured wave field components from Cluster. Figure 2 demonstrates that the resonant wave energy density at high θ-values (between the Gendrin and resonance cone angles) is 5–10 times larger than for parallel waves throughout the region L = 3–6. Hence, very oblique waves are expected to play a crucial role in the scattering of electrons in this region of space.

Electron lifetimes during geomagnetic storms. The remarkable effectiveness of the resonant interaction of very oblique waves with keV to MeV electrons can modify particle scattering and energization processes substantially in the radiation belts as compared with conventional theoretical estimates obtained for quasi-parallel waves alone. This effect should be most pronounced during moderately disturbed periods where oblique waves are more ubiquitous. To estimate the effects of oblique waves on resonant electron scattering during the course of a geomagnetic storm, we use here parameterizations of lower-band chorus wave magnetic intensity and θ distributions as functions of D_{st} devised on the basis of the same wave data set[18,27]. The

disturbance storm time D_{st} index is widely used to study the magnitude and internal variability of geomagnetic storms[28,29]. Two typical profiles $D_{st}(t)$ are considered (see top panel in Fig. 3), corresponding to storm types #1 and #2 (refs 28,30). The storm type #1 has a relatively long (~ 1.5 day) early recovery phase between $D_{st} = -100$ and -75 nT followed by a rapid increase of D_{st} back to -10 nT, while the second type has a much shorter early recovery phase followed by a much more prolonged stay (~ 3 days) around -50 nT.

The evolution of the lifetime τ_L of energetic electrons during the course of these two storms has been calculated numerically for various energies ranging from 100 eV to 1 MeV. Figure 3 first demonstrates the important variations of τ_L with D_{st}. Such strong variations can be explained by the combination of two main effects: (1) lifetimes increase when wave intensity decreases (both with and without oblique waves)[10,31] and (2) the wave–particle coupling Φ is significantly stronger for very oblique waves than for quasi-parallel waves over a very wide energy range (see Supplementary Fig. 2), leading to a reduction of lifetimes as the amount of very oblique waves increases during not-too-disturbed geomagnetic conditions[18,19,27]. The number of contributing resonances can moreover increase up to 10-fold for very oblique waves (see discussion of Supplementary Fig. 2 in Methods section).

When considering a realistic wave-normal angle distribution, the first clear consequence of the additional presence of very oblique waves is a general reduction of lifetimes during the storm. Most remarkably, however, such a reduction is much less significant during the early recovery period corresponding to $D_{st} < -75$ nT. The latter range actually corresponds to high parallel wave amplitudes. Very oblique waves are then almost absent, probably due to their quick damping by intense injections of hot electrons during the main phase of strong storms. Thus, an extended storm phase such that $D_{st} < -75$ nT, with intense waves and small losses, is particularly propitious for the strong energization of electrons. Later on, the competition between the opposite effects of a rapidly decreasing wave intensity and an increasing amount of oblique waves as D_{st} increases, results in a local minimum of τ_L near $D_{st} \sim -60$ nT during the early recovery phase. Finally, during nearly quiet periods with $D_{st} \sim -10$ nT at the end of storms, electron losses to the atmosphere are significantly increased by oblique waves, especially at very low energy. The remarkable difference between τ_L calculated for

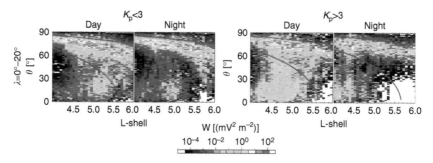

Figure 2 | Distribution of the energy of resonant whistler waves in the radiation belts. The distribution of the wave energy density Θ^2 of resonant waves (in $mV^2 m^{-2}$) is displayed in the (L,θ) space. The most effective resonant wave–particle interaction corresponds to a condition $\tan \alpha \tan \theta \approx 1$ (where α is the particle pitch angle) for electron energy < 2 MeV (ref. 31). This condition has been used to plot Θ^2 in this figure. Data are shown for one range of magnetic latitude $|\lambda| \in [0°, 20°]$, for day and night sectors, and for low ($K_p < 3$) and high ($K_p > 3$) geomagnetic activity. Red curves show the position of Gendrin θ_g and resonance cone θ_r angles. Both angles are calculated with the mean frequency of spacecraft observations, making use of precise plasma density measurements from IMAGE[39]. In the present figure, the wave refractive index has been limited to < 100 in agreement with rough but conservative upper bounds due to Landau damping by average levels of hot electrons[19]. Three frequency channels have been taken into account: 2,244.9, 2,828.4 and 3,563.6 Hz, covering almost the full range from 2 to 4 kHz. Each channel is used in the corresponding L-shell range to measure only waves in whistler-mode frequency range.

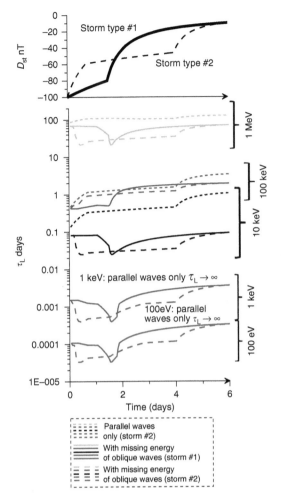

Figure 3 | Variation of electron lifetimes in the course of geomagnetic storms. Two different temporal profiles of the D_{st} index are shown in the top panel, corresponding to two different types of storms. Bottom panels show the corresponding variation of electron lifetimes τ_L calculated for $L \sim 5$ at different energies (each colour corresponding to one energy). Dotted lines correspond to τ_L for parallel chorus whistler-mode waves (τ_L is infinite for 100 eV and 1 keV in this case) in the case of storm type #2. Solid (storm type #1) and dashed (storm type #2) lines show τ_L when using a realistic distribution of wave-normal angle. Five energies are shown: 100 eV, 1 keV, 10 keV, 100 keV and 1 MeV (for $E \leq (>)$1 keV, lifetimes are numerically calculated for electrons of equatorial pitch angle <60°(85°)).

parallel waves alone and with a realistic θ-distribution reaches indeed one order of magnitude for 10 keV electrons, while at lower energies (≤ 1 keV), only very oblique waves are still able to resonantly scatter electrons towards the loss-cone.

Discussion

Such results definitely show that a precise knowledge of the actual distribution of wave energy as a function of propagation angle θ is a key factor for accurately modelling the evolution of relativistic as well as low-energy electron fluxes under the influence of resonant wave–particle interactions. As noted above, this θ distribution is tightly controlled by the density and temperature of hot electrons. The large energy stored in very oblique waves can be readily tapped by sufficiently hot electrons newly injected from the outer magnetosphere and lead to their further heating via Landau damping. It, therefore, represents an accelerating factor of change for this important population[25] of particles.

More generally, the intrinsic variability of hot electron injections with geomagnetic activity[21] probably explains the observed variation of wave obliquity[18,27]. As the latter is able to fine-tune the precipitation of very low energy (especially 0.1 to 10 keV) electrons, the presence of a large amount of very oblique waves could have unexpected and major consequences on the ionospheric conductivity and on the nightside upper atmosphere ionization level at various altitudes, potentially affecting the whole magnetosphere–ionosphere coupling[7,8,32].

Beside determining electron precipitations into the atmosphere, whistler-mode waves are also responsible for the rapid energization of ~ 10 keV to 1 MeV electrons to multi-MeVs in the radiation belts during geomagnetic storms[3,5,33]. To first order, the effective energization depends mainly on the dimensionless product $D_{EE}\tau_L$ of the energy diffusion rate D_{EE} and lifetime, because a longer τ_L leaves more time for electron acceleration to proceed[18,27,34]. Moreover, the important dependence of the energy diffusion rate on the wave magnetic intensity (strongly increasing with $-D_{st}$) is almost fully compensated in this factor $D_{EE}\tau_L$ by the inverse dependence of the lifetime on the wave intensity. Since D_{EE} varies also much more weakly with wave obliquity than τ_L[35], it is the important variation of the lifetime with wave obliquity that should mainly determine the variation of the effective energization level of electrons. Thus, the comparison of lifetimes calculated with and without very oblique waves in Fig. 3 directly demonstrates the often dramatic change in energization level between these two cases.

Furthermore, the results displayed in Fig. 3 suggest that two storms with the same maximal strength but with different temporal profiles may lead to different effects on energetic electron fluxes, because of the different lifetime reductions dictated by the varying amount of very oblique waves. A storm (close to type #1) with a prolonged early recovery phase at $D_{st} < -75$ nT followed by a quick return to $D_{st} > -20$ nT should take advantage of high parallel wave intensity and weak overall losses to strongly energize electrons. Conversely, another storm (close to type #2) with a shorter initial period at $D_{st} < -75$ nT followed by a much slower recovery back to $D_{st} > -20$ nT should generally involve much stronger electron losses induced by larger amounts of very oblique waves during the early recovery phase (up to 1.5 day in Fig. 3), associated with a smaller magnetic wave intensity—efficiently reducing electron energization during that period. Later, significant losses to the atmosphere combined with modest wave intensity should nearly prevent any substantial acceleration. This could help to answer one outstanding question in radiation belt physics—why some geomagnetic storms correspond to global electron energization, while other storms with the same magnitude of D_{st} variation do not[29].

Excluding oblique waves from consideration would actually make the dimensionless energization factor $D_{EE}\tau_L$ almost constant and independent of the D_{st} time profile, as it does not depend on the bounce-averaged wave intensity. Only the consideration of an additional dimension of the system, corresponding to wave obliquity, gives a chance to obtain a significant variation of particle acceleration efficiency with D_{st} and, as a result, immediately produces a difference in particle acceleration for different $D_{st}(t)$ profiles. This effect allows to separate precipitation-dominated storms with a fast early recovery further slowing down, from acceleration-dominated storms with a slow initial recovery later on speeding up.

The surprisingly high level of very oblique wave energy discovered in Figs 1 and 2 and the strong concomitant increase of the wave–particle coupling strength have revealed that the wave obliquity, regulated by low-energy electrons injected from the plasmasheet, represents a new and important lever governing the variations of energetic electron fluxes. It indicates one possible answer to the problem of often-noted discrepancies between modern radiation belt models and observations[9,36]. The consideration of only parallel waves mostly restricts the space of wave model parameters to a single parameter—the wave amplitude. However, the distributions of wave amplitudes with geomagnetic activity and spatial location are well documented[10,16] and included in modern codes. In this study, we have clearly shown that there exists at least one additional model parameter—wave obliquity, which can control both the energization of electrons, their precipitation into the atmosphere and even the energy range of precipitated particles. The revelation of this hidden parameter and of the corresponding missing energy of very oblique waves should provide new opportunities to better understand and forecast the observed variations of energetic electron fluxes in the radiation belts as well as the global dynamics of the magnetosphere–ionosphere coupling.

Methods

Evaluation of the wave energy density. The energy density W of whistler-mode waves in the Earth's magnetosphere is determined by wave electric field $\mathbf{E}$ and magnetic field $\mathbf{B}$ vectors through the relation:

$$W \approx \frac{1}{8\pi}\left(\mathbf{E}^* \frac{d(\omega\hat{\varepsilon})}{d\omega}\mathbf{E} + \mathbf{B}^2\right) \quad (1)$$

where $\mathbf{E}^*$ is the conjugate vector to $\mathbf{E}$, ω is the wave frequency and $\hat{\varepsilon}(\omega)$ the tensor of absolute permittivity. Using the dispersion relation for electromagnetic whistler-mode waves in a cold magnetized plasma, electric field components can be further expressed as a function of $\mathbf{B}$. One gets $W = B^2(1 + W_E)/8\pi$ where W_E depends only on the wave characteristics: its frequency ω, wave-normal angle θ (which defines the wave obliquity with respect to the geomagnetic field), and refractive index $N = kc/\omega$ (where k is the wave vector). W steeply increases with N, which is itself a rapidly growing function of θ. The refractive index N (as well as θ) can be determined either solely from full three-component wave magnetic field measurements on a given spacecraft, or else by complete wave magnetic and electric field measurements. The dominant contributions to the wave energy distribution can be further assessed on the basis of either method. However, wave electric field measurements on Cluster satellites are often noisy, at least much more than magnetic field measurements, limiting their use in practice to some case studies. Therefore, we have chosen to resort to the just-discussed method of determination of the full wave energy density on the basis of measurements of the wave magnetic components alone. Nevertheless, the accuracy and reliability of this method must first be demonstrated.

To this aim, we have compared the crucial N values obtained by the two methods in a series of Cluster observations of chorus waves displayed in Supplementary Fig. 1. The comparison of panels (a) and (b) shows clearly that wave activity can be identified not only in magnetic field fluctuations, but also in the concomitant variations of the electric field. Most of the wave-power is concentrated around ~3 kHz—the ratio of wave frequency to electron equatorial gyrofrequency is $\omega/\Omega_{c0} \sim 0.35$. Waves can be considered as very oblique when θ is comprised between the Gendrin angle $\theta_g \approx \arccos(2\omega/\Omega_{c0})$ (which corresponds to wave propagation at a group velocity independent of frequency[37]) and the so-called resonance cone angle $\theta_r \sim \arccos(2\omega/\Omega_{c0})$ (the upper bound on θ where the cold plasma refractive index N of whistler waves goes to infinity[38]). For events in Supplementary Fig. 1, we have $\theta_g \sim 55$–$65°$ and $\theta_r \sim 75$–$85°$. Most observed whistler-mode waves are such that $\theta \in [60°, 85°]$ and can, therefore, be classified as very oblique chorus waves. A substantial part of the wave energy density (see panel (d)) consists of such oblique waves. The large ratio $W/W_B \gg 1$ shows that most of the energy density then comes from the wave electric field. More importantly, evaluations of the wave refractive index N from three-component measurements of the sole wave magnetic field yield values very similar to calculations making use of both magnetic and electric field components, attesting the reliability of the former method (compare panels (e) and (f)). The discrepancy does not exceed 25% on average, showing that this method can be safely used for evaluating the wave energy density.

However, only some part of the total wave energy density can actually interact resonantly with trapped electrons[12]. This resonant part is determined by the wave–particle coupling efficiency Φ (ref. 2) which depends on resonance harmonic number n, electron energy and pitch angle, as well as on the wave field components[13,26]. The resonance condition

$$\omega\gamma - ck\sqrt{\gamma^2 - 1}\cos\theta\cos\alpha = -n\Omega_c \quad (2)$$

provides the necessary relation between particle energy (Lorentz factor γ), pitch angle α and wave obliquity θ. As a result, one gets a normalized estimate $\Theta^2 = \Phi^2 B^2/N^2$ of the resonant wave energy[13].

Evaluation of wave-particle coupling and diffusion rates. To demonstrate the peculiarities of electron resonant interaction with very oblique waves, additional numerical calculations of the wave–particle coupling efficiency Φ (averaged over latitude) have been performed as a function of wave propagation angle θ and geomagnetic activity index D_{st}, for various electron energies ranging from 100 eV to 1 MeV. Here, as well as for Fig. 3, we use usual values of the mean frequency $\omega_m/\Omega_{c0} \sim 0.35$ and frequency width $\Delta\omega/\Omega_{c0} \sim 0.2$ of lower-band chorus waves[19] and a ratio $\Omega_{pe}/\Omega_{ce} = 5$ corresponding to $L \sim 5$. Supplementary Fig. 1 shows that during not-too-disturbed geomagnetic conditions ($D_{st} > -60$ nT), wave–particle coupling Φ is clearly stronger for very oblique waves than for quasi-parallel waves over a wide energy range. For a given level of wave intensity, the available range of variation of the wave–particle coupling efficiency Φ at small equatorial pitch angles (near the loss-cone where particles are precipitated in the atmosphere) is so large that it could presumably explain any observed fluctuations of electron flux by fluctuations of the wave obliquity only and associated variations of electron scattering. In addition to the increase of Φ for a given resonance, the number of such contributing resonances can moreover increase 10-fold for oblique waves (see Supplementary Fig. 2).

The efficiency of charged particles resonant interaction with waves is determined by diffusion rates proportional to the weighting factor $\Phi_n^2 = \Theta^2 g_\theta(\theta)g_\omega(\omega)$ where g_θ and g_ω are normalized distributions of θ and wave frequency. To calculate the g_θ normalization, one should determine resonant k and ω for given particle pitch angle and energy. Then, an integration over θ must be performed. The upper limit of this integration is determined by the maximum value of the refractive index N_{Max}. The latter is imposed by the presence of both thermal effects in the dispersion relation and Landau damping by 100–500 eV suprathermal electrons of oblique waves near the resonance cone angle[19,22]. Using typically observed parameters for the thermal and suprathermal electron population at $L \sim 5$, it has been shown that one could take $N_{Max} \sim 120$ to 300 for lower-band chorus waves from low to high latitudes during periods of quiet to moderately disturbed geomagnetic activity, with N_{Max} varying as the inverse of the frequency ω and increasing with latitude[19]. It led us to use here (in Supplementary Fig. 2, and Fig. 3 in main text) a rough but realistic limit $N_{Max} \sim \min(300, 36\Omega_{ce}/\omega)$ corresponding to a predominant effect of Landau damping.

References

1. van Allen, J. A. & Frank, L. A. Radiation around the Earth to a radial distance of 107,400 km. *Nature* **183**, 430–434 (1959).
2. Schulz, M. & Lanzerotti., L. J. *Particle Diffusion in the Radiation Belts* (Springer, 1974).
3. Horne, R. B. Plasma astrophysics: acceleration of killer electrons. *Nat. Phys.* **3**, 590–591 (2007).
4. Horne, R. B., Lam, M. M. & Green, J. C. Energetic electron precipitation from the outer radiation belt during geomagnetic storms. *Geophys. Res. Lett.* **36**, L19104 (2009).
5. Thorne, R. M. *et al.* Rapid local acceleration of relativistic radiation-belt electrons by magnetospheric chorus. *Nature* **504**, 411–414 (2013).
6. Thorne, R. M. Energetic radiation belt electron precipitation—a natural depletion mechanism for stratospheric ozone. *Science* **195**, 287–289 (1977).
7. Brasseur, G. P. & Solomon., S. *Aeronomy of the Middle Atmosphere: Chemistry and Physics of the Stratosphere and Mesosphere*, (Springer, 2005).
8. Seppälä, A., Lu, H., Clilverd, M. A. & Rodger, C. J. Geomagnetic activity signatures in wintertime stratosphere wind, temperature, and wave response. *J. Geophys. Res.* **118**, 2169–2183 (2013).
9. Horne, R. B. *et al.* Space weather impacts on satellites and forecasting the Earth's electron radiation belts with SPACECAST. *Space Weather* **11**, 169–186 (2013).
10. Shprits, Y. Y., Meredith, N. P. & Thorne, R. M. Parameterization of radiation belt electron loss timescales due to interactions with chorus waves. *Geophys. Res. Lett.* **34**, L11110 (2007).
11. Omura, Y., Katoh, Y. & Summers, D. Theory and simulation of the generation of whistler-mode chorus. *J. Geophys. Res.* **113**, A04223 (2008).
12. Kennel, C. F. & Wong, H. V. Resonant particle instabilities in a uniform magnetic field. *J. Plasma Phys.* **1**, 75 (1967).
13. Lyons, L. R., Thorne, R. M. & Kennel, C. F. Electron pitch-angle diffusion driven by oblique whistler-mode turbulence. *J. Plasma Phys.* **6**, 589–606 (1971).
14. Hayakawa, M., Yamanaka, Y., Parrot, M. & Lefeuvre., F. The wave normals of magnetospheric chorus emissions observed on board GEOS 2. *J. Geophys. Res.* **89**, 2811–2821 (1984).
15. Meredith, N. P. *et al.* Global model of lower band and upper band chorus from multiple satellite observations. *J. Geophys. Res.* **117**, A10225 (2012).
16. Agapitov, O. *et al.* Statistics of whistler mode waves in the outer radiation belt: Cluster STAFF-SA measurements. *J. Geophys. Res.* **118**, 3407–3420 (2013).
17. Li, W. *et al.* Characteristics of the Poynting flux and wave normal vectors of whistler-mode waves observed on THEMIS. *J. Geophys. Res.* **118**, 1461–1471 (2013).
18. Artemyev, A. V., Agapitov, O. V., Mourenas, D., Krasnoselskikh, V. & Zelenyi, L. M. Storm-induced energization of radiation belt electrons: effect of wave obliquity. *Geophys. Res. Lett.* **40**, 4138–4143 (2013).

19. Li, W. *et al.* Evidence of stronger pitch angle scattering loss caused by oblique whistler-mode waves as compared with quasi-parallel waves. *Geophys. Res. Lett.* **41,** 6063–6070 (2014).

20. Chen, L., Thorne, R. M., Li, W. & Bortnik, J. Modeling the wave normal distribution of chorus waves. *J. Geophys. Res.* **118,** 1074–1088 (2013).

21. Li, W. *et al.* Global distributions of suprathermal electrons observed on THEMIS and potential mechanisms for access into the plasmasphere. *J. Geophys. Res.* **115,** A00J10 (2010).

22. Horne, R. B. & Sazhin., S. S. Quasielectrostatic and electrostatic approximations for whistler mode waves in the magnetospheric plasma. *Planet. Space Sci.* **38,** 311–318 (1990).

23. Ginzburg, V. L. & Rukhadze, A. A. *Waves in Magnetoactive Plasma* 2nd revised edition (Nauka, 1975).

24. Trakhtengerts., V. Y. A generation mechanism for chorus emission. *Ann. Geophys.* **17,** 95–100 (1999).

25. Fu, X. *et al.* Whistler anisotropy instabilities as the source of banded chorus: Van Allen probes observations and particle-in-cell simulations. *J. Geophys. Res.* **119,** 8288–8298 (2014).

26. Glauert, S. A. & Horne, R. B. Calculation of pitch angle and energy diffusion coefficients with the PADIE code. *J. Geophys. Res.* **110,** A04206 (2005).

27. Mourenas, D., Artemyev, A. V., Agapitov, O. V. & Krasnoselskikh, V. Consequences of geomagnetic activity on energization and loss of radiation belt electrons by oblique chorus waves. *J. Geophys. Res.* **119,** 2775–2796 (2014).

28. McPherron, R. L. in *Magnetic Storms* 98 (eds Tsurutani, B. T., Gonzalez, W. D., Kamide, Y. & Arballo, J. K.) 131–147 (American Geophysical Union Geophysical Monograph Series, 1997).

29. Reeves, G. D., McAdams, K. L., Friedel, R. H. W. & O'Brien, T. P. Acceleration and loss of relativistic electrons during geomagnetic storms. *Geophys. Res. Lett.* **30,** 1529 (2003).

30. Denton, M. H. *et al.* Geomagnetic storms driven by ICME- and CIR-dominated solar wind. *J. Geophys. Res.* **111,** A07S07 (2006).

31. Mourenas, D., Artemyev, A. V., Ripoll, J.-F., Agapitov, O. V. & Krasnoselskikh, V. V. Timescales for electron quasi-linear diffusion by parallel and oblique lower-band chorus waves. *J. Geophys. Res.* **117,** A06234 (2012).

32. Gkioulidou, M. *et al.* Effect of an MLT dependent electron loss rate on the magnetosphere-ionosphere coupling. *J. Geophys. Res.* **117,** A11218 (2012).

33. Mozer, F. S. *et al.* Direct observation of radiation-belt electron acceleration from electron-volt energies to megavolts by nonlinear whistlers. *Phys. Rev. Lett.* **1130,** 035001 (2014).

34. Horne, R. B. *et al.* Wave acceleration of electrons in the Van Allen radiation belts. *Nature* **437,** 227–230 (2005).

35. Mourenas, D., Artemyev, A., Agapitov, O. & Krasnoselskikh, V. Acceleration of radiation belts electrons by oblique chorus waves. *J. Geophys. Res.* **117,** A10212 (2012b).

36. Kim, K.-C., Shprits, Y., Subbotin, D. & Ni, B. Relativistic radiation belt electron responses to GEM magnetic storms: Comparison of CRRES observations with 3-D VERB simulations. *J. Geophys. Res.* **117,** A08221 (2012).

37. Gendrin, R. Le guidage des whistlers par le champ magnetique. *Planet. Space Sci.* **5,** 274 (1961).

38. Helliwell., R. A. *Whistlers and Related Ionospheric Phenomena* (Stanford Univ. Press, 1965).

39. Denton, R. E. *et al.* Distribution of density along magnetospheric field lines. *J. Geophys. Res.* **111,** A04213 (2006).

Acknowledgements

The work of F.S.M. and O.V.A. was supported by NASA Grant NNX09AE41G. The work of V.V.K. was supported by CNES grant Cluster Co-I DW.

Author contributions

All the authors contributed to all the aspects of this work.

Additional information

11

Determining volcanic eruption styles on Earth and Mars from crystallinity measurements

Kellie T. Wall[1], Michael C. Rowe[2], Ben S. Ellis[3], Mariek E. Schmidt[4] & Jennifer D. Eccles[2]

Both Earth and Mars possess different styles of explosive basaltic volcanism. Distinguishing phreatomagmatic eruptions, driven by magma–water interaction, from 'magmatic' explosive eruptions (that is, strombolian and plinian eruptions) is important for determining the presence of near-surface water or ice at the time of volcanism. Here we show that eruption styles can be broadly identified by relative variations in groundmass or bulk crystallinity determined by X-ray diffraction. Terrestrial analogue results indicate that rapidly quenched phreatomagmatic ejecta display lower groundmass crystallinity ($<$35%) than slower cooling ejecta from strombolian or plinian eruptions ($>$40%). Numerical modelling suggests Martian plinian eruptive plumes moderate cooling, allowing 20–30% syn-eruptive crystallization, and thus reduce the distinction between eruption styles on Mars. Analysis of Mars Curiosity rover CheMin X-ray diffraction results from Gale crater indicate that the crystallinity of Martian sediment (52–54%) is similar to pyroclastic rocks from Gusev crater, Mars, and consistent with widespread distribution of basaltic strombolian or plinian volcanic ejecta.

[1]School of the Environment, Washington State University, Webster Physical Science Building, Room 1228, Pullman, Washington 99164, USA. [2]School of Environment, University of Auckland, Commerce A Building, Private Bag 92019, Auckland 1142, New Zealand. [3]Institute of Geochemistry and Petrology, ETH Zurich, 8092 Zurich, Switzerland. [4]Department of Earth Sciences, Brock University, 500 Glenridge Avenue, Saint Catharines, Ontario, Canada L2S 3A1. Correspondence and requests for materials should be addressed to M.C.R. (email: Michael.rowe@auckland.ac.nz).

Predominant styles of volcanic eruptions on Mars are likely to differ as compared with those on Earth, with theoretical considerations suggesting that highly explosive, finer grained and more widely distributed mafic volcanism may be more prevalent in the Martian environment[1-5]. Despite the absence of a present-day liquid hydrosphere on Mars, magma–water driven (that is, phreatomagmatic) eruptions are also possible through interaction with ice or subsurface water[6]. Both direct and indirect observations of volcanic deposits on Mars are now possible, but unequivocally constraining Martian eruption styles remains challenging.

Factors affecting explosive mafic volcanism on Earth, such as overburden pressure in the magma chamber, geometry of the conduit, viscosity of the magma (influenced by microlite crystallization and bubble content), and volatile content and degassing history[7-9] are analogous to those operating on Mars, despite differences in eruption scale between these two rocky planets. Widespread and layered equatorial deposits including the Medusae Fossae Formation and kilometre-thick Vallis Marineris interior deposits have been postulated to at least partially comprise volcanic ash or pyroclastic flow deposits from explosive mafic eruptions[10-12]. The Mars Science Laboratory (MSL) Curiosity rover will have a unique opportunity to test models of layered deposit formation during its future traverse of Mount (Mt.) Sharp (officially termed Aeolis Mons), a 5.3-km-tall sequence of layered deposits in the centre of Gale Crater, that fits the criteria for these types of deposits[13] and potentially contains a continuous marker bed that is traceable around much of Mt. Sharp[14]. To date, Curiosity has encountered mainly volcaniclastic and isochemically altered basaltic sediments along its traverse to Yellowknife Bay[15]. Given the high likelihood that Curiosity will continue to encounter volcanically derived deposits, we present here a novel methodology to distinguish between types of mafic explosive eruptions on rocky planets (for example, plinian, strombolian and phreatomagmatic).

We have explored the hypothesis that groundmass crystallinity (C_{gm}) measured by X-ray diffraction (XRD) analysis may be correlated to eruption style and potentially used as a remote tool for constraining Martian eruption dynamics. C_{gm} is defined as the volume percent of a crystalline component in a mixed crystalline–amorphous material, which is a function of magma cooling history. Traditionally, the microlite content in basaltic volcanic rocks has been linked to pre-eruptive factors such as magma ascent rate and degassing history, particularly in the shallow conduit[16,17]. However, crystallization of basaltic melt can occur extremely rapidly, and significant groundmass crystallization may continue after eruption within more slowly cooling clasts[18]. Magma–water interaction driven by marine, ground- or surficial water results in extremely rapid cooling ($>10^6$ K s^{-1}; ref. 19), whereas studies of magmatic airfall products indicate maximum cooling rates of ≤ 10 K s^{-1} (refs 18,20). Phreatomagmatism favours the formation of quenched basaltic glass, with groundmass crystallinities significantly lower than those observed in air-cooled magmas.

Here we show how groundmass crystallinities, and by extension random bulk (that is, whole rock) crystallinities, determined from XRD analysis of a suite of terrestrial basaltic deposits enable development of a criterion for constraining eruption styles. We demonstrate from analysis of terrestrial volcanism that products of phreatomagmatic eruptions predominantly have groundmass and bulk (groundmass + phenocryst) crystallinities below ~40%, whereas crystallinities from strombolian and plinian basaltic eruptions are typically >40%. Remote determination of crystallinity of volcanic samples by the CheMin instrument, an XRD instrument on board the MSL Curiosity rover[21], therefore provides a novel approach for identifying magma–water interactions in the record of explosive volcanism on Mars, and potentially evaluating the evolution of the Martian hydrosphere as preserved in the volcanic record.

Results

Eruption style from crystallinity measurements. Eruptive styles of the explosive basaltic samples we have analysed include deposits from phreatomagmatic, basaltic plinian and strombolian (including fire fountaining) eruptions (Figs 1 and 2; and Table 1). Basaltic lava flows from the Columbia River flood basalts and Mt. Etna were also analysed to help verify the analytical technique (Fig. 1 and Table 1). Crystallinity is determined for both groundmass (C_{gm}) and random bulk sampling (C_{wr}) when possible. XRD

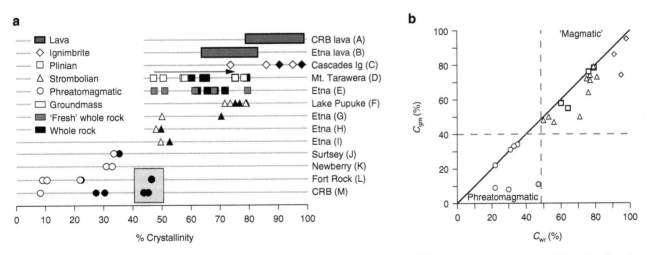

Figure 1 | Crystallinity of natural samples. (a) Calculated crystallinity of volcanic deposits from XRD analysis plotted versus sample location. Samples are broadly organized by volcanic eruption style. The range of groundmass crystallinity (C_{gm}) of lava flows are indicated by purple bars. Increasing crystallinity with clast size for the 1886 Mt. Tarawera plinian eruption is indicated by a black arrow. The blue box indicates phreatomagmatic whole-rock samples that are severely altered/weathered. **(b)** Comparison of hand-picked C_{gm} to bulk 'random sample' crystallinity (C_{wr}). Dashed grey lines delineate the approximate division between phreatomagmatic and 'magmatic' eruption styles. The severely altered/weathered whole-rock Fort Rock (Sample L) phreatomagmatic material containing fresh groundmass glass is indicated in blue. Sample locations and descriptions are keyed to Table 1 by their ID labels indicated in parentheses.

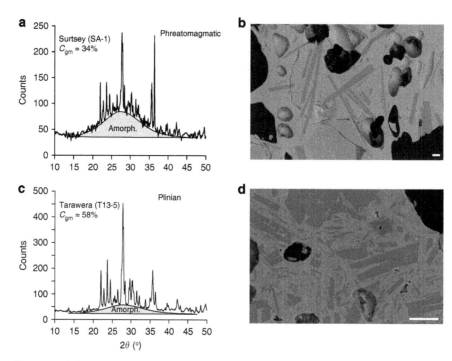

Figure 2 | Phreatomagmatic versus plinian eruptive products. Representative groundmass diffraction patterns (**a,c**) and scanning electron microscope back-scatter images (**b,d**) for a phreatomagmatic eruption (Surtsey) and basaltic plinian eruption (Tarawera). Grey regions in **a,c** represent the area under the diffraction pattern associated with the amorphous (Amorph., glass) component from 10–40° 2θ. Bragg peaks, identified here by X-ray count rates above background, represent the diffraction of X-rays off various lattice planes within minerals in the sample, and therefore represent the crystalline component. White scale bars, 20 μm (**b,d**).

crystallinity measurements qualitatively agree with textural observations, both macroscopically and from back-scatter electron imaging, which shows that the analysed material ranges from translucent and glassy (8% crystallinity; Columbia River basalt hyaloclastite) to dull (>50% crystallinity; Tarawera) (Figs 1 and 2 and Table 1). Prior investigations of this technique have documented the correlation between XRD results and crystallinity obtained from back-scatter electron image processing[22].

The crystallinity of phreatomagmatic materials (negligible to moderate weathering) investigated is generally low, ranging from 8 to 35% (Fig. 1). This contrasts with other 'magmatic' eruption styles that have C_{wr} and C_{gm} values >40% (47–80%). Lava crystallinity is generally high, with C_{wr} from 77 to 98% (C_{gm} >64%; Fig.1). Groundmass crystallinities are a function of cooling rate; thin, smaller lava flows cool more rapidly and, as such, have lower C_{gm} than slowly cooled massive lava flows. Particle size can variably affect interpretation of the crystallinity results. For example, groundmass chips were hand-picked from 125 and 250 μm sieve fractions to avoid microphenocrysts in one tuffaceous deposit (Fort Rock, OR). From the 125- to 250-μm-size fractions, we observed an increase in C_{gm} from 9 to 22%. This difference may reflect more complete exclusion of microphenocrysts in the 125-μm-sieved fraction, and may imply that the smaller, hand-picked size fraction more accurately represents C_{gm}. A bulk crystallinity (C_{wr}) measurement of the same 125-μm-sieved fraction as the groundmass separate (that is, comparable to the Curiosity rover sampling methodology[21]) has the same crystallinity ($C_{wr} = 22\%$) as the whole rock—note that this is also the same value as C_{gm} from the 250-μm-sieved fraction. The identical C_{wr} between the 125-μm-size fraction and the bulk rock analysis may imply that, unlike C_{gm}, C_{wr} is relatively insensitive to the analysed size fraction.

Clast size controls much of the variation seen in C_{gm} in samples representing strombolian and plinian eruption styles. Experimental studies have observed that larger clasts tend to have higher crystallinities as a result of slow air-cooling processes[18]. The low C_{gm} of phreatomagmatic samples in this study support the general model that rapid quenching vastly reduces post-eruption microlite growth. Taking into account differences in clast size and microphenocryst content, groundmass crystallinities of phreatomagmatic materials (<35% C_{gm}) are distinguishable from magmatic volcanic products (>40% C_{gm}). However, fresh bulk samples appear to be more distinctive, with C_{wr} of phreatomagmatic materials (<40%) significantly lower than other magmatic (>50%) volcanic deposits. The presence of phenocrysts causes a shift to higher crystallinity but appears to preserve the overall differences in eruption style in fresh to moderately altered samples (Fig. 1b). Therefore, crystallinity appears to be a robust proxy of cooling rate, with 'magmatic' explosive clasts showing evidence of greater microlite nucleation, owing to these materials remaining hotter for longer periods following fragmentation as compared with phreatomagmatic material.

The relationship between cooling rate and crystallinity is well illustrated by a basaltic ignimbrite[23] from the Cascade volcanic arc (Table 1). Crystallinity increases from 74 to 95% in weakly to moderately to densely welded ignimbrite. While basaltic ignimbrites are relatively rare in the geologic record on Earth, in silicic deposits the degree of welding is primarily controlled by bulk composition, emplacement temperature and accumulation rate. Densely welded ignimbrite deposits sinter following deposition and cool as a single unit rather than as discrete clasts. Such welding thus results in slower cooling rates and allows for a higher degree of groundmass crystallization[22].

Weathering and alteration effects on crystallinity. The results of this study show that it is possible to understand and quantify the effect of weathering and alteration on crystallinity. Obtaining accurate crystallinity measurements for basaltic volcanic

Table 1 | Sample descriptions and crystallinity measurements.

Sample	Location	Fig. 1 ID	Deposit type	Weathering	Clast size (cm)*	Trial	Crystallinity (%) Bulk	Groundmass[†]
Phreatomagmatic								
SA-1	Surtsey, Iceland	J	Basaltic sand	Mild	0.05-0.1		35	34
NL E	Newberry Volcano, OR, USA	K	Accretionary lapilli[‡]	Mild	<0.05	1	ND[§]	31
		K				2	ND	33
Tuff 1-B	CRB[||], OR, USA	M	Hyaloclastite tuff	Severe	<0.05		44	ND
		M			0.05		45	ND
Tuff 1-1m	CRB, OR, USA	M	Hyaloclastite tuff	Moderate	0.05		27	ND
		M			1		30	8[¶]
FR-12-91A	Fort Rock, OR, USA	L	Tuff	Mild	0.0125		22	9
		L			0.025		ND	22
		L			NA		22	ND
FR-12-97B	Fort Rock, OR, USA	L	Tuff	Severe	0.05		47	11[¶]
Plinian								
122 BC 1	Mt. Etna, Italy	E	Scoria (first phase)	Severe	2-3	1	72 (47[#])	ND
		E				2	64 (61[#])	ND
122 BC 2	Mt. Etna, Italy	E	Scoria (second phase)	Severe	2-3	1	62 (51[#])	ND
		E				2	68 (80[#])	ND
		E				3	72 (68[#])	ND
TB 1	Trailbridge, Oregon Cascades, USA	C	Ignimbrite (densely welded)	Moderate	NA		98	95
TB 2	Trailbridge, Oregon Cascades, USA	C	Ignimbrite (mod. welded)	Moderate	NA		91	86
TB 3	Trailbridge, Oregon Cascades, USA	C	Ignimbrite (lightly welded)	Moderate	NA		95	74
T13-5	Mt. Tarawera, New Zealand (summit)	D	Scoria	Mild	0.05-0.1		ND	47
		D			0.05-0.1		ND	51[¶]
		D			0.5		65	ND
		D			1		60	58
		D			1.5		76	76
		D			2		64	55
		D			4		79	78
Strombolian								
LP13-1	Lake Pupuke, New Zealand	F	Scoria	Mild	0.025		77	74
		F			0.05		75	72
		F			1		79	79
May 08 LF	Mt. Etna, Italy	I	Scoria	Negligible	2-5	1	56	47
		I				2	53	50
12SE07	Mt. Etna, Italy	H	Scoria	Negligible	1-2		50	48
3ET08	Mt. Etna, Italy	G	Scoria	Negligible	2-6		71	50
Lava flow				Negligible				
1SE04	Mt. Etna, Italy	B	Lava	Negligible	NA	1	77	71
		B				2	81	73
3SE04	Mt. Etna, Italy	B	Lava	Negligible	NA		76	64
Pom 1	CRB, WA, USA	A	Lava (top)	Moderate	NA		98 (93[#])	ND
Pom 2	CRB, WA, USA	A	Lava (flow interior)	Moderate	NA		95 (90[#])	ND
Pom 3	CRB, WA, USA	A	Lava (colonnade)	Moderate	NA		94 (93[#])	ND
LRP 1	CRB, WA, USA	A	Lava (interior)	Moderate	NA		87 (87[#])	ND
LRP 2	CRB, WA, USA	A	Lava (bottom)	Moderate	NA		91 (88[#])	ND
LRP 3	CRB, WA, USA	A	Lava (cooling surface)	Moderate	NA		81 (79[#])	ND
Ban 245	CRB, WA, USA	A	Lava (top)	Moderate	NA		85[#]	ND
Ban 290	CRB, WA, USA	A	Lava (interior)	Moderate	NA		90[#]	ND
Ban 315	CRB, WA, USA	A	Lava (bottom)	Moderate	NA		89[#]	ND
Butte 60	CRB, WA, USA	A	Lava (top)	Moderate	NA		80[#]	ND
Butte 115	CRB, WA, USA	A	Lava (interior)	Moderate	NA		87[#]	ND
Butte 190	CRB, WA, USA	A	Lava (bottom)	Moderate	NA		98[#]	ND

*NA, not applicable.
[†]Groundmass was picked from 500 µm to 1 mm size fraction of crushed sample, except for samples of particle size <500 µm as indicated.
[‡]The Newberry accretionary lapilli contain cores of scoria. Care was taken to isolate the juvenile ashy exteriors.
[§]ND, not determined.
[||]CRB denotes Columbia River basalt
[¶]Select groundmass cleaned with dilute nitric acid to reduce clay particulates in weathered samples.
[#]For bulk analyses of moderately to severely weathered samples, weathering surfaces were chipped away leaving fresher material (indicated in parentheses). Anomalously high crystallinity for 122 BC 2 trial 2 may result from uneven distribution of phenocrysts with the clast which was split for the analysis. Where both fresh and weathered crystallinities are presented, only 'fresh' analyses appear in Fig. 1.

material becomes more challenging when significant quantities of alteration products are present, which is a scenario that is not unexpected on the Martian surface. Spectral data of bright regions on Mars suggest a prevalence of poorly crystalline or amorphous material, such as palagonite or nanophase iron oxides (npOx;[24]). Poorly crystalline materials can have diffraction patterns that are not easily differentiated from those of true amorphous materials (for example, basaltic glass; Fig. 3). While basaltic glass creates one amorphous hump between $\sim$15 and 35° 2θ, palagonite, hisingerite, ferrihydrite and poorly crystalline aluminosilicates produce a low-angle rise at 4–10° 2θ, with a weaker amorphous hump at (variable) higher

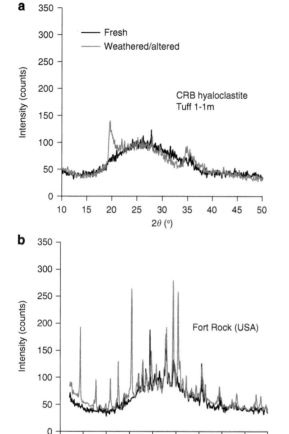

Figure 3 | Effects of alteration. Comparison between relatively fresh and weathered/altered samples from the same localities for phreatomagmatic samples. (**a**) Phreatomagmatic Columbia River basalt hyaloclastite fresh-picked glass (black line) and palagonitized bulk analysis (grey line). (**b**) Whole-rock analyses of weathered (grey line) and fresh (black line) tuffaceous material from Fort Rock (USA). An increase in peak intensity relative to that of the amorphous hump between 10–40° 2θ in weathered/altered samples corresponds to increased crystallinity in both samples.

angles (Fig. 3)[25,26]. In low proportions, the broad low-angle peak characteristic of such materials may not be discernible. Weathering processes and low-temperature hydrothermal alteration may affect the interpretation of XRD patterns since devitrification of the glass component will increase crystallinity, and alteration of crystalline components to clays and other partially amorphous components will yield lower crystallinity measurements. In addition, the precipitation of calcite and zeolites will result in an increased crystallinity.

Our results demonstrate that significant fractions of poorly crystalline alteration products in severely weathered/altered volcanic deposits can mask the original whole-rock crystallinity; however, the original C_{gm} may still be preserved within groundmass glass fragments despite seemingly pervasive alteration. In light to moderately altered deposits, the whole-rock and groundmass crystallinities still reflect the broad distinctions between phreatomagmatic (<35%) and other magmatic (>35%) deposits (Fig. 1b). For example, in hyaloclastite tuffs within the Columbia River flood basalts, unaltered material from sample Tuff 1-1m is only ~8% crystalline, whereas a bulk analysis of the moderately palagonitized sample is ~30% crystalline, which are both clearly identifiable within the

phreatomagmatic crystallinity range (Figs 1 and 3). Similar variations may be observed in tuff deposits from Fort Rock, OR, where C_{wr} increases from 22 to 47% with weathering; however, groundmass glass preserves nearly identical (~9–11%) crystallinities. More severe palagonitization of samples will blur the boundaries between these fields (Fig. 1 and Table 1). However, in such cases, the alteration products will be at high enough proportions for their presence to be identifiable in the diffraction patterns (Fig. 3) necessitating screening of results before interpretation. Bulk analyses of the severely weathered 122 BC plinian deposits from Mt. Etna generally yield higher crystallinities (by only 5–15%) than analyses of their fresher equivalents, whereas groundmass separates suggest a decrease from 46 to 36% as a result of greater weathering. These results suggest that in terms of measured crystallinities, alteration and weathering have varying and significant effects depending on the original crystal/glass content and style of alteration within volcanic products, but that C_{gm}, even in severely altered samples, and random bulk crystallinity in light to moderately altered samples still preserves the broadly defined eruption styles.

Discussion

To successfully transfer this technique from Earth to Mars, several important considerations must be addressed, including sampling strategies, environmental factors and alteration processes. As previously demonstrated (Fig. 1b), random bulk and groundmass crystallinities in fresh to moderately altered eruptive products record the same distinction between phreatomagmatic and 'magmatic' eruption styles. This observation is critical to the applicability of this technique as a remote tool because it is not possible to separate groundmass from crystals using current remote sampling techniques[21], and as a result, the Curiosity rover analyses are of bulk samples at the sub-125-μm sieve fraction. Hence, although the variations between eruption styles are defined by groundmass crystallization, this translates to the results of the random bulk (whole rock) analysis, and is therefore comparable to the MSL Curiosity rover sampling methodology[21]. Note that this does not discount the possibility that extremely high phenocryst contents in a phreatomagmatic deposit could skew results into the 'magmatic' crystallinity range. However, such deposits are rare on Earth (Fig. 1b).

While the same styles of basaltic explosive eruptions are interpreted to exist on Mars (for example, phreatomagmatic, strombolian and plinian), differences in the Martian environment will impact eruption dynamics and ultimately crystallization of erupted materials[1]. Of particular significance to this study is the effect of lower atmospheric pressure on Mars (600 Pa average surface pressure) compared with Earth (101.3 kPa at sea level). A lower atmospheric pressure results in a 300 times decrease in the density of the erupted gas, decreasing the size of transportable particles on Mars by a factor of ~100 compared with Earth[1]. However, the lower atmospheric pressure also results in a 1.5 times increase in gas velocity and an increase in magma fragmentation such that explosive eruptions result in a greater dispersal of finer pyroclastic material on Mars compared with Earth with particles transported in eruption clouds predominantly sub-centimetre in diameter[1].

The overall decrease in clast size in Martian explosive volcanism particularly affects groundmass crystallization in air-cooled pyroclasts since cooling times are proportional, and cooling rates are inversely proportional, to the particle radius[27]. Assuming a spherical cooling model, particles < ~1 cm rapidly re-equilibrate with their surroundings and remain in thermal equilibrium with the erupted gas in which they are entrained[1,27]. The cooling rate of small (<1 cm) erupted particles therefore is controlled by the cooling rate of the eruption plume they are in equilibrium with. Numerical simulations of a near-vent gas thrust model transitioning to a convective plume[27,28] modified for Martian atmospheric pressure and temperature[29], eruption velocities and vent diameters[1], and magma composition[30] (Supplementary Table 1) are coupled with cooling rate–crystallization rate relationships above an estimated glass transition temperature of ~730 °C (refs. 18,31) to reconstruct syn-eruptive groundmass crystallization during plinian eruptions on Mars (see Supplementary Discussion 1 for details of numerical modelling). Modelled total syn-eruptive groundmass crystallization varies from 20 to 30% depending on particle sizes (1–5 mm radius), vent geometry and eruption velocities, for cooling times from ~25 to 200 s. Numerical simulations indicate that despite the lower atmospheric pressure and smaller particle sizes, cooling is buffered within the eruption plume, still allowing for significant groundmass crystallization. As rapid cooling rates and times associated with phreatomagmatic eruptive processes are not greatly affected by the change to Martian conditions there remains a measureable, if reduced, difference in crystallization between eruption styles, with syn-eruptive groundmass crystallization during plinian style increasing C_{gm} by 20–30% compared with similar magma erupting phreatomagmatically. Larger clasts (> ~1 cm radius) that are modelled to rapidly fall out of the eruption plume have syn-eruptive C_{gm} of >30%, resulting from slow conductive cooling of the clast interiors.

Over geologic timescales, glass is a metastable material subject to alteration processes such as hydration and devitrification[32]. Although alteration of amorphous glass may result in the formation of crystalline materials (for example, zeolites), a by-product of both hydrothermal alteration and chemical weathering of glass on Mars is the formation of npOx. NpOx is an amorphous Fe-bearing constituent that may include haematite, goethite, ferrihydrite, hisingerite, schwertmanite, akaganeite and the Fe^{3+}-rich particles that pigment iddingsite and palagonite, and is presently not distinguished from glass based on remote XRD analysis[33–39]. Alteration of glass is considered to be one of the primary means of npOx formation in Martian volcanic deposits[33–37]. However, npOx may also form from the dissolution of jarosite, a hydrous, K- and Fe-bearing sulfate[40]. While severe alteration of glass in Earth-based analogues results in an increase in crystallinity (Fig. 1), from the assumption that npOx is predominantly derived from glass, the presence of significant amorphous npOx as an alteration product of glass on Mars should mitigate the effect of alteration on crystallinity measurements and the interpretation of eruptive processes.

Although there is an absence of XRD data for Martian volcanic rocks, miniature thermal emission spectrometer (Mini-TES) analysis of pyroclastic rocks at Home Plate in Gusev crater, Mars, indicate a significant component of a material similar to basaltic glass (20–45%; refs 30,34). In contrast, Mössbauer spectroscopy results indicate a high proportion (27–29%) of npOx (absent from the Mini-TES results), suggesting that npOx may appear spectrally similar to basaltic glass using Mini-TES[30]. However, from the assumption that amorphous npOx is primarily derived from basaltic glass, the proportion of npOx may provide a means of estimating primary glass content even in severely altered volcanic rocks. A C_{wr} of 55–80% from Mini-TES analysis of pyroclastic rocks in Gusev crater is consistent with a strombolian or plinian 'magmatic' eruption style, based on our methodology, rather than water/ice-driven phreatomagmatism[30,34]. However, high proportions of secondary silicate phases (up to ~25%) identified in some of the Home Plate mini-TES results may imply severe alteration and an underestimation of the glass content.

New XRD data from the MSL Curiosity rover[38] processed using the calibration routine of this study indicate that the

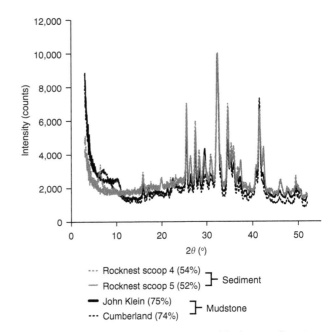

Figure 4 | Mars XRD patterns. XRD patterns of Rocknest sediment (scoops 4 and 5) and Yellowknife Bay mudstone rocks (John Klein and Cumberland) from Gusev crater, Mars, collected from the CheMin instrument on the Curiosity rover[38,39]. Bragg peaks in the diffraction pattern represent the crystalline component while the broad hump between ~10 and 40° 2θ is characteristic of the amorphous component. Increasing crystallinity is reflected by relatively higher peak intensity compared with the broad hump of the amorphous component. Calculated bulk crystallinity from this study is indicated in parentheses.

sediment making up the Rocknest sand shadow has a crystallinity of 52–54% (Fig. 4) (compared with independent estimates from XRD[38] and mass balance[33] approaches, yielding 73 ± 14% and 55% crystallinity, respectively). A calculation of the amorphous component chemical composition from a combination of CheMin and alpha particle X-ray spectrometer analysis on the Rocknest sand shadow sediment indicates a low-SiO$_2$ (37.2 wt%) component[33]. This is consistent with deconvolution of XRD patterns that indicate a best-fit to the CheMin XRD data with a 'Gusev-composition basaltic glass[38].' Although multiple lines of evidence suggest the presence of a basaltic glass component, the origin and nature of this amorphous component is unknown[39]. Both volcanic eruptions and impacts have the potential to widely distribute glass on the surface of Mars[41,42]. Recent experimental work suggests a correlation between glass composition and the shape of the diffuse amorphous peak as analysed by XRD[43]. Several different types of impact glass are predicted on the Martian surface including proximal, low-SiO$_2$ (45–60 wt%) and distal tektite-like, high-SiO$_2$ (60–75 wt%) glass[41]. While distally deposited high-SiO$_2$ glass should produce an XRD pattern distinct from basaltic volcanic glass, low-SiO$_2$ proximal impact glass may be indistinguishable from basaltic volcanic glass[43]. As both basaltic impact glass and volcanic glass have the potential for distribution on the Martian surface, for now both remain a potential source of the amorphous component in the Rocknest sand shadow.

The Rocknest sand shadow crystallinity estimate (52–54%) is significantly lower than the calculated crystallinity of Martian mudstones (John Klein and Cumberland) at Yellowknife Bay (Gale Crater)[39] (74–75%) using the same procedure, and is dominated by a basaltic mineral assemblage of plagioclase, olivine and pyroxene (Fig. 4)[33], indicating a potential source distinctive from that of the sediments comprising the mudstone rocks. With

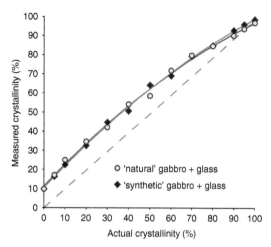

Figure 5 | Crystallinity calibration curves. Measured crystallinities, determined by XRD analysis, plotted against the samples' known crystallinities. Known crystallinities are determined from mixing proportions of gabbro and glass. Both the 'synthetic' gabbro powder series and the 'natural' gabbro powder series yield best-fit curves, blue and purple lines respectively, that nearly align, and are both close to the 1:1 line (dashed grey line).

an understanding that the source of the amorphous component is debatable, based on the methodology developed here and the similarity in crystallinity and mineralogy between the Rocknest sand shadow sediment and the basaltic pyroclastic rocks of Gusev crater, we suggest that X-ray results are consistent with a relatively high proportion of basaltic glass indicating potentially widely distributed strombolian or plinian volcanic eruptive material on the surface of Mars[30,33]. This interpretation is consistent with the greater anticipated dispersal of explosive volcanic deposits on Mars[1] and prior observations regarding the widespread similarity between Martian soil and sediment[33]. In summary, application of our calibrated crystallinity technique provides a broadly applicable remote tool for quantifying the amorphous content of both sediments and rocks, and for potentially characterizing juvenile volcanic eruption styles and magma–water interaction on Earth and Mars.

Methods

X-ray diffraction. All analyses in this study were conducted using a Siemens D500 X-ray diffractometer at Washington State University with a CuKα source ($\lambda = 1.54184$ Å) operated at 35 kV and 30 mA. Samples (1–2 g) were powdered to a uniform particle size of $\sim 20\,\mu$m, allowing for consistent and smooth sample preparation on glass plates (~ 0.1 g of material was used for sample mounts to ensure no interference from the substrate glass plate). Analyses were primarily conducted from 10 to 50° 2θ at 0.03° steps and with 3.0 s dwell times (on select samples, data were collected down to 2° 2θ to examine for low-angle peaks associated with alteration products; Fig. 3). Two incident beam slits and one diffracted beam slit were each positioned at 1°, followed by a diffracted beam graphite crystal monochronometer (0.05°). Using MDI JADE 8 software, diffractograms were processed with a 35-pt Savitzky–Golay smoothing filter to reduce background. Integrated counts were calculated above both cubic spline and linear backgrounds. The area above the linear background includes both the amorphous curve and crystalline peaks, whereas the integrated area above the cubic spline curve, applied to the base of the crystalline peaks, effectively eliminates the amorphous component and represents only the amount of crystalline material (Fig. 2). Thus, the ratio of this 'crystalline' area over 'total' area, multiplied by 100%, provides a relative percent crystallinity (RC(%)) of a sample (equation 1; ref. 22).

$$\frac{A_{CS}}{A_{L}} \times 100 = RC(\%) \qquad (1)$$

where A_{CS} and A_{L} represent the integrated counts above the cubic spline and linear backgrounds, respectively.

Crystallinity calibration. In order to determine crystallinity values from RC(%), a calibration curve was constructed using samples of known crystallinity. Powder

mixtures were created with varying weight percents of 100% crystalline material and 100% amorphous material (a glass plate) in order to mimic basalts of varying crystallinities (Fig. 5). Two mixture series were created with one using a 'natural' gabbro and the other using a 'synthetic' gabbro, the latter created by powdering and mixing by weight single coarse samples of plagioclase (50%), orthopyroxene (40%) and olivine (10%). The two calibration curves deviate only slightly at high crystallinities, suggesting that the natural gabbro is probably not completely crystalline. Both curves record artificially high relative crystallinities ($\sim 10\%$) at low true crystallinity, which is a result of increased background interference associated with low count rates for amorphous materials (Fig. 5). The 'synthetic' gabbro calibration was used in all subsequent calculations. The calibration curve is defined in terms of crystallinity as follows:

$$C(\%) = -\sqrt{\frac{-RC(\%)}{0.003} + 42500} + 198 \qquad (2)$$

where C(%) and RC(%) are the calibrated crystallinity and relative crystallinity, respectively. The calibrated methodology applied here for determining crystallinity is not a new approach. However, the application of this methodology as a new remote tool for determining volcanic eruption style is novel.

Analysis of natural samples. Whole-rock and groundmass crystallinities were determined from 17 eruptions of various styles. While the results discussed here focus on C_{gm} for remote applications, it is important to relate whole rock and C_{gms} since the physical separation of these different crystallinity measurements is not feasible during remote studies. Locations and descriptions of all samples are listed in Table 1. Multiple analyses were conducted from each sample to investigate groundmass versus whole-rock crystallinity as well as the effects of weathering. In addition, a limited number of repeat analyses were conducted to test reproducibility and homogeneity of samples, indicated by trial number (1, 2 and 3) in Table 1. A number of lava flows were also analysed to (1) provide a relative check of cooling rates and (2) to confirm that the amorphous substrate the samples are mounted on does not influence analyses. Massive Columbia River flood basalt lavas are nearly completely crystalline, with crystallinities varying from 80 to 98%, thus verifying that the sample mount substrate does not compromise the crystallinity calculations. In comparison with the slowly cooled Columbia River flood basalts, relatively glassy samples of smaller lava flows from Mt. Etna have substantially reduced crystallinities of 60–72%. XRD data from Mars sediment samples taken at the Rocknest sand shadow in Gale Crater (scoops 4 and 5; refs 33,38) and mudstones from Yellowknife Bay[39] were processed using our calibrated technique to determine the proportion of amorphous material. In contrast to the Siemens D500 that operates in reflection geometry mode with a CuKα X-ray source, the CheMin instrument operates in transmission geometry with a CoKα X-ray source[38]. While the absolute positions of peaks will vary as a result of the differences between wavelengths of CuKα and CoKα X-rays, overall differences in intensity will be relatively minor. As a result, crystallinities estimated from our calibrated approach, despite the differences in instruments, are expected to produce comparable results irrespective of the XRD instrument.

References

1. Wilson, L. & Head, J. W. Mars: review and analysis of volcanic eruption theory and relationships to observed landforms. *Rev. Geophys.* **32**, 221–263 (1994).
2. Glaze, L. S. & Baloga, S. A. Volcanic plume heights on Mars: limits of validity for convective models. *J. Geophys. Res.* **107**, 5086 (2002).
3. Wilson, L. & Head, J. W. Explosive volcanic eruptions on Mars; tephra and accretionary lapilli formation, dispersal and recognition in the geologic record. *J. Volcanol. Geotherm. Res.* **163**, 83–97 (2007).
4. Mouginis-Mark, P. J., Wilson, L. & Head, J. W. Explosive volcanism of Hecates Tholus, Mars: investigation of eruption conditions. *J. Geophys. Res.* **87**, 9890–9904 (1982).
5. Crown, D. A. & Greeley, R. Volcanic geology of Hadriaca Patera and the eastern Hellas region of Mars. *J. Geophys. Res.* **98**, 3431–3451 (1993).
6. Broz, P. & Hauber, E. Hydrovolcanic tuff rings and cones as indicators for phreatomagmatic explosive eruptions on Mars. *J. Geophys. Res.* **118**, 1656–1675 (2013).
7. Carey, S. & Sparks, R. S. J. Quantitative models of the fallout and dispersal of tephra from volcanic eruption columns. *Bull. Volcanol.* **48**, 109–125 (1986).
8. Roggensack, K., Hervig, R. L., McKnight, S. B. & Williams, S. N. Explosive basaltic volcanism from Cerro Negro volcano: influence of volatiles on eruptive style. *Science* **277**, 1639–1642 (1997).
9. Houghton, B. F. *et al.* The influence of conduit processes on changes in style of basaltic Plinian eruptions: Tarawera 1886 and Etna 122 BC. *J. Volcanol. Geotherm. Res.* **137**, 1–14 (2004).
10. Scott, D. H. & Tanaka, K. L. Ignimbrites of Amazonis Planitia region of Mars. *J. Geophys. Res.* **87**, 1179–1190 (1982).
11. Bradley, B. A., Sakimoto, S. E. H., Frey, H. & Zimbelman, J. R. Medusae Fossae Formation: new perspectives from Mars Global Surveyor. *J. Geophys. Res.* **107**, 5058 (2002).

12. Hynek, B. M., Phillips, R. J. & Arvidson, R. E. Explosive volcanism in Tharsis region: Global evidence in the martian geologic record. *J. Geophys. Res.* **108**, 5111 (2003).

13. Kite, E. S., Lewis, K. W., Lamb, M. P., Newman, C. E. & Richardson, M. E. Growth and form of the mound in Gale Crater, Mars: slope wind enhanced erosion and transport. *Geology* **41**, 543–546 (2013).

14. Milliken, R. E., Grotzinger, J. P. & Thompson, B. J. Paleoclimate of Mars as captured by the stratigraphic record in Gale Crater. *Geophys. Res. Lett.* **37**, L04201 (2010).

15. Stolper, E. M. et al. The petrochemistry of Jake_M: a martian mugearite. *Science* **341**, 1239463 (2013).

16. Hammer, J. E. & Rutherford, M. J. An experimental study of the kinetics of decompression-induced crystallization in silicic melt. *J. Geophys. Res.* **107**, 2021 (2002).

17. Szramek, L., Gardner, J. E. & Larsen, J. Degassing and microlite crystallization of basaltic andesite magma erupting at Arenal Volcano, Costa Rica. *J. Volcanol. Geotherm. Res.* **157**, 182–201 (2006).

18. Szramek, L., Gardner, J. E. & Hort, M. Cooling induced crystallization of microlite crystals in two basaltic pumice clasts. *Am. Mineral.* **95**, 503–509 (2010).

19. Zimanowski, B., Büttner, R., Lorenz, V. & Häfele, H. G. Fragmentation of basaltic melt in the course of explosive volcanism. *J. Geophys. Res.* **102**, 803–814 (1997).

20. Wallace, P. J., Dufek, J., Anderson, A. T. & Zhang, Y. Cooling rates of Plinian-fall and pyroclastic-flow deposits in the Bishop Tuff: inferences from water speciation in quartz-hosted glass inclusions. *Bull. Volcanol.* **65**, 105–123 (2003).

21. Blake, D. et al. Characterization and calibration of the CheMin mineralogical instrument on Mars Science Laboratory. *Space Sci. Rev.* **170**, 341–400 (2012).

22. Rowe, M. C., Ellis, B. S. & Lindeberg, A. Quantifying crystallization and devitrification of rhyolites via X-ray diffraction and electron microprobe analysis. *Am. Mineral.* **97**, 1685–1699 (2012).

23. Filiberto, J., Gross, J. & Treiman, A. H. Basaltic pyroclastic deposits on Earth and Mars: constraints for robotic exploration of martian pyroclastic deposits. *41st Lunar and Planetary Science Conference*, Abstract #1936 (The Woodlands, Texas, USA, 2010).

24. Morris, R. V., Golden, D. C., Bell, III J. F., Lauer, Jr. H. V. & Adams, J. B. Pigmenting agents in Martian soils: Inferences from spectral, Mössbauer, and magnetic properties of nanophase and other iron oxides in Hawaiian palagonitic soil PN-9. *Geochim. Cosmochim. Acta* **57**, 4597–4609 (1993).

25. Achilles, C. N., Morris, R. V., Chipera, S. J., Ming, D. W. & Rampe, E. B. X-ray diffraction reference intensity ratios of amorphous and poorly crystalline phases: Implications for CheMin on the Mars Science Laboratory Mission. *44th Lunar and Planetary Science Conference*, Abstract #3072 (The Woodlands, Texas, USA, 2013).

26. Rampe, E. B. et al. Detecting nanophase weathering products with CheMin: Reference intensity ratios of allophane, aluminosilicate gel, and ferrihydrite. *44th Lunar and Planetary Science Conference* Abstract #1188 (The Woodlands, Texas, USA, 2013).

27. Hort, M. & Gardner, J. Constraints on cooling and degassing of pumice during Plinian volcanic eruptions based on model calculations. *J. Geophys. Res* **105**, 25981–26001 (2000).

28. Woods, A. W. Moist convection and the injection of volcanic ash into the atmosphere. *J. Geophys. Res.* **98**, 179627–17636 (1993).

29. Magalhães, J. A., Schofield, J. T. & Seiff, A. Results of the Mars Pathfinder atmospheric structure investigation. *J. Geophys. Res.* **104**, 8943–8955 (1999).

30. Squyres, S. W. et al. Pyroclastic activity at Home Plate in Gusev Crater, Mars. *Science* **316**, 738–742 (2007).

31. Cashman, K. V. Relationship between plagioclase crystallization and cooling rate in basaltic melts. *Contib. Mineral. Petrol.* **113**, 126–142 (1993).

32. Lofgren, G. Experimental devitrification rate of rhyolite glass. *Geol. Soc. Am. Bull* **81**, 553–560 (1970).

33. Blake, D. F. et al. Curiosity at Gale Crater, Mars: characterization and analysis of the Rocknest sand shadow. *Science* **341**, 1239505 (2013).

34. Schmidt, M. E. et al. Spectral, mineralogical, and geochemical variations across Home Plate, Gusev Crater, Mars indicate high and low temperature alteration. *Earth Planet. Sci. Lett.* **281**, 258–268 (2009).

35. Morris, R. V. et al. Phyllosilicate-poor palagonitic dust from Mauna Kea Volcano (Hawaii): a mineralogical analogue for magnetic Martian dust? *J. Geophys. Res* **106**, 5057–5083 (2001).

36. Morris, R. V. et al. Mossbauer mineralogy of rock, soil, and dust at Gusev crater, Mars: Spirit's journey through weakly altered olivine basalt on the plains and pervasively altered basalt in the Columbia Hills. *J. Geophys. Res.* **111**, E2S13 (2006).

37. Morris, R. V. et al. Iron mineralogy and aqueous alteration from Husband Hill through Home Plate at Gusev Crater, Mars: results from the Mössbauer instrument on the Spirit Mars Exploration Rover. *J. Geophys. Res.* **113**, E12S42 (2008).

38. Bish, D. L. et al. X-ray diffraction results from Mars Science Laboratory: mineralogy of Rocknest at Gale Crater. *Science* **341**, 1238932 (2012).

39. Vaniman, D. T. et al. Mineralogy of a mudstone at Yellowknife Bay, Gale Crater, Mars. *Science* **343**, 1243480 (2014).

40. Elwood Madden, M. E. et al. Jarosite dissolution rates and nanoscale mineralogy. *Geochim.Cosmochim. Acta* **91**, 306–321 (2012).

41. Schultz, P. H. & Mustard, J. F. Impact melts and glasses on Mars. *J. Geophys. Res.* **109**, E01001 (2004).

42. Moroz, L. V. et al. Spectral properties of simulated impact glasses produced from martian soil analogue JSC Mars-1. *Icarus* **202**, 336–353 (2009).

43. Craig, M. A. et al. XRD patterns of glassy impactites: amorphous curve fitting and composition determination with implications for Mars. *44th Lunar and Planetary Science Conference*, Abstract #2319 (The Woodlands, Texas, USA, 2013).

Acknowledgements

This project was funded by the Washington State University (WSU) College of Arts and Sciences' Grants for Undergraduate Scholars, and by the NASA Space Grant Undergraduate Scholarship in Science and Engineering. We thank Matthew Nikiczuk, Rick Conrey, John Wolff, Klarissa Davis, Phil Shane, Marco Viccaro and Peter Kokelaar for samples and field assistance, and Anton Gulley for assistance with MATLAB coding. The WSU Geoanalytical Laboratory kindly provided for access to sample preparation facilities and the XRD data (C. Knaack).

Author contributions

K.T.W. performed all analyses and data processing. K.T.W. and M.C.R. developed interpretations and techniques. B.S.E. developed volcanological interpretations. M.E.S. consulted on Martian geology and J.D.E. devised crystallization models for clasts in plinian eruptions. K.T.W. and M.C.R. wrote the manuscript with contributions from all co-authors.

Additional information

12

Transition to a Moist Greenhouse with CO_2 and solar forcing

Max Popp[1,2,†], Hauke Schmidt[1] & Jochem Marotzke[1]

Water-rich planets such as Earth are expected to become eventually uninhabitable, because liquid water turns unstable at the surface as temperatures increase with solar luminosity. Whether a large increase of atmospheric concentrations of greenhouse gases such as CO_2 could also destroy the habitability of water-rich planets has remained unclear. Here we show with three-dimensional aqua-planet simulations that CO_2-induced forcing as readily destabilizes the climate as does solar forcing. The climate instability is caused by a positive cloud feedback and leads to a new steady state with global-mean sea-surface temperatures above 330 K. The upper atmosphere is considerably moister in this warm state than in the reference climate, implying that the planet would be subject to substantial loss of water to space. For some elevated CO_2 or solar forcings, we find both cold and warm equilibrium states, implying that the climate transition cannot be reversed by removing the additional forcing.

[1] Max Planck Institute for Meteorology, Bundesstrasse 53, Hamburg 20146, Germany. [2] Program in Atmospheric and Oceanic Sciences, Princeton University, 300 Forrestal Road, Sayre Hall, Princeton, New Jersey 08544, USA. † Present address: NOAA's Geophysical Fluid Dynamics Laboratory, Princeton, New Jersey, USA. Correspondence and requests for materials should be addressed to M.P. (email: mpopp@princeton.edu).

Water-rich planets such as Earth lose water by photo-dissociation of water vapour in the upper atmosphere and the subsequent escape of hydrogen. On present-day Earth, the loss occurs very slowly, because the mixing ratio of water vapour in the upper atmosphere is very low. But significant loss of water could occur over geological timescales if the surface temperature were around 70 K warmer than it is today[1-3]. For these high surface temperatures, the tropopause is expected to climb to high altitudes. As a consequence, the cold trapping of water vapour at the tropopause becomes ineffective, because the mixing ratio of water vapour increases with the rising tropopause. Steady states in which the mixing ratio in the upper atmosphere is sufficiently high for a water-rich planet to lose most of its water inventory in its lifetime are known as Moist-Greenhouse states[4]. A planet in this state would eventually become uninhabitable as all water is lost to space. For an Earth-like planet around a Sun-like star, a Moist Greenhouse would be attained if the mixing ratio in the upper atmosphere exceeds $\sim 0.1\%$ (ref. 1). For comparison, the mixing ratio in Earth's stratosphere is presently around two orders of magnitude smaller.

Moist-Greenhouse states were found and described in several studies with one-dimensional models[1-3,5,6] and have recently been found for terrestrial planets with three-dimensional models in different setups[7-9]. However, not all three-dimensional studies found stable Moist-Greenhouse states[10-12]. Instead the climate of these models would destabilize into a Runaway Greenhouse, a self-reinforcing water-vapour feedback-loop, before the Moist Greenhouse is attained. A few studies applied large forcing but the employed models became numerically unstable before the Moist-Greenhouse regime was attained[13-15]. Therefore, it remains unclear whether planets would attain a Moist-Greenhouse state before a Runaway Greenhouse occurs, especially for planets on an Earth-like orbit, where the only two previous studies with state-of-the-art general circulation models (GCM) gave contradicting results[12,9]. Moreover, all three-dimensional studies investigating Moist-Greenhouse states only applied solar forcing without considering greenhouse-gas forcing. Several studies have applied strong greenhouse-gas forcing, but either did not run their simulations to sufficiently high temperatures[16-19] or did not investigate the emergence of a Moist Greenhouse[20]. Greenhouse-gas forcing has long been assumed to be ineffective at causing Moist-Greenhouse states, because the greenhouse effect of any additional greenhouse gas would eventually be rendered ineffective by the increasing greenhouse effect of water vapour with increasing temperatures. Furthermore, large greenhouse-gas forcing would lead to a cooling of the upper atmosphere, which would push the Moist-Greenhouse limit to much higher surface temperatures[5]. However, if clouds are considered, these arguments may not apply, because clouds themselves can contribute to the climate becoming unstable[21].

Here we compare for the first time with a state-of-the-art GCM, namely ECHAM6 (ref. 22), how effective solar and CO_2 forcing are at causing a transition to a Moist Greenhouse. We couple the atmosphere to a slab ocean and choose an aqua-planet setup (fully water-covered planet) in perpetual equinox. This idealized framework is better suited than a present-day Earth setting to understand the involved dynamics while preserving the major feedback mechanisms of the Earth[23]. It also avoids conceptual problems with the representation of land-surface processes at high temperatures. We turn off sea ice in order to investigate the possibility of solely cloud-induced multiple steady states that were recently found in a one-dimensional study[21]. We modify the model such that it can deal with surface temperatures of up to 350 K (see Methods).

Thus we show that cloud-radiative effects (CRE) destabilize a present-day Earth climate as readily with CO_2 as with solar forcing. The changes in CRE are a consequence of the weakening of the large-scale circulation with increasing global-mean surface temperature (gST). However, the resulting climate transition does not lead to a Runaway Greenhouse, but instead a new regime of warm steady state with gST above 330 K is attained. This warm regime differs substantially in its dynamics from a present-day Earth-like climate and, most importantly, the upper atmosphere exceeds the Moist-Greenhouse limit in this regime. Hence a planet in such a state would lose water at a fast rate to space. Furthermore, there is hysteresis in the warm regime and removing the imposed forcing does therefore not necessarily cause a transition back to an Earth-like climate.

Results

Simulations with increased TSI. To assess the dependence of the climate state on total solar irradiance (TSI) for fixed CO_2 levels (at 354 p.p.m. volume mixing ratio), we apply a total of five different TSI-values that range from the present-day value on Earth (S_0) to 1.15 times that value. We find two regimes of steady states that are separated by a range of gST for which stable steady states are not found (Fig. 1). The regime of steady states with gST of up to ~ 298 K exhibits similar features as present-day Earth climate, such as a large pole-to-equator surface-temperature contrast (Fig. 2a) and a similar meridional distribution of cloud cover (Fig. 2b). Hence this regime of steady state can be considered to be Earth-like. In contrast, the warm regime of steady states with gST above 334 K is characterized by a considerably smaller pole-to-equator surface-temperature difference and a substantially different meridional distribution of cloud cover (Fig. 2a,b). This illustrates that the dynamics in the warm regime are quite different from the present-day Earth regime. Most importantly, the mixing ratio of water vapour at the uppermost level at 0.01 hPa is considerably higher in the warm regime and exceeds the Moist-Greenhouse limit (Fig. 2c). Therefore, a planet in such a state would be losing water to space at a fast rate. The minimum TSI required to cause a climate transition from the Earth-like to the warm regime lies between 1.03 S_0 and 1.05 S_0, whereas the maximum TSI to cause a climate transition from the warm back to the Earth-like regime lies

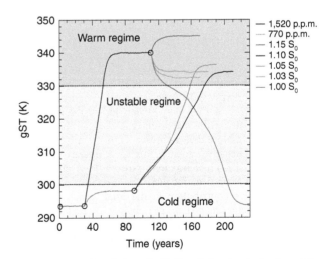

Figure 1 | Temporal evolution of gST. The circles denote the four states from which new simulations are started. For both a TSI of 1.00 S_0 and 1.03 S_0 as well as for a CO_2 concentration of 770 p.p.m., two simulations with different initial conditions are performed. The lines are interpolated from annual means.

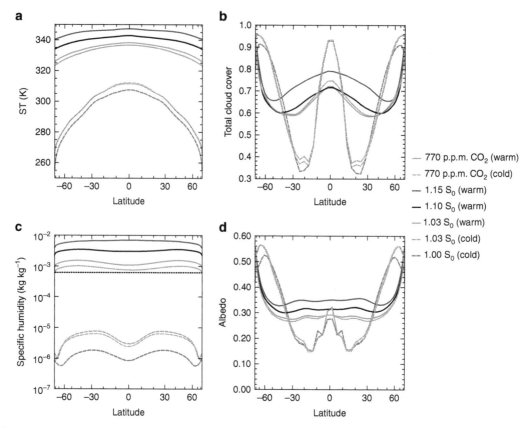

Figure 2 | Zonal means in steady state. (**a**) Surface temperature (ST); (**b**) the total cloud cover; (**c**) the specific humidity at the top level; and (**d**) the effective albedo. The effective albedo is defined as the ratio of the zonal and temporal means of the reflected solar radiation divided by the zonal and temporal means of the incoming solar radiation. The black horizontal line in **c** indicates the Moist-Greenhouse limit[1]. The temporal mean is taken over a period of 30 years. The horizontal axes are scaled with the sine of the latitude.

between 1.00 S_0 and 1.03 S_0 (Fig. 1). Consequently, there are two different stable steady states for a TSI of 1.03 S_0. Since sea ice is turned off in our model, this double steady state is entirely a consequence of atmospheric processes. In the warm regime, the cloud albedo increases at all latitudes with TSI, thus providing an efficient way to stabilize the climate against increased radiative forcing (Fig. 2d).

Energetics of the climate transition. To understand the processes governing the climate transition from the cold to the warm regime, we focus now on the transient simulation with a TSI of 1.05 S_0. The climate instability is evidenced by an increase in the total (shortwave plus longwave) net (downward minus upward component of the) global-mean top-of-the-atmosphere (TOA) radiative flux with increasing gST for gST between 300 and 330 K (Fig. 3). Note that a positive TOA radiative flux is net downward and a negative radiative flux is net upward. Therefore, the aqua-planet takes up energy for a positive TOA radiative flux and loses energy for a negative flux. The instability is caused by the cloud-radiative contribution to the total radiative flux that increases with increasing gST for gSTs below 330 K. The clear-sky contribution to the total radiative flux is decreasing with increasing gST and does thus not contribute to the climate instability. This decrease in clear-sky contribution is caused by an increase in clear-sky contribution to the outgoing longwave radiation, which is upward and thus decreases the net downward flux. At gST above 330 K, the cloud-radiative contribution decreases again with increasing gST. This allows together with the clear-sky contribution to attain a new steady state. Thus, clouds

destabilize the climate at lower gST and then stabilize again at higher gST. This change in sign of the cloud feedback is also responsible for the existence of the bistability.

Dynamics of the climate transition. The changes in cloud-radiative contribution to the total net TOA radiative flux (henceforth simply referred to as CRE) are caused by the weakening of the large-scale circulation (Fig. 4a) and the increase of water vapour in the atmosphere with increasing gST. The weakening of the circulation causes tropical convection to spread more evenly around the tropics, with less convection occurring around the equator and more convection occurring in the subsidence region. As a consequence, deep convective clouds with low cloud-top temperatures become more frequent in the subsidence region of the Hadley circulation (Fig. 4b). This in turn leads to a very strong increase in longwave CRE in this region, which dominates the increase in shortwave CRE (Fig. 5a,b). However, as the gST increases further and the specific humidity in the atmosphere increases, the clouds become thicker and thus more reflective, whereas the longwave CRE does not increase as fast anymore leading to a decrease in total CRE in the tropics for gST above 320 K. This decrease in tropical CRE with increasing gST contributes to the stabilization of the climate at gST above 330 K (Fig. 5c). In general, the changes in total CRE dominate the changes in clear-sky radiative effect in the tropics (Fig. 5c,d). In the extra-tropics, the weakening of the large-scale circulation leads to a steady decrease in cloud cover everywhere except at very high latitude (Fig. 4a,b). Therefore, the shortwave CRE increases in the extra-tropics (Fig. 5a). Since the tropopause deepens with increasing surface temperatures (not shown), the difference between the temperature at the surface and at the cloud tops increases as well and leads also to an increase in longwave CRE despite the decrease in cloud cover (Fig. 5b). Whereas the changes in the clear-sky radiative effect dominate the changes in CRE in most of the extra-tropics for gST up to 315 K, the changes in CRE increasingly dominate the extra-tropical response at gST above. This supports the idea that at high gST changes in CRE are

more important than changes in clear-sky radiative effect and dominate the climate response. In general, the weak large-scale circulation in the warm regime leads to a much more uniform meridional distribution of cloud condensate at all levels than in the cold regime (Fig. 6). Note that there is a decrease in global-mean convective precipitation but a slight increase in total precipitation from the cold to the warm regime (not shown). This may indicate that in the warm regime convection is overall less frequent but more intense, such that a significant fraction of condensate is not converted to convective precipitation but detrained to form large-scale precipitation. This trend continues as surface temperatures increase further in the warm regime.

Simulations with increased CO_2 concentrations. We start our comparison of CO_2-induced to solar forcing by increasing CO_2 concentrations to 770 p.p.m., while keeping the TSI fixed to 1.00 S_0. This corresponds to an equivalent adjusted forcing as is caused by an increase of TSI from 1.00 S_0 to 1.03 S_0. The adjusted forcing is defined to be the temporal and global mean of the energy uptake over the first year of simulation. The results suggest that the increase of the CO_2 concentrations leads to an equivalent warming and a similar meridional distribution of surface temperatures and clouds as the increase in TSI does (Fig. 2a,d). Since the aqua-planet warms by 4.57 K for an increase in CO_2 concentrations from 354 to 770 p.p.m., the climate sensitivity of the aqua-planet for a doubling of CO_2 concentrations is 4.08 K, if a $\log_2$ scaling is assumed. Starting from the final state of the simulation with a TSI of 1.03 S_0, we then increase the CO_2 concentrations to 1,520 p.p.m. and set the TSI back to 1.00 S_0 (Fig. 1). The combined effect leads to an adjusted forcing that is equivalent to increasing the TSI to 1.05 S_0. In this case the aqua-planet undergoes a climate transition into the warm regime (Fig. 1). Thus, the aqua-planet can as readily be forced to transition from the Earth-like to the warm regime by increasing CO_2 concentrations as by increasing the TSI. When starting in the warm regime, a reduction of CO_2 concentrations to 770 p.p.m. does not cause the planet to fall back into the Earth-like regime, but the aqua-planet remains in the warm regime. Therefore, the aqua-planet also exhibits a bistability of the climate for a TSI of 1.00 S_0 and a CO_2 concentration of 770 p.p.m.

Overall, the results suggest that the aqua-planet behaves similarly for solar forcing and CO_2-induced forcing (Figs 1,2 and 6). The most notable difference is that the steady-state gST in the warm regime is ~2 K lower for CO_2-induced forcing. The

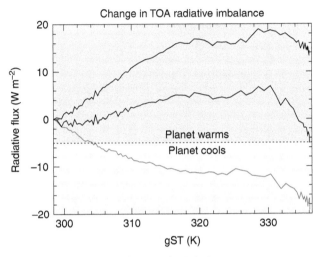

Figure 3 | Change of TOA radiative flux as a function of gST. The black line shows the change in global and annual mean of the total (shortwave plus longwave) net (downward minus upward component of the) TOA radiative flux, the blue line shows the cloud-radiative contribution and the red line the clear-sky contribution from the transient simulation with a TSI of 1.05 S_0. Note that a positive slope indicates an unstable state, because any warming/cooling would lead to an increase/decrease in energy uptake by the aqua-planet and thus to an additional warming/cooling. The cloud-radiative and clear-sky contributions sum up to the net TOA radiative flux. The changes are calculated by subtracting the global and annual means over the first year of the respective quantities. The horizontal line corresponds to the negative value of the initial total net TOA radiative flux. Hence, for a steady state to be attained, the total net TOA radiative flux must touch or intersect the horizontal line.

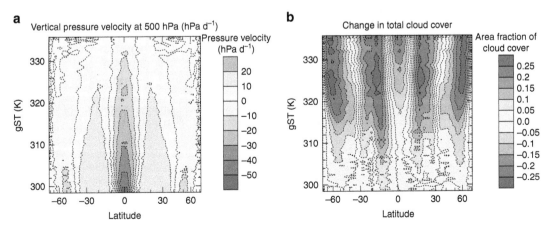

Figure 4 | Large-scale circulation and cloud cover during the climate transition. (**a**) Zonal and annual mean of the vertical pressure velocity as a function of latitude and gST for the transient period of the simulation with a TSI of 1.05 S_0. (**b**) Same as in **a** but for the change in zonal and annual mean of cloud cover from the first year of simulation.

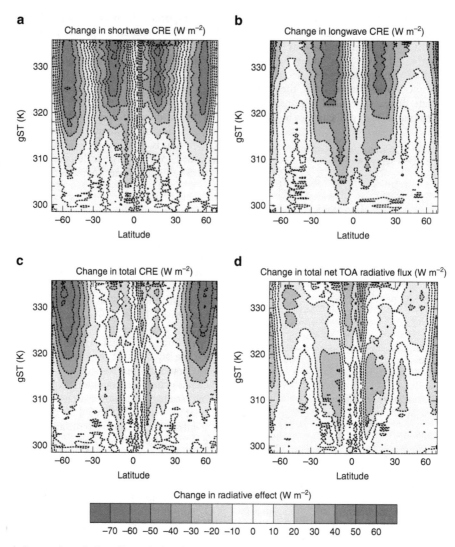

Figure 5 | Zonal means of changes in radiative effects during the climate transition. (**a**) Change of zonal and annual mean of the shortwave CRE; (**b**) change of the longwave CRE; (**c**) change of the total CRE; and (**d**) the change of the total net TOA radiative flux. All panels use the same colour bar. The horizontal axes are scaled with the sine of the latitude.

likely reason for this is that the thermal absorption by water vapour overlaps with the thermal absorption by CO_2 in the warm moist atmosphere, which renders the greenhouse effect of CO_2 less effective. However, since the climate instability in our simulations is caused by CRE at a gST at which the atmosphere is not yet sufficiently opaque to cancel the greenhouse effect of CO_2, CO_2-induced forcing can as easily cause a climate transition to the Moist Greenhouse as solar forcing does.

Sensitivity experiments. Two of the assumptions made in this study could potentially have a large influence on the results and may contribute to the differences in the results between our study and previous ones[17,12,9,20]. These assumptions concern the treatment of ozone and oceanic heat transport of the model. For these two cases, we show with sensitivity experiments that the qualitative nature of the results is not changed by the assumptions (Supplementary Figs 1–3). The neglect of sea ice should have no influence on the qualitative results, because there is no sea ice in the warm regime and because the sea-ice albedo feedback is positive and would thus favour a climate instability in the cold regime. However, the absence of sea ice may explain why our control simulation is warmer than present-day Earth.

Discussion

A recent study using the same model but in a different version found that the Earth's climate remains stable for CO_2 concentrations of at least 4,480 p.p.m. (ref. 17), whereas our study suggests that such concentrations would lead to a climate transition. Studies of Earth with other GCMs also found the climate to remain stable for higher CO_2 concentrations than we do[16,18]. However, the initial climate of our aqua-planet is $\sim 6\,K$ warmer than the one of present-day Earth. Such a warming would be attained by a quadrupling of CO_2 in the different version of our model used in ref. 17. By a simple estimate, this other study would thus have explored CO_2 concentrations of up to a fourth of 4,480 p.p.m.; hence, 1,120 p.p.m., if the simulations were started from a climate similar to ours. Therefore, if we account for the difference in the initial climates, the results of the two studies are not in contradiction. Indeed, the climate of the model version used in ref. 17 was recently shown to become unstable when the CO_2 concentrations were increased from 4,480 to 8,960 p.p.m. (eventually leading to numerical failure of their model)[19]. Nonetheless, the forcing required to cause a climate transition would certainly be higher on present-day Earth than on our aqua-planet, even with our version of the model. Several other studies of Earth have found lower climate sensitivities to

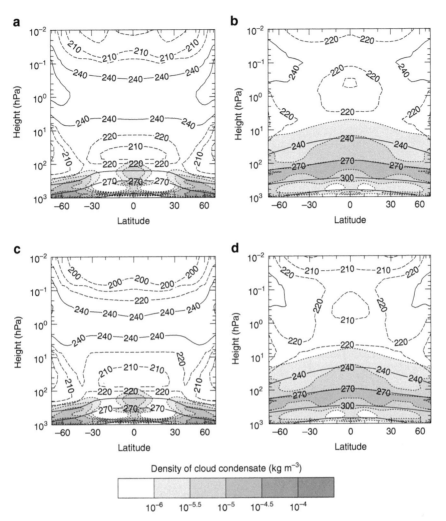

Figure 6 | Zonal means of cloud condensate in steady state. (a) Temporal mean over the last 30 years of simulation of the cloud condensate for a TSI of 1.03 S_0 in the cold regime and **(b)** same quantity for the same TSI but in the warm regime. **(c)** Same quantity obtained with a TSI of 1.00 S_0 but with atmospheric CO_2 concentrations of 770 p.p.m. in the cold regime **(d)** with the same CO_2 concentration and TSI in the warm regime. The contours denote temperatures in Kelvin, with the solid lines denoting the contours for 240, 270, 300 and 330 K and the dashed lines for the contours of 200, 210 and 220 K. The vertical axes are the height in terms of pressure of dry air and the horizontal axes are the latitudes scaled with their sines. The pressure of dry air is defined to be the pressure the atmosphere would have at a given level if no water vapour was present. Since the total mass of dry air does not change between simulations, the mean of the dry surface pressure is constant across simulations.

relatively large CO_2 forcing than we do which supports this notion[16,18,20].

Two studies recently investigated climates at gST above 330 K with state-of-the-art GCMs for Earth-like planets[12,9]. Wolf and Toon[9] used a modified version of the Community Atmosphere Model version 4 (CAM4) and Leconte et al.[12] a modified version of the Laboratoire de Météorologie Dynamique Generic (LMDG) climate model to investigate the climate response to strong solar forcing and both studies also found a region of increased climate sensitivity[12,9]. Therefore, a region of gST with increased climate sensitivity surrounded by regions of lower climate sensitivity appears to be a robust result, despite some differences in the magnitude of the region (Fig. 7c,e). We calculated the climate sensitivity here following[9], which yields a small climate sensitivity in the cold and in the warm regimes because the method uses the instantaneous forcing. If we use the adjusted forcing, the climate sensitivity is considerably larger, because the fast atmospheric adjustments reduce the initial TOA radiative imbalance quickly. Our model encounters the region of high climate sensitivity for smaller values of TSI (Fig. 7a), because our control climate is the

warmest and because the climate instability is encountered at ~300 K, whereas the region of increased climate sensitivity starts at ~310–315 K in the two other studies. Our steady-state albedo is a monotonically increasing function of both TSI and gST, in contrast to the two other studies where the albedo decreases with both increasing TSI and gST until either the Moist-Greenhouse state is attained or a Runaway Greenhouse occurs (Fig. 7b,d). However, in our transition from the cold to the warm regime the albedo decreases as well with increasing gST between 300 and 320 K. Only one of the two studies found a stabilizing cloud feedback similar to ours and a moist stratosphere at high gST (ref. 9), whereas the other one found that the cloud feedback is rather destabilizing and that the stratosphere remains dry[12]. In general, our results of the simulations with increased solar forcing are qualitatively similar to the ones found in ref. 9 in that we find two regimes of steady states, in that the warm regimes have a similar temperature structure (Fig. 8a), in that the cloud albedo increases with gST and most importantly in the existence of a stable Moist-Greenhouse regime. Similarly to ref. 9, the troposphere in the warm regime is characterized in our model by a particular

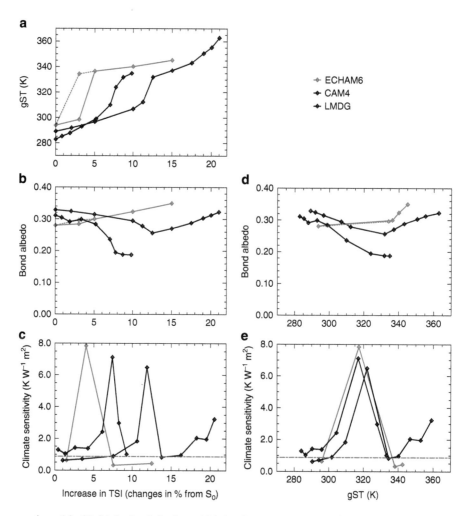

Figure 7 | Model inter-comparison. (**a**) gST; (**b**) the Bond albedo; and (**c**) the climate sensitivity as a function of the increase in TSI in per cents of S_0. The red solid lines and marks denote the results obtained with ECHAM6 when increasing TSI, whereas the dashed lines and mark denote the results obtained with ECHAM6 but when decreasing TSI. The black lines and marks denote the results obtained with CAM4 in ref. 9 and the blue lines and marks denote the results obtained with LMDG in ref. 12. (**d**) Bond albedo and (**e**) the climate sensitivity as a function of the gST. The red dashed-dotted line denotes the climate-sensitivity parameter for doubling CO_2 and is calculated from the simulation in which CO_2 concentrations are increased from 354 to 770 p.p.m. using $\log_2$ scaling. Note that the Bond albedo is equal to the global-mean of the effective albedo.

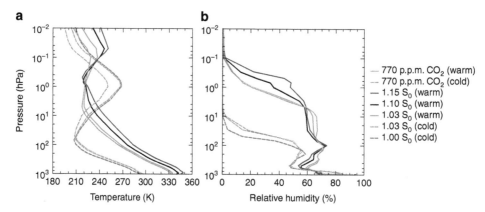

Figure 8 | Atmospheric profiles of temperature and relative humidity. (**a**) Global-mean vertical profiles of gST and (**b**) of relative humidity for the different steady states. The vertical axes are the height in terms of pressure of dry air.

radiative-convective equilibrium with a temperature inversion close to the surface, a somewhat drier region above and a more humid region up to the tropopause. A similar structure has also been found and discussed in two one-dimensional studies[5,21].

Wolf and Toon[9] argues that the change to the aforementioned radiative-convective equilibrium is crucial for the emergence of stable Moist-Greenhouse states, but our results suggest that the weakening of the large-scale circulation is equally important by allowing the radiative-convective regime to spread over the entire tropics. This spread of the convective region over a large fraction of the planet, namely the tropics, also explains some of the similarities between the three- and the one-dimensional models in the warm regime. Compared with ref. 9, the dry region is more humid in our warm regime and in general more humid than in our cold regime (Fig. 8b, Supplementary Fig. 4). Given the different dynamical cores, radiative transfer schemes, convection schemes and cloud schemes, it is, however, remarkable how many similarities the three models share.

In the two aforementioned studies of large solar forcing, ozone was entirely removed from the whole atmosphere in all simulations[12,9], whereas we do include ozone in our calculations. Since ozone has a strong warming influence in the middle atmosphere due to its strong absorption of solar radiation, the inclusion of ozone in our model may explain why our atmosphere is moister in the upper levels than theirs for similar values of gSST. To test this hypothesis, we perform a simulation with a TSI of 1.10 S_0 without atmospheric ozone. Both the upper atmospheric temperatures and specific humidity are lower but still above the Moist-Greenhouse limit in the experiments where ozone is removed (Supplementary Fig. 1). Thus, the specific humidity is still considerably higher than in our cold regime and especially than in the simulations of Wolf and Toon[9] and Leconte et al.[12] at similar gSST. In general, removing ozone does not appear to alter the results substantially.

One study investigated strong CO_2 forcing and gST above 330 K with the Fast Atmosphere-Ocean Model (FAOM) developed at the Goddard Institute for Space Studies and did not find any region of strongly increased climate sensitivity[20]. The climate sensitivity increases somewhat but not nearly as much as with our or with the other two state-of-the-art models used for solar forcing. As a consequence, considerably higher values of CO_2 are required in that study compared with ours to attain a gST of 330 K. Furthermore, the control climate is colder in that study, and it takes around one doubling of CO_2 to attain the gST of our control simulation. The humidity at the top level of FAOM remains roughly one order of magnitude below the Moist-Greenhouse limit in the warmest steady states found in ref. 20. This may be partly due to the CO_2 cooling of the upper atmosphere, but could also be a consequence of not running the model to sufficiently high temperatures. We perform a sensitivity experiment, where we increase CO_2 concentrations to 9,000 p.p.m. in order to assess whether increased CO_2 concentrations could cause a substantial drying of our upper atmosphere (not shown). The upper atmosphere-specific humidity stays; however, well above the Moist-Greenhouse limit also in that case. Some of the differences between FAOM and ECHAM6 may simply be caused by the use of different setups and parameterizations. But FAOM is a simplified model designed for fast computation and uses simplified cloud physics, which may be the cause for the absence of a region of strongly increased climate sensitivity. So, whereas our version of ECHAM6 is rather on the low side of CO_2 concentrations required to cause the gST to rise above 330 K, FAOM likely is on the high side of the concentrations.

To conclude, we have demonstrated with a state-of-the-art climate model that a water-rich planet might lose its habitability as readily by CO_2 forcing as by increased solar forcing through a transition to a Moist Greenhouse and the implied long-term loss of hydrogen. We confirm previous results that a region of increased climate sensitivity exists and show that the climate is unstable in our model in that region due to positive cloud feedbacks caused by a weakening of the large-scale circulation. We also demonstrate that there is hysteresis and that once a transition to the Moist-Greenhouse regime has occurred, the process may not simply be reversed by removing the additional forcing.

Methods

General setup. We employ a modified version of the GCM ECHAM6 (ref. 22) in an aqua-planet setting in which the whole surface is covered by a 50-m-deep mixed-layer ocean. We run the model with a spectral truncation of T31, which corresponds to a Gaussian grid with a grid-point spacing of 3.75°. The atmosphere is resolved vertically by 47 layers up to a pressure (of dry air) of 0.01 hPa. The oceanic heat transport is prescribed by a sinusoidal function of latitude. There is no representation of sea ice in our model and as a consequence water may be colder than the freezing temperature. The orbit of the aqua-planet is perfectly spherical with a radius of 1 AU. The obliquity of the aqua-planet is 0°. For simplicity, a year is set to be 360 days. The rotation velocity of the aqua-planet corresponds to present-day Earth.

Parameterized oceanic heat transport. To account for the meridional oceanic heat transport, we introduce an additional energy flux from the mixed-layer ocean to the atmosphere, which mimics the divergence of a prescribed meridional oceanic heat flux. This energy flux is commonly called q-flux. Imposing a q-flux is necessary to prevent the atmosphere from having to take over all the meridional energy transport, which would result in an amplified atmospheric large-scale circulation. However, since the oceanic circulation depends also on atmospheric properties, the imposed q-flux may be inaccurate in a warming climate. The weak atmospheric large-scale circulation suggests that the oceanic heat transport would be weaker in the warm regime and that thus the absolute values of the q-flux are too large. To assess whether the imposed q-flux affects the results in the warm regime, a number of simulations without q-flux are performed. Despite small differences in the large-scale circulation (Supplementary Fig. 2), the warm steady states exist and are stable even without q-flux (Supplementary Fig. 3).

Treatment of ozone. If the tropopause climbs, regions with high ozone concentrations may come to lie in the troposphere of the model, because ozone concentrations are prescribed to climatological values. High ozone concentrations could, however, not occur in the presence of tropospheric water vapour concentrations. Therefore, we limit the tropospheric ozone concentrations to a volume mixing ratio of 1.5×10^{-7}. As a consequence, ozone is taken out of the atmosphere, if the tropopause rises to levels where the climatological values would exceed this limit. This process is reversible if the tropopause descends again.

Modifications to the grid-point physics. Our version of the model incorporates several changes to the grid-point physics, such that we obtain a more accurate representation of several physical processes in warm climates[21]. The grid-point physics include representation of surface exchange, turbulence and vertical diffusion[24,25], gravity-wave drag[26,27], radiative transfer and radiative heating[28,29], convection[30,31], cloud cover[32] and cloud microphysics[33]. In summary, these changes are the inclusion of the mass of water vapour when calculating the total pressure and the omission of all approximations where small specific humidities are assumed (as for example in the calculation of density). The pressure effects of water vapour are not considered for the horizontal transport. So, the model is in sorts a hybrid model, with water vapour adding to the total pressure for local effects but not so for the large-scale transport. A detailed description of the modified model thermodynamics can be found in the appendix of ref. 34. We will give here a short overview of the modified radiative transfer scheme as well as of the convection, cloud-cover and cloud-microphysical schemes.

Radiative transfer. The radiative transfer scheme has recently been described and evaluated extensively in ref. 21, but as a courtesy to the reader we will repeat some of the major features here. It is based on the Rapid Radiative Transfer Model[28,29], but includes some small modifications. It uses the correlated-k method to solve the radiative transfer equations in the two-stream approximation. The k-coefficients are calculated from the HITRAN (1996 and 2000) database using a line-by-line radiative transfer model[28,29]. The water vapour continuum is based on CKD_v2.4. The shortwave radiation spectrum is divided into 14 bands, and the longwave radiation spectrum is divided into 16 bands. Since the lookup-tables of the molecular absorption coefficients are designed for a limited range of temperatures only, an exponential extrapolation for temperatures up to 400 K for the longwave radiation scheme is performed. The same extrapolation scheme is also applied to the lookup-tables for the absorption coefficients of the water vapour self-broadened continuum in the shortwave radiation scheme, but the original linear extrapolation scheme is kept for all the other absorption coefficients. The lookup-tables for the bandwise spectrally integrated Planck function and the derivative thereof with respect to temperature have been extended to 400 K. Furthermore the water vapour self-broadened continuum is introduced in the upper atmosphere radiation calculations, to account for the increase in water vapour with increasing gST. The effect of pressure broadening by water vapour on the molecular absorption coefficients is neglected, as is scattering by water vapour. The thus modified radiation scheme is not as accurate as line-by-line radiative transfer models or models using recalculated k-coefficients for the higher temperatures, but still sufficiently accurate for the task at hand as has recently been demonstrated[21].

Convective scheme. ECHAM6 uses a mass-flux scheme for cumulus convection[30], with modifications for penetrative convection to the original scheme[31]. The contribution of cumulus convection to the large-scale budgets of heat, moisture and momentum is represented by an ensemble of clouds consisting of updrafts and downdrafts in a steady state. Depending on moisture convergence at the surface and depth of the convection cell, the model will either run in penetrative, mid-level, or shallow convection mode. The scheme allows for the formation of precipitation, but not for radiatively active convective clouds. Instead, detrained water is passed to the cloud-microphysical scheme which creates or destroys cloud condensate in a further step.

Cloud-cover scheme. ECHAM6 uses the Sundqvist scheme for fractional cloud cover[32]. This cloud scheme has been tuned for the use with present-day Earth's climate[33,35]. However, the scheme is well suited for simulations of cloud cover in warm climates as it diagnoses cloud cover directly from relative humidity, which should be crucial to cloud formation irrespective of the temperature.

Cloud-microphysical scheme. The cloud-microphysical scheme is described in detail in ref. 33. The scheme consists of prognostic equations for the vapour, liquid, and ice phases. There are explicit microphysics for warm-phase, mixed-phase and ice clouds. The cloud-condensation-nuclei concentration follows a prescribed vertical profile which is typical for the present-day Earth's maritime conditions. Since we have no estimate of the aerosol load in a hypothetical warm climate, we assume that this profile is also a reasonable choice for warmer climates. The microphysics do not require changes for the use of the scheme in warm climates, since cloud formation in a warm-phase cloud, in which potential changes may occur, is not directly temperature-dependent (at least not in the range of temperatures we consider).

Cloud-radiative interactions. Clouds are represented in the radiative transfer calculations, assuming the so-called maximum-random-overlap assumption. Under this assumption cloud layers are assumed to be maximally overlapping if they are adjacent to one another, and randomly overlapping if they are separated by a clear layer. The absorptivity of clouds depends on their combined optical depths, the gas in which they are embedded, and the interstitial aerosol. The microphysics to determine the optical properties of the cloud particles involve the liquid water and ice paths, cloud-drop radii, as well as liquid water and ice content. Cloud scattering is represented as a single-scattering albedo by assuming Mie scattering from cloud droplets in the shortwave calculations, but is neglected in the longwave calculations. Clouds are not considered in the radiative transfer routines if the specific mass of the cloud condensate does not exceed 10^{-7} kg per kg of air.

Special settings for high-temperature simulations. To run the model at high temperatures, a few special settings are necessary. The time step is reduced from 2,400 to 600 s, and the radiation time step is reduced from 7,200 to 2,400 s, except for the simulation with a TSI of 1.15 S_0, where the time step is reduced to 360 and the radiation time step to 1,440 s. Nonetheless, we sometimes encounter problems with resolved waves propagating to the top levels of the model, where their amplitude grows and where they may be reflected. To avoid frequent model failure due to these effects, we introduce Rayleigh friction to the vorticity and the divergence as well as increased horizontal diffusion in the top six layers (above ~ 0.75 hPa). The time constant of the Rayleigh friction is (10800 s) at the sixth layer and is increased by a factor of 3.2 per layer towards the top and is hence increased by a factor of $\sim 1{,}000$ in the top layer. The horizontal diffusion is increased by a factor of 3.2 per layer. The values of the time constant of the Rayleigh friction and the magnitude of horizontal diffusion are determined by trial and error. Since we investigate a large range of climates, it is difficult to find suitable values for these time constants, and despite these modifications the model fails occasionally. In these cases, however, the runs can be continued by slightly changing their trajectory, which is achieved by reducing the factor of multiplication per level for the Rayleigh friction from 3.2 to 3.199 for 30 days.

References

1. Kasting, J. F. Runaway and moist greenhouse atmospheres and the evolution of earth and venus. *Icarus* **74**, 472–494 (1988).
2. Kasting, J. F., Whitmire, D. P. & Reynolds, R. T. Habitable zones around main-sequence stars. *Icarus* **101**, 108–128 (1993).
3. Kopparapu, R. K. *et al.* Habitable zones around main-sequence stars: new estimates. *Astrophys. J.* **765**, 131 (2013).
4. Kasting, J. F., Pollack, J. B. & Ackerman, T. P. Response of Earth's atmosphere to increases in solar flux and implications for loss of water from venus. *Icarus* **57**, 335–355 (1984).
5. Wordsworth, R. D. & Pierrehumbert, R. T. Water loss from terrestrial planets with CO2-rich atmospheres. *Astrophys. J.* **778**, 154 (2013).

6. Ramirez, R. M., Kopparapu, R. K., Lindner, V. & Kasting, J. F. Can increased atmospheric CO2 levels trigger a runaway greenhouse? *Astrobiology* **14**, 714–731 (2014).
7. Abe, Y., Abe-Ouchi, A., Sleep, N. H. & Zahnle, K. J. Habitable zone limits for dry planets. *Astrobiology* **11**, 443–460 (2011).
8. Yang, J., Cowan, N. B. & Abbot, D. S. Stabilizing cloud feedback dramatically expands the habitable zone of tidally locked planets. *Astrophys. J. Lett.* **771**, L45 (2013).
9. Wolf, E. & Toon, O. The evolution of habitable climates under the brightening sun. *J. Geophys. Res. Atmos.* **120**, 5775–5794 (2015).
10. Ishiwatari, M., Takehiro, S., Nakajima, K. & Hayashi, Y. Y. A numerical study on appearance of the runaway greenhouse state of a three-dimensional gray atmosphere. *J. Atmos. Sci.* **59**, 3223–3238 (2002).
11. Ishiwatari, M., Nakajima, K., Takehiro, S. & Hayashi, Y.-Y. Dependence of climate states of gray atmosphere on solar constant: From the runaway greenhouse to the snowball states. *J. Geophys. Res.* **112**, D13120 (2007).
12. Leconte, J., Forget, F., Charnay, B., Wordsworth, R. & Pottier, A. Increased insolation threshold for runaway greenhouse processes on earth-like planets. *Nature* **504**, 268–271 (2013).
13. Boer, G. J., Hamilton, K. & Zhu, W. Climate sensitivity and climate change under strong forcing. *Clim. Dyn.* **24**, 685–700 (2005).
14. Heinemann, M. Warm and sensitive paleocene-eocene climate. *Reports on Earth System Science 70/2009* (Max-Planck-Institute for Meteorology, 2009).
15. Wolf, E. T. & Toon, O. Delayed onset of runaway and moist greenhouse climates for earth. *Geophys. Res. Lett.* **41**, 167–172 (2014).
16. Hansen, J. *et al.* Efficacy of climate forcings. *J. Geophys. Res.* **110**, D18104 (2005).
17. Meraner, K., Mauritsen, T. & Voigt, A. Robust increase in equilibrium climate sensitivity under global warming. *Geophys. Res. Lett.* **40**, 5944–5948 (2013).
18. Wolf, E. T. & Toon, O. B. Hospitable archean climates simulated by a general circulation model. *Astrobiology* **13**, 656–673 (2013).
19. Bloch-Johnson, J., Pierrehumbert, R. T. & Abbot, D. S. Feedback temperature dependence determines the risk of high warming. *Geophys. Res. Lett.* **42**, 4973–4980 (2015).
20. Russell, G. L., Lacis, A. A., Rind, D. H. & Colose, C. Fast atmosphere-ocean model runs with large changes in CO2. *Geophys. Res. Lett.* **40**, 5787–5792 (2013).
21. Popp, M., Schmidt, H. & Marotzke, J. Initiation of a runaway greenhouse in a cloudy column. *J. Atmos. Sci.* **72**, 452–471 (2015).
22. Stevens, B. *et al.* The atmospheric component of the MPI-M earth system model: ECHAM6. *J. Adv. Model. Earth Syst.* **5**, 146–172 (2013).
23. Medeiros, B., Stevens, B. & Bony, S. Using aquaplanets to understand the robust responses of comprehensive climate models to forcing. *Clim. Dyn.* **44**, 1957–1977 (2015).
24. Brinkop, S. & Roeckner, E. Sensitivity of a general circulation model to parameterizations of cloudturbulence interactions in the atmospheric boundary layer. *Tellus A* **47**, 197–220 (1995).
25. Giorgetta, M. A. *et al.* The atmospheric general circulation model ECHAM6 - model description. *Technical Report 135* (Max-Planck-Institute for Meteorology, 2013).
26. Hines, C. O. Doppler-spread parameterization of gravity-wave momentum deposition in the middle atmosphere. Part 1: basic formulation. *J. Atmos. Sol.-Terr. Phys.* **59**, 371–386 (1997).
27. Hines, C. O. Doppler-spread parameterization of gravity-wave momentum deposition in the middle atmosphere. Part 2: Broad and quasi mono-chromatic spectra, and implementation. *J. Atmos. Sol.-Terr. Phys.* **59**, 387–400 (1997).
28. Mlawer, E. J., Taubman, S. J., Brown, P. D., Iacono, M. J. & Clough, S. A. Radiative transfer for inhomogeneous atmospheres: RRTM, a validated correlated-k model for the longwave. *J. Geophys. Res.* **102**, 16663–16682 (1997).
29. Iacono, M. J. *et al.* Radiative forcing by long-lived greenhouse gases: calculations with the AER radiative transfer models. *J. Geophys. Res.* **113**, D13103 (2008).
30. Tiedtke, M. A comprehensive mass flux scheme for cumulus parameterization in large-scale models. *Mon. Wea. Rev.* **117**, 1779–1800 (1989).
31. Nordeng, T.-E. Extended versions of the convective parametrization scheme at ECMWF and their impact on the mean and transient activity of the model in the tropics. *Technical Memorandum 206, ECMWF* (European Centre for Medium-Range Weather Forecasts, Shinfield Park, Reading, RG2 9AX, UK, 1994).
32. Sundqvist, H., Berge, E. & Kristjansson, J. E. Condensation and cloud parameterization studies with a mesoscale numerical weather prediction model. *Mon. Weather Rev.* **117**, 1641–1657 (1989).
33. Lohmann, U. & Roeckner, E. Design and performance of a new cloud microphysics scheme developed for the ECHAM general circulation model. *Clim. Dyn.* **12**, 557–572 (1996).

34. Popp, M. Climate instabilities under strong solar forcing. *Reports on Earth System Science 152/2014* (Max-Planck-Institute for Meteorology, 2014).

35. Mauritsen, T. *et al.* Tuning the climate of a global model. *J. Adv. Model. Earth Syst.* **4,** M00A01 (2012).

Acknowledgements

We thank Thorsten Mauritsen for a thorough internal review. We thank Eric Wolf for an insightful discussion and for making the data from CAM4 available to us. We thank Jérémy Leconte for making the data from LMDG available to us. We thank Dorian Abbot and two anonymous reviewers for very constructive reviews of this manuscript. We thank the Max Planck Society for the Advancement of Science for financial support.

Author contributions

J.M., H.S. and M.P. conceived the study. M.P. performed the changes to the model and conducted the modelling work. J.M., H.S. and M.P. all contributed substantially to the analysis and the discussion. M.P. wrote the manuscript with comments from J.M. and H.S.

Additional information

13

Early aqueous activity on the ordinary and carbonaceous chondrite parent bodies recorded by fayalite

Patricia M. Doyle[1,2,†], Kaori Jogo[1,2,†], Kazuhide Nagashima[1], Alexander N. Krot[1,2], Shigeru Wakita[3], Fred J. Ciesla[4] & Ian D. Hutcheon[5,‡]

Chronology of aqueous activity on chondrite parent bodies constrains their accretion times and thermal histories. Radiometric ^{53}Mn-^{53}Cr dating has been successfully applied to aqueously formed carbonates in CM carbonaceous chondrites. Owing to the absence of carbonates in ordinary (H, L and LL), and CV and CO carbonaceous chondrites, and the lack of proper standards, there are no reliable ages of aqueous activity on their parent bodies. Here we report the first ^{53}Mn-^{53}Cr ages of aqueously formed fayalite in the L3 chondrite Elephant Moraine 90161 as $2.4^{+1.8}_{-1.3}$ Myr after calcium-aluminium-rich inclusions (CAIs), the oldest Solar System solids. In addition, measurements using our synthesized fayalite standard show that fayalite in the CV3 chondrite Asuka 881317 and CO3-like chondrite MacAlpine Hills 88107 formed $4.2^{+0.8}_{-0.7}$ and $5.1^{+0.5}_{-0.4}$ Myr after CAIs, respectively. Thermal modelling, combined with the inferred conditions (temperature and water/rock ratio) and ^{53}Mn-^{53}Cr ages of aqueous alteration, suggests accretion of the L, CV and CO parent bodies $\sim$1.8 − 2.5 Myr after CAIs.

[1] Hawai'i Institute of Geophysics and Planetology, University of Hawai'i at Mānoa, Pacific Ocean Science & Technology (POST) Building, 1680 East-West Road, Honolulu, Hawai'i 96822, USA. [2] University of Hawai'i NASA Astrobiology Institute, Honolulu, Hawai'i 96822, USA. [3] Center for Computational Astrophysics, National Astronomical Observatory of Japan, 2-21-1 Osawa, Mitaka, Tokyo 181-8588, Japan. [4] Department of the Geophysical Sciences, University of Chicago, 5734 South Ellis Avenue, Chicago, Illinois 60637, USA. [5] Glenn Seaborg Institute, Lawrence Livermore National Laboratory, L-231, Livermore, California 94551, USA. † Present addresses: Department of Geological Sciences, University of Cape Town, Rondebosch 7701, South Africa (P.M.D.); Division of Polar Earth-System Sciences, Korea Polar Research Institute, 26 Songdomirae-ro, Yeonsu-gu, Incheon 406-840, Korea. (K.J.). ‡ Deceased. Correspondence and requests for materials should be addressed to P.M.D. (email: pdoyle@higp.hawaii.edu) or to A.N.K. (sasha@higp.hawaii.edu).

Most chondritic meteorites (chondrites) experienced aqueous alteration resulting in the formation of a diverse suite of secondary minerals, including phyllosilicates, magnetite ($FeFe_2O_4$), sulfides, carbonates (calcite ($CaCO_3$), dolomite ($CaMg(CO_3)_2$), breunnerite ($(Mg,Fe,Mn)CO_3$) and siderite ($FeCO_3$)), ferromagnesian olivines ($(Fe,Mg)_2SiO_4$) and Ca-rich pyroxenes ($Ca(Fe,Mg)Si_2O_6$; ref. 1). Mineralogical observations, isotopic data and thermodynamic analysis suggest that the alteration resulted from interactions between a rock and an aqueous solution in an asteroidal setting[2]. Therefore, dating minerals formed by aqueous alteration provides important constraints on the accretion ages of chondrite parent bodies. The chondrite accretion ages, the conditions of aqueous alteration (temperature and water/rock ratio), and the inferred hydrogen and oxygen isotopic compositions of chondritic water[3,4] can potentially be used to constrain chondrite accretion regions and to test the recently proposed Grand Tack dynamical model of the early Solar System evolution[5]. According to this model, the hydrated P-type, D-type and taxonomic C-complex asteroids, which are commonly associated with carbonaceous chondrites[6], accreted outside Jupiter's orbit (5–15 AU from the Sun; 1 AU is the distance between the Sun and the Earth), and were scattered and implanted into the main asteroid belt (located between 2 and 4 AU from the Sun) during the migration of Jupiter and Saturn within the first several million years of the Solar System formation.

The short-lived radionuclide ^{53}Mn, which decays to ^{53}Cr with a half-life of ~ 3.7 Myr and appears to have been uniformly distributed in the protoplanetary disk[7], is known to be a useful tool for dating aqueous alteration[1]. As yet, accurate ^{53}Mn–^{53}Cr ages of aqueous alteration are only known for the CM (Mighei-like) carbonaceous chondrites[8] containing aqueously formed Mn-rich, Cr-poor calcite suitable for *in situ* radiometric ^{53}Mn–^{53}Cr dating with secondary ion mass spectrometry (SIMS). SIMS measurements of manganese-chromium (Mn–Cr) isotope systematics of a mineral require a proper standard to determine a relative sensitivity factor ($RSF = (^{55}Mn^+/^{52}Cr^+)_{SIMS}/(^{55}Mn/^{52}Cr)_{mineral}$) to correct for the relative sensitivities between $^{55}Mn^+$ and $^{52}Cr^+$ ions and to calculate true $^{55}Mn/^{52}Cr$ ratios in the mineral. The initial ^{53}Mn/^{55}Mn ratio, $(^{53}Mn/^{55}Mn)_0$, of calcites in CM chondrites has been determined by using an RSF measured on a synthetic Mn- and Cr-doped calcite standard[8]. The inferred initial ^{53}Mn/^{55}Mn ratios, ranging from $\sim 2.7 \times 10^{-6}$ to $\sim 3.4 \times 10^{-6}$, correspond to calcite formation ages of $\sim 4 - 5$ Myr after the formation of calcium–aluminium-rich inclusions (CAIs), the oldest Solar System solids dated[9]. Carbonates, however, are virtually absent in weakly aqueously altered ordinary (H, L and LL) and several carbonaceous chondrite groups, including CV (Vigarano-like) and CO (Ornans-like). Instead, the least metamorphosed meteorites of these groups contain aqueously formed Mn-rich, Cr-poor fayalite ($Fa_{>90}$; Fayalite (Fa) number = atomic $Fe/(Fe + Mg) \times 100$ in olivine) suitable for *in situ* Mn–Cr isotope dating with SIMS[10–12]. As for carbonates, SIMS measurements of ^{53}Mn–^{53}Cr systematics in fayalite require a proper standard to determine the RSF and calculate $^{55}Mn/^{52}Cr$ ratios. Because natural fayalite contains virtually no chromium, it cannot be used as a standard for Mn-Cr isotope measurements; San Carlos olivine (Fa_{10}) was typically used instead[10–12]. However, it has been recently discovered that the RSF in olivine changes as a function of its fayalite content[13,14]. Therefore, all previously published ^{53}Mn–^{53}Cr ages of CV and CO chondritic fayalite, acquired with a San Carlos olivine standard[10–12], need to be corrected.

To obtain accurate ^{53}Mn–^{53}Cr ages of fayalite, a Mn- and Cr-doped fayalite (Fa_{99}) standard was synthesized (ref. 14 and Doyle *et al.*, manuscript in preparation). Here we report on the mineralogy, petrography and oxygen-isotope compositions of the fayalite-bearing assemblages and ^{53}Mn–^{53}Cr ages of fayalite in type 3 L, LL, CO and CV chondrites. Our data confirm the origin of fayalite during low-temperature aqueous alteration on the L, CV and CO chondrite parent bodies[10–12], and allow us to constrain the accretion ages and the accretion regions of these bodies.

Results

Petrographic evidence for *in situ* formation of fayalite. Nearly pure fayalite (Fa_{90-100}; Supplementary Table 1) was found in the aqueously altered ordinary (Semarkona (LL3.00), Elephant Moraine (EET) 90161 (L3.05), Meteorite Hills (MET) 00452 (L/LL3.05), MET 96503 (L3.1–3.2), Ngawi (LL3–6)) and carbonaceous (MacAlpine Hills (MAC) 88107 (CO3-like) and Asuka (A) 881317 (CV3)) chondrites. There are three major textural occurrences of fayalite in these meteorites: (1) fayalite with tiny inclusions of hedenbergite ($CaFeSi_2O_6$) and magnetite overgrows onto ferromagnesian chondrule olivines (Fa_{1-30}; Fig. 1a,b). There are sharp compositional boundaries between the fayalite and the chondrule olivines. These observations are consistent with low-temperature formation of the fayalite overgrowths and the lack of subsequent thermal metamorphism above ~ 300–$400\,^{\circ}$C that would have resulted in Fe–Mg interdiffusion between the adjacent fayalite and ferromagnesian olivines. (2) In the hydrated chondrite matrices, fayalite associates with hedenbergite, magnetite and Fe,Ni-sulfides, and often forms euhedral crystals that show no evidence for corrosion or replacement by the surrounding phyllosilicates. These observations indicate a local equilibrium between the fayalite and the phyllosilicates (Fig. 1c). In addition, in matrices of the ordinary chondrites (OCs) EET 90161, MET 96503 and MET 00452, fayalite replaces amorphous ferromagnesian silicates (Fig. 1a). (3) Fine-grained matrix-like rims around magnesian porphyritic chondrules in MAC 88107 and A-881317 are commonly crosscut by veins composed of fayalite, hedenbergite and magnetite (Fig. 1d), indicating *in situ* crystallization of this mineral paragenesis.

Oxygen isotopic compositions. Oxygen isotopic compositions of fayalite, magnetite, chondrule olivines and the bulk compositions of type 3 ordinary, CV and CO chondrites are plotted on a three-isotope oxygen diagram, $\delta^{17}O$ versus $\delta^{18}O$ (Fig. 2a and Supplementary Table 2). In Fig. 2b, the same data are plotted as deviation from the terrestrial fractionation line, $\Delta^{17}O = \delta^{17}O - 0.52 \times \delta^{18}O$.

Oxygen isotopic compositions of fayalite and magnetite in type 3 OCs plot along a mass-dependent fractionation line with a slope of ~ 0.5 (Fig. 2a). The compositions of fayalite and magnetite, having an average $\Delta^{17}O$ value of $+4.8 \pm 1.7\permil$ (2 s.d.), are significantly different from those of chondrule olivines[15] and bulk OCs[16], which both have $\Delta^{17}O$ values of $\sim +1\permil$ (Fig. 2b).

Oxygen isotopic compositions of fayalite and magnetite in A-881317 (CV3) plot along a mass-dependent fractionation line with a slope of ~ 0.5 (Fig. 2a) and have an average $\Delta^{17}O$ value of $-0.4 \pm 0.9\permil$ (Fig. 2b), consistent with the previously published data for fayalite and magnetite from the Kaba and Mokoia CV3 chondrites[17]. The bulk oxygen isotopic compositions of CV chondrites[18] and CV chondrule olivines[19] plot along the carbonaceous chondrite anhydrous mineral (CCAM) line with a slope of ~ 1 (Fig. 2a) and show a range of $\Delta^{17}O$ values from -6 to $-2\permil$ (Fig. 2b).

Oxygen isotopic compositions of fayalite and magnetite in MAC 88107 (CO3-like) plot along a mass-dependent fractionation line with a slope of ~ 0.5 (Fig. 2a) and have an average $\Delta^{17}O$ value of $-1.6 \pm 0.9\permil$ (Fig. 2b), whereas chondrule olivines plot

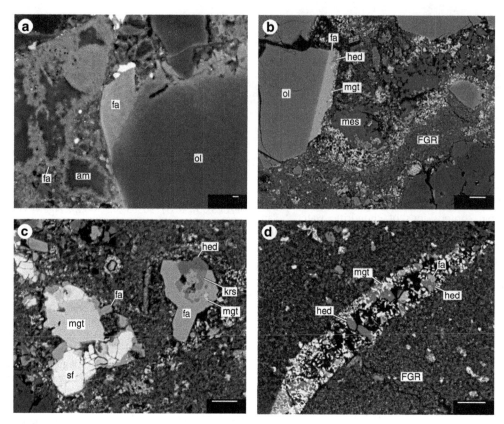

Figure 1 | Textural occurrences of fayalite-bearing assemblages in ordinary and carbonaceous chondrites. Backscattered electron images of fayalite-bearing assemblages in (**a**) ordinary and (**b** − **d**) carbonaceous chondrites. Shown are (**a**) two occurrences of fayalite (fa) in the OC EET 90161 (L3.05): fayalite overgrows onto an Mg-rich chondrule olivine (ol) and replaces amorphous (am) material; and (**b** − **d**) three occurrences of fayalite-hedenbergite (hed)-magnetite (mgt) assemblages in the CO3-like carbonaceous chondrite MAC 88107. (**b**) Ferroan porphyritic chondrule (type II) surrounded by a fine-grained rim (FGR). Chondrule mesostasis (mes) is replaced by phyllosilicates and partially leached out. The olivine phenocryst at the boundary with the mesostasis is overgrown by fayalite with inclusions of hedenbergite and magnetite. (**c**) Fe,Ni-sulfide (sf)-magnetite-fayalite-hedenbergite-kirschsteinite (krs) particles in the hydrated (phyllosilicate-bearing) matrix. (**d**) Fine-grained rim around magnesian porphyritic chondrule (type I) is crosscut by a vein composed of magnetite, fayalite and hedenbergite. Scale bars, 1 μm (**a**) and 10 μm (**b** − **d**).

along the CCAM line and show a range of $\Delta^{17}O$ values from -9 to -2‰ (Fig. 2b). The bulk oxygen isotopic composition of MAC 88107 also plots along the CCAM line and has a $\Delta^{17}O$ value of -4.8‰ (ref. 18).

^{53}Mn–^{53}Cr ages of fayalite formation. To obtain an olivine standard suitable for SIMS measurements of Mn–Cr isotope systematics of chondritic fayalite, we synthesized a suite of ferromagnesian olivines ranging from Fa_{31} to Fa_{99} (ref. 14; see Methods section and Supplementary Table 3). Our Mn–Cr isotope measurements of the synthesized olivine grains with the University of Hawai'i (UH) ims-1280 SIMS revealed that the RSF strongly depends on the fayalite content in olivine[14]. In olivine with Fa_{10-30}, the RSF increases from ~0.9 to ~1.5 as a function of fayalite content; in more ferroan olivines ($Fa_{>30}$), the RSF is nearly constant, ~1.6 (Fig. 3 and Supplementary Table 4). The observed differences in RSF values between fayalite (Fa_{90-100}) and San Carlos olivine (Fa_{10}) suggest that all previously reported ^{53}Mn–^{53}Cr dating of fayalite in carbonaceous chondrites[10–12] are in error and yield systematically young ages[14].

With the matrix-matched standard (Fa_{99}), we measured Mn–Cr isotope systematics of fayalite in EET 90161, A-881317 and MAC 88107 (Table 1 and Fig. 4). The excess of ^{53}Cr in fayalite from these meteorites correlates with the $^{55}Mn/^{52}Cr$ ratio, indicating *in situ* decay of ^{53}Mn. The inferred initial $^{53}Mn/^{55}Mn$ ratios in fayalite from EET 90161, A-881317 and

MAC 88107 are $(4.35 \pm 1.23) \times 10^{-6}$, $(3.07 \pm 0.44) \times 10^{-6}$ and $(2.58 \pm 0.21) \times 10^{-6}$, respectively.

The ^{53}Mn–^{53}Cr chronometer provides a relative chronology, that is, the ^{53}Mn–^{53}Cr age of mineral formation is given relative to a reference material: $\Delta t_{\text{mineral-reference}} = 1/\lambda 53_{Mn} \times \ln[(^{53}Mn/^{55}Mn)_{\text{reference}}/(^{53}Mn/^{55}Mn)_{\text{mineral}}]$, where $\lambda 53_{Mn}$ is the decay constant of ^{53}Mn. In cosmochemistry, the commonly used reference material is CV CAIs, which are the oldest Solar System solids, with the U-corrected Pb–Pb absolute age of $4,567.30 \pm 0.16$ Ma (ref. 9) that is considered to represent the age of the Solar System. The initial $^{53}Mn/^{55}Mn$ ratio in CV CAIs has, however, not been measured owing to a lack of primary CAI minerals having high Mn/Cr ratios and the post-crystallization disturbance of their Mn–Cr isotope systematics[20]. Instead, the initial $^{53}Mn/^{55}Mn$ ratio in CV CAIs can be calculated from the difference in the U-corrected Pb–Pb ages of CAIs and the U-corrected Pb–Pb ages angrites (basaltic meteorites) with the measured initial $^{53}Mn/^{55}Mn$ ratio (Supplementary Methods). Using the measured $(^{53}Mn/^{55}Mn)_0$ in the D'Orbigny angrite of $(3.24 \pm 0.04) \times 10^{-6}$ (ref. 21), its U-corrected Pb–Pb absolute age of $4,563.4 \pm 0.3$ Ma (ref. 22) and the U-corrected Pb–Pb absolute age of CV CAIs[9], the estimated $(^{53}Mn/^{55}Mn)_0$ ratio in the CV CAIs is ~6.8×10^{-6}. Therefore, the inferred $(^{53}Mn/^{55}Mn)_0$ in fayalites from EET 90161, A-881317 and MAC 88107 correspond to the formation ages of $2.4^{+1.8}_{-1.3}$, $4.2^{+0.8}_{-0.7}$ and $5.1^{+0.5}_{-0.4}$ Myr after the CV CAIs, respectively (Fig. 5 and Supplementary Table 5).

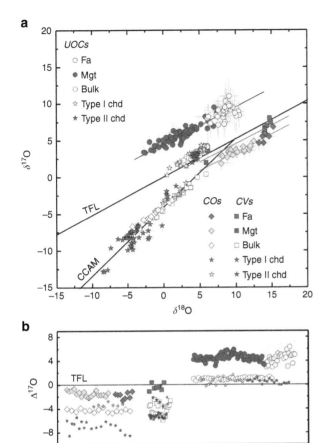

Figure 2 | Oxygen isotopic compositions of fayalite-bearing assemblages and chondrule olivines. (**a**) Three-isotope oxygen diagram (where $\delta^{17,18}O = [(^{17,18}O/^{16}O_{sample})/(^{17,18}O/^{16}O_{SMOW}) - 1] \times 1,000$ and SMOW is standard mean ocean water) showing the compositions of fayalite (Fa), magnetite (Mgt), chondrule (chd) olivine fragments, and bulk compositions for EET 90161 and other unequilibrated OCs (UOCs), A-881317 (CV3), MAC 88107 (CO3-like) and other CO and CV carbonaceous chondrites. The terrestrial fractionation line (TFL) and CCAM line are shown for reference. (**b**) $\Delta^{17}O$ values of fayalite, magnetite, chondrule fragments and bulk compositions of UOCs, CO and CV chondrites. The data are plotted arbitrarily on the x axis to show individual compositions. The difference in $\Delta^{17}O$ observed between chondrule olivines and the secondary minerals (fayalite and magnetite) indicate that they are in isotopic disequilibrium, which is consistent with the low-temperature formation of fayalite and magnetite. Uncertainties shown are 2σ. Fayalite and magnetite data are from EET 90161, Semarkona (LL3.00), MET 00452 (L/LL3.05) and Ngawi (L3-6 breccia) (Supplementary Table 2). Bulk meteorite and chondrule compositions from refs 15–19.

Thermal modelling of L, CO and CV chondrite parent bodies. Mineral thermometry suggests that L, CV and CO chondrite parent bodies reached peak metamorphic temperatures of ~950, ~600 and ~600 °C, respectively[23]. Thermal evolution of the L-, CV- and CO-like bodies with different radii heated only by decay of the short-lived radionuclide ^{26}Al, having a half-life of ~0.7 Myr, is modelled with reference to the peak metamorphic temperatures and the ages of fayalite formation in the L, CV and CO chondrites (see Methods section).

Our calculations indicate that L chondrite-like bodies with radii of 20 − 40 km need to accrete within ~1.6–1.8 Myr after CV CAIs, respectively, in order to reach a peak metamorphic temperature of 950 °C (Fig. 6a). The CV and CO chondrite-like bodies with radii of 20–50 km need to accrete within ~2.4–2.6 Myr and ~2.1–2.4 Myr after CV CAIs, respectively,

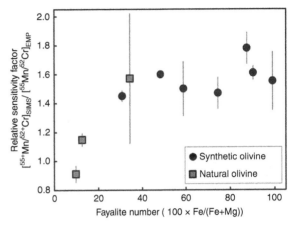

Figure 3 | Mn-Cr RSF in ferromagnesian olivines. Mn–Cr RSF, determined with the UH Cameca ims-1280 using a primary beam ~5 μm in diameter, as a function of fayalite number. The RSFs reported for samples Fa12-90 are an average of 2–7 analyses. Four to five analyses were collected daily from either Fa99 or Fa10, and the RSFs reported are an average of 18 and 20 analyses, respectively. The RSFs for Fa10 and Fa99 collected during individual sessions are shown in Doyle *et al.* (manuscript in preparation) and described in the Supplementary Discussion. Uncertainties are 2 s.d., with uncertainty in Fa number smaller than the symbols.

to reach a peak metamorphic temperature of 600 °C (Fig. 6b). The CO chondrites have lower bulk aluminium contents than the CV chondrites (Supplementary Table 6). As a result, the CO chondrite parent body heated by decay of ^{26}Al must have accreted earlier than the CV chondrite parent body in order to reach the same peak metamorphic temperature. We note that owing to the balance between the heat generated by decay of ^{26}Al and heat lost through radiation, the maximum temperature reached by a body larger than 30 km in radius largely depends on its accretion time and not its size (Fig. 6a,b).

Temperature evolution at different depths of the L, CV and CO chondrite-like bodies with different sizes and accretion ages are shown in Fig. 6c–f. Regions ~26.5 and ~36.5 km from the centre of L chondrite-like bodies with radii of 30 and 40 km, respectively, would have conditions (100 − 200 °C) suitable for the formation of fayalite (Fig. 6c,e). In a body 30 km in radius, the fayalite-forming region would not have been heated above 300 °C, and would have avoided Fe–Mg diffusion between fayalite and ferromagnesian olivines, consistent with our observations of fayalites in EET 90161. In a body 40 km in radius, the fayalite-forming region would have reached up to 350 °C ~0.5 Myr after fayalite formation, which would have resulted in some Fe–Mg interdiffusion in fayalite. The CV and CO chondrite-like bodies, each with a 50-km radius, have regions ~43 − 47 km from the centre of the bodies, which experienced metamorphic temperatures in the range of 100 − 300 °C (Fig. 6d,f). These calculations indicate that secondary fayalite is likely to have formed and survived only near the peripheral parts of the L, CV and CO chondrite parent bodies.

Discussion

The oxygen-isotope systematics of fayalite and magnetite in type 3 ordinary and carbonaceous chondrites indicate that these minerals are in isotopic disequilibrium with the chondrule olivines, precluding a high-temperature origin of fayalite and magnetite during chondrule formation[24]. Instead, the large differences in $\delta^{18}O$ values between the coexisting fayalite and magnetite (up to 10‰ in A-881317 and up to 8‰ in MAC 88107)

Table 1 | Mn–Cr isotope compositions of fayalite in EET 90161 (L3.05), MAC 88107 (CO3-like) and A-881317 (CV3).

Chondrite	Sample	Spot no.	$^{55}Mn/^{52}Cr$	2σ	$\delta^{53}Cr$ (‰)	2σ	$^{53}Cr/^{52}Cr$	2σ
EET 90161 (L3.05)	23-20b	1	4,090	660	124	99	0.128	0.011
	20-3d	1	4,175	253	169	80	0.133	0.009
	13-x	1	2,188	144	88	115	0.123	0.013
	10-58	2	39	1	1	8	0.114	0.001
	10-58	3	590	24	26	76	0.116	0.009
	27-40b	1	169	10	28	22	0.117	0.003
	28-46	2	693	39	22	34	0.116	0.004
	20-3	2	2,418	372	100	64	0.125	0.007
	23-23	1	883	129	27	43	0.116	0.005
	7-48	1	1,123	71	31	31	0.117	0.003
	3-18	1	99	5	2	13	0.114	0.002
	9-52	2	1,386	202	68	30	0.121	0.003
MAC 88107 (CO3)	45b	1	10,074	955	179	68	0.134	0.008
	63a	1	58,122	7,041	1,157	248	0.245	0.028
	63a	2	48,964	7,404	1,066	278	0.234	0.032
	82	1	4,903	349	108	43	0.126	0.005
	82	2	296	10	12	9	0.115	0.001
	82	3	190	6	2	8	0.114	0.001
	63b	1	30,154	3,709	710	174	0.194	0.020
	63b	2	33,955	4,017	677	197	0.190	0.022
	73	1	26,320	1,781	630	138	0.185	0.016
	73	2	25,520	2,314	643	139	0.186	0.016
	54	1	8,799	1,370	195	56	0.136	0.006
	54	2	16,497	2,924	504	108	0.171	0.012
	53	1	29,181	5,661	622	150	0.184	0.017
	53	2	26,666	4,795	631	164	0.185	0.019
	70	1	20,630	3,653	524	125	0.173	0.014
A-881317 (CV3)	1-3	1	10,533	4,207	302	104	0.148	0.012
	1-3	2	16,185	6,348	373	127	0.156	0.014
	1-3	3	10	4	−5	7	0.113	0.001
	1-3	4	4,225	1,636	121	53	0.127	0.006
	1_3	5	20,428	7,584	529	141	0.174	0.016
	4-1	1	12,278	4,810	334	83	0.151	0.009
	4-1	2	8,387	3,285	279	71	0.145	0.008
	4-1	3	11,159	4,382	268	107	0.144	0.012
	3-1	1	14,676	5,703	454	99	0.165	0.011
	3-1	2	14,372	5,586	352	91	0.153	0.010
	3-1	3	7,089	2,749	161	59	0.132	0.007
	6_2	1	3,037	1,087	43	33	0.118	0.004
	6_2	2	22,706	8,287	632	136	0.185	0.015

A-881317, Asuka 881317; EET, Elephant Moraine; MAC, MacAlpine Hills.

are consistent with a low-temperature origin of these minerals in equilibrium with an aqueous solution[25,26].

The mineralogy, petrography and oxygen isotopic compositions, therefore, support the formation of fayalite-bearing assemblages in type 3 ordinary, CO and CV chondrites during fluid–rock interactions on their respective parent bodies[1,10–12]. Thermodynamic analysis of the gas–solution–rock system[27,28] shows that nearly pure fayalite ($Fa_{>90}$) is stable at low temperatures ($\sim 100 - 200\,^{\circ}C$) and low water/rock mass ratios ($\sim 0.1 - 0.2$), and, therefore, these conditions are inferred for aqueous alteration on the ordinary, CV and CO chondrite parent bodies. These alteration conditions are different from those recorded by the more extensively aqueously altered but less metamorphosed CM and CI (Ivuna-like) carbonaceous chondrites, which have experienced alteration at temperatures of ~ 20–$80\,^{\circ}C$ (refs 18,29) and < 150–$210\,^{\circ}C$ (refs 18,30), and water/rock mass ratios of $\sim 0.3 - 0.6$ and > 0.8 (ref. 18), respectively. The CI and CM chondrites contain abundant phyllosilicates, carbonates and magnetite, but lack fayalite, which is unstable under these alteration conditions[28].

The formation ages of fayalite in L, CO and CV chondrites and the conditions of fayalite stability[28] can be linked to the metamorphic histories and the accretion ages of their parent bodies (Fig. 6). The L chondrites define a metamorphic sequence of petrologic types from 3 to 6, and reached peak metamorphic temperatures of $\sim 950\,^{\circ}C$ (ref. 23). In contrast, the CV and CO chondrites define a narrower range of petrologic types, 3.1 to > 3.6 and 3.0 to 3.8, respectively, and reached peak metamorphic temperatures of $\sim 600\,^{\circ}C$ (refs 23,31,32). As it is not known how well the CV and CO chondrite parent bodies are sampled by the recovered meteorites, the estimated peak metamorphic temperatures provide only a lower limit on the maximum temperatures experienced by these bodies.

It is generally accepted that ^{26}Al was the major heat source responsible for thermal metamorphism and aqueous alteration of early accreted planetesimilas[33]. Assuming ^{26}Al was homogeneously distributed in the protoplanetary disk with an initial abundance corresponding to the canonical $^{26}Al/^{27}Al$ ratio of 5.25×10^{-5} (refs 34–36), we modelled the thermal evolution of several chondrite parent bodies accreting 1.6–4.0 Myr after CV

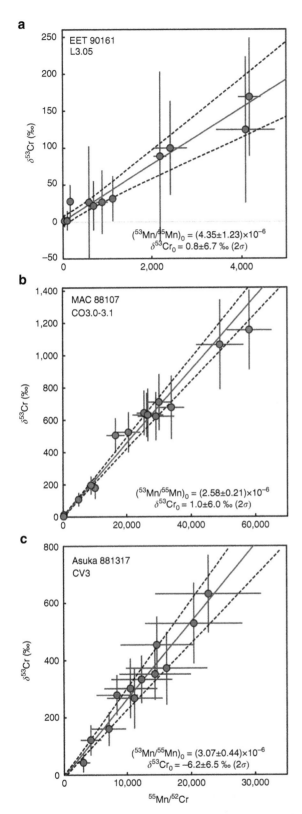

Figure 4 | Mn-Cr evolutionary diagrams of fayalite in ordinary and carbonaceous chondrites. Mn-Cr evolutionary diagrams of fayalite in (**a**) EET 90161, (**b**) MAC 88107 and (**c**) A-881317. $\delta^{53}Cr = [(^{53}Cr/^{52}Cr)_{fayalite}/(^{53}Cr/^{52}Cr)_{terrestrial})-1] \times 1,000$, where $(^{53}Cr/^{52}Cr)_{terrestrial} = 0.113459$. Excesses of ^{53}Cr in the EET 90161, MAC 88107 and A-881317 fayalites are well correlated with $^{55}Mn/^{52}Cr$, indicative of *in situ* decay of ^{53}Mn to ^{53}Cr. The initial $(^{53}Mn/^{55}Mn)_0$ and $^{53}Cr/^{52}Cr$ ratios ($\delta^{53}Cr_0$) are indicated in the diagrams. Uncertainties shown are 2σ.

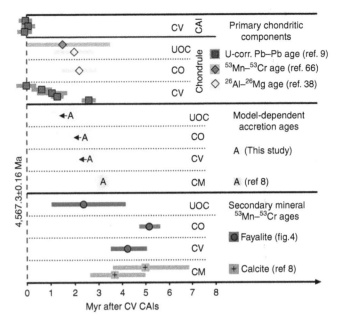

Figure 5 | Chronology of aqueous alteration and accretion of ordinary and carbonaceous chondrite parent bodies. Chronology of aqueous alteration of unequilibrated OC (UOC: L and LL) and carbonaceous (CO, CV and CM) chondrite parent bodies. Relative ages of fayalite in UOC, CO and CV chondrites, and calcites in CM chondrites[8] are anchored to the D'Orbigny angrite and shown relative to the age of CV CAIs[9] (Methods section). Also shown are the model accretion ages of UOC, CO and CV (this study), and CM ([8]) chondrites, and U-corrected Pb-Pb ([9]), ^{53}Mn-^{53}Cr (ref. 66) and ^{26}Al-^{26}Mg (ref. 38) ages of chondrules in CV, UOC and CO chondrites, relative to the CV CAIs[9].

CAIs (Supplementary Table 6). Numerical modelling of an L chondrite-like body with a radius of ~ 30–$40\,km$ (Fig. 6c,e) suggests that this body accreted with an initial $^{26}Al/^{27}Al$ ratio of $\sim 9 \times 10^{-6}$, corresponding to $\sim 1.8\,Myr$ after CV CAIs. Fayalite could have precipitated $\sim 0.6\,Myr$ later in the outer portion of this body, which may never have experienced temperatures above $300\,°C$ (Fig. 6c,e). Similarly, the CV and CO chondrite-like bodies with radii of $50\,km$ (Fig. 6d,f) accreted with an $^{26}Al/^{27}Al$ ratio of $\sim 4 \times 10^{-6}$ corresponding to $\sim 2.5\,Myr$ after CV CAIs. Fayalite could have precipitated ~ 1.5–$2.5\,Myr$ later in the outer portions of these bodies, which may never have experienced temperatures above $300\,°C$.

The modelled accretion age of the CV chondrite parent body is in agreement with the ^{26}Al-^{26}Mg formation age of $> 2.6\,Myr$ for secondary grossular in the CAIs from the CV3 Allende chondrite[37]. The modelled accretion ages of the L and CO chondrite parent bodies are consistent with the ^{26}Al-^{26}Mg ages of chondrules from the type 3 ordinary and CO chondrites, 1.98 ± 0.82 and $2.19 \pm 0.72\,Myr$ after CV CAIs (ref. 38 and references therein), respectively, but require rapid accretion of chondrules after their formation into their parent bodies. The rapid accretion of the chondrite parent bodies after chondrule formation may explain the mineralogical, chemical and isotopic differences between the chondrite groups[39]. The short time interval between the accretion of L, CV and CO chondrite parent bodies and the formation ages of aqueously produced fayalite suggest that the water ices were incorporated into the accreting parent bodies rather than being added later, during regolith gardening.

The inferred water/rock mass ratios (~ 0.1–0.2) in ordinary, CO and CV chondrites are much lower than the solar value of 1.2 (ref. 40). The estimated low permeability of chondritic

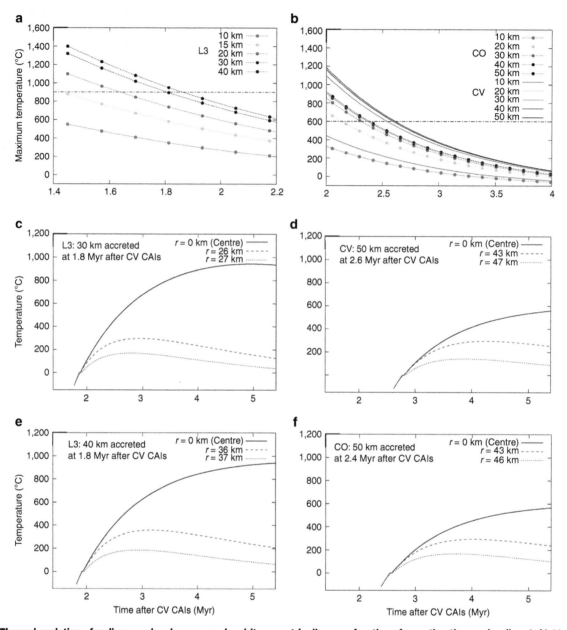

Figure 6 | Thermal evolution of ordinary and carbonaceous chondrite parent bodies as a function of accretion time and radius. (a,b) Maximum temperature as a function of time after CV CAIs modelled for L, CV and CO chondrite-like parent bodies ranging from 10 to 50 km in radius. **(c – f)** Temperature evolution diagrams at different depths shown for **(c,e)** L chondrite-like parent bodies 30 and 40 km in radius; **(d)** a 50-km radii CV-like parent body; and **(f)** a 50 km radii CO chondrite-like parent body.

meteorites[41] and the lack of variations in bulk chemical compositions between meteorites with different degrees of aqueous alteration[1,2,42] imply that the alteration occurred in a chemically closed system, with the fluid flow being restricted to 10–100 µm, which is consistent with rare occurrences of short veins in aqueously altered chondrites (Fig. 1d). We suggest that the inferred low abundance of water ices in ordinary, CO and CV chondrite parent bodies resulted from their accretion close to the snow line, possibly slightly inside it, where only the relatively large ice-bearing particles could have avoided instantaneous evaporation and survived to be accreted to the parent body[43].

The location of the snow line in the protoplanetary disk is uncertain; it likely did not reside at a single location, but rather migrated with time as the luminosity of the proto-Sun, the mass transport rate through the disk, and the disk opacity all evolved with time. In viscous disks, where mass accretion occurs throughout the disk, energy from internal dissipation can keep the inner disk warm for much of the lifetime of the disk. As mass accretion rates diminished with time, the viscous dissipation would slow, causing the snow line to migrate inwards with time. While specific details vary with the assumed disk structure and viscosity, models[44–47] suggest that the snow line would be located beyond 5 AU early in disk evolution, but is likely to have been present at 2–3 AU in the 1.8–2.5-Myr-old disk, which corresponds to the accretion ages of ordinary, CV and CO chondrite parent bodies. Thus, this would suggest that the ordinary, CO and CV chondrite parent bodies most likely accreted in the inner part of the Solar System, close to the current position of the main asteroid belt, rather than experiencing large radial excursions that may have occurred during planet migration[5]. This conclusion is inconsistent with the Grand Tack dynamical model of the Solar System evolution that predicts implantation of hydrated carbonaceous chondrite-like planetesimals, formed beyond Jupiter, into the main asteroid belt[5,48].

Methods

Synthesis experiments. To measure the Mn–Cr RSF in fayalite, ferromagnesian olivines covering a wide compositional range (Fa_{31-99}) were synthesized, as described by Doyle *et al.* (manuscript in preparation). The methodology is summarized as follows: pre-dried Fe_2O_3, MgO, SiO_2, $MnCO_3$, Cr_2O_3 and NiO were mixed in stoichiometric proportions and ground by hand under ethanol. The respective powders were made into a slurry with a polyvinyl alcohol solution and attached to platinum loops. The loops were suspended from a platinum chandelier within a 1-atm vertical gas mixing furnace in which the temperature was monitored using an S-type thermocouple. Mixtures of H_2 and CO_2 were used to control oxygen fugacity (fO_2), which was monitored using a SIRO2 C700+ solid electrolyte oxygen sensor. The samples were at dwell temperatures ranging from 1,200 to 1,500 °C for up to 19 h (Supplementary Table 3), after which they were quenched into water.

Petrography and major element analysis. The products of the synthesis runs were mounted in epoxy resin, polished and characterized using the UH JEOL JXA-8500F field emission electron microprobe (EMP) operating at 20 kV accelerating voltage, 50 nA beam current and 1 – 5 micron beam sizes. Counting times on both peak (and background) were 90 s (and 45 s) for Mg, Cr and Si, 30 s (and 15 s) for Ni and Mn, and 20 s (and 10 s) for Fe. Minerals with known chemical compositions were used as standards. The compositions of the olivine samples are described using their fayalite numbers. Liquidus olivine grains (Fa_{31-99}) obtained from partial melts are typically > 100 μm in size and are compositionally uniform; they contain ~0.7 ± 0.1 wt% MnO and ~0.06 ± 0.03 wt% Cr_2O_3.

The mineralogy and petrography of EET 90161, Semarkona, MET 00452, MET 96503, Ngawi, MAC 88107 and A-881317 were studied in transmitted and reflected light with an optical microscope and in backscattered electrons with the UH EMP. Quantitative chemical analyses were performed with the UH JEOL JXA-8500F EMP operated at 15 kV accelerating voltage, 15–20 nA beam current, and ~1 μm beam size using five wavelength spectrometers. Minerals with known chemical compositions were used as standards. Matrix effects were corrected using the ZAF correction method[49].

Oxygen-isotope systematics. Oxygen isotopic compositions of fayalite, magnetite and chondrule olivines were measured *in situ* by SIMS with the UH Cameca ims-1280 ion microprobe using two measurement protocols, depending on grain size.

For grains > 10 μm, an ~0.8–1.2 nA Cs^+ primary ion beam was focused to a diameter of ~5 μm and rastered over a ~10 × 10 μm^2 area for pre-sputtering (120 s). After pre-sputtering, the raster size was reduced to ~7 × 7 μm^2 for automated centring of the secondary ion beam followed by data collection. An energy window of ~40 eV was used. A normal-incident electron flood gun was used for charge compensation with homogeneous electron density over a region ~70 μm in diameter. Three oxygen isotopes ($^{16}O^-$, $^{18}O^-$ and $^{17}O^-$) were measured in multicollection mode using multicollection Faraday cups and an axial electron multiplier (EM). The mass-resolving power for $^{17}O^-$ and for $^{16}O^-$ and $^{18}O^-$ were set to ~5,500 and ~2,000, respectively. $^{16}OH^-$ signal was monitored in every measured spot and was typically $< 10^6$ counts per second, compared with typical $^{17}O^-$ count rates of 2×10^5 counts per second. Contribution of $^{16}OH^-$ onto $^{17}O^-$ was corrected based on a peak/tail ratio. The correction was typically <0.1‰ (~0.5‰ at most). Instrumental mass fractionation effects for fayalite, magnetite and chondrule olivine were corrected by analysing terrestrial fayalite, terrestrial magnetite and San Carlos olivine standards, respectively. The standards were analysed repeatedly before and after each run. Reported errors (2σ) include both the internal measurement precision and the external reproducibility (~0.5–1.4‰ (2σ) in both $\delta^{17}O$ and $\delta^{18}O$) of standard data obtained during a given session.

For grains < 10 μm in size, an ~20–30 pA primary Cs^+ beam was focused to ~1–2 μm. The three oxygen isotopes, $^{16}O^-$, $^{17}O^-$ and $^{18}O^-$, were measured in multicollection mode using a multicollection Faraday cup, an axial EM and a multicollection EM, respectively. The internal and external reproducibility on the multiple analyses of the standards was ~1–3‰ (2σ) for both $\delta^{17}O$ and $\delta^{18}O$.

Small grains (< 10 μm) are generally very difficult to identify under the optical microscope of the ims-1280 ion microprobe. To measure such small grains, we used the UH JEOL-5900LV scanning electron microscope to mark the regions of interest. This is done by focusing the electron beam, centring the grain of interest in the field of view and then increasing the magnification so that the electron beam is effectively a spot on the centre of the grain. The electron beam removes adsorbed water on the carbon coating in such a way that it appears as a dark spot in $^{16}O^-$ scanning ion image in the ion probe. Scanning $^{16}O^-$ images are obtained by rastering the ~20 pA ion beam over an ~10 × 10 to 50 × 50 μm^2 square. After the mark is found using $^{16}O^-$ scanning ion imaging, we centre the point of interest under the ion beam and then turn off the primary beam to prepare for measurement. The sputter rate of the rastered beam is sufficiently low that it does not completely remove the sample's conductive coating during ion imaging. A conductive pathway therefore remains available for the duration of the oxygen-isotope measurements.

After analysis, the location of each probe spot was re-imaged by electron microscopy to check for beam overlap between phases and to identify large cracks or impurities that may have affected the result.

Mn-Cr isotope systematic. Mn–Cr isotope data were collected *in situ* with the UH Cameca ims-1280 ion microprobe using two measurement protocols, depending on grain sizes.

A $^{16}O^-$ primary beam with a 100 pA current was focused into a spot ~5 μm in diameter and used to collect data from terrestrial and meteoritic olivines (Fa_{10-34}), liquidus-phase synthetic olivines (Fa_{31-99}) and fayalites (Fa_{95-100}) from the carbonaceous chondrites A-881317 and MAC 88107. Owing to the small size of the fayalite grains (≤4–5 μm) in the OC EET 90161, their Mn-Cr isotope data were collected using a $^{16}O^-$ primary beam with a 65–75 pA current focused into a spot ~3 μm in diameter.

The positive secondary ions were accelerated with 10 kV. A 50 eV energy window was used. Three chromium ions, $^{50}Cr^+$, $^{52}Cr^+$ and $^{53}Cr^+$, were measured simultaneously using two multicollection EMs and an axial EM, respectively. Subsequently $^{55}Mn^+$ was measured on the axial EM by peak-jumping. The mass-resolving power was set to ~4,500 for $^{50}Cr^+$ and $^{52}Cr^+$, and ~6,000 for $^{53}Cr^+$ and $^{55}Mn^+$. These settings were sufficient to separate the $^{50}Cr^+$, $^{52}Cr^+$, $^{53}Cr^+$ and $^{55}Mn^+$ ions from interfering species, including $^{52}CrH^+$. Isotopic data were typically collected over 125 cycles (EET 90161) and 100 cycles (A-881317 and MAC 88107). The chromium count rates collected during the first 25 cycles were often higher than during the remaining cycles, so the first 25 cycles were removed due to possible contamination. Corrections were made for both the EM background noises and dead times.

Chromium-isotope and $^{55}Mn/^{52}Cr$ ratios were calculated from the total number of counts to suppress systematic bias caused by low count rates of chromium[50,51]. The measured $^{53}Cr/^{52}Cr$ ratios were corrected for instrumental mass fractionation determined by repeat analyses of synthetic fayalite (Fa_{99}; ref. 14), which was assumed to have a terrestrial $^{53}Cr/^{52}Cr$ ratio of 0.113459 (ref. 52). We could not do an internal mass fractionation correction because the interferences (^{50}V and ^{50}Ti) on ^{50}Cr were too large. The measured $^{55}Mn/^{52}Cr$ ratios were reduced using an RSF (defined here as $(^{55}Mn^+/^{52}Cr^+)_{SIMS}/(^{55}Mn/^{52}Cr)_{EMPA}$) determined on synthetic fayalite (Fa_{99}; ref. 14; Fig. 3, Supplementary Table 4 and Supplementary Discussion) that was measured daily and bracketed the unknown. The Fa_{99} $(^{55}Mn/^{52}Cr)_{EMPA}$ ratio was calculated assuming 83.789 % of Cr is ^{52}Cr (ref. 53). As the ion yields changed with time, a time-averaged RSF was used. The measurement durations of the bracketing Fa_{99} standards were matched to that of the unknown if (on the rare occasion) the meteoritic fayalite analysis was shortened (for example, owing to the base of the grain being breached). The reported uncertainties (2σ) in chromium-isotope ratio and $^{55}Mn/^{52}Cr$ ratio include both the internal precision of an individual analysis and the external reproducibility for standard measurements during a given analytical session.

After analysis, the location of each probe spot was re-imaged by electron microscopy to check for beam overlap between phases and to identify large cracks or impurities that may have affected the result.

Age anchors. We used the U-isotope-corrected Pb-Pb absolute age of the CV CAIs (4,567.30 ± 0.16 Ma; ref. 9) as the age of the Solar System. Owing to the presence of isotopically anomalous chromium, absence of primary minerals with Mn/Cr ≫ 1 and evidence for a late-stage disturbance of ^{53}Mn–^{53}Cr systematics in CV CAIs[20], the initial $^{53}Mn/^{55}Mn$ ratio of the Solar System is not known. In this paper, relative ^{53}Mn–^{53}Cr ages obtained in fayalites as well as those in calcites reported in literature[8] are anchored to the D'Orbigny angrite for which $(^{53}Mn/^{55}Mn)_0$ and U-corrected Pb–Pb absolute ages are known[21,22].

Angrites are a small but diverse group of igneous meteorites that are divided into two subgroups: coarse-grained plutonic and fine-grained quenched angrites (ref. 54 and references therein). The mineralogy, bulk chemical and isotopic compositions suggest that both subgroups originated on the same parent body. Crystallization of the quenched angrites predates that of the plutonic angrites by ~7 Myr (refs 55,56). The quenched angrites are useful as age anchors for multiple short-lived chronometers (ref. 56) as (1) having crystallized from a melt, they were isotopically homogenized at the time of their original formation; (2) they formed sufficiently early in the history of the Solar System to contain measurable excesses of the daughter products of several short-lived radionuclides; and (3) they crystallized and cooled rapidly at the time of their formation, effectively closing the isotope systems of long- and short-lived chronometers at the same time.

Among the quenched angrites dated, we used D'Orbigny as an age anchor for calculating model ages for the following reasons: (1) it is relatively unmetamorphosed, and therefore preserves the chronological records largely undisturbed[57]; (2) it has a precise U-isotope-corrected Pb–Pb absolute age (4,563.37 ± 0.25 Ma; ref. 22); and (3) very well-defined ^{53}Mn–^{53}Cr isochron ($(^{53}Mn/^{55}Mn)_0$ ratio of $(3.24 \pm 0.04) \times 10^{-6}$; ref. 21). Anchoring the fayalite data to other plutonic anchors (Supplementary Fig. 1) could result in a 1.4 Myr age difference, as discussed in the Supplementary Methods.

The calculated ^{53}Mn–^{53}Cr ages of fayalite were subsequently compared with the U-corrected Pb–Pb absolute age (4,567.30 ± 0.16 Ma) of the CV CAIs[9]. We recalculated the previously reported model ages of secondary calcites in CM carbonaceous chondrites[8] relative to the D'Orbigny angrite anchor (Supplementary Table 5 and Supplementary Fig. 2).

Thermal modelling of chondrite parent bodies. To model the thermal evolution of the L, CV and CO chondrite-like bodies consistent with the peak metamorphic temperatures, the temperature range of fayalite formation and the inferred [53]Mn–[53]Cr ages of fayalite crystallization, we considered spherically symmetric, instantaneously accreting bodies that were heated by decay of [26]Al only. A heat-conduction equation,

$$\rho c \frac{\partial T}{\partial t} = \frac{1}{r^2} \frac{\partial}{\partial r}\left(r^2 K \frac{\partial T}{\partial r}\right) + A e^{-\lambda t} \tag{1}$$

where ρ is density, c is specific heat, T is temperature, t is time, r is radius from the centre, K is thermal conductivity, A is initial radiogenic heat generation rate per unit volume and λ is the decay constant of [26]Al, is solved numerically using a finite difference method and an explicit integral method[58].

We assumed that (1) [26]Al was the only heat source of a chondrite parent body, and was uniformly distributed in the protoplanetary disk with the canonical [26]Al/[27]Al ratio of 5.25×10^{-5}; (2) some regions of these parent bodies reached $100 - 200\,^{\circ}\text{C}$ (temperature range of fayalite formation[28]) at the time of fayalite formation; and (3) that these regions avoided subsequent heating above $300\,^{\circ}\text{C}$ (to preclude Fe–Mg diffusion in secondary fayalite). Supplementary Table 6 (part 1) defines parameters specific to the parent bodies from which the L, CV and CO chondrites were derived, such as the fayalite formation ages (Figs 4 and 5) and peak metamorphic temperatures experienced by these parent bodies[23,31,32]. We note that using an initial [26]Al/[27]Al ratio of 5.25×10^{-5}, rather than 5×10^{-5}, corresponds to 0.05 Myr, which is well within the uncertainty of our isotope measurements ($0.4 - 1.8$ Myr).

The initial parameters (assumed to be constant) of the L, CV and CO chondrite parent bodies are detailed in Supplementary Table 6 (part 2; refs 28,59). In addition, we used the physical properties (thermal conductivity, specific heat and density) for ice, water[60,61] and rock[62,63], detailed in Supplementary Table 6 (part 3).

The latent heat of ice melting has been included, and we numerically calculated the effect of ice melting on the thermal evolution. Hydration reactions were not included, as the primary goal is to determine whether or not the parent bodies could reach the formation temperature range of fayalite. The thermal model also does not include the potential influence of an overlying regolith to reduce the thermal diffusivity. Two previous estimates of [26]Al heating of meteorite parent bodies[64,65] reported much higher temperatures than those calculated here, most likely owing to the differences in the assumed physical properties such as the thermal diffusivity (previous studies assumed values of <0.01) and the water/rock ratio (assumed to be zero in the previous studies). We note that increasing the water/rock ratio not only increases the amount of energy required to heat the parent body to a given temperature (ice has a higher heat capacity than rock) but also reduces the amount of [26]Al (per unit mass) available to produce heat as water does not contain any [26]Al.

Parent bodies with radii of 20, 30, 40 and/or 50 km were modelled for the L, CV and CO chondrite parent bodies (Fig. 6). The size of the bodies selected may also represent larger bodies as the maximum temperature reached by a parent body larger than 30 km in radius is largely independent of its size, and there is little difference in the accretion times of the larger bodies (Fig. 6a,b).

References

1. Krot, A. N. et al. in *Meteorites and the Early Solar System II.* (eds Lauretta, D. S. & McSween, Jr. H. Y.) 525–553 (University of Arizona Press, 2006).
2. Zolensky, M. E., Krot, A. N. & Benedix, G. Record of low-temperature alteration in asteroids. *Rev. Mineral. Geochem.* **68,** 429–462 (2008).
3. Alexander, C. M. O. D. et al. The provenances of asteroids, and their contributions to the volatile inventories of the terrestrial planets. *Science* **337,** 721–723 (2012).
4. Krot, A. N. et al. Origin of asteroidal water: constraints from isotopic compositions of aqueously formed minerals. *Meteorit. Planet. Sci.* **48** issue s1, Supplement, id.5161 (2013).
5. Walsh, K. J., Morbidelli, A., Raymond, S. N., O'Brien, D. P. & Mandell, A. M. A low mass for Mars from Jupiter's early gas-driven migration. *Nature* **475,** 206–209 (2011).
6. Burbine, T. H., McCoy, T. J., Meibom, A., Gladman, B. & Keil, K. Meteoritic parent bodies: Their number and identification. in *Asteroids III.* (eds Bottke, Jr. W. F. et al.) 653–667 (The University of Arizona Press, 2002).
7. Trinquier, A., Birck, J.-L., Allegre, C. J., Göpel, C. & Ulfbeck, D. [53]Mn–[53]Cr systematics of the early Solar System revisited. *Geochim. Cosmochim. Acta* **72,** 5146–5163 (2008).
8. Fujiya, W., Sugiura, N., Hotta, H., Ichimura, K. & Sano, Y. Evidence for the late formation of hydrous asteroids from young meteoritic carbonates. *Nat. Commun.* **3,** 1–6 (2012).
9. Connelly, J. N. et al. The absolute chronology and thermal processing of solids in the Solar protoplanetary disk. *Science* **338,** 651–655 (2012).
10. Hutcheon, I. D., Krot, A. N., Keil, K., Phinney, D. L. & Scott, E. R. D. [53]Mn–[53]Cr dating of fayalite formation in the CV3 chondrite Mokoia: evidence for asteroidal alteration. *Science* **282,** 1865–1867 (1998).
11. Jogo, K., Nakamura, T., Noguchi, T. & Zolotov, M. Y. Fayalite in the Vigarano CV3 carbonaceous chondrite: occurrences, formation age and conditions. *Earth Planet. Sci. Lett.* **287,** 320–328 (2009).
12. Krot, A. N. et al. Evidence for low-temperature growth of fayalite and hedenbergite in MacAlpine Hills 88107, an ungrouped carbonaceous chondrite related to the CM-CO clan. *Meteorit. Planet. Sci.* **35,** 1365–1386 (2000).
13. McKibbin, S. J., Ireland, T. R., Amelin, Y., O'Neill, H. S. C. & Holden, P. Mn-Cr relative sensitivity factors for secondary ion mass spectrometry analysis of Mg-Fe-Ca olivine and implications for the Mn-Cr chronology of meteorites. *Geochim. Cosmochim. Acta* **110,** 216–228 (2013).
14. Doyle, P. M., Nagashima, K., Jogo, K. & Krot, A. N. Relative sensitivity factor defined for [53]Mn–[53]Cr chronometry of secondary fayalite. *Lunar Planet. Sci.* **46,** #1792 (2013).
15. Kita, N. T. et al. High precision SIMS oxygen three isotope study of chondrules in LL3 chondrites: Role of ambient gas during chondrule formation. *Geochim. Cosmochim. Acta* **74,** 6610–6635 (2010).
16. Clayton, R. N., Mayeda, T. K., Goswami, J. N. & Olsen, E. J. Oxygen isotope studies of ordinary chondrites. *Geochim. Cosmochim. Acta* **55,** 2317–2337 (1991).
17. Choi, B.-G., Krot, A. N. & Wasson, J. T. Oxygen isotopes in magnetite and fayalite in CV chondrites Kaba and Mokoia. *Meteorit. Planet. Sci.* **35,** 1239–1248 (2000).
18. Clayton, R. N. & Mayeda, T. K. Oxygen isotope studies of carbonaceous chondrites. *Geochim. Cosmochim. Acta* **63,** 2089–2104 (1999).
19. Rudraswami, N. G., Ushikubo, T., Nakashima, D. & Kita, N. T. Oxygen isotope systematics of chondrules in the Allende CV3 chondrite: High precision ion microprobe studies. *Geochim. Cosmochim. Acta* **75,** 7596–7611 (2011).
20. Papanastassiou, D. A., Wasserburg, G. J. & Bogdanovski, O. The [53]Mn–[53]Cr system in CAIs: an update. *Lunar Planet. Sci.* **36,** #2198 (2005).
21. Glavin, D. P., Kubny, A., Jagoutz, E. & Lugmair, G. W. Mn-Cr isotope systematics of the D'Orbigny angrite. *Meteorit. Planet. Sci.* **39,** 693–700 (2004).
22. Brennecka, G. A. & Wadhwa, M. Uranium isotope compositions of the basaltic angrite meteorites and the chronological implications for the early Solar System. *Proc. Natl Acad. Sci. USA* **109,** 9299–9303 (2012).
23. Huss, G. R., Rubin, A. E. & Grossman, J. N. in *Meteorites and the Early Solar System II.* (eds Lauretta, D. S. & McSween, Jr. H. Y.) 567–586 (University of Arizona Press, 2006).
24. Lauretta, D. S. & Buseck, P. R. Opaque minerals in chondrules and fine-grained chondrule rims in the Bishunpur (LL3.1) chondrite. *Meteorit. Planet. Sci.* **38,** 59–79 (2003).
25. Zhen, Y.-F. Calculation of oxygen isotope fractionation in metal oxides. *Geochim. Cosmochim. Acta* **55,** 2299–2307 (1991).
26. Zhen, Y.-F. Calculation of oxygen isotope fractionation in anhydrous silicate minerals. *Geochim. Cosmochim. Acta* **57,** 1079–1091 (1993).
27. Krot, A., Petaev, M. I. & Bland, P. A. Multiple formation mechanisms of ferrous olivine in CV carbonaceous chondrites during fluid-assisted metamorphism. *Antarct. Meteorite Res.* **17,** 154–171 (2004).
28. Zolotov, M. Y., Mironenko, M. V. & Shock, E. L. Thermodynamic constraints on fayalite formation on parent bodies of chondrites. *Meteorit. Planet. Sci.* **41,** 1775–1796 (2006).
29. Guo, W. & Eiler, J. M. Temperatures of aqueous alteration and evidence for methane generation on the parent bodies of the CM chondrites. *Geochim. Cosmochim. Acta* **71,** 5565–5575 (2007).
30. Berger, E. L., Zega, T. J., Keller, L. P. & Lauretta, D. S. Evidence for aqueous activity on comet 81P/Wild 2 from sulfide mineral assemblages in Stardust samples and CI chondrites. *Geochim. Cosmochim. Acta* **75,** 3501–3513 (2011).
31. Bonal, L., Quirico, E., Bourot-Denise, M. & Montagnac, G. Determination of the petrologic type of CV3 chondrites by Raman spectroscopy of included organic matter. *Geochim. Cosmochim. Acta* **70,** 1849–1863 (2006).
32. Bonal, L., Bourot-Denise, M., Quirico, E., Montagnac, G. & Lewin, E. Organic matter and metamorphic history of CO chondrites. *Geochim. Cosmochim. Acta* **71,** 1605–1623 (2007).
33. Grimm, R. E. & McSween, Jr. H. Y. Heliocentric zoning of the asteroid belt by aluminum-26 heating. *Science* **259,** 653–655 (1993).
34. Jacobsen, B., Yin, Q.-Z., Moynier, F., Amelin, Y., Krot, A. N., Nagashima, K., Hutcheon, I. D. & Palme, H. [26]Al-[26]Mg and [207]Pb-[206]Pb systematics of Allende CAIs: canonical solar initial [26]Al/[27]Al ratio reinstated. *Earth Planet. Sci. Lett.* **272,** 353–364 (2008).
35. Kita, N. T. et al. [26]Al-[26]Mg isotope systematics of the first solids in the early solar system. *Meteorit. Planet. Sci.* **48,** 1383–1400 (2013).
36. Kruijer, T. S., Kleine, T., Fischer-Gödde, M., Burkhardt, C. & Wieler, R. Nucleosynthetic W isotope anomalies and the Hf-W chronometry of Ca-Al-rich inclusions. *Earth Planet. Sci. Lett.* **403,** 317–327 (2014).

37. Jacobsen, B. et al. Formation of the short-lived radionuclide ^{36}Cl in the protoplanetary disk during late-stage irradiation of a volatile-rich reservoir. Astrophys. J. Lett. 731, L28–L31 (2011).
38. Kita, N. T. & Ushikubo, T. Evolution of protoplanetary disk inferred from ^{26}Al chronology of individual chondrules. Meteorit. Planet. Sci. 47, 1108–1119 (2012).
39. Alexander, C. M. O'D., Grossman, J. N., Ebel, D. S. & Ciesla, F. J. The formation conditions of chondrules and chondrites. Science 320, 1617–1619 (2008).
40. Lodders, K. Solar System abundances and condensation temperatures of the elements. Astrophys. J. 591, 1220–1247 (2003).
41. Bland, P. A. et al. Why aqueous alteration in asteroids was isochemical: High porosity ≠ high permeability. Earth Planet. Sci. Lett. 287, 559–568 (2009).
42. Rubin, A. E., Trigo-Rodrıguez, J. M., Huber, H. & Wasson, J. T. Progressive aqueous alteration of CM carbonaceous chondrites. Geochim. Cosmochim. Acta 71, 2761–2782 (2007).
43. Cyr, K. E., Sears, W. D. & Lunine, J. I. Distribution and evolution of water ice in the solar nebula: Implications for solar system body formation. Icarus 135, 537–548 (1998).
44. Ciesla, F. J. & Cuzzi, J. N. The evolution of the water distribution in a viscous protoplanetary disk. Icarus 181, 178–204 (2006).
45. Kennedy, G. M. & Kenyon, S. J. Planet formation around stars of various masses: The snow line and the frequency of giant planets. Astrophys. J. 673, 505–512 (2008).
46. Min, M., Dullemond, C. P., Kama, M. & Dominik, C. The thermal structure and the location of the snow line in the protosolar nebula: Axisymmetric models with full 3-D radiative transfer. Icarus 212, 416–426 (2011).
47. Martin, R. G. & Livio, M. On the evolution of the snow line in protoplanetary disks. Mon. Not. R Astron. Soc. 425, L6–L9 (2012).
48. Walsh, K. J., Morbidelli, A., Raymond, S. N., O'Brien, D. P. & Mandell, A. M. Populating the asteroid belt from two parent source regions due to the migration of giant planets—'The Grand Tack'. Meteorit. Planet. Sci. 47, 1941–1947 (2012).
49. Armstrong, J. T. in Microbeam Analysis. (ed. Newbury, D. E.) 239–246 (San Francisco Press, 1988).
50. Ogliore, R. C., Huss, G. R. & Nagashima, K. Ratio estimation in SIMS analysis. Nucl. Instr. Meth. Phys. Res. B 269, 1910–1918 (2011).
51. Telus, M., Huss, G. R., Ogliore, R. C., Nagashima, K. & Tachibana, S. Recalculation of data for short-lived radionuclide systems using less-biased ratio estimation. Meteorit. Planet. Sci. 47, 2013–2030 (2012).
52. Papanastassiou, D. A. Chromium isotopic anomalies in the Allende meteorite. Astrophys. J. 308, L27–L30 (1986).
53. Rosman, K. J. R. & Taylor, P. D. P. Isotopic compositions of the elements 1997. Pure Appl. Chem. 70, 217–235 (1998).
54. Krot, A. N., Keil, K., Scott, E. R. D., Goodrich, C. A. & Weisberg, M. K. in Meteorites and Cosmochemical Processes. 2nd edn Vol. 1 (ed. Davis, A. M.) Treatise on Geochemistry 1–63 (Elsevier, 2014).
55. Amelin, Y. U–Pb ages of angrites. Geochim. Cosmochim. Acta 72, 221–232 (2008).
56. Kleine, T., Hans, U., Irving, A. J. & Bourdon, B. Chronology of the angrite parent body and implications for core formation in protoplanets. Geochim. Cosmochim. Acta 84, 186–203 (2012).
57. Mittlefehldt, D. W., Killgore, M. & Lee, M. T. Petrology and geochemistry of D'Orbigny, geochemistry of Sahara 99555, and the origin of angrites. Meteorit. Planet. Sci. 37, 345–369 (2002).
58. Wakita, S. & Sekiya, M. Thermal evolution of icy planetesimals in the solar nebula. Earth Planets Space 63, 1193–1206 (2011).
59. Hutchison, R. Meteorites: A petrologic, chemical and isotopic synthesis (Cambridge University Press, 2007).
60. National Astronomical Observatory of Japan. Chronological Scientific Tables (Maruzen Publishing Co., Ltd., 2010, in Japanese).
61. Murphy, D. M. & Koop, T. Review of the vapour pressures of ice and supercooled water for atmospheric applications. Q. J. R. Meteorol. Soc. 131, 1539–1565 (2005).
62. Yomogida, K. & Matsui, T. Physical-properties of ordinary chondrites. J. Geophys. Res. 88, 9513–9533 (1983).
63. Opeil, C. P., Consolmagno, G. J. & Britt, D. T. The thermal conductivity of meteorites: new measurements and analysis. Icarus 208, 449–454 (2010).
64. Lee, T., Papanastassiou, D. A. & Wasserburg, G. J. Demonstration of ^{26}Mg excess in Allende and evidence for ^{26}Al. Geophys. Res. Lett. 3, 109–112 (1976).
65. Hutcheon, I. D. & Hutchison, R. Evidence from the Semarkona ordinary chondrite for ^{26}Al heating of small planets. Nature 337, 238–241 (1989).
66. Yin, Q. Z., Jacobsen, B., Moynier, F. & Hutcheon, I. D. Toward consistent chronology in the early Solar System: high-resolution ^{53}Mn-^{53}Cr chronology for chondrules. Astrophys. J. 662, L43–L46 (2007).

Acknowledgements

This study is dedicated to Dr. Ian Douglas Hutcheon who was the pioneer in using ^{53}Mn-^{53}Cr systematics in fayalite for dating aqueous alteration of chondritic meteorites. He passed away on March 26, 2015, after submission of the final revised version of the manuscript. The Meteorite Working Group and National Institute of Polar Research (Japan) are thanked for the loan of meteorites from the Antarctic Meteorite Collections. We also thank Dr T. Nakamura from Tohoku University (Japan) for providing terrestrial magnetite and fayalite standards for oxygen-isotope measurements, and Drs W. Fujiya and B. Jacobsen for their helpful discussions. Thermal modelling was carried out on the PC cluster at the Center for Computational Astrophysics, National Astronomical Observatory of Japan. This material is based upon work supported by the National Aeronautics and Space Administration (NASA) Cosmochemistry grants NNX12AH69G (A.N.K., PI) and NNH10AO48I (I.D.H., PI), and by the NASA Astrobiology Institute under Cooperative Agreement No. NNA09DA77A issued through the Office of Space Science, with additional funding from South African National Research Foundation grant no. 88191 (P.M.D., PI). Portions of this work were performed under the auspices of the US Department of Energy by Lawrence Livermore National Laboratory under contract DE-AC52-07NA27344.

Author contributions

A.N.K. initiated and oversaw the study. P.M.D. designed and performed the olivine synthesis experiments. P.M.D., K.J., KN and A.N.K. collected EMP and SIMS data. S.W. performed the thermal modelling. A.N.K., P.M.D. and K.N. wrote the manuscript, including discussions with S.W., K.J., F.J.C. and I.D.H.

Additional information

14

Lonsdaleite is faulted and twinned cubic diamond and does not exist as a discrete material

Péter Németh[1,2], Laurence A.J. Garvie[3,4], Toshihiro Aoki[5], Natalia Dubrovinskaia[6], Leonid Dubrovinsky[7] & Peter R. Buseck[2,4]

Lonsdaleite, also called hexagonal diamond, has been widely used as a marker of asteroidal impacts. It is thought to play a central role during the graphite-to-diamond transformation, and calculations suggest that it possesses mechanical properties superior to diamond. However, despite extensive efforts, lonsdaleite has never been produced or described as a separate, pure material. Here we show that defects in cubic diamond provide an explanation for the characteristic d-spacings and reflections reported for lonsdaleite. Ultrahigh-resolution electron microscope images demonstrate that samples displaying features attributed to lonsdaleite consist of cubic diamond dominated by extensive {113} twins and {111} stacking faults. These defects give rise to nanometre-scale structural complexity. Our findings question the existence of lonsdaleite and point to the need for re-evaluating the interpretations of many lonsdaleite-related fundamental and applied studies.

[1] Institute of Materials and Environmental Chemistry, Research Centre for Natural Sciences, Hungarian Academy of Sciences, Budapest 1117, Hungary. [2] Department of Chemistry and Biochemistry, Arizona State University, Tempe, Arizona 85287-1604, USA. [3] Center for Meteorite Studies, Arizona State University, Tempe, Arizona 85287-6004, USA. [4] School of Earth and Space Exploration, Arizona State University, Tempe, Arizona 85287-1404, USA. [5] LeRoy Eyring Center for Solid State Science, Arizona State University, Tempe, Arizona 85287-1704, USA. [6] Material Physics and Technology at Extreme Conditions, Laboratory of Crystallography, University of Bayreuth, Bayreuth D-95440, Germany. [7] Bayerisches Geoinstitut, Universität Bayreuth, Bayreuth D-95440, Germany. Correspondence and requests for materials should be addressed to P.N. (email: nemeth.peter@ttk.mta.hu).

The allotropes of carbon display a wide diversity of structures that include the three-dimensional (3D) diamond and graphite, two-dimensional (2D) graphene and curved nanotubes and fullerenes. Within this diversity are materials with extraordinary properties, paramount being cubic diamond, which has the highest known hardness and thermal conductivity. Diamond is reported to have a number of polytypes, of which lonsdaleite (also called hexagonal diamond) has received particularly intense attention. Lonsdaleite was first described almost 50 years ago from the Canyon Diablo iron meteorite[1,2]. Its formation was attributed to shock-induced transformation of graphite within the meteorite upon impact with Earth, and its occurrence was used as an indicator of shock[1-3]. It has since been reported from several meteorites as well as from terrestrial sediments and has been attributed to asteroidal impacts, both extraterrestrial and on Earth[4-7].

Lonsdaleite was proposed to have a wurtzite (ZnS)-type structure with space group $P6_3/mmc$ ($a = 0.251$ and $c = 0.412$ nm) and with all structural positions occupied by carbon[1-3,8]. Observations and theoretical studies suggested a structural relationship among graphite, cubic diamond and lonsdaleite and an important role of the latter during the graphite-to-diamond transition[3,9-14]. Furthermore, an area centred around 18 GPa and 1,400 K in the pressure–temperature diagram for carbon was attributed to a phase called 'retrievable hexagonal-type diamond'[10], which corresponds to the conditions where lonsdaleite has been reported[3,11,15].

Lonsdaleite has also received much attention because of its potentially superior mechanical properties, such as compressive strength, hardness and rigidity, thought to rival or exceed those of cubic diamond[16,17]. However, these exceptional properties have not been proven experimentally because of the inability to synthesize lonsdaleite as a pure phase. It has been reported to form during static compression of graphite[3,9,13,15-18]; high-pressure–high-temperature treatment of powdered diamond, graphite and amorphous carbon[19]; explosive detonation and shock compression of graphite[11,20] and diamond[21]; and chemical vapour deposition of hydrocarbon gases[22]; however, in all cases the synthesis product also contained cubic diamond, graphite or both.

Published powder X-ray diffraction (XRD) patterns of lonsdaleite show peaks of cubic diamond plus extra, very broad and poorly resolved maxima at 0.218, 0.193, 0.151 and 0.116 nm that have been indexed using a hexagonal unit cell[1-3,14,15,20]. However, these maxima either occur on the shoulders of diamond peaks (Fig. 1a,b) or match those of graphite, but well-resolved X-ray reflections for lonsdaleite have not been reported. Selected-area electron diffraction (SAED), Raman, electron energy-loss spectroscopy (EELS) and high-resolution transmission electron microscopy have also been used for identification of lonsdaleite[4-7,14,15,18,23,24]; however, interpretation of data is ambiguous (Supplementary Note 1). In spite of the many diffraction and spectroscopic studies, unambiguous data that prove the existence of lonsdaleite as a distinct material have not been reported.

Here we provide a new explanation, based on faulted and twinned cubic diamond, for the diffraction features attributed to lonsdaleite. Because we question its existence, we will hereafter refer to it as 'lonsdaleite' to indicate scepticism that it exists as a discrete material. XRD and SAED patterns, EELS data and scanning transmission electron microscope (STEM) images of natural samples (Supplementary Fig. 1) from the Canyon Diablo meteorite, the type specimen from which 'lonsdaleite' was first described[1,2], and of synthetic material (Supplementary Fig. 1) prepared under conditions where 'retrievable hexagonal-type diamond' was reported[10], support our argument. The finding calls for re-evaluation of previous reports regarding 'lonsdaleite.'

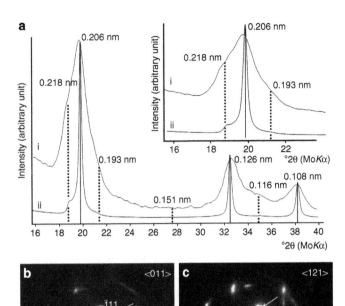

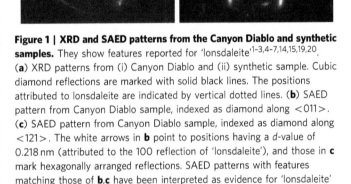

Figure 1 | XRD and SAED patterns from the Canyon Diablo and synthetic samples. They show features reported for 'lonsdaleite'[1-3,4-7,14,15,19,20]. (**a**) XRD patterns from (i) Canyon Diablo and (ii) synthetic sample. Cubic diamond reflections are marked with solid black lines. The positions attributed to lonsdaleite are indicated by vertical dotted lines. (**b**) SAED pattern from Canyon Diablo sample, indexed as diamond along <011>. (**c**) SAED pattern from Canyon Diablo sample, indexed as diamond along <121>. The white arrows in **b** point to positions having a d-value of 0.218 nm (attributed to the 100 reflection of 'lonsdaleite'), and those in **c** mark hexagonally arranged reflections. SAED patterns with features matching those of **b,c** have been interpreted as evidence for 'lonsdaleite' projected along <010> and <001>[4-7,14,15,20], respectively.

Results

'Lonsdaleite' diffraction features. The XRD patterns of Canyon Diablo diamonds and the synthetic material display the poorly resolved diffraction maxima attributed[1-3,14,15,19,20] to 'lonsdaleite' (Fig. 1a). The SAED patterns of numerous grains consist of either spotty rings with streaking and smeared intensities or reflections arranged nearly hexagonally (Figs 1b,c and 2a). These patterns are consistent with diamond projected along <011> or <121>, respectively, although the streaking and hexagonally arranged 111 reflections are incompatible with defect-free single-crystal cubic diamond. Such data have been interpreted as evidence for 'lonsdaleite' projected along <010> and <001>[4-7,14,15,20]. The circularly integrated intensity profiles of these SAED patterns (Supplementary Fig. 2) match the profile of the XRD pattern and show peaks for cubic diamond together with poorly resolved maxima at the d-spacings attributed to 'lonsdaleite.' In order to understand the structural features that give rise to these reflections and their d-spacings, we imaged these samples using a state-of-the-art ultrahigh-resolution STEM.

The STEM images of the samples differ from well-ordered cubic diamond by displaying prominent chevron patterns and features that arise from multiple twins and stacking faults (Figs 2–4, Supplementary Fig. 3). The twins in the Canyon Diablo sample

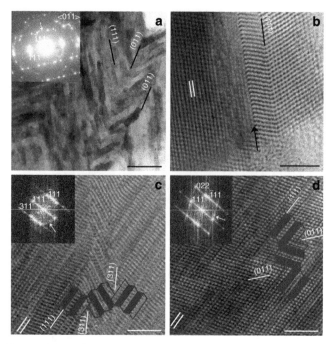

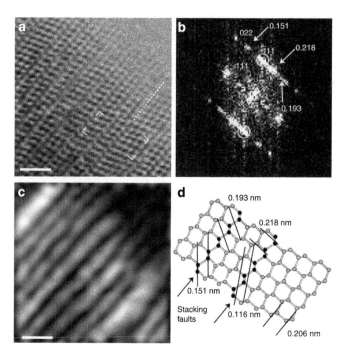

Figure 2 | STEM images provide evidence for stacking faults and multiple types of twins. These defects give rise to the diffraction features attributed to 'lonsdaleite.' (**a**) Low-magnification STEM image and corresponding SAED pattern (inset) from the synthetic sample. Black lines and symbols in the image indicate twin planes. In **a** multiple {111} twins (indicated by black and white bands) give rise to {011} twins. The white arrow in the insert points to the position having a d-value of 0.218 nm that has been attributed to the 100 reflection of 'lonsdaleite.' (**b**) A (111) twin (black line) and a (111) stacking fault (indicated by a black arrow) in the synthetic sample. The 0.206-nm {111} fringes are marked with white parallel lines here and elsewhere. (**c**) {113} twins with corresponding FFT (insert) for the Canyon Diablo sample. (**d**) {011} twins with corresponding FFT (insert) for the Canyon Diablo sample. {111} stacking faults result in streaking of reflections in the SAED pattern and FFTs. In **c,d**, multiple {111} twins (indicated by shaded overlays) give rise to the {113} and {011} twins. These twins result in the characteristic SAED ring patterns (Figs 1b and 2a) as they give rise to additional streaking, respectively, at 59° (white arrow on FFT in **c**) and 71° (white arrow on FFTs in **d**) to the strong {111} streaking. Scale bars mark 200 nm for **a** and 2 nm for **b–d**.

divide the grains into domains that are 1- to 3-nm wide (Fig. 2), and the stacking faults produce domains two to four layers across (0.4–0.8 nm wide; Fig. 3). The synthetic sample shows a similar domain structure, although the domains are wider (2–20 nm), and there are fewer stacking faults. The C K-edge EELS data from the structurally complex areas are consistent with sp^3-bonded carbon and are indistinguishable from cubic diamond (Supplementary Fig. 4). A question thus arises about how to explain the reflections and d-spacings of 'lonsdaleite?'

Explanation of the 'lonsdaleite' diffraction features. SAED patterns and STEM images along <011> and <121> of both the natural and synthetic samples demonstrate their structural complexity. The STEM images reveal two mechanisms that explain the features attributed to 'lonsdaleite.' Images along <011> show extensive {111} twinning (mechanism I), which results in abundant twin interfaces and stacking faults. These twins occur together with two new types of twins: {113} and {011} (Figs 2 and 3). Although {113} diamond twins were proposed from molecular dynamics calculations[25], they have not been reported experimentally. Observing the {113} twins requires

Figure 3 | Stacking faults provide an explanation for the reflections and d-spacings of 'lonsdaleite.' (**a**) <011> STEM image from the Canyon Diablo sample; one of many {111} stacking faults is indicated by the dotted white line. (**b**) FFT calculated from **a**. White arrows indicate spacings (0.218, 0.193 and 0.151 nm) that have been attributed to 'lonsdaleite'. (**c**) Amplitude image calculated from the {−111} set of diamond reflections (white circles in **b**). Bright regions indicate domains two to four layers across separated by {111} stacking faults. (**d**) Structure model of the region marked with white corners in **a**. Stacking faults (black layers of atoms) result in d-spacings that are absent in single-crystal diamond and give rise to the broad X-ray and electron-diffraction features. Scale bars mark 1 nm for **a,c**.

resolving the 0.126-nm spacings, which was only possible with the ultrahigh resolution provided by an aberration-corrected microscope. The other new type of twin is {011} and it can be recognized from prominent chevron patterns in low-magnification images (Fig. 2a). We interpret the chevron twins as evidence for local regions lacking the symmetry of cubic diamond. The local symmetry loss is the result of multiple {111} twins and abundant stacking faults (Fig. 2a,d). The {111} stacking faults and twins interrupt the {111} periodicity (Figs 2 and 3) and result in broad diffraction maxima at the d-spacings of 'lonsdaleite' on XRD (Fig. 1a) and SAED patterns (Figs 1b and 2a) as well as fast Fourier transforms (FFTs) (Figs 2c,d and 3b).

The STEM images along <121> show evidence for another mechanism (mechanism II) that can explain the reported reflections and d-spacing of 'lonsdaleite.' The {113} twinning visible in the <121> projections results in prominent domains elongated parallel to {113} and recognized by their conspicuous {111} and perpendicular {022} fringes (Fig. 4, Supplementary Fig. 3). The domains are bordered by paired {113} twins. Adjacent {113} twins result in regions that are two-layer thick (Fig. 4d) and produce hexagonally arranged reflections in FFTs and SAED patterns (Figs 1c and 4b, Supplementary Fig. 3). These twins result in a broad diffraction maximum at 0.216 nm, with hexagonally arranged carbon atoms across the twin boundary (Fig. 4c,d, Supplementary Fig. 3c).

Consistent with prior work, discrete 'lonsdaleite' grains were not found; however, the above results provide a new interpretation of the observed diffraction features previously attributed to 'lonsdaleite.' We found its reported features on multiple XRD and

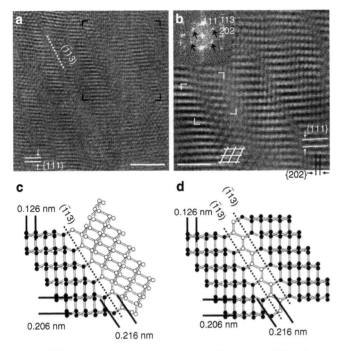

Figure 4 | Twins provide an alternate explanation for the diffraction features of 'lonsdaleite.' (**a**) <121> STEM image from the Canyon Diablo sample; one of many {113} twins is indicated by the dotted white line. (**b**) Background-filtered image calculated from the region marked by black corners in **a**. Intersecting white lines in **b** mark domains separated by regions that exhibit hexagonal fringes. The FFT in the insert shows hexagonally arranged reflections (black arrows) that are absent in single-crystal diamond. (**c**) Structure model of the {113} diamond twin. The structure across the twin consists of hexagonally arranged carbon atoms. Black and white atoms indicate twin domains. (**d**) Structure model of the region marked with white corners in **b**. The figure shows two layers of hexagonally arranged carbon atoms with 0.216-nm spacings, equivalent to two 0.108-nm {113} planes, which matches the d-value of 100 reflection attributed to 'lonsdaleite.' Scale bars mark 2 nm for **a** and 1 nm for **b**.

SAED patterns; however, detailed STEM investigations showed no evidence for 'lonsdaleite.' The local arrangements of carbon atoms across the stacking faults and twin boundaries give rise to diffraction intensities with the d-spacings reported for 'lonsdaleite' (Figs 3d and 4a, Supplementary Fig. 3). The subnanometre domains produce asymmetric broadening of diffraction peaks in XRD and SAED patterns. These observations are consistent with the proposed 111 diamond peak-broadening caused by stacking faults[26,27], and are analogous to what occurs in face-centred cubic metals such as Co[28].

Discussion

We provide new insights into the pressure–temperature phase diagram of carbon. The STEM images and diffraction patterns of the synthetic sample prepared through direct conversion from graphite and under conditions corresponding to 'retrievable hexagonal-type diamond'[10] show features consistent with cubic diamond twins and faults (Fig. 2a,b, Supplementary Fig. 3). Therefore, we suggest that the region centred around 18 GPa and 1,400 K, actually corresponds to the field of cubic diamond containing defects in the form of twins and stacking faults. Furthermore, our results imply that graphite transforms to diamond without intermediate 'lonsdaleite,' which is consistent with the recent findings of shock-produced interstratified graphite and diamond[29].

The abundant stacking faults and twins in the impact-produced diamonds from the Canyon Diablo meteorite result from shock metamorphism of graphite[30]. Similar nanometre-scale structural complexity occurs in the synthetic sample, which was prepared through static compression. Since both samples contain the same structural features, the origin of the observed defect structure is not necessarily shock. In fact, extensive diamond twins have been reported from highly strained, mechanically twinned pink diamonds, which are thought to form as a result of plastic deformation in the mantle[31]. These features indicate that deformation may also produce diamond having high concentrations of planar defects.

Theoretical predictions suggest outstanding mechanical properties for 'lonsdaleite'[16,17]. These predictions are based on the presumption of a periodic structure; however, a material with such periodicity has not been found. We demonstrate that the samples that show the characteristic diffraction features of 'lonsdaleite' display extensive diamond twins and stacking faults that divide the grains into subnanometre domains. Grain-size reduction is a powerful mechanism for improving the hardness of many materials, diamond among them[32–34]. Decreasing the domains' size through defects may similarly contribute to strengthening of the materials. Thus, materials with the structural complexity similar to that of the studied Canyon Diablo meteorite samples can be candidates for synthetic products that possess exceptional mechanical properties for technological applications.

Methods

Materials. Diamond-bearing regions from the Canyon Diablo meteorite were reacted with dilute HCl for 2 days. The resulting insoluble material was washed several times with distilled water, dried and concentrated by centrifuging in a lithium heteropolytungstate solution (density 2.9 g ml^{-1}) followed by washing in distilled water. As a result of these procedures, black adamantine grains with rhombohedral and cubic forms having size from 0.05 to 1 mm were obtained (Supplementary Fig. 1). A 0.12 × 0.10 × 0.08-mm grain was mounted on a glass capillary and X-rayed with a Bruker SMART APEX single-crystal diffractometer employing MoK$_\alpha$ radiation, a graphite monochromator and a 2,000-K CCD detector (Arizona State University, USA). A data set was obtained by rotating the omega goniometer by one-degree per frame and by collecting intensities by scanning the frames for 1-min per/frame. The intensities of the data set were integrated and converted to a one-dimensional X-ray pattern.

A synthetic sample was prepared in a 5,000-ton multi-anvil press (Supplementary Fig. 1). The sample assembly consisted of a MgO (+5 weight% Cr$_2$O$_3$) octahedron (18-mm edge length) containing a LaCrO$_3$ heater. A cylinder of high purity graphite was used as a starting material. The cylinder was enclosed in a Ta capsule, pressurized to 19 GPa and heated at 2,473 K for 5 min, the conditions that correspond to those reported by[7], where 'retrievable hexagonal-type diamond' occurs. The temperature was monitored with a W$_3$Re/W$_{25}$Re thermocouple located axially with respect to the heater, with a junction close to the Ta capsule. Powder XRD from the synthetic sample was obtained using a high-brilliance Rigaku diffractometer (MoK$_\alpha$ radiation) equipped with Osmic focusing X-ray optics and Bruker Apex CCD detector (University of Bayreuth, Germany). The diffraction patterns were processed using the Fit2D software[35].

Transmission electron microscopy (TEM) and image processing. Samples were crushed with a pestle and mortar. A droplet (ca. 2 µl) of the crushed grains in suspension in water was placed on a lacy-C-coated Cu TEM grid and dried. Bright-field STEM images, SAED patterns and EELS data were acquired from electron-transparent areas of the residue protruding into the holes of the lacy-C grid using an aberration-corrected JEOL ARM200F scanning TEM (Arizona State University, USA; 200 keV, 0.08-nm point resolution).

FFTs obtained from the STEM images and background-filtered images were calculated using the Gatan Digital Micrograph 3.5 software. The amplitude image of Fig. 3c was generated from Fig. 3a following the method described in refs 36,37, using routines written for the Digital Micrograph software and applying 0.06-nm^{-1}-size Lorentzian masks for the {−111} set of diamond reflections. The background-filtered image of Fig. 4b was obtained by applying 0.06-nm^{-1}-size Lorentzian masks for the hexagonally arranged reflections.

References

1. Frondel, C. & Marvin, U. B. Lonsdaleite, a hexagonal polymorph of diamond. *Nature* **217**, 587–589 (1967).
2. Hanneman, R. E., Strong, H. M. & Bundy, F. P. Hexagonal diamonds in meteorites: Implications. *Science* **155**, 995–997 (1967).
3. Bundy, F. P. & Casper, J. S. Hexagonal diamond - a new form of carbon. *J. Chem. Phys.* **46**, 3437–3446 (1967).
4. Le Guillou, C., Rouzaud, J. N., Remusat, L., Jambon, A. & Bourot-Denise, M. Structures, origin and evolution of various carbon phases in the ureilite Northwest Africa 4742 compared with laboratory-shocked graphite. *Geochim. Cosmochim. Acta* **74**, 4167–4185 (2010).
5. Kennett, D. J. et al. Shock-synthesized hexagonal diamonds in Younger Dryas boundary sediments. *Proc. Natl Acad. Sci. USA* **106**, 12623–12628 (2009).
6. Israde-Alcantara, I. et al. Evidence from central Mexico supporting the Younger Dryas extraterrestrial impact hypothesis. *Proc. Natl Acad. Sci. USA* **109**, E738–E747 (2012).
7. Kvasnytsya, V. et al. New evidence of meteoritic origin of the Tunguska cosmic body. *Planet. Space Sci.* **84**, 131–140 (2013).
8. Ergun, S. & Alexander, L. E. Crystalline forms of carbon: a possible hexagonal polymorph of diamond. *Nature* **195**, 765–767 (1962).
9. Utsumi, W. & Yagi, T. Formation of hexagonal diamond from room temperature compression of graphite. *Proc. Jpn. Acad. B Phys.* **67**, 159–164 (1991).
10. Bundy, F. P. et al. The pressure-temperature phase and transformation diagram for carbon; updated through 1994. *Carbon. N. Y.* **34**, 141–153 (1996).
11. Erskine, D. J. & Nellis, W. J. Shock-induced martensitic phase-transformation of oriented graphite to diamond. *Nature* **349**, 317–319 (1991).
12. Khaliullin, R. Z., Eshet, H., Khne, T. D., Behler, J. & Parrinello, M. Nucleation mechanism for the first graphite-to-diamond phase transition. *Nat. Mater.* **10**, 693–697 (2011).
13. Le Guillou, C., Brunet, F., Irifune, T., Ohfuji, H. & Rouzaud, J. N. Nanodiamond nucleation below 2273 K at 15 GPa from carbons with different structural organizations. *Carbon* **45**, 636–648 (2007).
14. Nakamuta, Y. & Toh, S. Transformation of graphite to lonsdaleite and diamond in the Goalpara ureilite directly observed by TEM. *Am. Mineral.* **98**, 574–581 (2013).
15. Isobe, F., Ohfuji, H., Sumiya, H. & Irifune, T. Nanolayered diamond sintered compact obtained by direct conversion from highly oriented graphite under high pressure and high temperature. *J. Nanomater.* **2013**, 380136 (1–6) (2013).
16. Pan, Z., Sun, H., Zhang, Y. & Chen, C. Harder than diamond: superior indentation strength of wurtzite BN and lonsdaleite. *Phys. Rev. Lett.* **102**, 055503 (2009).
17. Quingkun, L., Yi, S., Zhiyuan, L. & Yu, Z. Lonsdaleite – a material stronger and stiffer than diamond. *Scripta Mater.* **65**, 229–232 (2011).
18. Kulnitskiy, B., Perezhogin, I., Dubitsky, G. & Blank, V. Polytypes and twins in the diamond-lonsdaleite system formed by high-pressure and high-temperature treatment of graphite. *Acta Crystallogr. Sect. B Struct. Sci.* **69**, 474–479 (2013).
19. Dubrovinskaia, N., Dubrovinsky, L., Langenhorst, F., Jacobsen, S. & Liebske, C. Nanocrystalline diamond synthesized from C60. *Diam. Rel. Mater.* **14**, 16–22 (2005).
20. Kurdyumov, A. V., Britun, V. F., Yarosh, V. V., Danilenko, A. I. & Zelyavskii, V. B. The influence of shock compression conditions on the graphite transformations into lonsdaleite and diamond. *J. Superhard Mater.* **34**, 19–27 (2012).
21. He, H. K., Sekine, T. & Kobayashi, T. Direct transformation of cubic diamond to hexagonal diamond. *Appl. Phys. Lett.* **81**, 610–612 (2002).
22. Misra, A. et al. Hexagonal diamond synthesis on h-GN strained films. *Appl. Phys. Lett.* **89**, 071911 (2006).
23. Daulton, T. L., Eisenhour, D. D., Bernatowicz, T. J., Lewis, R. S. & Buseck, P. R. Genesis of presolar diamonds: comparative high-resolution transmission electron microscopy study of meteoritic and terrestrial nano-diamonds. *Geochim. Cosmochim. Acta* **60**, 4853–4872 (1996).
24. Smith, D. C. & Godard, G. UV and VIS Raman spectra of natural lonsdaleite: towards a recognized standard. *Spectrochim. Acta Mol. Biomol. Spectros* **73**, 428–435 (2009).
25. Morris, J. R., Fu, C. L. & Ho, K. M. Tight-binding study of tilt grain boundaries in diamond. *Phys. Rev. B* **54**, 132–138 (1996).
26. Treacy, M. M. J., Newsam, J. M. & Deem, M. W. A general recursion method for calculating diffracted intensities from crystals containing planar faults. *Proc. R. Soc. A-Math. Phys. Eng. Sci.* **433**, 499–520 (1991).
27. Fayette, L. et al. Analysis of the fine structure of the Raman line and of X-ray reflection profiles for textured CVD diamond films. *Diam. Rel. Mater.* **4**, 1243–1250 (1995).
28. Yu, X. H., Zhang, J. Z., Zhang, Y. Y., Wang, L. P. & Zhao, Y. S. Comparative studies of yield strength and elastic compressibility between nanocrystalline and bulk cobalt. *J. Appl. Phys.* **111**, 113506 (2012).
29. Garvie, A. J. L., Németh, P. & Buseck, P. R Transformation of graphite to diamond via a topotactic mechanism. *Am. Mineral.* **99**, 531–538 (2014).
30. Lipschutz, M. & Anders, E. The record in the meteorites-IV: origin of diamonds in iron meteorites. *Geochim. Cosmochim. Acta* **24**, 83–105 (1961).
31. Gaillou, E. et al. Spectroscopic and microscopic characterization of color lamellae in natural pink diamond. *Diam. Rel. Mater.* **19**, 1207–1220 (2010).
32. Irifune, T., Kurio, A., Sakamoto, S., Inoue, T. & Sumiya, H. Ultrahard polycrystalline diamond from graphite. *Nature* **421**, 599–600 (2003).
33. Sumiya, H. & Irifune, T. Indentation hardness of nano-polycrystalline diamond prepared from graphite by direct conversion. *Diam. Rel. Mater.* **13**, 1771–1776 (2004).
34. Dubrovinsky, L., Dubrovinskaia, N., Prakapenka, V. & Abakumov, A. Implementation of micro-ball nanodiamond anvils for high-pressure studies above 6 Mbar. *Nat. Commun.* **3**, 1163 (2012).
35. Hammersley, A. P. ESRF International Report ESRF98HA01T, FIT2D V9.129 (Reference Manual Version 3.1, Grenoble, France, 1998).
36. Hÿtch, M. J. & Potez, L. Geometric phase analysis of high-resolution electron microscopy images of antiphase domains: example Cu3Au. *Philos. Mag. A* **76**, 1119–1138 (1997).
37. Hÿtch, M. J., Snoeck, E. & Kilaas, R. Quantitative measurement of displacement and strain fields from HREM micrographs. *Ultramicroscopy* **74**, 131–146 (1998).

Acknowledgements

We are grateful to the staff and for use of the facilities in the John M. Cowley Center for High Resolution Electron Microscopy at the Arizona State University. We thank Thomas Groy for X-raying the Canyon Diablo sample. P.N. acknowledges support from the Hungarian Scientific Research Fund and Hungarian Economic Development Centre grant HUMAN_MB08-1-2011-0012 and the János Bolyai Research Scholarship. L.A.J.G and P.R.B. were supported by NASA Origins of Solar Systems grant NNX11AK58G. N.D. thanks Deutsche Forschungsgemeinschaft (DFG) for financial support through the Heisenberg Program and the DFG Project No. DU 954-8/1 and the Federal Ministry of Education and Research (Germany) for the BMBF grant No. 5K13WC3 (Verbundprojekt O5K2013, Teilprojekt 2, PT-DESY).

Author contributions

P.N. and L.A.J.G. conceived the project and P.N. took the lead on data analysis and manuscript writing. L.A.J.G. performed the TEM sample preparation; T.A. conducted the ultrahigh-resolution microscopy; N.D. and L.D. prepared the synthetic diamond sample using high pressure and high temperature technique and characterized it by means of powder XRD. All authors contributed to interpreting the data and the writing of the paper.

Additional information

15

High-temperature water–rock interactions and hydrothermal environments in the chondrite-like core of Enceladus

Yasuhito Sekine[1], Takazo Shibuya[2,3], Frank Postberg[4,5], Hsiang-Wen Hsu[6], Katsuhiko Suzuki[3], Yuka Masaki[3], Tatsu Kuwatani[7], Megumi Mori[8], Peng K. Hong[9], Motoko Yoshizaki[10], Shogo Tachibana[8] & Sin-iti Sirono[11]

It has been suggested that Saturn's moon Enceladus possesses a subsurface ocean. The recent discovery of silica nanoparticles derived from Enceladus shows the presence of ongoing hydrothermal reactions in the interior. Here, we report results from detailed laboratory experiments to constrain the reaction conditions. To sustain the formation of silica nanoparticles, the composition of Enceladus' core needs to be similar to that of carbonaceous chondrites. We show that the presence of hydrothermal reactions would be consistent with NH_3- and CO_2-rich plume compositions. We suggest that high reaction temperatures ($>50\,°C$) are required to form silica nanoparticles whether Enceladus' ocean is chemically open or closed to the icy crust. Such high temperatures imply either that Enceladus formed shortly after the formation of the solar system or that the current activity was triggered by a recent heating event. Under the required conditions, hydrogen production would proceed efficiently, which could provide chemical energy for chemoautotrophic life.

[1] Department of Earth and Planetary Science, University of Tokyo, Bunkyo 113-0033, Japan. [2] Laboratory of Ocean-Earth Life Evolution Research, Japan Agency for Marine-Earth Science and Technology, Yokosuka 237-0061, Japan. [3] Research and Development Center for Submarine Resources / Project Team for Next-Generation Technology for Ocean Resources Exploration, Japan Agency for Marine-Earth Science and Technology, Yokosuka 237-0061, Japan. [4] Institut für Geowissenschaften, Universität Heidelberg, Heidelberg 69120, Germany. [5] Institut für Raumfahrtsysteme, Universität Stuttgart, Stuttgart 70569, Germany. [6] Laboratory for Atmospheric and Space Physics, University of Colorado, Boulder, Colorado 80303, USA. [7] Department of Solid Earth Geochemistry, Japan Agency for Marine-Earth Science and Technology, Yokosuka 237-0061, Japan. [8] Department of Natural History Science, Hokkaido University, Sapporo 060-0810, Japan. [9] The University Museum, University of Tokyo, Bunkyo 113-0033, Japan. [10] Department of Earth and Planetary Science, Tokyo Institute of Technology, Meguro 152-8551, Japan. [11] Graduate School of Environmental Science, Nagoya University, Nagoya 464-8601, Japan. Correspondence and requests for materials should be addressed to Y.S. (email: sekine@eps.s.u-tokyo.ac.jp).

Water-rich plumes of vapour and ice particles with sodium salts erupting from warm fractures near the south pole of Saturn's icy moon Enceladus suggest the presence of a liquid water reservoir in the interior[1–4]. Recent work combining Cassini measurements and experimental results shows that some of the observed plume materials are associated with ongoing hydrothermal activity in the interor[5]. Nanometre-sized silica particles with a confined size range detected by the Cassini Cosmic Dust Analyser are found to have originated from Enceladus[5]. Supported by the results of hydrothermal experiments, it is indicated that these particles originated from nanosilica colloids that formed when silica saturation was reached upon cooling of hydrothermal fluids[5]. The presence of these particles provides tight constraints on the particular conditions of the interior ocean; that is, the presence of high-temperature reactions ($\geq \sim 90\,^\circ C$), moderate salinity ($\leq \sim 4\%$), and alkaline seawater ($pH = 8.5–10.5$)[5]. Products of ongoing hydrothermal reactions would have been transported upwards from an interior ocean located at a depth of $\sim 30\,km$ beneath the surface at Enceladus' south pole[6,7] and would have then been ejected into the plume[8].

One major difference between hydrothermal reactions on Enceladus and those currently occurring on Earth is the plausible presence of abundant primordial volatiles (for example, NH_3 and CO_2) provided from icy planetesimals that formed the Saturnian system[9], and these volatiles are abundant in the gas component of Enceladus' plumes[3,10]. However, detailed laboratory investigations on the fate of these volatiles under hydrothermal conditions within Enceladus have not been performed. In addition, Na^+ is a major constituent in Enceladus' alkaline ocean[1,11,12], as Cassini has detected sodium salts, such as $NaHCO_3$ and $NaCl$, in the plume's ice grains[1,2].

Another possible difference between hydrothermal reactions on Enceladus and those currently occurring on Earth is the rock composition. Given the presence of olivine and pyroxene in comets[13,14], these primitive crystalline silicates would also have been two of the most abundant constituents in the building blocks of Enceladus. If the rocks of Enceladus have not experienced large-scale silicate melting throughout its history, the composition of the core would have been chondritic[11], containing abundance of these primitive minerals. On the other hand, if Enceladus' rocky core has experienced silicate melting in the early stages of its evolution, a more ultramafic, olivine-rich rocks would have been formed within the core, similar to Earth's upper mantle and as proposed for the interior of Ceres[15]. The low-density rocky core suggested by Cassini's data[6,7] (for example, $\sim 2.4–2.5\,g\,cm^{-3}$ for an H_2O mantle with $60\,km$ thickness) is consistent with the presence of hydrous minerals with significant porosity, suggesting the widespread occurrence of water–rock interactions in the core and a supply of aqueous fluid to the subsurface ocean.

To constrain the conditions of hydrothermal reactions on Enceladus, the present study provides the results of further hydrothermal experiments, temporal variations in fluid composition and microscope observations of rock residues collected after the experiments. The experimental results are compared with chemical equilibrium calculations. Based on the results of the experiments, the present study constrains both the reactions of primordial volatiles and the composition of the rock core within Enceladus. Although the previous study[5] shows that the minimum temperatures of hydrothermal reactions required for the formation of silica nanoparticles depend on a pH change of fluids, it does not discuss the mechanisms or possible range of the change. In the present study, we discuss the range of pH changes and required temperature conditions based on the detailed experiments. Finally, we propose thermal evolution

scenarios that could support ongoing hydrothermal activity within Enceladus.

Results

Hydrothermal simulations. In the experiments, we used two types of starting minerals with low and high Si contents: a powdered San Carlos olivine (olivine experiment), and a mixture of powdered orthopyroxene (opx) (orthoenstatite: 70 wt.%) and San Carlos olivine (30 wt.%) (opx experiment) (see Methods). The opx experiment ($Mg/Si = \sim 1.2$) simulates the alteration of a relatively Si-rich rocky core that has not experienced silicate melting, such as a parent body of carbonaceous chondrites in terms of Mg/Si ratios[16] ($Mg/Si = \sim 1.0–1.1$), whereas the olivine experiment ($Mg/Si = \sim 1.8$) simulates the alteration of a more ultramafic rocky core formed by large-scale silicate melting. An aqueous solution of NH_3 and $NaHCO_3$ was used for starting solution (see Methods). Using a steel-alloy autoclave (Supplementary Fig. 1), we simulate hydrothermal reactions within the rocky core of Enceladus by performing the experiments at pressure of $400\,bar$ ($\sim 150\,km$ below the water–rock boundary). The possibility of the occurrence of hydrothermal reactions at the ocean–rock interface of Enceladus' ocean will be discussed below in the Discussion section. The experimental conditions are summarized in Supplementary Table 1.

Dissolved gases and metals. The measured concentrations of dissolved gas species and metallic ions in the fluid samples of the experiments can be compared directly with Cassini's observations of Enceladus' plume compositions[1,3]. Based on chemical equilibrium[17,18], the lack of abundant N_2 in the plumes[19] might suggest the absence of hydrothermal activity, as it was proposed that N_2 should form by the decomposition of NH_3 at high temperatures[17,18] ($\geq 200\,^\circ C$). However, our experimental results indicate that no N_2 was produced from NH_3 and that NH_3 remains unaltered even at $300\,^\circ C$ (Fig. 1 and Supplementary Fig. 2) (N_2 production $< \sim 50\,\mu mol\,kg^{-1}$ H_2O: also see Supplementary Table 2 and Supplementary Note 1). These results indicate that the decomposition of NH_3 is kinetically inhibited and is not catalysed by olivine, pyroxene or their alteration minerals under our experimental conditions. Given high activation energy for reducing-oxidizing reactions of N_2, it has been suggested that catalysts would be required to promote these reactions at $500–1000\,^\circ C$ (refs 20–22). Typical catalysts attempted for decomposition of aqueous NH_3 are platinum group, transition metals or their oxides[20,21]. However, these catalytic decomposition reactions of aqueous NH_3 at high temperatures usually requires a significant amount of effective oxidants such as O_2 (refs 20–22), which is probably unavailable in Enceladus. Thus, our experimental results indicate that the lack of N_2 in the plumes[19] is not indicative of the absence of hydrothermal reactions.

Our results also show that the conversion of CO_2 to CH_4 is suppressed (Fig. 1) (also see Supplementary Fig. 2, Supplementary Table 2, and Supplementary Note 1), as reported previously[23,24]. Based on chemical equilibrium calculations, previous studies hypothetically discuss the conversion of CO_2 into CH_4 under hydrothermal conditions in Enceladus[17,18]. In fact, given the presence of metallic grains, such as Fe-Ni alloy, in meteorites, the conversions of CO_2 to CH_4 would have proceeded in Enceladus through Fischer–Tropsch-type reactions[25]. However, McCollom and Seewald[23] showed that these metallic catalysts were rapidly deactivated over time[23], suggesting the loss of catalytic activity under sub-to-supercritical conditions over geological timescales. Thus, we suggest that the presence of abundance of CO_2 in the

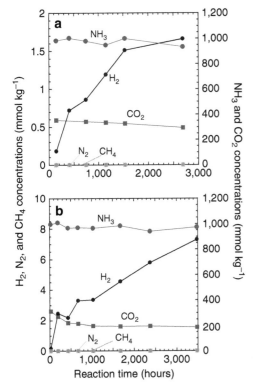

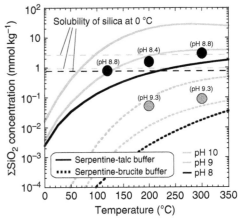

Figure 2 | Experimental and calculation results of dissolved silica concentrations. Results of ΣSiO_2 ($= SiO_{2(aq)} + HSiO_3^- + NaSiO_{3(aq)}$) are shown as a function of temperature at variable pH. Black and grey circles are the measured ΣSiO_2 in fluid samples of the opx and olivine experiments, respectively. The results of total silica concentrations in the opx experiments are also given in ref. 5. The numbers annotated to the experimental data are the calculated in situ pH values (see Methods). Solid lines are ΣSiO_2 values in chemical equilibrium at 400 bars according to the following reaction between serpentine and saponite/talc (serpentine–talc buffer): serpentine $+ 2SiO_{2(aq)} \leftrightarrow$ talc (saponite) $+ H_2O$. Dotted lines are ΣSiO_2 values in chemical equilibrium at 400 bars according to the following reaction between serpentine and brucite (serpentine–brucite buffer): serpentine $+ H_2O \leftrightarrow 3$brucite $+ 2SiO_{2(aq)}$. The concentrations of $HSiO_3^-$ and $NaHSiO_{3(aq)}$ were calculated for different pH values and at a constant Na^+ concentration (100 mmol kg^{-1}) using the equilibrium constants of the following reactions: $SiO_{2(aq)} + H_2O \leftrightarrow HSiO_3^- + H^+$ and $HSiO_3^- + Na^+ \leftrightarrow NaHSiO_{3(aq)}$. Horizontal broken lines show the solubility of amorphous silica at 0 °C and 100 bars for each pH value.

Figure 1 | Variations in the concentrations of dissolved gas species. Results of H_2, N_2, CH_4, ΣCO_2 ($= CO_{2(aq)} + CO_3^{2-} + HCO_3^-$) and ΣNH_3 ($= NH_{3(aq)} + NH_4^+$) during the experiments of (**a**) the opx experiment at 300 °C, and (**b**) the olivine experiment at 300 °C. Dissolved H_2 was generated through the oxidation of Fe(II) in olivine to magnetite and serpentine, which were observed in the rocks after the experiments. The decreasing ΣCO_2 is due to the formation of carbonate in the solid phase. The ΣNH_3 concentrations are high and almost constant during the experiments. Our results provide no evidence for CH_4 or N_2 production from CO_2 or NH_3, respectively (CH_4 production $< 5 \, \mu mol \, kg^{-1}$; N_2 production $< 50 \, \mu mol \, kg^{-1}$) (see Supplementary Note 1). The experimental data are given in the Supplementary Table 2.

plume[3] also does not indicate the absence of hydrothermal reactions in Enceladus.

Furthermore, Fe^{2+}, Mg^{2+} and Ca^{2+} become depleted in alkaline hydrothermal fluids (Supplementary Fig. 3 and Supplementary Table 3). These elements tend to be distributed in the rocky phase under alkaline hydrothermal conditions. Thus, the non-detection of these elements in the solid components of the plume[1] also supports the proposition that the solution compositions of Enceladus' ocean are controlled mainly by hydrothermal reactions involving alteration minerals under alkaline conditions.

Dissolved silica concentration. We observed considerable variability in the dissolved total silica concentrations ($\Sigma SiO_2 = SiO_{2(aq)} + HSiO_3^- + NaHSiO_{3(aq)}$) measured at the end of the experiments (Fig. 2). The presence of unaltered starting minerals in solid samples collected after the experiments (Supplementary Figs 4–6) and the observed continuous H_2 formation during the experiments (Fig. 1) indicate that mineral alteration was still occurring after 3–10 months of the reaction time. However, ΣSiO_2 in fluid samples reached steady levels within several months of reaction time (Supplementary Fig. 3). These steady-state levels of ΣSiO_2 in the opx experiments are always much higher than those in the olivine experiments (Fig. 2). For instance, at 300 °C, ΣSiO_2 in the opx experiment was ~ 30 times that in the

olivine experiment (Fig. 2). Fig. 2 also illustrates that ΣSiO_2 generally increases with reaction temperature. In the opx experiments, ΣSiO_2 at 300 °C was higher than that at 120 °C by a factor of ~ 4.

To understand the factors that determine the observed trends, mineralogical and chemical analyses of the hydrothermally altered solid samples were performed (Supplementary Figs 4–6 and Supplementary Table 4). The major alteration products of the olivine experiments were serpentine (chrysotile), along with brucite, magnetite and carbonate (magnesite and dolomite), which is consistent with previous experiments and calculations of terrestrial ultramafic-based hydrothermal vents[26,27] and with the proposed surface materials on Ceres[15]. In contrast, the alteration products of the opx experiments were dominated by serpentine (chrysotile) and saponite, along with talc, magnetite and carbonate (also see ref. 5), which are typical of carbonaceous chondrites[16]. Given the efficient oxidization of Fe(II) via high-temperature hydrothermal reactions[26,27] (≥ 150 °C), the molar ratio of Mg to Si in the starting minerals is important in determining the major compositions of the alteration minerals. Serpentine and brucite are dominant in hydrous minerals when the Mg/Si ratio is high (Mg/Si > 1.5), whereas serpentine and saponite/talc are the major hydrous silicates when the starting minerals are Si rich (Mg/Si < 1.5).

In geothermal fields on Earth, the ΣSiO_2 value of fluids is thought to be strongly influenced by reactions of alteration minerals, including serpentine, brucite and talc[28]. Fig. 2 shows the calculated equilibrium concentrations of ΣSiO_2 for the reaction between serpentine and talc (that is, serpentine $+ 2SiO_{2(aq)} \leftrightarrow$ talc (saponite) $+ H_2O$: serpentine–talc buffer) and between serpentine and brucite (that is, serpentine $+ H_2O \leftrightarrow 3$brucite $+ 2SiO_{2(aq)}$: serpentine–brucite buffer) (also see the caption of Fig. 2). Fig. 2 indicates that the calculated values of ΣSiO_2 for the two buffer systems are in good agreement with the measured values of ΣSiO_2 in the opx and olivine experiments, respectively. Because of the relatively slow formation of the alteration minerals in the experiments, the measured ΣSiO_2 would not be controlled by the buffer systems for the initial periods of reaction time (Supplementary Fig. 3). However, as the alteration reactions proceed, the measured ΣSiO_2 contents are determined by the chemical equilibrium of the buffer systems, regardless of the presence of unaltered starting minerals. These results strongly suggest that the ΣSiO_2 content of fluids in Enceladus' interior is also controlled by the buffer systems of the alteration minerals over geological timescales.

The formation of nanosilica colloids in a cooling silica-saturated solution explains Cassini's observations[5]. In this scenario, high-temperature fluids in chemical equilibrium with rocks of the core enter and mix with a low-temperature ocean[5]. Colloidal silica nanoparticles form upon cooling in the ocean when the ΣSiO_2 content of hydrothermal fluids exceeded the solubility of amorphous silica[5]. As in a previous associated study[5], the present study assumes that silica nanoparticles are generated in an ocean at $0\,°C$ (ref. 4), which provides a lower limit on the required temperature of hydrothermal reactions. Fig. 2 shows that ΣSiO_2 for the serpentine–talc buffer exceeds the solubility of silica at $0\,°C$ when the fluid temperature becomes sufficiently high[5] ($\geq 90\,°C$). In contrast, ΣSiO_2 for the serpentine–brucite buffer is much lower than the solubility of silica for fluid temperatures of $\leq 350\,°C$. These results suggest that to sustain high ΣSiO_2 contents sufficient to form silica nanoparticles, hydrous silicates on Enceladus should have been dominated by serpentine and saponite/talc, which are similar phases to those found in carbonaceous chondrites[16]. These results further imply that Enceladus' rocky core would not have experienced large-scale silicate melting and formation of more ultramafic rocks, if hydrothermal reactions took place in the core.

Discussion

A previous study indicates that to form silica nanoparticles, the temperature of fluids on Enceladus needs to exceed $\sim 90\,°C$ if fluid pH remains constant upon cooling[5]. However, fluid pH is highly likely to change upon cooling and mixing with seawater. Fig. 3 shows the minimum temperatures of hydrothermal reactions for the serpentine–talc buffer for different fluid pH values as a function of seawater pH, required to produce silica nanoparticles on Enceladus. Enceladus' seawater is suggested to be mildly alkaline (pH ~ 8.5–10.5), based on both the composition of emitted salt-rich grains[1,2] and the stable existence of silica nanoparticles[5]. On the other hand, pH values of pore water in the rocky core are only roughly constrained[5] (pH $> \sim 8.5$).

If the ocean-core system in Enceladus is chemically closed to other volatile reservoirs, such as the icy crust, the pH values of hydrothermal fluids and the ocean would be controlled by water–rock interactions[11]. In such a chemically closed system, fluid pH tends to increase upon cooling[11]. In fact, our experimental results show that the pH values of fluids range 8–9 at high temperatures (120–$300\,°C$) and increase to ~ 10 upon cooling to the room temperature ($\sim 15\,°C$) in the opx experiments (Supplementary Table 3). This is because the dissociation constant of H_2O to H^+ and OH^- has a maximum at 200–$300\,°C$, and because the conversion of NH_3 and H^+ to NH_4^+ tends to proceed at lower temperatures. Our experimental results show that the thermal decomposition of NH_3 to N_2 is efficiently inhibited even at high temperatures (Fig. 1), which, in turn, facilitates the increases in

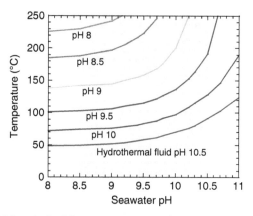

Figure 3 | Required minimum temperatures of hydrothermal fluids on Enceladus. Results are obtained from the cross-sections between the solubility of silica at $0\,°C$ and the ΣSiO_2 values determined by the serpentine–talc buffer as a function of seawater pH at $0\,°C$ for different values of hydrothermal fluid pH. In other words, the required minimum temperatures are determined when the ΣSiO_2 value for a given hydrothermal fluid exceeds the solubility of silica at $0\,°C$. The silica solubility is obtained for Na^+ concentration of $0.1\,\mathrm{mol\,kg}^{-1}$ and pressure 30 bar. The pH of Enceladus' seawater has been suggested to be in the range of 8.5–10.5 (refs 1,5) (the shaded area). The solid lines show the results when hydrothermal fluid pH values change to seawater pH values by cooling and mixing with oceanic water (see text).

fluid pH upon cooling. These results suggest that pH values of NH_3-containing fluids increase, possibly by one unit or more, upon cooling. If pH values increase by one unit upon cooling, required temperatures become $\sim 200\,°C$ for seawater pH of 8.5–10.5 (Fig. 3), as shown previously[5]. Given that the required temperatures may vary within $\sim 20\,°C$ depending on Na^+ and ΣCO_2 concentrations and pressure[5], hydrothermal activity at $\geq \sim 150$–$200\,°C$ is required to account for the formation of silica nanoparticles on Enceladus for a chemically closed, ocean-core system in Enceladus.

On the other hand, if Enceladus' ocean is chemically open to the icy crust through effective volatile exchanges[10], pH values of fluids and ocean are not determined simply by a change in dissolved species upon cooling. In fact, pH values of hydrothermal fluids may possibly be close to, or even higher than, those of oceanic water if there is a significant difference in ΣCO_2 concentrations in solutions between the ocean and hydrothermal fluids. Such differences in ΣCO_2 can occur when pore water contains a lower ΣCO_2 concentration than the seawater due to formation of carbonates and organic matter by interactions with the rocks; whereas a high ΣCO_2 concentration in the ocean is sustained by a supply from CO_2 clathrates in the icy crust[10]. In this case, pH values of the pore water could be moderately to strongly alkaline because of low abundances of CO_3^{2-} and HCO_3^-, and those of the ocean would be mildly alkaline buffered by a $NaHCO_3$ or $NaCO_3$ system[1,29]. Thus, the existence of a ΣCO_2 gradient in Enceladus would lower the minimum temperatures of water–rock interactions required for the formation of silica nanoparticles, compared with those for a closed system (Fig. 3). However, strongly alkaline solutions (pH $= 11$–13) buffered by a NaOH system[29] are unlikely to occur within Enceladus because of the presence of abundance of CO_2 in the interior[3]. In addition, given a plausible rapid water circulation in Enceladus inferred from the size of silica nanoparticles expelled by the plumes[5], a large difference in ΣCO_2 between the ocean and core tends to be mitigated over geological timescales. Thus, even when Enceladus' ocean is largely affected by CO_2 supply from the icy crust, we conclude that hydrothermal fluids should be

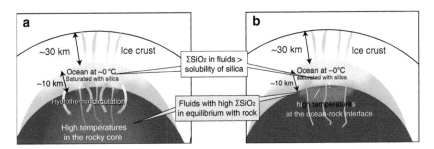

Figure 4 | Schematic illustration of hydrothermal circulations and the formation of silica on Enceladus. (a) Deep hydrothermal circulation would have occurred between a warm, and probably porous[5-7,45,70], rocky core and a cold ocean because of remnant heat from the early stages of Enceladus' evolution. **(b)** If heating has been induced by a recent heating event such as crustal overturn[41], orbital evolution[42] or an impact[43], hydrothermal reactions would have taken place at the ocean-rock interface. In this case, serpentinization and its associated heat production may have been important in sustaining high-temperature water-rock interactions.

moderately alkaline at most (pH ~ 8.5–10.5). In the extreme case of fluid pH of 10.5 and seawater pH of 8.5 within the range constrained by the previous studies[1,5], the minimum temperature required to form silica nanoparticles becomes $\sim 50\,°C$ (Fig. 3). A lower pH value of hydrothermal fluids requires a higher minimum temperature for the formation of silica nanoparticles (Fig. 3). Thus, we conclude that whether Enceladus' ocean is chemically open or closed, high-temperature water-rock reactions ($> \sim 50\,°C$) would be required in the interior.

It has been indicated that the silica nanoparticles of 2–8 nm in radius observed in Saturn's stream particles must have been formed by recent or ongoing hydrothermal activity on Enceladus, because the growth of several nm sized silica particles, for example, by Ostwald ripening, in the subsurface ocean would take months to several years at most[5]. Our experimental results provide additional supporting evidence for the presence of ongoing hydrothermal activity on Enceladus. We show that chemical equilibrium between dissolved silica and alteration minerals is achieved within months, even at a relatively low hydrothermal temperature of $120\,°C$ (Supplementary Fig. 3). These results suggest that if the interior of Enceladus had become completely cold ($\ll 100\,°C$) and hydrothermal reactions had ceased, then ΣSiO_2 in the ocean would have reached low levels over a geologically short time, as determined by the equilibrium at low temperatures (Fig. 2).

We estimated the dissolution rate of silica nanoparticles in an unsaturated solution. Previous work shows that the dissolution rate of silica nanoparticles in pure water at $0\,°C$ and pH 5.7 is $8.8 \times 10^{-15}\,\mathrm{cm\,s^{-1}}$ ($\sim 3\,\mathrm{nm\,year^{-1}}$)[30]. Under these conditions, even a several nm sized particle is estimated to dissolve in a few years. In an alkaline and NaCl-rich solution, the dissolution proceeds much more quickly (by one or two orders of magnitude) than that in pure water[30,31]. This suggests that silica nanoparticles would have readily dissolved over geological timescales after ΣSiO_2 in the ocean fell below the solubility of silica. Accordingly, the formation of silica nanoparticles is most likely sustained by geologically recent or ongoing hydrothermal activity.

Although the energy budget of Enceladus' current geological activity remains unclear[32,33], constraints derived from the observations of silica nanoparticles may help our understanding of the interior structure and thermal evolution of Enceladus. We show that the rocky core of Enceladus is most likely composed of Si-rich, carbonaceous chondritic rocks and that the core has not experienced large-scale silicate melting and therefore remains porous. This view is in agreement with the recent findings of a low-density core as inferred from the gravity data[6,7]. These results regarding the properties of the rocky core imply that the oceanic water could penetrate deep below the ocean-rock interface[34], resulting in deep hydrothermal circulation driven by remnant

heat of the early stages of Enceladus' evolution stored in the deep core (Fig. 4a). Thermal evolution models suggest that Enceladus' core reached high temperatures due to short-lived radiogenic heating and was dehydrated in the early stage of its evolution[35,36], if it formed within 4 million years (Myrs) of the formation of the solar system. Such radiogenic heat together with steady and episodic tidal dissipation heating could be retained in the deep core for a long time (on the order of 2 billion years (Gyrs) or more)[37]. In this case, exothermic re-serpentinization of the deep core would have subsequently occurred and could have kept the interior warm for longer[37]. A formation age of Enceladus, and thus the Saturnian system, within 4 Myrs of the formation age of the solar system is consistent with the formation age proposed for Iapetus[38] and with the typical lifetime of protoplanetary disks around the Sun-like stars[39,40].

However, given that a porous rocky core tends to lose remnant heat rapidly, especially if it is percolated by the oceanic water, it may be more likely that hydrothermal activity on Enceladus was triggered by a recent incidental heating event (for example, a catastrophic crustal overturn[41], an orbital evolution[42] or an impact[43]). The thickness of plume particles deposits on the small Saturnian satellites also implies that the duration of cryovolcanic activities on Enceladus would be as short as 10 Myrs (ref. 44). These incidental heating event could have increased the temperature near the ocean-rock interface (Fig. 4b). It is highly uncertain whether this event alone could have produced a sufficient amount of heat to cause hydrothermal activity, because such an event provides heat mainly in the icy shell rather than in the rocks of the seafloor. However, if Enceladus' rocky core is fragmented, the incidental events would have triggered effective tidal dissipation within the core[45], especially near the ocean-rock interface. In addition, if the ocean-rock interface had contained pristine minerals, such an event might have initiated ice melting and subsequent exothermic serpentinization. This in turn could have triggered a positive feedback between serpentinization, temperature increase and large tidal dissipation[32,45], possibly leading to hydrothermal reactions. Numerical simulations of Europa's ocean[46,47] show that hydrothermal plumes produce upwelling currents at a velocity of 1–$5\,\mathrm{cm\,s^{-1}}$. Although simulations for Enceladus' ocean are required to evaluate the intensity of upwelling currents, these results imply that hydrothermal plumes[46,47] and water convection[8,48] in the ocean could have transported nanoparticles from the seafloor to the plume source near the ice-ocean interface.

We propose that the temperature of ongoing hydrothermal reactions on Enceladus may be sufficiently high to cause effective Fe(II) oxidization associated with serpentinization, especially when the reactions occur in a relatively chemically closed system. However, if Enceladus has been warm since its formation, the rocky core might have already become completely serpentinized and oxidized by past water–rock reactions[34]. The available data on Enceladus' plumes are insufficient to determine whether hydrothermal activity is involved in ongoing serpentinization and Fe(II) oxidization. Our experimental results (Fig. 1 and Supplementary Fig. 2) and theoretical modelling[12] indicate that further evidence for ongoing serpentinization on Enceladus would be high levels of H_2 in the plumes ($\gg 1\,mmol\,kg^{-1}$), which may be testable via *in situ* measurements by Cassini and future missions. Although Cassini's observations during a recent series of low-velocity flybys of Enceladus show that the presence of substantial abundances of H_2 in the plume[10], it is still unclear whether the hydrogen is native or generated by reactions with titanium wall of the INMS antechamber[3]. Serpentinization could also support the emergence and survival of possible chemoautotrophic life on Enceladus through the provision of reducing power (that is, H_2) into CO_2-rich water, as proposed for early Earth and Mars[49,50]. Our experiments suggest that H_2 generation on Enceladus is as efficient as that in terrestrial ultramafic-hosted hydrothermal vents[27] (Fig. 2), where H_2-based microbial ecosystems are sustained[51].

Methods

Hydrothermal experiments. The apparatus and methodology of the hydro-thermal experiments conducted in the present study were based on the previous study simulating terrestrial hydrothermal vents[52,53]. Supplementary Fig. 1 shows the schematic diagram of the flexible gold reaction cell and steel (Inconel)-alloy autoclave used in the present study. The flexible cell consisted of a gold bag with a titanium head (Supplementary Fig. 1). The surface of the titanium head was oxidized by combustion before use to avoid catalytic reactions on the surface of metallic titanium. The inside wall of the sampling tube was coated with gold to avoid catalytic reactions. The flexible gold reaction cell was heated at 500 °C for 3 h in air to remove potential contamination of organic matter before each experiment.

Olivine used in the experiments of the present study was extracted from mantle peridotite originated from San Carlos ($Mg_{1.8}Fe_{0.2}SiO_4$). As natural olivine was used, it included minor minerals such as orthopyroxene, clinopyroxene and spinel, which have provided Al, Ca, and other elements in the fluids and alteration minerals (see the caption of Supplementary Fig. 6). In addition, trace amounts of transition elements, such as Ni and Mn, were also contained in olivine. Orthoenstatite crystals were synthesized by the flux method[54]. Special grade reagents of MgO and SiO_2 were mixed with enstatite stoichiometry, and added to the flux of special grade Li_2O_3, MoO_3, and V_2O_5 that was mixed in the proportions of 34.3, 55.9 and 9.8 wt.%, respectively. The weight ratio of the nutriment (that is, $MgO + SiO_2$) to the flux was 0.05. The mixture of nutriment and flux were heated in a platinum crucible at 970 °C for 100 h in the air, and then cooled to 730 °C at an average cooling rate of $2\,°C\,h^{-1}$ and to room temperature at $\sim 100\,°C\,h^{-1}$. Synthesized orthoenstatite crystals were separated from the solvent by washing in hot water. In the opx experiments, we mixed the synthesised opx with San Carlos olivine (the bulk composition of the opx-olivine mixtures: $Mg/Si = \sim 1.2$) to reproduce the Mg/Si ratios of carbonaceous chondrites ($Mg/Si = \sim 1.0$–1.1). Given the stoichiometry of alteration reactions, alteration mineral assemblages would change drastically at $Mg/Si = \sim 1.5$ (see the text); accordingly, we consider that the rock mixtures of the opx experiments can approximately simulate the reactions of carbonaceous chondritic rocks.

The initial concentrations of NH_3 and $NaHCO_3$ in the starting solution were 1.1×10^3 and $3.6 \times 10^2\,mmol\,kg^{-1}$ H_2O, or 2% and 0.7% relative to H_2O, respectively. These concentrations are comparable to volatile compositions of comets[55] (NH_3: $\sim 1\%$ relative to H_2O) and to Na^+ abundances observed in Enceladus' plumes[1] (~ 1–$3 \times 10^2\,mmol\,kg^{-1}$ H_2O), respectively. The ΣCO_2 ($= CO_{2(aq)} + CO_3^{2-} + HCO_3^-$) concentrations in the experiments may be 1/2 to 1/10 of the CO_2 abundance in comets[55]. However, the plume activities might have resulted in a loss of CO_2 throughout the history of Enceladus[8], and the change in ΣCO_2 would not change our conclusions significantly. Isotopic labelling was used for the species in the starting aqueous solution (that is, $^{15}NH_3$ with 1% of ^{15}N; $NaH^{13}CO_3$ with 10% of ^{13}C) to verify the products of hydrothermal reactions, such as CH_4 and N_2.

The starting minerals were powdered with an alumina mortar and ultrasonically cleaned with acetone first and then pure water, before use. The size of the powdered minerals was typically ~ 10–$200\,\mu m$. Mixtures of aqueous solution ($\sim 60\,g$) and starting minerals ($\sim 15\,g$) were introduced into the reaction cell. The initial water/rock ratio was fixed at ~ 4 in the experiments, because the water/rock ratio in submarine hydrothermal environments on Earth is considered to be limited to $< \sim 5$ (ref. 56). During the experiments, fluid samples of $\sim 2\,g$ were collected. Thus, the water/rock ratio decreased to ~ 3 at the end of the experiments because of fluid sampling.

During the experiments, the flexible cell collapses as fluids are removed during sampling, which allowed us to conduct online sampling of the fluids at a near constant temperature and pressure condition without vapour phase present[52,53]. Glass and Teflon vials purged with pure Ar gas were connected with the sampling valve and used to sample fluids in the flexible cell (Supplementary Fig. 1). The blank experiments (that is, without starting minerals or dissolved species) were performed in our previous study[57], which used the same experimental system as the present study. The H_2 concentration throughout the blank experiment is $\sim 0.008\,mmol\,kg^{-1}$ or less[57], which is negligible compared with the H_2 concentrations in our experiments (Fig. 1). More detailed descriptions on the experimental systems may be found in refs 52,57.

The experiments were conducted at a constant pressure of 400 bar, corresponding to the pressure of the interior of Enceladus' core. The temperature conditions of the olivine experiments were 200 and 300 °C, and those of the opx experiments were 120, 200 and 300 °C (Supplementary Table 1). In the opx experiments at 120 and 200 °C, we started the experiment at 120 °C for the first ~ 2 months, and then increased the temperature to 200 °C and continued the experiment for another ~ 1 month. The other experiments were performed at constant temperatures. The durations of the experiments were 2–10 months.

Chemical analyses of dissolved gas species were performed using a gas chromatograph (GC-2014 Shimadzu) and ion chromatograph (ICS-1600 DIONEX) at JAMSTEC, Japan Agency for Marine-Earth Science and Technology, and a gas chromatograph-mass spectrometer (GCMS-QP2010 Shimazdu) at the University of Tokyo. Inductively coupled plasma atomic emission spectroscopy (PerkinElmer) was also conducted at JAMSTEC to measure the concentrations of dissolved elements. Mineralogical and chemical analyses of the rocks were conducted using an X-ray diffraction spectrometer (X'PERT-PRO PANanlytical) and a scanning electron microscope with an electron probe microanalyser (JXA-8200 JEOL) at the University of Tokyo.

Equilibrium calculations. For calculating ΣSiO_2 determined by the equilibrium of secondary minerals and the solubility of amorphous silica at 0 °C, we used the equilibrium constants computed by the SUPCRT92 program[58]. Although Na-rich saponite was the major alteration mineral found in samples collected after the opx experiments, we used thermodynamic data for talc [$Mg_3Si_4O_{10}(OH)_2$] rather than for Na-rich saponite [$(Na)_{0.3}Mg_3(Si,Al)_4O_{10}(OH)_2$]. This is because of the absence of Na-rich saponite in the database[58] and the similarity in the chemical formulae of these minerals. We also used the thermodynamic data of amorphous silica for nanoparticles. This assumption provides a lower limit of the temperature required for hydrothermal reactions to form nanosilica, because silica nanoparticles are less stable than amorphous silica because of the difference in surface free energy.

In situ pH calculations. *In situ* pH ($pH_{in\,situ}$) was calculated with the Geochemist's Workbench computer code[59] based on pH at room temperature ($pH_{25°C}$) and concentrations of dissolved elements and species in fluids collected at the final sampling of the experiments (Supplementary Tables 2 and 3). In this $pH_{in\,situ}$ calculations, charge balance was constrained from the $pH_{25°C}$ value, while Na was used as the element compensating the imbalanced charge derived from analytical errors. The thermodynamic database required for this calculation was generated by the SUPCRT92 computer program[58], with thermodynamic data for mineral, aqueous species and complexes from refs 60–65. The B-dot activity model was used[66,67]. The temperature-dependent activity coefficient for aqueous CO_2 was derived from the empirical relationship of ref. 68 and the temperature-dependent activity of water in a NaCl solution was derived from the formulation of ref. 59. Cleverley and Bastrakov[69] provide useful temperature-dependent polynomial functions for both these last two parameters[69]. Although the calculations were carried out using a total pressure of 500 bars, pressure is a minor factor since the equilibrium constants are not sensitive to the modest changes in pressure.

References

1. Postberg, F. *et al.* Sodium salts in E-ring ice grains from an ocean below the surface of Enceladus. *Nature* **459**, 1098–1101 (2009).
2. Postberg, F., Schmidt, J., Hillier, J., Kempf, S. & Srama, R. A salt-water reservoir as the source of a compositionally stratified plume on Enceladus. *Nature* **474**, 620–622 (2011).
3. Waite, Jr J. H. *et al.* Liquid water on Enceladus from observations of ammonia and ^{40}Ar in the plume. *Nature* **460**, 487–490 (2009).
4. Schmidt, J., Brilliantov, N., Spahn, F. & Kempf, S. Slow dust in Enceladus' plume from condensation and wall collisions in tiger stripe fractures. *Nature* **451**, 685–688 (2008).
5. Hsu, H.-W. *et al.* Silica nanoparticles as an evidence of hydrothermal activities at Enceladus. *Nature* **519**, 207–210 (2015).
6. Iess, L. *et al.* The gravity field and interior structure of Enceladus. *Science* **344**, 78–80 (2014).
7. McKinnon, W. B. Effect of Enceladus's rapid synchronous spin on interpretation of Cassini gravity. *Geophys. Res. Lett.* **42**, 2137–2143 (2015).
8. Matson, D. L., Castillo-Rogez, J. C., Davies, A. G. & Johnson, T. V. Enceladus: a hypothesis for bringing both heat and chemicals to the surface. *Icarus* **221**, 53–62 (2012).

9. Mousis, O. *et al.* Formation conditions of Enceladus and origin of its methane reservoir. *Astrophys. J. Lett.* **701**, 39–42 (2009).

10. Bouquet, A., Mousis, O., Waite, J. H. & Picaud, S. Possible evidence for a methane source in Enceladus' ocean. *Geophys. Res. Lett.* **42**, 1334–1339 (2015).

11. Zolotov, M. Y. An oceanic composition on early and today's Enceladus. *Geophys. Res. Lett.* **34**, L23203 (2007).

12. Glein, C. R., Baross, J. A. & Waite, Jr J. H. The pH of Enceladus' ocean. *Geochim. Cosmochim. Acta* **162**, 202–219 (2015).

13. Nakamura, T. *et al.* Chondrulelike objects in short-period comet 81 P/Wild 2. *Science* **321**, 1664–1667 (2008).

14. Crovisier, J. *et al.* The spectrum of comet Hale-Bopp (C/1995 O1) observed with the infrared space observatory at 2.9 astronomical units from the sun. *Science* **275**, 1904–1907 (1997).

15. Milliken, R. E. & Rivkin, A. S. Brucite and carbonate assemblages from altered olivine-rich materials on Ceres. *Nat. Geosci.* **2**, 258–261 (2009).

16. Brearley, A. J. in *Meteorites and the Early Solar System II* (eds Lauretta, D. S. & McSween, H. Y.) 587–624 (University of Arizona Press, 2006).

17. Glein, C. R., Zolotov, M. Y. & Shock, E. L. The oxidation state of hydrothermal systems on early Enceladus. *Icarus* **197**, 157–163 (2008).

18. Matson, D. L., Castillo, J. C., Lunine, J. & Johnson, T. V. Enceladus' plume: compositional evidence for a hot interior. *Icarus* **187**, 569–573 (2007).

19. Hansen, C. J. *et al.* The composition and structure of the Enceladus plume. *Geophys. Res. Lett.* **38**, L11202 (2011).

20. Lee, D. K. Mechanism and kinetics of the catalytic oxidation of aqueous ammonia to molecular nitrogen. *Environ. Sci. Technol.* **37**, 5745–5749 (2003).

21. Oshima, Y., Inaba, K. & Koda, S. Catalytic supercritical water oxidation of coke works waste with manganese oxide. *Sekiyu Gakkaishi* **44**, 343–350 (2001).

22. Helling, R. K. & Tester, J. W. Oxidation of simple compounds and mixtures in supercritical water: carbon monoxide, ammonia, and ethanol. *Envion. Sci. Technol.* **22**, 1319–1324 (1988).

23. McCollom, T. M. & Seewald, J. S. Experimental constraints on the hydrothermal reactivity of organic acids and acid anions: I. Formic acid and formate. *Geochim. Cosmochim. Acta* **67**, 3625–3644 (2003).

24. McCollom, T. M., Lollar, B. S., Lacrampe-Couloume, G. & Seewald, J. S. The influence of carbon source on abiotic organic synthesis and carbon isotope fractionation under hydrothermal conditions. *Geochim. Cosmochim. Acta* **74**, 2717–2740 (2010).

25. Horita, J. & Berndt, M. E. Abiogenic methane formation and isotopic fractionation under hydrothermal conditions. *Science* **285**, 1055–1057 (1999).

26. Seyfried, Jr W. E., Foustoukos, D. I. & Fu, Q. Redox evolution and mass transfer during serpentinization: an experimental and theoretical study at 200 °C, 500 bar with implications for ultramafic-hosted hydrothermal systems at mid-ocean ridges. *Geochim. Cosmochim. Acta* **71**, 3872–3886 (2007).

27. McCollom, T. M. & Bach, W. Thermodynamic constraints on hydrogen generation during serpentinization of ultramafic rocks. *Geochim. Cosmochim. Acta* **73**, 856–875 (2009).

28. Frost, B. R. & Beard, J. S. On silica activity and serpentinization. *J. Petrol.* **48**, 1351–1368 (2007).

29. Zolotov, M. Y. Aqueous fluid composition in CI chondritic materials: chemical equilibrium assessments in closed systems. *Icarus* **220**, 713–729 (2012).

30. Icenhower, J. P. & Dove, P. M. The dissolution kinetics of amorphous silica into sodium chloride solutions: effects of temperature and ionic strength. *Geochim. Cosmochim. Acta* **64**, 4193–4203 (2000).

31. Brantley, S. L. in *Kinetics of Water-Rock Interaction* (eds Kubicki, J. D. & White, A. F.) 151–210 (Springer, 2008).

32. Spencer, J. R. *et al.* in *Saturn from Cassini-Huygens* (eds Dougherty, M. K., Esposito, L. W. & Krimigis, S. M.) 683–724 (Springer, 2005).

33. Shoji, D., Hussmann, H., Sohl, F. & Kurita, K. Non-steady state tidal heating of Enceladus. *Icarus* **235**, 75–85 (2014).

34. Vance, S. *et al.* Hydrothermal systems in small ocean planets. *Astrobiology* **7**, 987–1005 (2007).

35. Castillo-Rogez, J. C., Matson, D. L., Vance, S. D., Davies, A. G. & Johnson, T. V. in *Proceedings of the 38th Lunar and Planetary Science* Vol. 38, 2265 (League City, 2007).

36. Schubert, G., Anderson, J. D., Travis, B. J. & Palguta, J. Enceladus: present internal structure and differentiation by early and long-term radiogenic heating. *Icarus* **188**, 345–355 (2007).

37. Travis, B. J. & Schubert, G. Keeping Enceladus warm. *Icarus* **250**, 32–42 (2015).

38. Castillo-Rogez, J. *et al.* ^{26}Al-decay: heat production and a revised age for Iapetus. *Icarus* **204**, 658–662 (2009).

39. Wyatt, M. C. Evolution of debris disks. *Annu. Rev. Astron. Astrophys.* **46**, 339–383 (2008).

40. Pascucci, I. & Tachibana, S. in *Protoplanetary Dust: Astrophysical and Cosmochemical Perspectives* (eds Apai, D. & Lauretta, D. S.) 263–298 (Cambridge University Press, 2010).

41. O'Neill, C. O. & Nimmo, F. The role of episodic overturn in generating the surface geology and heat flow on Enceladus. *Nat. Geosci* **3**, 88–91 (2010).

42. Wisdom, J. Spin-orbit secondary resonance dynamics of Enceladus. *Astron. J.* **128**, 484–491 (2004).

43. Roberts, J. H. & Stickle, A. M. *Proceedings of the 46th Lunar and Planetary Science* Vol. 46, 1468 (The Woodlands, 2015).

44. Hirata, N., Miyamoto, H. & Showman, A. P. Particle deposition on the Saturnian satellites from ephemeral cryovolcanism on Enceladus. *Geophys. Res. Lett.* **41**, 4135–4141 (2014).

45. Roberts, J. H. The fluffy core of Enceladus. *Icarus* **258**, 54–66 (2015).

46. Vance, S. & Goodman, J. in *Europa.* (eds Pappalardo, R. T., McKinnon, W. B. & Khurana, K.) 459–482 (University of Arizona Press, 2006).

47. Goodman, J. C. & Lenferink, E. Numerical simulations of marine hydrothermal plumes for Europa and other icy worlds. *Icarus* **221**, 970–983 (2012).

48. Soderlund, K. M., Schmidt, B. E., Wicht, J. & Blankenship, D. D. Ocean-driven heating of Europa's icy shell at low latitudes. *Nat. Geosci.* **7**, 16–19 (2014).

49. Russell, M. J., Hall, A. J. & Martin, W. Serpentinization as a source of energy at the origin of life. *Geobiology* **8**, 355–371 (2010).

50. Schulte, M., Blake, D., Hoehler, T. & McCollom, T. Serpentinization and its implications for life on the early Earth and Mars. *Astrobiology* **6**, 364–376 (2006).

51. Kelley, D. S. *et al.* A serpentine-hosted ecosystem: the Lost City hydrothermal field. *Science* **307**, 1428–1434 (2005).

52. Shibuya, T. *et al.* Reactions between basalt and CO_2-rich seawater at 250 and 350 °C, 500 bars: Implications for the CO_2 sequestration into the modern oceanic crust and composition of hydrothermal vent fluid in the CO_2-rich early ocean. *Chem. Geol.* **359**, 1–9 (2013).

53. Seyfried, Jr W. E. Experimental and theoretical constraints on hydrothermal alteration processes at Mid-Ocean Ridges. *Annu. Rev. Earth Planet. Sci.* **15**, 317–335 (1987).

54. Tachibana, S., Tsuchiyama, A. & Nagahara, H. Experimental study of incongruent evaporation kinetics of enstatite in vacuum and in hydrogen gas. *Geochim. Cosmochim. Acta* **66**, 713–728 (2002).

55. Bocklelée-Morvan, D., Crovisier, J., Mumma, M. J. & Weaver, H. A. in *Comets II* (eds Festou, M. C., Keller, H. U. & Weaver, H. A.) 391–423 (University Arizona Press, 2004).

56. Wetzel, L. R. & Shock, E. L. Distinguishing ultramafic—from basalt-hosted submarine hydrothermal systems by comparing calculated vent fluid compositions. *J. Geophys. Res.* **105**, 8319–8340 (2000).

57. Yoshizaki, M. *et al.* H_2 generation by experimental hydrothermal alteration of komatiitic glass at 300 °C and 500 bars: A preliminary result from on-going experiment. *Geochem. J.* **43**, 17–22 (2009).

58. Johnson, J. W., Oelkers, E. H. & Helgeson, H. C. SUPCRT92: a software package for calculating the standard molal thermodynamic properties of minerals, gases, aqueous species, and reactions from 1 to 5000 bar and 0 to 1000 °C. *Comput. Geosci.* **18**, 899–947 (1992).

59. Bethke, C. M. *Geochemical and Biogeochemical Reaction Modeling* (Cambridge University Press, 2008).

60. Shock, E. L. & Helgeson, H. C. Calculation of the thermodynamic and transport properties of aqueous species at high pressures and temperatures: correlation algorithms for ionic species and equation of state predictions to 5 kb and 1000 °C. *Geochim. Cosmochim. Acta* **52**, 2009–2036 (1988).

61. Shock, E. L. & Koretsky, C. M. Metal-organic complexes in geochemical processes: Estimation of standard partial molal thermodynamic properties of aqueous complexes between metal cations and monovalent organic acid ligands at high pressures and temperatures. *Geochim. Cosmochim. Acta* **59**, 1497–1532 (1995).

62. Shock, E. L., Helgeson, H. C. & Sverjensky, D. A. Calculation of the thermodynamic and transport properties of aqueous species at high pressures and temperatures: standard partial molal properties of inorganic neutral species. *Geochim. Cosmochim. Acta* **53**, 2157–2183 (1989).

63. Shock, E. L., Sassani, D. C., Willis, M. & Sverjensky, D. A. Inorganic species in geologic fluids: correlations among standard molal thermodynamic properties of aqueous ions and hydroxide complexes. *Geochim. Cosmochim. Acta* **61**, 907–950 (1997).

64. Sverjensky, D. A., Shock, E. L. & Helgeson, H. C. Prediction of the thermodynamic properties of aqueous metal complexes to 1000 °C and 5 kb. *Geochim. Cosmochim. Acta* **61**, 1359–1412 (1997).

65. McCollom, T. M. & Shock, E. L. Geochemical constraints on chemolithoautotrophic metabolism by microorganisms in seafloor hydrothermal systems. *Geochim. Cosmochim. Acta* **61**, 4375–4391 (1997).

66. Helgeson, H. C. Thermodynamics of hydrothermal systems at elevated temperatures and pressures. *Am. J. Sci.* **267**, 729–804 (1969).

67. Helgeson, H. C. & Kirkham, D. H. Theoretical prediction of the thermodynamic behavior of aqueous electrolytes at high pressures and temperatures: II. Debye-Huckel parameters for activity coefficients and relative partial molal properties. *Am. J. Sci.* 1199–1261 (1974).

68. Drummond, S. E. *Boiling and Mixing of Hydrothermal Fluids: Chemical Effects on Mineral Precipitation* (PhD thesis: Pennsylvania State University, 1981).

69. Cleverley, J. S. & Bastrakov, E. N. K2GWB: utility for generating thermodynamic data files for the Geochemist's Workbench ® at 0–1000 °C and 1–5000 bar from UT2K and the UNITHERM database. *Comput. Geosci.* **31,** 756–767 (2005).

70. Collins, G. C. & Goodman, J. C. Enceladus' south polar sea. *Icarus* **189,** 72–82 (2007).

Acknowledgements

This study was supported by Grant-in-Aids for Scientific Research from the Ministry of Education, Culture, Sports, Science and Technology, Japan (23103003 and 15H05830), from the Japan Society for Promotion of Science (26707024 and 15H02142), from the Mitsubishi Foundation, from the JGC-S Scholarship Foundation, and from the Astrobiology Program of the National Institutes of Natural Sciences (NINS). Y.S., F.P. and H.-W.H. thank J. Castillo-Rogez, S. Charnoz and A. Crida for helpful discussions. Y.S. thanks K. Shimizu for his help in analyses of fluid samples and Y. Oshima, E. Shimoda and T. Sasaki for discussion on ammonia dissociation and catalytic reactions under hydrothermal conditions.

Author contributions

Y.S. performed the experiments and calculations simulating Enceladus and wrote the manuscript. T.S., K.S. and Y.M. designed the hydrothermal experiments and the analysis system. Y.S., T.S., F.P. and H.-W.H. produced the outline of the study. Y.S., T.K. and M.M. performed microscope analyses of the rock samples. S.T. synthesised the starting minerals for the experiments. S.S. calculated the lifetime of nanosilica in a solution. All the authors discussed and contributed intellectually to the interpretation of the results.

Additional information

Competing financial interests: The authors declare no competing financial interests.

Cracks in Martian boulders exhibit preferred orientations that point to solar-induced thermal stress

Martha-Cary Eppes[1], Andrew Willis[2], Jamie Molaro[3], Stephen Abernathy[1] & Beibei Zhou[2]

The origins of fractures in Martian boulders are unknown. Here, using Mars Exploration Rover 3D data products, we obtain orientation measurements for 1,857 cracks visible in 1,573 rocks along the Spirit traverse and find that Mars rock cracks are oriented in statistically preferred directions similar to those compiled herein for Earth rock cracks found in mid-latitude deserts. We suggest that Martian directional cracking occurs due to the preferential propagation of microfractures favourably oriented with respect to repeating geometries of diurnal peaks in sun-induced thermal stresses. A numerical model modified here with Mars parameters supports this hypothesis both with respect to the overall magnitude of stresses as well as to the times of day at which the stresses peak. These data provide the first direct field and numerical evidence that insolation-related thermal stress potentially plays a principle role in cracking rocks on portions of the Martian surface.

[1]Department of Geography and Earth Sciences, University of North Carolina at Charlotte, Charlotte, North Carolina 28223, USA. [2]Department of Electrical and Computing Engineering, University of North Carolina at Charlotte, Charlotte North Carolina 28223, USA. [3]Lunar and Planetary Laboratory, University of Arizona, Tucson, Arizona 85721, USA. Correspondence and requests for materials should be addressed to M.-C.E. (email: meppes@uncc.edu).

Physical weathering is the primary, non-tectonic mechanism for the mechanical breakdown of rock into smaller particles. On Mars, as on Earth, mechanical weathering is the precursor to sediment production and rock erosion, see, for example, refs 1–4, and can potentially influence chemical weathering and subsequent atmospheric feedbacks[5–8]. Hence, identifying the key drivers of weathering is therefore possibly tantamount to understanding the key drivers of landscape change on the Martian surface.

Rocks found in the Mars Exploration Rover (MER) images of the Martian surface commonly exhibit fractures that are visible without magnification (hereafter: 'cracks'; Fig. 1). In general, Mars cracks have similar characteristics to those observed in rocks found in Earth's deserts[2,9–11]. Various hypotheses have been proposed to explain non-Earth mechanical weathering, including damage by original ejecta emplacement[3], salt weathering[2,10,12] and thermal stress[13–16]. To date, however, no one has collected mechanical weathering data from extraterrestrial rocks themselves that might support, refute or quantify the relative importance of these processes.

Numerical modelling, field data and rock instrumentation convincingly target solar-induced thermal stress as a key mechanism responsible for physical weathering on Earth[17–21]. In particular, a majority of non-bedrock boulders found in Earth's mid-latitude deserts exhibit cracks with strongly preferred, roughly north–north-east orientations (Fig. 2), even when rock anisotropies such as bedding or foliation are taken into account[22–25]. These orientations are hypothesized to result from Earth's rotation[22,25], a geometric relationship which imparts cyclically occurring maximum thermal stresses[26] on boulders as they are directionally heated and cooled during the sun's daily east-to-west transit across the sky. Although other non-tectonic sources of fracture-inducing rock stress, such as freezing or salt precipitation, may also play a role in fracturing overall, these sources of stress by themselves should produce random crack orientations without the influence of the sun. It is unknown if rocks in other Earth deserts or in other celestial bodies exhibit preferred crack orientations.

Long-term orbital cycles on Mars are thought to be roughly analogous to those of Earth's Milankovitch cycles[27]. Although the Martian year is about 1.8 times as long as an Earth year, the Martian sidereal day is only about 40 min longer than that of Earth. Mars' obliquity, or axial tilt, is centred around roughly the same angle as Earth's[28], though known to vary by as much as ±20° more over intermediate timescales (~5–10 Myr ago). Although obliquity is thought to have varied significantly more and in chaotic ways over longer timescales[29], addressing how these variations might affect insolation-related cracking is beyond the scope of this paper.

Nevertheless, the overall similarity in geometry of insolation between Earth and Mars leads to the expectation that the resulting geometry of thermal stresses that arise, particularly near the equator where MER rovers are located, are also similar and might result in preferred orientations of cracks, albeit possibly at different angles. Herein, analysis of three-dimensional (3D) reconstructions of stereo pair photographs of the Martian surface and output from a two-dimensional (2D) model of diurnal thermoelastic stresses expected from simple insolation-related deformation reveals that Martian boulders do exhibit cracks with preferred orientations, and that calculated solar-induced thermal stresses for Martian rocks are consistent with solar-driven directional cracking.

Results

Identifying sampling bias potential in Spirit data products. We collected a detailed preliminary data set of rock and crack data (Methods and Supplementary Methods) to characterize potential sampling biases that might arise from making such 'field' measurements using MER Spirit PANCAM data products in combination with the 'ImageRover' software that we developed and validated (Supplementary Figs 1 and 2; Supplementary Table 1). We identified three important data set biases: (1) due to their visibility, cracks observed on Mars will be from relatively larger rocks (Supplementary Data set 1) compared with the data set derived from Earth rocks (Supplementary Data set 2); (2) images collected by the Spirit Rover were disproportionately collected with the PANCAM pointing parallel to the direction of rover motion (NW–SE; Fig. 3). Consequently, the azimuths of randomly chosen images such as those in this preliminary analysis reflect this directional predilection (Fig. 4a); and (3) because of the overall lower visibility of image-parallel-striking cracks in 2D MER images (Methods), the majority of cracks measured in

Figure 1 | Cracked rock on Mars. Example of a MER Spirit PANCAM image (data product 2p130443923eff0900p2555l7m1.img, Courtesy NASA/JPL-Caltech Planetary Data System) of a rock with visible cracks from the Martian surface. Image azimuth: 292°. Local True Solar Time: 10:45:08. Sol 46. Site 9. Solid arrows point to linear features that would meet our criteria for a crack (Methods). Dashed arrows point to features such as edges or wide voids that would not meet our criteria for a crack.

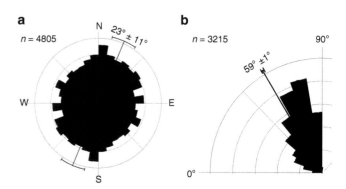

Figure 2 | Earth crack data compilation. Data compiled for this paper from other studies of Earth rock cracks in northern, and to a lesser extent southern, hemisphere, mid-latitude deserts (Methods and Supplementary Data set 2)[22–25,62]. Lines point to data vector mean with 95% confidence interval depicted by brackets. Data are plotted in 10° bins with frequency reported in %. (**a**) Circular bidirectional, histogram of crack strike orientations. Raleigh test P value <0.01. Rao's spacing test P value <0.01. (**b**) Circular histogram of dip angles from the same cracks where available.

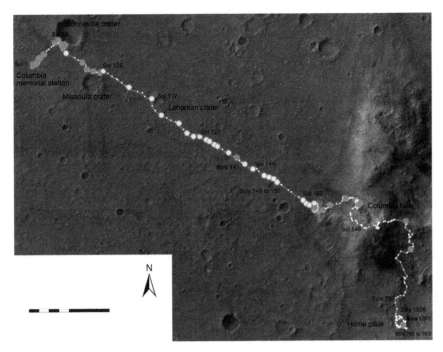

Figure 3 | Location map. Sites used for data collection for this study are marked on a map of the Spirit Traverse modified from an image by Ohio State University Mapping and GIS Laboratory[67] provided courtesy of NASA/JPL-CalTech Planetary Data System. The red lines and small white dots represent the path of the Spirit rover and its location by Sol. Larger light blue dots are located at sites where data were collected for the 133-rock data set (Supplementary Data set 1). Green dots are located at sites where data were collected for the 1,440-rock data set (Supplementary Data set 3). Orange dots are located at sites where data were collected for both data sets. Prominent Mars surface features are indicated by name on the map. Scale bar is 1,000 m.

any given image will be striking into, and out of, the photograph. Consequently, there is a strong correlation ($R^2 = 0.7$, Pearson $r = 0.9$, P value ≤ 0.01) between image azimuth and measured crack strike (Fig. 4b); the difference between image azimuth and measured crack strike was normally distributed about a mean of $1.7 \pm 27°$ (Fig. 4c).

Crack strike data measured for this preliminary analysis (Fig. 4d,e) therefore share the MER database NW–SE sampling bias of the images themselves (Fig. 4a) and cannot be considered in the context of weathering as sampled. Furthermore, because of the likely obscuration of a larger proportion of vertical to subvertical cracks compared with lower dip cracks (Methods), dip angles measured from this randomly chosen population of images are also therefore likely of a lower-angle on average compared with the actual population. The vector mean and 95% confidence interval of measured Mars crack dip angles was $39 \pm 3°$ (Fig. 4f and Supplementary Data set 1).

Nevertheless, crack dip direction measured from MER data should not be subject to visual bias. For example west- versus east-dipping cracks should be equally visible in a north–south-oriented photo. Although the time of the day might produce shadows in one direction over another, the images employed were obtained during mid-day hours (Local True Solar Time) with relatively high sun angles (Supplementary Data set 1), precluding bias due to preferential shadowing.

Dip directions observed in Mars rock cracks, as indicated by strikes collected with right-hand rule, were dominantly oriented to the north (75% of all dips; Fig. 4e). This promising result combined with our understanding of sampling biases of the Spirit database led us to the development of a crack measurement method, which would allow us to confidently examine the orientation of crack strikes for Martian rocks.

Martian rock crack strike orientation analysis. To accurately record rock crack strikes on the Martian surface, we employed a stratified sampling strategy to carefully mitigate effects due to observed sampling biases associated with rover traverse direction and crack visibility. Using randomly selected images whose azimuths fell within each of every 10° bin between 0 and 360°, we randomly selected 40 rocks, all >20 cm in maximum ImageRover-measured diameter (Methods; Supplementary Fig. 3). In all, we selected 1,440 rocks (40 rocks from each of 36 bins) located in 31 different sites along the Spirit traverse (Fig. 3; Supplementary Data set 3). Thus, the azimuths of the images for each of the 1,440 selected rocks are equally distributed around 360° (Fig. 5a,b), and this uniform distribution is statistically supported (Raleigh P value $= 0.995$; unity on a quantile–quantile plot in Fig. 5c). Both statistical tests are for the null hypothesis of uniformity, with Rao's testing for multiple modes and Rayleigh testing for a single mode. Because, as described above, crack strikes measured with ImageRover were found to almost equal the image azimuths in which they are found (Fig. 4), we asserted that all visible cracks in the 1,440 rocks will have strikes whose distribution is also normal and equal to that of image azimuths. Thus, if crack orientations on the Martian surface are random in all rocks, then the total numbers of visible cracks should be roughly equal in each of the 36 bins of data, and their overall distribution should mimic the uniform distribution of the image azimuths themselves.

We counted all visible cracks in each of the 1,440 rocks, and 1,635 cracks were observed in 875 rocks (Supplementary Data set 3). Overall, rocks exhibit about one crack per rock; however, this number is likely significantly smaller than the actual average given likely crack visibility limitations for MER images (Supplementary Methods). For each of the 1,635 cracks counted, we assigned it an orientation equal to that of its image azimuth (Fig. 6 and Supplementary Data set 3) to test if the numbers of visible cracks were uniformly distributed about 360°, as would be expected if crack orientations are random.

In the 1,440-rock data set, circular histograms and a quantile–quantile plot of the orientations of cracks reveals a non-uniform,

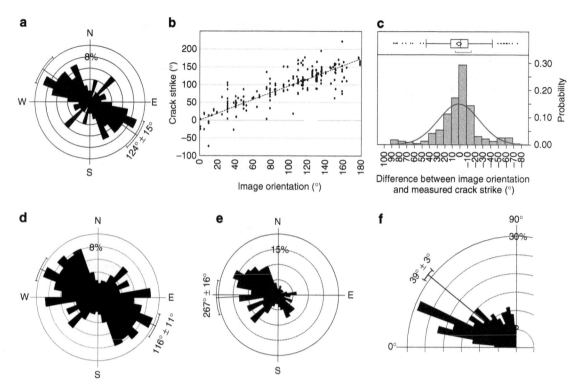

Figure 4 | 133-Rock data set. Graphs depicting orientation data collected using ImageRover for 133 rocks and their associated images and cracks. Lines and arcs on circular histograms indicate vector means and associated 95% confidence intervals. Circular histogram data are plotted in 10° bins with frequency reported in %. (**a**) Bidirectionally plotted circular histogram of the orientations of the azimuth of the random selection of images in which the cracks were measured ($n = 76$). (**b**) Graph showing the strong correlation between camera orientations and measured crack strike orientations ($n = 222$; $y = 0.9672x$; Pearson $r = 0.85$; Pearson P value ≤ 0.01; $R^2 = 0.71$). (**c**) Histogram, normal fit curve (red line) and box and whisker plot (diamond, mean; vertical line, median; red bracket, densest 50% of data; dots, outliers) of the difference between measured crack strike orientations and their associated image orientations ($n = 222$). (**d**) Bidirectionally plotted circular histogram of the strikes of the cracks ($n = 222$) measured directly from the 133 rocks using ImageRover. (**e**) Unidirectionally plotted circular histogram depicting same data as **b**. Because all crack strikes were measured using the right-hand rule, this graph indicates that measured cracks dip primarily to the northeast. (**f**) Unidirectionally plotted circular histogram depicting the dip angles measured for the cracks ($n = 222$).

preferred distribution of Mars rock crack strikes (Fig. 6a–c). The vector mean and 95% confidence interval for all 1,635 cracks is $46 \pm 20°$ with a Rayleigh P value of 0.018 and Rao's spacing test P value of <0.01.

To ensure that this preferred distribution was not related in some way to the visibility of cracks in the afternoon, given the slight preference towards afternoon time stamps in images of the entire data set (average time stamp $= 13{:}15 \pm 1$ h 35 min), we subsampled the 1,440-rock data set for 360 rocks (10 randomly selected rocks for each of the 36 bins from 0 to 360) to create a subset of rock images with an average time stamp of $\sim 11{:}15 \pm 1$ h, 7 min (Supplementary Methods). The vector mean orientation of cracks from this morning-derived subset of images was within 10° of that of the entire population and was also statistically non-random (Supplementary Fig. 4) suggesting no significant difference in crack visibility due to time of day that the image was taken within the range of the mid-day hours of our data set.

We also examined the orientations of images for the 40% of the 1,440 Martian rocks that exhibited no visible cracks with the idea that, if the majority of cracks are preferentially aligned then there are necessarily fewer cracks in other orientations. Since visible cracks share the orientations of images in which they are found, it would therefore be predicted that rocks would be more likely to have zero cracks in image orientations, relative to the viewer, roughly normal to those of the observed majority of cracks. As before, because the azimuths of the images for the 1,440 rocks are equally distributed about 360°, if there is no preference, then the

images associated with rocks with zero cracks should also be randomly distributed around 360°.

The image azimuths for the rocks with no cracks were found to have a preferred direction that was 75° offset from that of visible cracks (vector mean and 95% confidence interval: $121 \pm 38°$; Fig. 6d). Therefore, the implication of this result is that, relative to the viewer, there are in fact fewer visible cracks overall in the NW–SE directions, and thus fewer cracks striking in those directions, as would be predicted given that a majority of Martian rock cracks strike NE–SW. To ensure that cracks with NW–SE orientations are not less visible in afternoon hours, we also analysed the 360-rock morning-only subset of images for rocks with zero cracks. For this subset of data, the NW mode is still evident and prominent (Supplementary Fig. 4). Thus overall, these 'zero crack' data provide additional support for the major finding of our work, which is that cracks in rocks on the Martian surface are statistically non-uniform in their orientations.

Rock cracking and solar-induced thermal stresses on Mars. Fracture mechanics dictates that rocks will crack catastrophically when the applied stress exceeds a material-dependent critical threshold (for example, tensile strength, critical energy release rate of Griffith and so on[30]). Single cycles of rapid temperature change are known to exert such stresses on rocks and to cause cracking (thermal shock). Such thermal shock-related cracking is recognized to sometimes result in a tell-tale polygonal pattern of cracking, see, for example, ref. 17. Numerical models[13,15] indicate

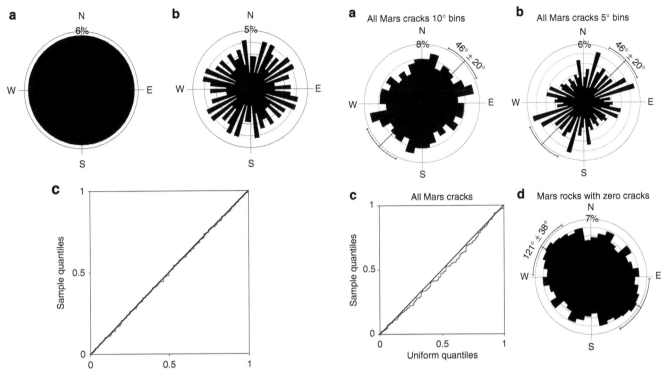

Figure 5 | Image azimuths 1,440-rock data set. Graphs of data derived from the orientations of images associated with each of the rocks in which cracks were counted ($n=1,440$). These graphs and their data visually and statistically demonstrate that the selected rock data set removes the image-direction sampling bias associated with MER traverse direction through stratified sampling of images whose azimuths occur in each of 36 10° bins. Frequency data are plotted as percentages. (**a**) Bidirectionally plotted circular histogram of image orientations, plotted in 10° bins ($n=1,440$ Raleigh test P value $=0.995$.). (**b**) Bidirectionally plotted circular histogram of image orientations, plotted in 5° bins ($n=1,440$ Raleigh test P value $=0.995$.) (**c**) Quantile–quantile plot of image orientations ($n=1,440$). Black line represents a perfectly uniform distribution. Red line is plotted from image azimuth data.

Figure 6 | Mars rock-crack data analysis. Graphs of crack data derived from 1,440 rocks selected along the MER Spirit traverse indicating that cracks exhibit preferred orientations. Lines and arcs on circular histograms indicate vector means and associated 95% confidence intervals. Frequencies are reported in %. Bidirectionally plotted circular histogram of the strikes of all visible cracks, plotted in (**a**) 10° and (**b**) 5° bins ($n=1,635$; Rayleigh test P value $=0.018$ and Rao's spacing test P value <0.01). (**c**) A quantile–quantile plot of the same data as **a**. Black line represents a perfectly uniform distribution. Red line is plotted from image azimuth data. Divergence of the plot of cracks with the 1:1 line indicates preferred orientations. (**d**) Bidirectionally plotted circular histogram of image orientations for rocks that exhibited no cracks, plotted in 10° bins ($n=565$; Raleigh test P value $=0.35$; Rao's spacing test P value <0.01).

that terrestrial-body (including Mars) rates of solar-induced rock surface temperature change are as fast as those hypothesized to lead to thermal shock, and polygonal cracking is observed in some MER images of bedrock, see, for example, ref. 2; however, we did not commonly observe these crack patterns in Martian boulders.

It is well established, however, that all rocks are characterized by flaws such as grain boundaries or pores that will act as foci for stresses and thus for crack propagation, see, for example, ref. 31. Such flaws can develop into cracks that slowly and predictably propagate via subcritical crack growth at stresses much lower (the limit is not well defined but is often cited as ~20% of the critical value[32]) than a material's critical threshold.

There is evidence to suggest that such processes contribute to mechanical weathering of rocks on Mars. For thermal stress cycles associated with diurnal insolation, subcritical crack growth can be facilitated by fatigue, whereby cyclic loading serves to preferentially weaken the material at crack tips and result in crack lengthening and rock degradation, see, for example, ref. 20. These thermally driven cracking processes are undeniably complicated and likely influenced by factors ranging from mineral axis thermal anisotropies, see, for example, ref. 33, to mineral translucence[34,35], to edge effects. Nevertheless, laboratory experiments produce thermal fatigue-related fracturing when

subjecting basalts to thermal cycling replicating Mars temperature extremes[16]. Furthermore, numerical models of solar-induced thermal stresses on Earth[36] and other planetary bodies[37] are well within the range of stresses expected to produce subcritical crack growth, and in some cases exceed the reported critical rock strength values of most rock types. Such stresses have not been calculated for the Mars surface.

Here we use a previously developed 2D finite-element model[38,39] of the thermoelastic behaviour of rock micro-structures (small sections of an infinite half-space) subjected to diurnal thermal forcing to simulate, for the first time, solar-induced thermal stresses that might develop on a polycrystalline rock (Supplementary Table 2) on the Martian surface, and determine if such stresses may be sufficient to induce crack propagation (Methods). The microstructures have properties typical of basalt, a common rock type abundant along the MER Spirit traverse.

The surface temperature of a microstructure at a longitude of 0° and the latitude of the Spirit traverse (15°S) was calculated over one solar day on the warmest day of the year (L_s 68) at that location. We calculate that rock surface temperature varies on this day from 186 to 192 K (Fig. 7). This range is substantially smaller than many reported values for Martian surface temperatures (up to 120 K, see, for example, ref. 40). Nevertheless this small

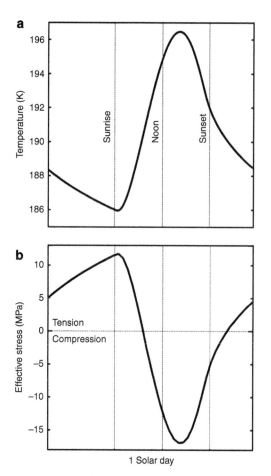

Figure 7 | Numerical model results. Results from our modification of a pre-existing model[38] to calculate solar-induced thermal stresses arising in the surface of a subaerially exposed Martian rock (see Methods for details). (**a**) Surface temperatures calculated over one solar day for a microstructure found in an idealized basalt located at a longitude of 0° and the latitude (15°S) of the Spirit traverse. (**b**) Calculated thermoelastic stresses induced within the microstructure throughout the solar day for the same rock using the temperatures derived for **a** above as inputs.

temperature range is consistent with THEMIS data, which shows decreasing range in temperature with increasing thermal inertia[41]. Bedrock has a higher thermal conductivity than unconsolidated sediment, resulting in four to five times higher thermal inertia[42,43]. Also, the thinner Martian atmosphere will serve to dampen diurnal warming effects compared with those of Earth. Although our calculated values represent an annual maxima[38], these diurnal temperature ranges calculated for a large slab are also likely lower than those which might arise in an individual, relatively small boulder, which will heat up more than flat ground due to its size and the influence of radiation from surrounding terrain.

The thermoelastic stresses induced within the microstructure throughout the solar day are also calculated (Methods). These stresses peak in the tensile regime at 12 MPa, and at 17 MPa in the compressive regime. These values represent the amount of idealized energy available for crack propagation, as the model does not account for the release of mechanical energy as propagation occurs.

Our modelling indicates that the magnitude of stress induced by solar-forcing within the microstructure is primarily controlled by rock surface temperature diurnal range and the Young's modulus and coefficient of thermal expansion of component mineral types[38,39]. This dependence is consistent with other types

of models developed for solar-induced thermal stresses in rocks on Earth, see, for example, ref. 36.

Molaro et al.[38] found that bodies close to the Sun and bodies that rotate slowly have the largest daily temperature ranges and thus highest stresses. Earth and Mars have nearly the same rotation rates; however, due to its closer proximity to the Sun, Earth is expected to have a larger diurnal temperature range. Maximum diurnal rock surface temperature fluctuations measured in a variety of climates and latitudes on Earth range from ~20 to ~60 °C, see, for example, ref. 44 and are therefore consistent with this prediction from our model. Consequently, rocks on Earth likely experience higher solar-induced microstructure stresses than on Mars and rock lifetimes on the two planets support this conclusion (see below). In addition, the strength of the materials in vacuum can be one to several times higher than in atmosphere[45]; thus, even comparable stresses between the two bodies may result in faster breakdown rates on Earth.

Similar modelling of grain-scale diurnal thermal stresses has not been completed for Earth. Nevertheless, calculations of maximum solar-induced macroscopic thermal stresses arising in Earth boulders yield values on the order of those calculated herein[26,36]. Because these macroscopic stresses would be predicted to be lower overall than grain-scale stresses, the other modelling results are consistent with our prediction of higher microstructure stresses for a rock on Earth. How such macrostresses might interact with microstresses to produce rock fractures is currently unknown.

It is non-trivial to relate calculated stresses to actual rock breakdown, as rock strength at the grain scale is generally not a well characterized material property and is dependent on a host of factors. Nonetheless, measurements of rock tensile strengths at larger scales are generally of a similar order of magnitude (10–100 s of MPa, see, for example, ref. 46) to stresses calculated by our model and within the cited 80% limit for subcritical crack growth. However, any given amount of external loading exerted on a large scale will result in more damage than for a small scale due to the former having generally more, and larger, pre-existing cracks, see, for example, ref. 30. Thus, rock strengths at grain scales are likely higher than those measured in a hand sample. However, the magnitudes of our modelled stresses likely represent a minimum since our model does not account for pre-existing defects, which will serve to amplify stresses, see, for example, ref. 45. Thus, we conclude that the results of our numerical model, at a minimum, are consistent with the suggestion that thermal stresses arising from simple diurnal insolation are sufficient to drive subcritical, fatigue-driven fracture growth in Martian rocks.

Linking macrocrack orientation and insolation. We hypothesize that cracks on Mars exhibit preferred orientations because the geometry of directional heating by the sun drives fatigue-related cracking in preferred directions. Both critical- and subcritical-fracture propagation direction in solid materials is strongly linked to the direction of the applied stress, see, for example, refs 30,47. In a rock with randomly oriented microfractures or zones of weakness (for example, different sides of an individual crystal grain), those heterogeneities whose orientations are favourable with respect to the orientation of the external loading will preferentially propagate and coalesce relative to cracks with other less-favourable orientations, see, for example, refs 48,49. Because subsequent fracture will favourably occur in directions where the density of pre-existing cracks is highest[49], if loading direction is consistent, a positive feedback develops whereby additional cracking is predicted to result in fracturing along already preferentially weakened planes. Furthermore, it has been explicitly demonstrated through laboratory experiments that directional thermal stress loading in particular can result in the preferential propagation of cracks of specific orientations with respect to the location of the heat source[50], and models of thermal stress arising in a hollow sphere subject to a rotating heat source indicate that the orientation of stresses are dictated by the orientation of heating[51]. Thus, if solar-induced thermal stresses on Mars are sufficient, as we have shown, to result in fracture propagation in rocks exposed on the Martian surface, it would then be expected that fracture growth would occur preferentially in directions dictated by insolation itself.

Calculated solar-induced, cyclic thermal stresses on Earth and other terrestrial bodies are not diurnally static, but instead peak at particular times of day: sunrise, mid-day or shortly thereafter and sunset[15,26]. Acoustic emission monitoring of cracking in three different ~25 cm diameter boulders, placed for multiple years in mid-latitude locations on Earth, indicates that >80% of all cracking events occur in the afternoon hours, with frequency peaks at times[18] that are virtually identical to the timing of peak solar-induced thermal stresses calculated using inputs from the rocks themselves[26].

Because the peak stresses that are driving this cracking are induced by a rotating heat source, they represent directional loading whose average direction should relate to the average sun orientation at those times of day. Assuming that cracks propagate in a direction that is directly related to the direction of external thermal stress loading as explained above, then arises the potential for a majority of cracks in any given rock to be oriented in a range of directions that reflect the orientation of stresses induced by the sun. The higher density of cracks with sun-related orientations could then result in their preferential extension by other processes like salt-shattering. That such preferentially oriented cracks might develop despite these other processes as well as complicating factors, such as mineral anisotropy or translucence at grain scales, suggests that rock-scale stresses related to the geometry of the boulder itself potentially play an important role in their formation.

To date, there is no existing analysis of the orientations of rock-scale stresses that arise during peak-solar-induced-stress times for a boulder sitting on a ground surface. However, high spatial- and temporal-resolution field measurements of rock surface temperatures reveal that there is a distinct gradient and geometry in average boulder surface temperatures that recurs throughout the day and/or year for rocks on Earth exposed to the sun, see, for example, refs 18,52. Furthermore, isotherms (zones of equal temperatures) measured on boulder surfaces on Earth at the times of day when cracking occurs exhibit strongly defined NE orientations[18]. The overall mean and modal orientation(s) of these isotherms is roughly the same as the vector mean of both Earth (Fig. 2) and Mars (Fig. 6) crack strike data.

Given the overall general similarity in geometry of insolation between Earth and Mars along the Spirit traverse, there is no reason to believe that similar isotherms, and thus similarly oriented stresses, would not also develop in Martian rocks. For example, the surface depth of a diurnal-temperature-wave on Earth and Mars are comparable (such 'skin depth' is proportional to the square root of the diurnal period). Even though overall Martian rock surface temperatures will be lower, the portions of a Martian rock in sunlight will still heat up relative to parts of the rock in shadows. We thus expect that specific geometries of peak stresses will recur on Mars as on Earth even if the magnitude of those stresses differs. This analogy leads to the expectation that Martian insolation-driven thermal stresses will also result in preferred propagation of favourably oriented cracks, ultimately resulting in a population of macrocracks with preferred

directions, as we have observed (see, for example, Figs 2a and 6a). We test this hypothesis by determining if peak stresses that arise in Martian rocks also occur at specific times of day.

Our thermoelastic model shows that the magnitude of stresses induced within microstructures on the Martian surface varies throughout the day with distinct peaks (Fig. 7). Peak tensile stresses occur just before sunrise, as the surface continues to radiate heat into space throughout the night, more so than occurs on Earth. Any atmosphere dampens this cooling by reradiating some of that infrared energy back into the surface. Thus, all else being equal, an airless surface would be able to reach a lower temperature than a Martian surface, or in turn Earth's surface. Nevertheless, similar sunrise peaks in calculated tensile stresses also occur for Earth boulders[26].

Our model's calculated peak compressional stresses on Mars occur in afternoon hours between noon and sunset. At this time of day, atmospheric radiation causes net heating to continue for longer on Mars than it would on an airless body, and thus this peak occurs slightly later on Mars[38,39]. This result suggests that, on Earth, surfaces should experience this peak state even later in the day, a prediction that is supported by observations of rock surface temperatures that peak in the late afternoon, see, for example, ref. 18.

The mid-afternoon timing of peak compressional stresses developing on Mars is similar to those calculated for compressional stresses arising in the surface of a large isotropic rock sphere due to solar-induced thermal cycles on Earth[26]. While compressional stresses are less likely to produce fracture, the Earth model predicts a roughly simultaneously occurring peak in interior tensile stresses as the warm boulder surface pulls away from the relatively cool interior during these hours. Our model does not explicitly calculate interior stresses that may develop in a Martian boulder. However, the similarities between the two models and the similar skin depths between the two planets suggests that Martian interior tensile stresses will peak coincidentally with exterior compressional stresses, in the afternoon hours, as on Earth. Thus overall, the temporal peaks in stresses that our model calculates suggest that Martian rock cracking will also preferentially occur in the mid-afternoon and morning hours.

Our model predicts that a broad, flat Martian rock surface will go back into tension shortly after sunset. However, we expect variability in the timing of this transition and magnitude of tensile stresses when considering boulders, which have different sizes and shapes. Variability in the overall timing of stresses will also arise from the influence of atmospheric radiation; thus, we expect that the times of peak stresses on Earth and Mars to be somewhat different, and to possibly lead to subtle differences in crack orientations.

Discussion

The strikes of Mars rock cracks along the Spirit traverse strike more easterly with a somewhat greater variance (Fig. 6) than those from our compilation of published strike data derived from 4,805 cracks in rocks from Earth's mid-latitude deserts, whose vector mean and 95% confidence interval is $23 \pm 11°$ (Raleigh and Rao's P values <0.01; Fig. 2). A 0.003 P value for a circular-data χ^2-test[53] indicates that the two populations are statistically distinct. Also, observed Mars crack dip directions (Fig. 4) are opposite those of many observed on Earth[23], and there is an overall higher prominence of an E-W mode of orientations in the Mars data.

For a variety of reasons, it might be expected that Earth and Mars cracks may differ in crack orientations and variance. For example, the fact that the Mars crack data set was derived from a low-latitude, southern hemisphere location, combined with the higher angle of average Martian obliquity[29] would lead to the expectation of different angles of insolation and thus thermal stress-driven cracking. Furthermore, in the most general terms, solar-induced thermal stresses that likely arise in any given rock are complex, and the magnitude of the influence of such stress on rock cracking is likely dependent on a host of factors that we do not explicitly address including latitude[44], topographic shading[13], rock composition[34,37] and/or rock shape and size[26]. This complexity is highlighted by the fact that calculated maxima in solar-related thermal stresses occur at different times of day for different topographic locations and different size rocks[13,15,26,37]. These complexities also lead to the expectation that dominant crack orientations will vary somewhat with rock size, composition, shape and overall location. We see evidence of this variability in our compilation of Earth field data which shows relatively large variance (Fig. 2) compared with that of individual data sets[23-25]. On Earth, this variance can be explained by the wide variety of landscapes in which these data are derived, as well as by the fact that other mechanical weathering processes are likely active on Earth's surface.

On Mars, it might therefore be expected that such other processes are more limited and thus crack orientations should exhibit less variance; however, there are other mechanical weathering processes likely acting on the Martian surface (for example, wind abrasion, impact shattering) that could introduce variance into the observed Mars crack orientation data. For example, in addition to insolation, wind can also induce rapid rock surface temperature changes on Earth[54]. Strong afternoon-dominant winds are expected from numerical models of the terrain along the Spirit traverse, and mid-day winds have been noted from images of rock cutting dust[55]. Although thermal advection effects are complicated on Mars due to the lower density of its atmosphere, in general surface winds might serve to advect heat off the rock surface at a time of day when thermal stress is already high, amplifying the potential for thermal stress in afternoon hours and increasing variance in potential stress fields that develop.

Furthermore, the range of rock sizes for the Martian data set (20–400 cm) was much higher than that of the Earth data sets (generally < 50 cm; Supplementary Data set 2), potentially resulting in greater variance in the times of day when cracking occurs and thus in crack orientations. In particular, sunset-coincident peaks in solar-induced thermal stress are not predicted in Earth-based models for relatively large rocks like those measured here[26]. Instead, calculated peak stresses occur during the early mid-afternoon. Less sunset cracking in the overall larger rocks of the data set might therefore account for the more easterly orientation of the Mars data set compared with Earth.

Finally, very low erosion rates for geomorphic surfaces along the Spirit traverse (0.03 nm per year)[1] attest to the slow rates of cracking that must occur in observed Mars rocks compared with Earth, because there are still large rocks present on these extremely old surfaces. In addition to the lower expected amplitude of Martian stresses relative to Earth's, these slow rates are likely also due to the lack of water which, through a variety of processes, propagates fractures at higher rates, see, for example, refs 45,56. Nevertheless, the size, and variance in the size, of rocks in the cratered plains of Gusev notably decreases with lower measured thermal inertia[1], a proxy for the age of the surface where younger surfaces have higher thermal inertia. This trend is similar to the diminution of boulder size on surfaces of increasing age observed in desert pavements in arid environments on Earth where, through time, poorly sorted gravel-boulder dominated surfaces mechanically weather without significant transport into well-sorted 1–3 cm size pebbles, see, for example, refs 57,58. Thus, the observation that exposed Mars rocks get

smaller through time suggests that when surfaces are not buried by sand, thermal stress weathering processes on Mars ultimately result in a distribution of broken rocks similar to those observed in desert pavements of Earth, albeit at a slower rate. All of these observations point to the potential long-term importance of insolation in mechanical weathering processes acting on both of these terrestrial planets.

Methods

Collecting crack data from MER data products. The MER Spirit and Opportunity missions have derived numerous data products from the Martian surface which are accessible on the internet (http://an.rsl.wustl.edu/mer/). In particular, 1,024 × 1,024 pixel stereo images collected by mission PANCAMs[59] are merged by NASA to create associated 3D data files. These files are composed of georeferenced *xyz* coordinates that represent the scene visible in the PANCAM 2D images, providing the data from which geometric measurements of the Martian surface may be made (summarized in Supplementary Methods). As of the time of publication, however, there had been no means by which to readily derive distance or orientation data from these data products.

We developed software, 'ImageRover,' (http://visionlab.uncc.edu/index.php/mer-imagerover-app), to provide an interface for visualization and geometric analysis of NASA's MER PANCAM 3D data products (Supplementary Fig. 1; Supplementary Methods). ImageRover employs MER mission 2D and 3D data directly from their publicly available repository (for example, NASA's Planetary Data System; http://pdsimg.jpl.nasa.gov/Atlas/MER/mer1po_0xxx/data/) and renders views of the Martian surface that enable user-friendly extraction of measurements. Specifically, a user is able to measure (1) point-to-point lengths between user-defined end-points and (2) orientations of user-defined planes (Supplementary Fig. 1). Crack plane azimuth orientations (strike) are expressed relative to Mars north–south–east–west (0–359°) and crack plane slope (dip) is expressed relative to horizontal (0–90°) in the same way that a geologist would report a right-hand rule strike and dip on Earth. All results published in this article are compiled from data available before July 2013 which include 29,714 EFF/XYZ stereoscopic images from the Spirit rover recorded during sols 1–2,209 (the entire mission).

To validate the accuracy of ImageRover calculations made from the MER data products, we compared crack length and strike and dips measured on 3D data of Earth boulders using ImageRover to the same measurements made using a tape measure and a compass in the field (Supplementary Fig. 2, Supplementary Table 1, Supplementary Methods). Overall, the differences between the measurements made using ImageRover and the same measurements made in the field were found to fall within errors associated with making the measurements with a compass. For example, the difference between measured biaxial strikes on individual cracks using the two methods are comparable to reported errors associated with making these measurements in the field alone; that is, ~9° for steeply dipping beds to ~30–50° for shallowly dipping beds[60]. Thirty-two per cent of the dip directions measured using ImageRover were opposite of that measured in the field particularly for high- and low-angle dips (Supplementary Methods), rendering strikes measured using the right-hand rule off by ±180° in Supplementary Table 1, but within the stated errors above when considered as biaxial data.

Mars rock and crack measurements overview. Using our software and MER Spirit data, we examined cracks on 1,573 rocks in total from 59 different sites along the Spirit traverse (Fig. 3), which comprises a range of volcanic rock types and geomorphic surfaces[61]. For all data collected, we examined only larger rocks (> 10 and > 20 cm) within a similar field of view within the image (that is, far distant rocks visible in images were avoided) to ensure good visibility of as many cracks as possible on rock surfaces. Also, to avoid thermal complexities associated with slope aspect or topography-induced shading[13], we avoided collecting data on visibly significant ground slopes or immediately adjacent to significant topography. To avoid making measurements on tectonically induced fractures, all data were collected for individual boulders; no data were collected for bedrock outcrops. Although it has been suggested that solar-induced directional cracking may vary with latitude[22,25], we limited our sampling to the Spirit traverse which, at the time of data collection, had overall larger numbers of data products with images of rocks that fit our sampling criteria.

Cracks were defined as any narrow lineation of pixels that strongly contrasted with adjacent features within a rock mass (Fig. 1). To minimize any bias in visualization of cracks of certain orientations due to shadowing effects, the time of day of all images employed in the study was limited to mid-day hours (~ 10:00 to ~ 15:00 hours) when the sun would be relatively high overhead (Supplementary Methods).

To produce a data set of crack orientations that were free from sampling bias associated with observation direction, we took advantage of observed relationships between image orientations and crack orientations. We first analysed the data set and our methods for potential sampling bias by measuring crack strikes and dips in 133 rocks using ImageRover. The results of this analysis (above), combined with the extremely time intensive nature of searching for, and analysing, appropriate rocks with sufficiently complete 3D data available, led us to the development of methods specifically aimed at maximizing the numbers of rocks analysed while minimizing all sampling bias inherent to the Spirit data products (see below). Finally 1,440 rocks and their associated cracks were visualized in ImageRover and analysed for crack orientations.

Identifying sampling bias potential for Spirit data products. To characterize potential sampling biases that might arise from making field measurements using MER Spirit PANCAM data products, we collected a preliminary data set of rock and crack data (Supplementary Methods for details; Supplementary Data set 1). In 36 randomly selected MER Spirit sites, we randomly selected 76 images for which we used ImageRover software to measure the length and strike and dip of all visible cracks found in all imaged rocks that were >10 cm in diameter (Supplementary Data set 1). The size of these analysed cracked rocks ranged from 10 cm to ~4 m with an average size of 0.6 ± 0.7 m; however, our validation tests (Supplementary Methods) indicated that this average likely speaks to crack visibility in large rocks more than to the average size of a cracked rock. Thus crack data from Martian rocks are necessarily derived from a population of relatively large rocks compared with the compilation of similar data from Earth (Supplementary Data set 2).

Our analysis of randomly chosen images along the Spirit traverse also revealed a strong directional sampling bias of MER Spirit data products due to Spirit's predilection towards recording images in particular directions relative to the motion of the Rover. Specifically, the vast majority of Spirit images for which 3D data are available were recorded when the PANCAMs were pointing either in the direction of the motion of the rover (image azimuth = SE) or directly behind the rover (image azimuth = NW). Thus, the azimuths of any group of randomly selected images such as those chosen for our preliminary analysis (Fig. 4a) reflect this bias.

Another expected bias in the Spirit data set relates to the visibility of cracks of varying locations on the rock and orientations with respect to the orientation of the camera or viewer. In our field validation tests on boulders in North Carolina, USA, we noted that sunlit-rock-face-cracks having orientations oblique, or parallel to, the viewpoint of a distant observer (3–5 m) were less clear, or not visible at all, compared with cracks on rock faces oriented normal to the viewpoint of the observer.

We noted that this disparity in crack visibility transfers to 2D images and results in a smaller count, using our criteria, of those cracks that strike unfavourably relative to the azimuth orientation of the image. In particular, cracks that strike parallel to a 2D image (normal to the camera orientation) are overall less visible in any given image compared with those that strike into, and out of, the image for a combination of reasons. First, if they are vertical or subvertical, image-parallel-striking cracks will likely not be exposed in any rock face that is parallel to the image azimuth itself, giving such cracks a much lower chance of being observed. Furthermore, since the majority of these image-parallel-/viewer normal-striking cracks will be located on rock faces that are oblique to the orientation of the photo, they have a much higher likelihood of being obscured by the protruding portions of the rock itself unless the rock is perfectly spherical. Such an obscuration would be particularly relevant to the MER data set and for the blocky, irregular rocks of the Martian surface because (1) any rock protrusion would reduce the contrast and/or consistent linear visibility for cracks located on the side of the rock behind the protrusion and/or (2) such protrusions would result in a hole in the 3D data where the crack is located behind them. Both of these outcomes would preclude measuring these image-parallel-striking cracks, because we defined cracks as any continuous lineation that starkly contrasted with adjacent pixels. Furthermore measurements can only be made on a feature that 3D data are complete in the area of interest. As a result, the majority of any measureable cracks in a given image will be striking in and out of the image itself. Analysis of the orientations of the 222 ImageRover-measured crack strikes and dips visible in the 133 rocks confirmed that the measured crack strikes were typically nearly equal to the image azimuth (Fig. 4b,c).

Mars crack data collection. Starting at MER Spirit site 2, we examined all images in order of the data product file listing via JPL. We searched for images that met the following criteria in order of image azimuth bin from 0 to 360° via 10° bins: only images with similar clarity of rock detail in the 2D and 3D images (that is, similar focal lengths in the bottom half of the image) and relatively level visible foreground surface were chosen. As such, we ensured that all the data were collected from images of comparable resolution and visual distance. If such an image was not available for the necessary 10° bin orientation at the chosen site, we went on to the next site. As such, we employed a stratified sampling strategy to select a random representation of differently oriented images along the Spirit Traverse (Figs 3 and 5 and Supplementary Data set 3).

For each selected image, a 50-box grid overlay provided a random reference from which to select rocks. The closest rocks greater than 20 cm diameter to each of the bottom 10 grid line intersections were chosen (Supplementary Fig. 3). If no rock was present at an interval, that interval was skipped, and we looked for the closest rock to the next grid point. All the above steps were repeated until we had identified 10 rocks from a suite of images that all fell within a given 10° image azimuth bin. We then moved on to the next azimuth bin and repeated the image/ rock selection process starting with whichever site and image we had ended on

previously. We repeated these steps four times per 10° image azimuth bin for a total of 40 rocks per bin and 1,440 total rocks from 31 different sites along the Spirit traverse.

Finally, for each of these 1,440 rocks, we counted the total number of cracks visible in each rock using the criteria defined above. If a rock exhibited no cracks, then it was assigned a '0'. As such, we counted the total number of cracks per rock visible in typical large boulders on the Martian surface. The statistically robust correlation between crack and image orientation, combined with the equally distributed azimuths of the 1,440-rock images, enabled the confident use of image azimuth as a proxy for the orientation of any visible cracks in the image (Fig. 4; main text). Therefore, for each crack counted, we assigned it an orientation equal to the orientation of the image that it was found in (Fig. 6). We also plotted the image orientation for all rocks (565 of the 1,440) that did not exhibit visible cracks (Fig. 6d).

Earth desert crack data compilation. We compiled all available crack strike ($n = 4,805$) and dip ($n = 3,215$) data that we could find for surface boulders in Earth's deserts[22–25,62] (Fig. 2 and Supplementary Data set 2). Crack dip data were collected for only two of these studies[23,62]. If raw data were not available in the published work, we obtained the data from the first author directly. From each study, we included all the crack data available in our compilation. We did not attempt to account for rock type, anisotropies, surface age or other possible variables that would influence crack properties. As such, the data set comprises a variety of rock sizes (from ~1 cm to ~2 m) from a variety of desert locations ranging from about 40°N latitude to 30°S latitude. The majority of crack data is derived from study sites in the southwestern United States[22,23,25,62], with about 450 cracks from the eastern Libyan Plateau in Egypt[24] and about 150 cracks from the Gobi and Strzelecki deserts[22]. Thus, most, but not all, data are from the northern hemisphere. Rock types included basalt, intrusive igneous rocks, meta-volcanic and meta-sedimentary rocks, carbonates and, to a lesser extent, clastic sedimentary rocks. In all studies, individual boulders (no bedrock data are reported) were chosen using selected spacing intervals, designated areas on specified geomorphic surfaces, and/or certain size criteria. Crack orientation data were collected on each selected boulder. For some studies[24,25], only one orientation measurement was reported for all parallel cracks on an individual boulder. Thus, about one-fourth of the compiled data set might be underreporting certain orientations in the context of 'all cracks'. The minimum size boulder examined, as well as the minimum length of crack measured, varied somewhat for each of the studies, with a cutoff of 5 cm for rocks and 2 cm for cracks for the majority of the data. Rock size data are included in the compilation if available.

Mars solar-induced thermal stress model. In this study, we utilized the model employed by Molaro et al.[38] to simulate grain-scale thermoelastic stresses produced on Martian surfaces. This model is composed of two parts: a one-dimensional thermal model that calculates the temperatures in a macroscopic column of material throughout the solar day on a given body, and a 2D finite-element model[63], which calculates thermoelastic deformation in a microstructure over the same period. The thermal model is used to calculate time-dependent solar and conductive heat fluxes at the surface and some depth, which can then be applied as boundary conditions in the 2D model, which uses those heat fluxes to calculate thermoelastic deformation over time. The idea behind coupling these two models together is that the 2D model investigates a 'zoomed in' version of the thermal model.

The original study did not include the effects of an atmosphere, and thus had to be modified for our purposes to make calculations for rocks on the Martian surface. It solves the heat equation, which is given by:

$$c_p \rho \frac{dT}{dt} + \nabla Q = 0 \qquad (1)$$

where c_p is the specific heat capacity, ρ is the density, T is the temperature, k is the thermal conductivity and Q is the heat flux. Following the example of Aharonson and Schorghofer[64], we included terms for atmospheric extinction (2), emission (3) and scattering (3) for a flat surface:

$$Q_{sol} = \frac{S}{R^2}(1-A)e^{\frac{\tau}{\max(0.04 \text{ or } \sin(\theta))}} \qquad (2)$$

$$Q_{IR} = f_{IR}\frac{S}{R^2}\sin(\theta_{noon}) \qquad (3)$$

$$Q_{scat} = \frac{1}{2}f_{scat}\frac{S}{R^2} \qquad (4)$$

where S is the solar constant, R is the solar distance, A is the albedo and θ is the solar elevation angle. τ is the optical depth at the Martian surface, which is taken to be 0.2 (refs 65,66). The extinction term (1) is proportional to the path length through the atmosphere, except at times of the day when it is limited by the radius of curvature of the planet. The emission term (2) approximates the atmospheric radiation as a fraction of the noontime solar insolation, and is kept constant throughout the day. The scattering term (4) assumes that half of the scattered light is lost to space. The emission factor (f_{IR}) has a value of 0.04, and the scattering factor (f_{scat}) has a value of 0.02. We refer the reader to Aharonson and Schorghofer[64] for additional details on this model. We calculated the temperature of a Martian surface over one solar day during the warmest day of the year (L_s 68) at a longitude of 0° and a latitude of 15°S.

The 2D model used was Finite Element Analysis of Microstructures (OOF2)[63], a finite-element modelling programme designed to help scientists simulate the behaviour of microstructures. OOF2 calculates (1) and the force-balance (2) equation, given by:

$$M\frac{d^2u}{dt^2} + \nabla \cdot \sigma = 0 \qquad (5)$$

where M is the mass density tensor, u is the displacement field and σ is the stress tensor. It utilizes the time-dependent heat flux from the thermal model as boundary conditions to calculate stresses over one solar day. Additional details on this model can be found in Molaro et al.[38], as we did not make any modifications to what was used in their original study. The microstructure modelled can be thought of as a small piece in an infinite half-space, where its top represents a surface open to space, the bottom represents 5 mm in depth. While the temperature was set to be periodic across horizontal boundaries, the displacement across horizontal boundaries is fixed. In this sense, the stresses cannot be directly compared with those calculated for boulders, as this model does not include the effects of boulder shape or surface curvature. We used a randomized grid of hexagons, each of which represents a pyroxene (75% of the total) or plagioclase (25% of the total) grain with a diameter of 360 microns. This composition approximates basaltic rock, which is abundant along the Spirit traverse. Values for the material parameters can be found in the Supplementary Table 2.

The stress parameter presented herein is the von Mises stress (σ_e, also sometimes called the equivalent stress. It is calculated from the principal stresses:

$$\sigma_e = \sqrt{\frac{(\sigma_1 - \sigma_2)^2 + (\sigma_2 - \sigma_3)^2 + (\sigma_1 - \sigma_3)^2}{2}} \qquad (6)$$

Because the top of the microstructure is a free surface, most of the stress induced in our microstructure is along the x axis ($\sigma_e \approx |\sigma_{xx}|$). Note that while σ_e always has a positive value, an artificial sign correction has been applied to Fig. 7 to visually separate the tensile and compressional regimes modelling.

References

1. Golombek, M. et al. Geology of the Gusev cratered plains from the Spirit rover traverse. J. Geophys. Res.-Planet. 111, 1–27 (2006).
2. Thomas, M., Clarke, J. & Pain, C. Weathering, erosion and landscape processes on Mars identified from recent rover imagery, and possible Earth analogues. Aust. J. Earth Sci. 52, 365–378 (2005).
3. Grant, J. et al. Surficial deposits at Gusev crater along Spirit rover traverses. Science 305, 807–810 (2004).
4. Ollier, C. Weathering (Longman Group, 1984).
5. Larsen, I. J. et al. Rapid soil production and weathering in the Southern Alps, New Zealand. Science 343, 637–640 (2014).
6. Yokoyama, T. & Matsukura, Y. Field and laboratory experiments on weathering rates of granodiorite: separation of chemical and physical processes. Geology 34, 809–812 (2006).
7. Kirchner, J. W., Riebe, C. S., Ferrier, K. L. & Finkel, R. C. Cosmogenic nuclide methods for measuring long-term rates of physical erosion and chemical weathering. J. Geochem. Expl. 88, 296–299 (2006).
8. Millot, R., Gaillardet, J., Dupré, B. & Allègre, C. J. The global control of silicate weathering rates and the coupling with physical erosion: new insights from rivers of the Canadian Shield. Earth Planet. Sci. Lett. 196, 83–98 (2002).
9. Ashley, J. W. et al. Evidence for mechanical and chemical alteration of iron-nickel meteorites on Mars: process insights for Meridiani Planum. J. Geophys. Res.-Planet. 116, 1–22 (2011).
10. Jagoutz, E. Salt-induced rock fragmentation on Mars: the role of salt in the weathering of Martian rocks. Adv. Space Res. 38, 696–700 (2006).
11. Chan, M. A., Yonkee, W. A., Netoff, D. I., Seiler, W. M. & Ford, R. L. Polygonal cracks in bedrock on Earth and Mars: implications for weathering. Icarus 194, 65–71 (2008).
12. Clarke, J. in Arid Zone Geomorphology: Process, Form and Change in Drylands 3rd edn (ed. Thomas, D. S. G.) 61–82 (Wiley-Blackwell, 2011).
13. Leask, H. & Wilson, L. in 34th Annual Lunar Planet. Sci. Conf. 1320 (Houston, TX, USA, 2003).
14. Delbo, M. et al. Thermal fatigue as the origin of regolith on small asteroids. Nature 508, 233–236 (2014).
15. Molaro, J. & Byrne, S. Rates of temperature change of airless landscapes and implications for thermal stress weathering. J. Geophys. Res.-Planet. 117, E10011 (2012).
16. Viles, H. et al. Simulating weathering of basalt on Mars and Earth by thermal cycling. Geophys. Res. Lett. 37, L18201 (2010).
17. Hall, K. & Thorn, C. E. Thermal fatigue and thermal shock in bedrock: an attempt to unravel the geomorphic processes and products. Geomorphology 206, 1–13 (2014).

18. Warren, K., Eppes, M.-C., Swami, S., Garbini, J. & Putkonen, J. Automated field detection of rock fracturing, microclimate, and diurnal rock temperature and strain fields. Geosci. Instrum. Method. Data Syst. 3, 371–406 (2013).
19. McKay, C. P., Molaro, J. L. & Marinova, M. M. High-frequency rock temperature data from hyper-arid desert environments in the Atacama and the Antarctic dry valleys and implications for rock weathering. Geomorphology 110, 182–187 (2009).
20. Gómez-Heras, M., Smith, B. J. & Fort, R. Surface temperature differences between minerals in crystalline rocks: implications for granular disaggregation of granites through thermal fatigue. Geomorphology 78, 236–249 (2006).
21. Hall, K. The role of thermal stress fatigue in the breakdown of rock in cold regions. Geomorphology 31, 47–63 (1999).
22. Eppes, M. C., McFadden, L. D., Wegmann, K. W. & Scuderi, L. A. Cracks in desert pavement rocks: further insights into mechanical weathering by directional insolation. Geomorphology 123, 97–108 (2010).
23. Eppes, M. et al. in AGU Fall Meeting Abstracts 0745 (San Francisco, CA, USA, 2010).
24. Adelsberger, K. A. & Smith, J. R. Desert pavement development and landscape stability on the Eastern Libyan Plateau, Egypt. Geomorphology 107, 178–194 (2009).
25. McFadden, L., Eppes, M., Gillespie, A. & Hallet, B. Physical weathering in arid landscapes due to diurnal variation in the direction of solar heating. Geol. Soc. Am. Bull. 117, 161–173 (2005).
26. Shi, J. Study of Thermal Stresses in Rock Due to Diurnal Solar Exposure (MS thesis, University of Washington, 2011).
27. Schorghofer, N. Temperature response of Mars to Milankovitch cycles. Geophys. Res. Lett. 35, L18201 (2008).
28. Forget, F., Haberle, R., Montmessin, F., Levrard, B. & Head, J. Formation of glaciers on Mars by atmospheric precipitation at high obliquity. Science 311, 368–371 (2006).
29. Laskar, J. et al. Long term evolution and chaotic diffusion of the insolation quantities of Mars. Icarus 170, 343–364 (2004).
30. Anderson, T. L. Fracture Mechanics: Fundamentals and Applications (CRC press, 2005).
31. Atkinson, B. K. Fracture Mechanics of Rock (Academic Press, 1987).
32. Atkinson, B. K. Subcritical crack growth in geological materials. J. Geophys. Res.-Solid Earth 89, 4077–4114 (1984).
33. Goudie, A. S. & Viles, H. A. The thermal degradation of marble. Acta Univ. Carol. Geogr. 35, 7–16 (2000).
34. Hall, K. Natural building stone composed of light-transmissive minerals: Impacts on thermal gradients, weathering and microbial colonization. A preliminary study, tentative interpretations, and future directions. Environ. Earth Sci. 62, 289–297 (2011).
35. Hall, K., Meiklejohn, I., Sumner, P. & Arocena, J. Light penetration into Clarens sandstone and implications for deterioration of San rock art. Geoarchaeology 25, 122–136 (2010).
36. Holzhausen, G. R. Origin of sheet structure, 1. Morphology and boundary conditions. Eng. Geol. 27, 225–278 (1989).
37. Molaro, J. & Byrne, S. in 44th Lunar Planet. Sci. Conf. 1790 (Arizona, USA, 2013).
38. Molaro, J. & Byrne, S. Grain-scale thermoelastic stresses and spatiotemporal temperature gradients on airless bodies, implications for rock breakdown. J. Geophys. Res.-Planet. 120, 255–277 (2015).
39. Molaro, J. & Byrne, S. Thermoelastic Grain-Scale Stresses on Airless Bodies and Implications for Rock Breakdown #46 (American Astronomical Society, DPS meeting, 2014).
40. Mellon, M. T. Small-scale polygonal features on Mars: seasonal thermal contraction cracks in permafrost. J. Geophys. Res.-Planet. 102, 25617–25628 (1997).
41. Mellon, M. T., Jakosky, B. M., Kieffer, H. H. & Christensen, P. R. High-resolution thermal inertia mapping from the Mars global surveyor thermal emission spectrometer. Icarus 148, 437–455 (2000).
42. Fergason, R. L., Christensen, P. R. & Kieffer, H. H. High-resolution thermal inertia derived from the Thermal Emission Imaging System (THEMIS): thermal model and applications. J. Geophys. Res.-Planet. 111, E12004 (2006).
43. Edwards, C., Bandfield, J., Christensen, P. & Fergason, R. Global distribution of bedrock exposures on Mars using THEMIS high-resolution thermal inertia. J. Geophys. Res.-Planet. 114, E11001 (2009).
44. Sumner, P., Nel, W. & Hedding, D. Thermal attributes of rock weathering: zonal or azonal? A comparison of rock temperatures in different environments. Polar Geogr. 28, 79–92 (2004).
45. Kranz, R. L. Microcracks in rocks: a review. Tectonophysics 100, 449–480 (1983).
46. Ersoy, A. & Atici, U. Performance characteristics of circular diamond saws in cutting different types of rocks. Diam. Relat. Mater. 13, 22–37 (2004).
47. Amitrano, D. & Helmstetter, A. Brittle creep, damage, and time to failure in rocks. J. Geophys. Res.-Solid Earth 111, B11201 (2006).

48. Zhao, Y. Crack pattern evolution and a fractal damage constitutive model for rock. *Int. J. Rock Mech. Min. Sci.* **35,** 349–366 (1998).

49. Nara, Y. & Kaneko, K. Sub-critical crack growth in anisotropic rock. *Int. J. Rock Mech. Min. Sci.* **43,** 437–453 (2006).

50. Widhalm, C., Tschegg, E. & Eppensteiner, W. Anisotropic thermal expansion causes deformation of marble claddings. *J. Perform. Constr. Fac.* **10,** 5–10 (1996).

51. Tanigawa, Y. & Takeuti, Y. Three-dimensional thermoelastic treatment in spherical region and its application to solid sphere due to rotating heat source. *J. Appl. Math. Mech./Z. Angew. Math. Mech.* **63,** 317–324 (1983).

52. Hall, K. & André, M.-F. New insights into rock weathering from high-frequency rock temperature data: an Antarctic study of weathering by thermal stress. *Geomorphology* **41,** 23–35 (2001).

53. Batschelet, E. *Circular Statistics in Biology* Vol. 371 (Academic Press, 1981).

54. Molaro, J. L. & McKay, C. P. Processes controlling rapid temperature variations on rock surfaces. *Earth Surf. Proc. Land.* **35,** 501–507 (2010).

55. Greeley, R. *et al.* Wind-related processes detected by the Spirit rover at Gusev Crater, Mars. *Science* **305,** 810–813 (2004).

56. Moores, J. E., Pelletier, J. D. & Smith, P. H. Crack propagation by differential insolation on desert surface clasts. *Geomorphology* **102,** 472–481 (2008).

57. Mushkin, A., Sagy, A., Trabelci, E., Amit, R. & Porat, N. Measuring the time and scale-dependency of subaerial rock weathering rates over geologic time scales with ground-based lidar. *Geology* **42,** 1063–1066 (2014).

58. McFadden, L. D., Ritter, J. B. & Wells, S. G. Use of multiparameter relative-age methods for age estimation and correlation of alluvial fan surfaces on a desert piedmont, eastern Mojave Desert, California. *Quat. Res.* **32,** 276–290 (1989).

59. Bell, J. *et al.* Mars exploration rover Athena panoramic camera (PANCAM) investigation. *J. Geophys. Res.-Planet.* **108,** 8063 (2003).

60. Cruden, D. & Charlesworth, H. Errors in strike and dip measurements. *Geol. Soc. Am. Bull.* **87,** 977–980 (1976).

61. Crumpler, L. *et al.* Mars exploration Rover geologic traverse by the Spirit rover in the plains of Gusev crater, Mars. *Geology* **33,** 809–812 (2005).

62. Doyle, S. *Remotely Mapping Surfce Roughness on Alluvial Fans: an Approach for Understanding Depositional Processes* (MS thesis, Univ. New Mexico, 2013).

63. Langer, S. A., Fuller, Jr E. R. & Carter, W. C. OOF: image-based finite-element analysis of material microstructures. *Comput. Sci. Eng.* **3,** 15–23 (2001).

64. Aharonson, O. & Schorghofer, N. Subsurface ice on Mars with rough topography. *J. Geophys. Res.-Planet.* **111,** E11007 (2006).

65. Vincendon, M., Langevin, Y., Poulet, F., Bibring, J. P. & Gondet, B. Recovery of surface reflectance spectra and evaluation of the optical depth of aerosols in the near-IR using a Monte Carlo approach: application to the OMEGA observations of high-latitude regions of Mars. *J. Geophys. Res.-Planet.* **112,** E08S13 (2007).

66. Dundas, C. M. & Byrne, S. Modelling sublimation of ice exposed by new impacts in the Martian mid-latitudes. *Icarus* **206,** 716–728 (2010).

67. NASA. *Spirit Traverse Map (Sol 1506).* Available at http://mars.nasa.gov/mer/mission/tm-spirit/images/MERA_A1506_2_br2.jpg (2008).

Acknowledgements

The data presented in this paper are available in Supplementary Materials defined in the text. All MER data products depicted and employed are in the public domain and provided courtesy of NASA-JPL-CalTech; Planetary Data System. The ImageRover software used in this study is available at http://visionlab.uncc.edu/index.php/mer-ima-gerover-app. This work was funded by NASA ROSES Mars Data Analysis Program award # NNX09AI43G. We would like to thank Alan Gillespie and Les McFadden for helpful reviews of a previous version of this paper as well as the contributions of students Jennifer Aldred, Rebecca Adamick, and Jungphil Kwon. Dr Katherine Adelsberger generously provided unpublished raw data from her published work.

Author contributions

The idea for this research, as well as the design, collection, analysis and interpretation of all MER data were completed or overseen by M.-C.E., who wrote the manuscript with input from A.W. and J.M. A.W., with M.-C.E.'s input, completed and/or oversaw the design and writing of ImageRover software. J.M. designed and wrote the numerical model presented and provided the majority of its interpretation within the manuscript. S.A. contributed to MER data collection and analysis. B.Z. contributed to ImageRover development.

Additional Information

17

A new type of solar-system material recovered from Ordovician marine limestone

B. Schmitz[1,2], Q.-Z. Yin[3], M.E. Sanborn[3], M. Tassinari[1], C.E. Caplan[2] & G.R. Huss[2]

From mid-Ordovician ∼470 Myr-old limestone >100 fossil L-chondritic meteorites have been recovered, representing the markedly enhanced flux of meteorites to Earth following the breakup of the L-chondrite parent body. Recently one anomalous meteorite, Österplana 065 (Öst 65), was found in the same beds that yield L chondrites. The cosmic-ray exposure age of Öst 65 shows that it may be a fragment of the impactor that broke up the L-chondrite parent body. Here we show that in a chromium versus oxygen-isotope plot Öst 65 falls outside all fields encompassing the known meteorite types. This may be the first documented example of an 'extinct' meteorite, that is, a meteorite type that does not fall on Earth today because its parent body has been consumed by collisions. The meteorites found on Earth today apparently do not give a full representation of the kind of bodies in the asteroid belt ∼500 Myr ago.

[1] Astrogeobiology Laboratory, Department of Physics, Lund University, 221 00 Lund, Sweden. [2] Hawai'i Institute of Geophysics and Planetology, University of Hawai'i at Manoa, Honolulu, Hawaii 96822, USA. [3] Department of Earth and Planetary Sciences, University of California at Davis, Davis, California 95616, USA. Correspondence and requests for materials should be addressed to B.S. (email: birger.schmitz@nuclear.lu.se).

The ordinary chondrites are the dominant type of meteorite falling on Earth today, representing about 85% of the total flux. They are divided in three subgroups, H (42% of ordinary chondrite falls), L (47%) and LL (11%), depending on their iron and metal content. In the 1960s, it was realized that the L chondrites show a major peak in K–Ar degassing ages at ~ 500 Myr ago[1,2], an age recently better constrained to ~ 470 Myr ago[3]. These meteorites apparently underwent a massive collisional event at this time, involving the breakup of their parent body. In contrast, the typical degassing age of most meteorite types goes back to the young solar system, that is, > 3.5 Gyr ago. In marine limestone that formed on Earth around 470 Myr ago, there is an evidence for an at least two orders of magnitude increase in the flux of L-chondritic meteorites and micrometeorites (for a review, see ref. 4). The prime evidence is the more than 100 fossil L chondrites recovered during the industrial quarrying of Ordovician limestone in southern Sweden[4,5]. The meteorites are almost completely replaced by secondary minerals, but spinel grains, and often also petrographic textures, are preserved. Mineralogical, chemical, isotopic and petrographic studies have shown that mid-Ordovician fossil extraterrestrial material is indeed of L-chondritic origin[4]. Recently, one 8-cm-large meteorite was found, containing spinel grains very different from all the other meteorites[6]. On the basis of ^{21}Ne cosmic-ray exposure ages of its spinel grains this meteorite, Österplana 065, previously referred to as the 'mysterious object', appears to have been liberated as a meteoroid at the same time as the other fossil meteorites[6-8]. It may thus represent a piece of the impactor that broke up the L-chondrite parent body[6]. The breakup of this body and the following enhanced flux of extraterrestrial material, including km-sized asteroids, coincide with marked evolutionary changes in Earth's invertebrate life, the so-called Great Ordovician Biodiversification Event[9]. Unravelling why the L-chondrite parent body broke up at this time, may lead to a better understanding of possible large-scale astronomical perturbations affecting both Earth and the solar system[4].

Österplana 065 was preliminarily classified as 'winonaite-like', based on oxygen isotopic and elemental composition of its spinels and by comparison with a winonaite clast in the Villabeto de la Peña ordinary (L6) chondrite[6,10]. In order to further clarify the origin of Öst 65 here we apply chromium isotopic analyses of whole-rock samples of Öst 65 plus three fossil L chondrites (Österplana 018, 029 and 032, or informal names Ark 018, Gol 001 and Bot 003, respectively), as well as clast and host material from Villabeto de la Peña. We also perform Cr-isotopic analyses of separated chrome-spinel grains from two recent meteorites, the winonaite NWA 725 and the L6 ordinary chondrite Lundsgard. In addition, we present refined oxygen three-isotopic analyses for the chrome spinels of Öst 65, and new data on the appearances and distribution of relict minerals in Öst 65. We show that Öst 65 has no documented analogue among the known meteorites that have fallen on Earth in recent time. The ^{54}Cr values of Öst 65 are similar to those of ordinary chondrites, whereas oxygen isotopes show its affinity with some rare primitive achondrites.

Results

Isotopic results. The fossil meteorite Öst 65 is shown in Fig. 1, and its stratigraphic position in the Thorsberg quarry, together with previously established ^{21}Ne cosmic-ray exposure ages[6,7], is presented in Fig. 2. Our new ε^{54}Cr and Δ^{17}O data are compared to similar data for many of the major groups of meteorites known today in Fig. 3. For the detailed results of

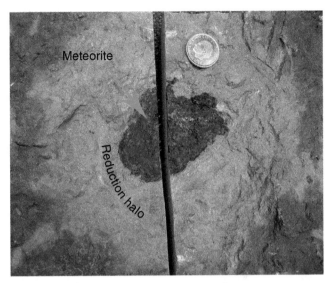

Figure 1 | The Österplana 065 fossil meteorite from the Glaskarten 3 bed. The meteorite is $8 \times 6.5 \times 2$ cm large. It is surrounded by a grey reduction halo, in the otherwise red limestone. Oxygen was consumed when the meteorite weathered on the sea floor. The coin in the image has a diameter of 2.5 cm.

our isotopic analyses, see Tables 1 and 2. The spinel grains from Lundsgard and NWA 725 yield ε^{54}Cr values that are the same as previously analyzed L6 chondrites and winonaites, respectively. The samples of the fossil L chondrites yield ε^{54}Cr results in agreement with an ordinary chondrite origin, that is, in the range -0.34 to -0.24. This indicates that whole-rock analyses of fossil meteorites provide reliable Cr-isotopic results. This is in line with previous elemental analyses of whole-rock fossil L chondrites showing that they have retained most of their original chromium[11]. The Villabeto de la Peña host and clast give ε^{54}Cr values confirming the previous classifications based on oxygen isotopes alone[10]. The two whole-rock samples of Öst 65 yield ε^{54}Cr values of -0.26 to -0.23, which are within the uncertainties of previously measured ordinary chondrites[12]. The new oxygen-isotope analyses confirm previous Δ^{17}O values for Öst 65 spinels just below the terrestrial fractionation line. The data plot with the winonaites, acapulcoites and lodronites, clearly separated from the chromites of the typical L-chondrite fossil meteorites. Although the oxygen isotopes support a relationship to the winonaites (as suggested in ref. 6), acapulcoites or lodronites, the ε^{54}Cr results for Öst 65 rule out such a relationship. Our combined ε^{54}Cr and Δ^{17}O analyses of Öst 65 suggest that it originated from an asteroid of a kind not represented in our collections of recent meteorites (Fig. 3).

Mineralogy and petrology. Mineralogical and petrological data support the conclusion that Öst 65 represents a so far unknown type of meteorite. Our data rule out that Öst 65 is a breccia with clasts of, for example, winonaite and ordinary chondrite origin. The relict texture is unbrecciated and no relict chondrule texture is observed. We have dissolved several fragments from different parts of Öst 65, and have not been able to recover one single grain of the type of chromite that is so common in equilibrated ordinary chondrites. Such chromite ($>$ca. 30 µm) has a narrow element compositional range, making it easy to identify[4]. Our studies of polished sections show that the type of chrome-spinel grains we used for oxygen isotopic analyses are distributed throughout the meteorite. There are no MgAl spinels (>5 µm) in Öst 65. The only type of inclusion found in

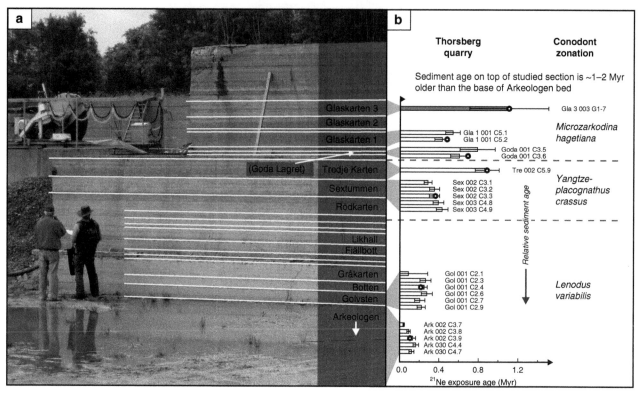

Figure 2 | Thorsberg quarry and cosmic-ray exposure ages of fossil meteorites. (a) The quarried profile in the Thorsberg quarry with the names of individual beds. (b) Cosmic-ray exposure ages for some of the mid-Ordovician fossil meteorites. Meteorite cosmic-ray exposure ages increase with stratigraphic height, in accordance with sedimentation rates as established from conventional biochronology[7,8]. The data indicate that all the fossil meteorites, including Öst 65, originate from a breakup event that took place about 200 Ka before the lowermost beds in the quarry formed[6]. The details and the analytical background of the [21]Ne exposure ages are discussed in ref. 6. In the figure the Österplana 065 meteorite is represented by its 'field' name, Gla3 003, at the top of the section studied. Our fossil meteorites obtain official names, approved by the Meteoritical Society, following the convention of naming meteorites after the locality at which they have been found, that is, in this case Österplana. But in order to emphasize that meteorites in different beds fell at different times they also receive an informal 'field' name. For Öst 65 the name Gla3 003, implies that it is the third meteorite found in the bed Glaskarten 3.

the chrome spinels from Öst 65 has a Ti- and Cr-rich composition (similar to olkhonskite). These inclusions are quite common in the spinels of Öst 65 (Fig. 4a), but no silicate inclusions have been found although searched extensively. The chrome-spinel grains are sometimes associated with 10–30 μm large rutile grains (Fig. 4b). To our knowledge there is no known meteorite with the chrome-spinel–rutile assemblage that we observe in Öst 65.

The majority of the chrome-spinel grains in Öst 65 show abundant planar deformation features (Fig. 4c), something not observed in the many fossil L chondrites. The planar features appear similar to shock lamellae in chromites from shock melt veins in recent meteorites[13]. The Öst 65 spinels are also heavily fractured (Fig. 4a,c), which provides independent evidence for severe shock. Chromites of recently fallen L chondrites show increasing levels of fracturing with increasing shock stage[14].

Discussion

The meteorite Öst 65 accompanies > 100 L chondrites recovered in the sediments of the Swedish quarry. Although single random meteorites are possible, one has to consider that Öst 65 represents on the order of one per cent of the meteorites that have been found on the mid-Ordovician sea floor. This indicates that Öst 65 may represent one of the dominant types of meteorites arriving on Earth 470 Myr ago. Although we cannot rule out that a meteorite similar to Öst 65 may eventually be found among

meteorites falling on Earth today, we can be quite certain that the Öst 65 type does not represent one per cent of the 52,600 classified recent meteorites. At the most there could be a few overlooked specimens.

The discovery in the Ordovician record of a meteorite type that is apparently not falling on the Earth today has important implications. For example, it is possible that the asteroid that produced Öst 65 no longer exists, and there is no source for such meteorites today. The asteroid belt has been evolving through collisions over the history of the solar system, and many of the original asteroids have undoubtedly been destroyed. The record of fossil meteorites on Earth (or on other planetary bodies) provides the only evidence for their former existence and the only way to investigate the collisional evolution of the solar system. The diversity of meteorite parent bodies even relatively late in solar-system history, ~ 500 Myr ago, may have been significantly greater than today. We know, for example, that remnants of Earth building blocks are not present in our meteorite collections. Such meteorites are extinct now but clearly built the Earth in the distant past.

The cosmic-ray exposure age of Öst 65 is ~ 1 Myr and is within the range of those of the fossil L chondrites (Fig. 2)[6–8]. This associates Öst 65 with the breakup of the L-chondrite parent body. The high level of shock metamorphism in the chrome-spinel grains (Fig. 4c) is consistent with the Öst 65 parent body having experienced a major collision. It is possible that a collision between the L-chondrite parent body and the Öst 65 parent body resulted in almost complete destruction of the Öst 65 parent

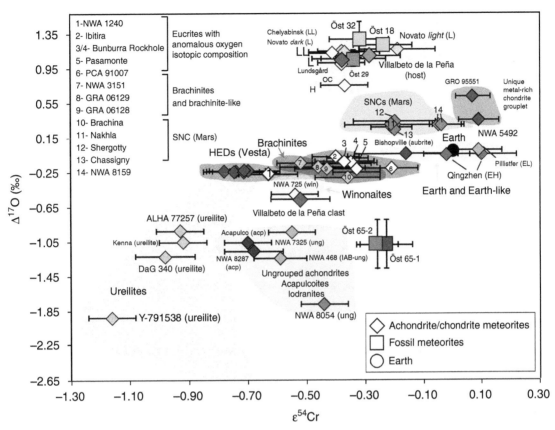

Figure 3 | Oxygen and chromium isotopic composition of meteorites. Comparison of $\Delta^{17}O$ versus $\varepsilon^{54}Cr$ of achondrites, and ordinary chondrites fallen on Earth in recent time with Österplana 065 (Öst 65) and fossil L-chondritic meteorites Österplana 018, 029 and 032 (Öst 18, Öst 29 and Öst 32 in figure). The fields for ordinary chondrites (OC), Mars (SNCs), Earth and earth-likes, Vesta (HEDs), brachinites, ureilites, winonaites (win), acapulcoites (acp)/lodranites/ungrouped achondrites (ung) and a newly identified unique metal-rich chondrite are marked with select representative samples with available data. Symbol colours indicate meteorite type or grouping. Note that the carbonaceous chondrites and affiliated achondrites plot outside the field, with highly positive $\varepsilon^{54}Cr$. The $\varepsilon^{54}Cr$ literature values are from refs 12, 18, 19, 24-33, and $\Delta^{17}O$ literature values are from refs 10, 12, 25, 30, 34-43.

Table 1 | Chromium isotopic results for recent and fossil meteorites.*

Meteorite name & sample	$\varepsilon^{53}Cr$	$\varepsilon^{54}Cr$
Lundsgard	+0.21 ± 0.04	−0.38 ± 0.10
NWA 725	+0.05 ± 0.04	−0.54 ± 0.08
Öst 65 sample 1	+0.24 ± 0.03	−0.23 ± 0.07
Öst 65 sample 2	+0.23 ± 0.04	−0.26 ± 0.09
Öst 32	+0.21 ± 0.04	−0.32 ± 0.10
Öst 29	+0.19 ± 0.05	−0.34 ± 0.11
Öst 18	+0.31 ± 0.04	−0.24 ± 0.10
Villabeto de la Peña clast	+0.06 ± 0.03	−0.52 ± 0.10
Villabeto de la Peña host	+0.26 ± 0.04	−0.29 ± 0.11

*Errors for $\varepsilon^{53}Cr$ and $\varepsilon^{54}Cr$ values are 2σ SD.

asteroid while only disrupting the L-chondrite parent body into several smaller bodies and debris. These smaller bodies continue to feed meteorites, but there is nothing left to provide Öst 65 meteorites to Earth today.

We do not know whether the Öst 65 parent body was one of the original asteroids or already a fragment of a larger object. Iron meteorites come from parent bodies that experienced melting and differentiation that separated the Fe–Ni metal from the silicate components. Trace-element data indicate that iron meteorites may originate from up to 60 different parent bodies[15]. With a few potential exceptions, we have not been able to identify the silicate fractions that accompanied the Fe–Ni metal in the original parent

bodies. The silicates were probably stripped from tougher iron cores of the asteroids early in solar-system history. The iron cores of these parent bodies were much more resistant to collisional degradation, whereas the outer silicate parts have been consumed over time in this process[16]. Because Öst 65 does not exhibit chondrite texture and has other characteristics that do not match most chondrites, the Öst 65 parent body might represent some of this missing silicate. Winonaite meteorites have very similar oxygen-isotope compositions to the silicate inclusions in IAB and IIICD iron meteorites, implying that they may have originated on the same parent asteroid[15,17]. The 'forensic' power of paired $\Delta^{17}O$-$\varepsilon^{54}Cr$ isotope systematics[18,19] could be applied to silicate and oxide inclusions of many types of iron meteorites in order to identify their genetic link with other planetary materials. Studies of fossil meteorites in the Earth sedimentary record may identify other classes of 'extinct' meteorites, some of which could be linked to known classes of iron meteorites.

The Öst 65 meteorite is significant because it demonstrates that ~500 Myr ago we may have had different meteorites falling to Earth than what we see today. Some may be extinct, whereas others, such as the L chondrites, still fall on Earth. Apparently there is potential to reconstruct important aspects of solar-system history by looking down in Earth's sediments, in addition to looking up at the skies[4].

Methods
Chromium isotopic analysis. Aliquots for Cr-isotopic composition measurements were prepared using bulk material for seven of the samples: Öst 65-1, Öst

Table 2 Oxygen isotopic results for Öst 65 chrome-spinel grains.*

Grain and spot	$^{17}O/^{16}O$	$^{18}O/^{16}O$
Grain 4, spot 1	0.00038301 ± 0.00000017	0.0020113 ± 0.0000009
Grain 4, spot 2	0.00038305 ± 0.00000020	0.0020110 ± 0.0000009
Grain 5, spot 1	0.00038329 ± 0.00000016	0.0020117 ± 0.0000009
Grain 8, spot 1	0.00038310 ± 0.00000019	0.0020119 ± 0.0000011
Grain 8, spot 2	0.00038317 ± 0.00000021	0.0020118 ± 0.0000010
Grain 9, spot 1	0.00038312 ± 0.00000020	0.0020127 ± 0.0000010
Grain 10, spot 1	0.00038306 ± 0.00000017	0.0020119 ± 0.0000012
Weighted mean and 2σ SD	0.00038311 ± 0.00000007	0.0020117 ± 0.0000004
Reduced χ^2	1.226	1.178

Grain and spot	$\delta^{17}O$	$\delta^{18}O$	$\Delta^{17}O$
Grain 4, spot 1	0.33 ± 0.43	3.02 ± 0.46	−1.24 ± 0.50
Grain 4, spot 2	0.44 ± 0.51	2.92 ± 0.46	−1.07 ± 0.57
Grain 5, spot 1	1.06 ± 0.42	3.24 ± 0.44	−0.62 ± 0.47
Grain 8, spot 1	0.57 ± 0.49	3.36 ± 0.46	−1.18 ± 0.55
Grain 8, spot 2	0.75 ± 0.56	3.28 ± 0.51	−0.96 ± 0.62
Grain 9, spot 1	0.62 ± 0.52	3.74 ± 0.52	−1.32 ± 0.59
Grain 10, spot 1	0.47 ± 0.45	3.37 ± 0.60	−1.28 ± 0.55
Weighted mean and 2σ SD	0.61 ± 0.18[†]	3.25 ± 0.18	−1.08 ± 0.21[†]
Reduced χ^2	1.266	1.178	0.933

*Data are from measurement spots that are free of cracks, inclusions, overlaps onto other phases and so on that could have caused shifts in the oxygen-isotope ratios.
†Errors do not include the systematic uncertainty from the deadtime correction (∼ ±20‰). Including that uncertainty increases the errors on the mean values for $\delta^{17}O$ and $\Delta^{17}O$ to ±0.27 and ±0.29, respectively. The larger errors are used in Fig. 3. The errors on individual measurements are only minimally affected by the uncertainty in the deadtime.

65-2, Öst 18, Öst 29 and Öst 32, a winonaite clast fraction from Villalbeto de la Peña and Villalbeto de la Peña L-chondrite host matrix. For the Villalbeto de la Peña clast, the clast fraction was visually inspected for adhered matrix material and any visible matrix was removed by hand. Bulk powders were generated for each sample by crushing a fusion crust-free piece using an agate mortar and pestle. For two of the samples, Lundsgard and NWA 725, Cr-isotopic composition measurements were made using separated chromite grains instead of bulk powders. The bulk powders were homogenized and an aliquot of the powder (30–50 mg) was placed into a PTFE Parr bomb capsule with a 2:1 concentrated HF-HNO$_3$ acid solution. For Lundsgard and NWA 725 the chromite grains were placed directly into their respective PTFE capsules with the 2:1 HF-HNO$_3$ acid solution. The capsules were sealed in stainless steel jackets and heated in an oven for 96 h at 190 °C to ensure complete dissolution of all refractory phases. After dissolution, the sample solutions were dried down and treated with concentrated HNO$_3$ and 6 N HCl to eliminate fluorides generated during the dissolution processes.

Chromium was removed from the sample matrix utilizing a 3-column chemistry procedure described in ref. 20. High-precision isotope ratios were measured in the purified Cr fractions using a Thermo Triton Plus thermal ionization mass spectrometer at the University of California at Davis. The Cr fractions were loaded onto outgassed tungsten filaments after being mixed with an Al-boric acid-silica gel activator solution. Each filament was loaded with 3 µg of Cr and four filaments were loaded for each sample for a total load of 12 µg. Each four filament set was bracketed with filaments loaded with the NIST SRM 979 Cr standard. The signal intensity for each filament was set to 10 V (± 15%) for ^{52}Cr with 10^{11} Ω resistors. Each filament analysis consisted of 1,200 ratios with 8 s integrations times. A gain calibration was made at the beginning of each filament, and the baseline was measured and the amplifiers were rotated between each block of 25 ratios. The mass fractionation was corrected using an exponential law and a $^{50}Cr/^{52}Cr$ ratio of 0.051859 (ref. 21). The $^{53}Cr/^{52}Cr$ and $^{54}Cr/^{52}Cr$ ratios are expressed in ε-notation (that is, parts per 10,000 deviation from the NIST SRM 979 Cr standard).

Oxygen isotopic analysis. Individual chromite grains, 50–100 µm in diameter, separated from Öst 65, were mounted in resin in 0.25-inch-diameter stainless steel bullets. Somewhat larger grains of Stillwater chromite standard were mounted in similar bullets. All grains were ground and polished to give a flat cross section at least 50 µm across. The grains were imaged with secondary electrons and measurement spots were selected to avoid cracks, surface imperfections, inclusions and so on. The spots were marked using the scanning electron microscope so that the spots could be seen by scanning ion imaging to accurately position the beam for measurement.

Oxygen-isotope compositions were measured using the University of Hawai'i Cameca ims-1280 ion microprobe. The detailed methodology is provided in ref. 22. Here, we provide details specific for these measurements. A 1 nA Cs$^+$ primary ion beam, focused to ∼10 µm, with a total impact energy of 20 keV was used. Spots were presputtered for 60 s using a 5-µm raster, after which the raster was reduced to 3 µm for measurement. Masses $^{16}O^-$, $^{17}O^-$ and $^{18}O^-$ were measured simultaneously in multicollection mode. $^{16}O^-$ and $^{18}O^-$ were measured by multicollector Faraday cups with low-mass resolving power (MRP ∼ 2,000), while $^{17}O^-$ was measured using the axial monocollector electron multiplier with MRP ∼ 5,600, sufficient to separate the $^{16}OH^-$ interference. The relatively high $^{17}O^-$ count rate produces a loss in gain due to aging of the first dynode of the electron multiplier. To minimize this effect, the $^{17}O^-$ signal was measured for only 4 s in each cycle, after which the beam was deflected into a monocollector Faraday cup for 10 s (procedure originally proposed by Kita et al.[23]). Signals for $^{16}O^-$ and $^{17}O^-$ collected during the 4-s interval were used to determine the $^{17}O/^{16}O$ ratio and the $^{16}O^-$ and $^{18}O^-$ signals collected during the 10-s interval were used to determine the $^{18}O/^{16}O$ ratio. Each measurement consisted of 30 cycles.

Measured data were corrected for detector background and deadtime (electron multiplier). The background corrections for the Faraday cups were made using a smoothed fit to the backgrounds measured during presputtering. A tail correction was made for the $^{16}OH^-$ interference. The contribution of the $^{16}OH^-$ tail at mass ^{17}O was estimated from measurement of the low-mass tail of the ^{17}O peak to be ∼6 × 10^{-6} of the $^{16}OH^-$ interference. The height of the $^{16}OH^-$ interference was measured at the end of each measurement. Even with our protocol to minimize changes in the gain of the electron multiplier, the gain did change with time. We checked the pulse-height distribution at the beginning of each day and set the gain to our 'standard' condition. We did not check the gain again during the day, as we have learned that the act of checking the multiplier gain introduces a transient instability in the multiplier gain that compromises the data. The gain of the monocollector electron multiplier during the day was monitored by measuring the Stillwater chromite standard for five-six measurements at the beginning of each day, after five-six measurements of the unknown, and at the end of the day. The Stillwater measurements permitted us to model and account for the gain drift in the multiplier during each measurement day.

The uncertainties for the individual measurement data reported in Table 2 include the counting-statistical uncertainties from the individual measurements, the standard deviation of the standard measurements (including uncertainties in the drift correction in the case of ^{17}O), and the uncertainties in the background and tail corrections. The mean values for Öst 65 are the error-weighted means and the standard errors for the set of measurements. The uncertainties do not include a systematic uncertainty associated with the deadtime correction for ^{17}O. The deadtime of 30 ns has an uncertainty of ± 1 nanosecond, which results in an uncertainty on $\delta^{17}O$ and $\Delta^{17}O$ of 0.2‰. When combined with the statistical errors for individual $\delta^{17}O$ and $\Delta^{17}O$ measurements, this uncertainty makes little difference. But this error cannot be incorporated into the calculation of the weighted mean and standard error, because it is a systematic error that is not reduced by making more measurements. It must be applied after the mean values and standard errors are calculated. When combined quadratically with the standard errors on the mean values, the total uncertainty for $\delta^{17}O$ becomes ± 0.27‰ and for $\Delta^{17}O$ becomes ± 0.29‰.

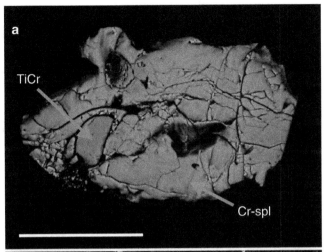

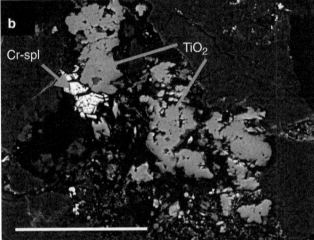

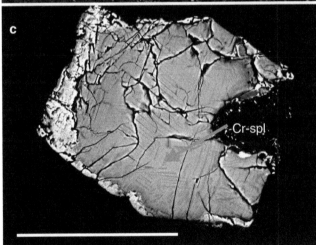

Figure 4 | Back-scattered electron images of mineral grains from Österplana 065. (**a**) Chrome-spinel grain with inclusion of a TiCr-mineral, possibly olkhonskite ($TiO_2 = 56$ wt%, $Cr_2O_3 = 30$ wt%, $FeO = 9$ wt%). (**b**) Rutile and chrome spinel in polished section of Öst 65. (**c**) Chrome-spinel grain with apparent shock deformation lamellae. Scale bars, 50 μm.

References

1. Anders, E. Origin, age and composition of meteorites. *Space Sci. Rev.* **3**, 583–714 (1964).
2. Keil, K., Haack, H. & Scott, E. R. D. Catastrophic fragmentation of asteroids: evidence from meteorites. *Planet. Space Sci.* **42**, 1109–1122 (1995).
3. Korochantseva, E. V. *et al.* L-chondrite asteroid breakup tied to Ordovician meteorite shower by multiple isochron ^{40}Ar–^{39}Ar dating. *Meteorit. Planet. Sci.* **42**, 113–130 (2007).
4. Schmitz, B. Extraterrestrial spinels and the astronomical perspective on Earth's geological record and evolution of life. *Chem. Erde.* **73**, 117–145 (2013).
5. Schmitz, B., Tassinari, M. & Peucker-Ehrenbrink, B. A rain of ordinary chondritic meteorites in the early Ordovician. *Earth Planet. Sci. Lett.* **194**, 1–15 (2001).
6. Schmitz, B. *et al.* A fossil winonaite-like meteorite in Ordovician limestone: a piece of the impactor that broke up the L-chondrite parent body? *Earth Planet. Sci. Lett.* **400**, 145–152 (2014).
7. Heck, P. R., Schmitz, B., Baur, H., Halliday, A. N. & Wieler, R. Fast delivery of meteorites to Earth after a major asteroid collision. *Nature* **430**, 323–325 (2004).
8. Heck, P. R., Schmitz, B., Baur, H. & Wieler, R. Noble gases in fossil micrometeorites and meteorites from 470 Myr old sediments from southern Sweden and new evidence for the L-chondrite parent body breakup event. *Meteorit. Planet. Sci.* **43**, 517–528 (2008).
9. Schmitz, B. *et al.* Asteroid breakup linked to mid-Ordovician biodiversification event. *Nat. Geosci.* **1**, 49–53 (2008).
10. Bischoff, A. *et al.* Reclassification of Villalbeto de la Peña—occurrence of a winonaite-related fragment in a hydrothermally metamorphosed L-chondritic breccia. *Meteorit. Planet. Sci.* **48**, 628–640 (2013).
11. Schmitz, B. *et al.* Determining the impactor of the Ordovician Lockne crater: oxygen and neon isotopes in chromite versus sedimentary PGE signatures. *Earth Planet. Sci. Lett.* **306**, 149–155 (2011).
12. Jenniskens, P. *et al.* Fall, recovery, and characterization of the Novato L6 chondrite breccia. *Meteorit. Planet. Sci.* **49**, 1388–1425 (2014).
13. Chen, M., Shu, J., Xie, X. & Mao, H. K. Natural $CaTi_2O_4$-structured $FeCr_2O_4$ polymorph in the Suizhou meteorite and its significance in mantle mineralogy. *Geochim. Cosmochim. Acta* **67**, 3937–3942 (2003).
14. Alwmark, C. *et al.* A 3-D study of mineral inclusions in chromites from ordinary chondrites using synchrotron microtomography—method and applications. *Meteorit. Planet. Sci.* **46**, 1071–1081 (2011).
15. Goldstein, J. I., Scott, E. R. D. & Chabot, N. L. Iron meteorites: crystallization, thermal history, parent bodies, and origin. *Chem. Erde* **69**, 293–325 (2009).
16. Burbine, T. H., Meibom, A. & Binzel, R. P. Mantle material in the main belt: battered to bits? *Meteoritics* **31**, 607–620 (1996).
17. Benedix, G. K., McCoy, T. J., Keil, K. & Love, G. A petrologic study of the IAB iron meteorites: constraints on the formation of the IAB-winonaite parent body. *Meteorit. Planet. Sci.* **35**, 1127–1141 (2000).
18. Sanborn, M. E., Yin, Q.-Z. & Irving, A. J. in *45th Lunar and Planetary Science Conference*, The Woodlands, TX, USA. Abstract #2032 (2014).
19. Sanborn, M. E. & Yin, Q.-Z. in *45th Lunar and Planetary Science Conference*, The Woodlands, TX, USA. Abstract #2018 (2014).
20. Yamakawa, A., Yamashita, K., Makishima, A. & Nakamura, E. Chemical separation and mass spectrometry of Cr, Fe, Ni, Zn, and Cu in terrestrial and extraterrestrial materials using thermal ionization mass spectrometry. *Anal. Chem.* **81**, 9787–9794 (2009).
21. Shields, W. R., Murphy, T. J., Catanzaro, E. J. & Garner, E. L. Absolute isotopic abundance ratios and the atomic weight of a reference sample of chromium. *J. Res. Natl Bur. Stand.* **70A**, 193–197 (1966).
22. Makide, K. *et al.* Oxygen- and magnesium-isotope compositions of calcium–aluminum-rich inclusions from CR2 carbonaceous chondrites. *Geochim. Cosmochim. Acta* **73**, 5018–5050 (2009).
23. Kita, N. T., Ushikubo, T., Fu, B., Spicuzza, M. J. & Valley, J. W. in *38th Lunar and Planetary Science Conference*, League City, TX, USA. Abstract #1338 (2007).
24. Yamakawa, A., Yamashita, K., Makishima, A. & Nakamura, E. Chromium isotope systematics of achondrites: chronology and isotopic heterogeneity of the inner solar system bodies. *ApJ* **720**, 150–154 (2010).
25. Benedix, G. K. *et al.* in *45th Lunar and Planetary Science Conference*, The Woodlands, TX, USA. Abstract #1650 (2014).
26. Shukolyukov, A. & Lugmair, G. W. in *37th Lunar and Planetary Science Conference*, League City, TX, USA. Abstract #1478 (2006).
27. Ueda, T., Yamashita, K. & Kita, N. Chromium isotopic study of ureilites. *Meteorit. Planet. Sci.* **41**, A176 (2006).
28. Sanborn, M. E. *et al.* Re-evaluation of anomalous metal-rich lodranite Northwest Africa 468 based on combined chromium and oxygen isotopes. *Meteorit. Planet. Sci.* **49**, A352 (2014).
29. Sanborn, M. E., Yin, Q.-Z., Zipfel, J. & Palme, H. Investigating the genetic relationship between NWA 5492 and GRO 95551 using high-precision chromium isotopes. *Meteorit. Planet. Sci.* **50**, A5159 (2015).
30. Popova, O. P. *et al.* Chelyabinsk airburst, damage assessment, meteorite recovery, and characterization. *Science* **342**, 1069–1073 (2013).
31. Trinquier, A., Birck, J.-L. & Allegre, C. Widespread ^{54}Cr heterogeneity in the inner solar system. *ApJ* **655**, 1179–1185 (2007).

32. Sanborn, M. E. *et al.* Chromium isotopic studies of ungrouped achondrites NWA 7325, NWA 2976, and NWA 6704. *Meteorit. Planet. Sci.* **48**, A5220 (2013).

33. Sanborn, M. E. & Yin, Q.-Z. in *46th Lunar and Planetary Science Conference*, The Woodlands, TX, USA. Abstract #2241 (2015).

34. Clayton, R. N. & Mayeda, T. K. Oxygen isotope studies of achondrites. *Geochim. Cosmochim. Acta* **60**, 1999–2017 (1996).

35. Clayton, R. N., Mayeda, T. K. & Rubin, A. E. Oxygen isotopic compositions of enstatite chondrites and aubrites. *J. Geophys. Res.* **89**, C245–C249 (1984).

36. Clayton, R. N., Mayeda, T. K., Goswami, J. N. & Olsen, E. J. Oxygen isotope studies of ordinary chondrites. *Geochim. Cosmochim. Acta* **55**, 2317–2337 (1991).

37. Rubin, A. E., Kallemeyn, G. W. & Wasson, J. T. A AB-complex iron meteorite containing low-Ca clinopyroxene: northwest Africa 468 and its relationship to lodranites and formation by impact melting. *Geochim. Cosmochim. Acta* **66**, 3657–3671 (2002).

38. Scott, E. R. D., Greenwood, R. C., Franchi, I. A. & Sanders, I. S. Oxygen isotopic constraints on the origin and parent bodies of eucrites, diogenites, and howardites. *Geochim. Cosmochim. Acta* **73**, 5835–5853 (2009).

39. Irving, A. J. *et al.* in *45th Lunar and Planetary Science Conference*, The Woodlands, TX, USA. Abstract #2465 (2014).

40. Weisberg, M. K. *et al.* Petrology and oxygen isotopes of NWA 5492, a new metal-rich chondrite. *Meteorit. Planet. Sci.* **47**, 363–373 (2012).

41. Day, J. M. D. *et al.* Origin of felsic achondrites Graves Nunataks 06128 and 06129, and ultramafic brachinites and brachinite-like achondrites by partial melting of volatile-rich primitive parent bodies. *Geochim. Cosmochim. Acta* **81**, 94–128 (2012).

42. Heck, P. R. *et al.* A single asteroidal source for extraterrestrial Ordovician chromite grains from Sweden and China: high-precision oxygen three-isotope SIMS analysis. *Geochim. Cosmochim. Acta* **74**, 497–509 (2010).

43. Greenwood, R. C., Franchi, I. A., Gibson, J. M. & Benedix, G. K. Oxygen isotope variation in primitive achondrites: the influence of primordial, asteroidal, and terrestrial processes. *Geochim. Cosmochim. Acta* **94**, 146–163 (2012).

Acknowledgements

We thank quarry owners Göran, Sören and Stig Thor for enduring support with the meteorite search; K. Deppert and P. Eriksson for support at Lund University; R. Weihard for steadfast support in Lidköping; A. Cronholm and F. Terfelt for technical assistance; and L. Ferrière, P. Heck, J. Holstein, U. Kolitsch, M. Meier, S. Rout and D. Topa for discussions. Two anonymous referees and D. Lauretta provided helpful comments. B.S. is funded by a European Research Council-Advanced Grant (213000). Q.-Z.Y. thanks K. Ziegler for providing materials from the Villalbeto de la Peña meteorite, and acknowledges NASA (NNX14AM62G) and the UC Office of the President (UC Lab Fees Award ID# 12-LR-237921) for partial support of this work. G.R.H. was supported by NASA (NNX14AI19G). This is Hawai'i Institute of Geophysics and Planetology Publication #2188 and School of Ocean and Earth Science and Technology Publication #9612.

Author contributions

All authors contributed equally to the study.

Additional information

Competing financial interests: The authors declare no competing financial interests.

18

A potential hidden layer of meteorites below the ice surface of Antarctica

G.W. Evatt[1], M.J. Coughlan[1], K.H. Joy[2], A.R.D. Smedley[3], P.J. Connolly[3] & I.D. Abrahams[1]

Antarctica contains some of the most productive regions on Earth for collecting meteorites. These small areas of glacial ice are known as meteorite stranding zones, where upward-flowing ice combines with high ablation rates to concentrate large numbers of englacially transported meteorites onto their surface. However, meteorite collection data shows that iron and stony-iron meteorites are significantly under-represented from these regions as compared with all other sites on Earth. Here we explain how this discrepancy may be due to englacial solar warming, whereby meteorites a few tens of centimetres below the ice surface can be warmed up enough to cause melting of their surrounding ice and sink downwards. We show that meteorites with a high-enough thermal conductivity (for example, iron meteorites) can sink at a rate sufficient to offset the total annual upward ice transport, which may therefore permanently trap them below the ice surface and explain their absence from collection data.

[1] School of Mathematics, University of Manchester, Manchester M13 9PL, UK. [2] The School of Earth, Atmospheric and Environmental Sciences, University of Manchester, Manchester M13 9PL, UK. [3] Centre for Atmospheric Science, The School of Earth, Atmospheric and Environmental Sciences, University of Manchester, Manchester M13 9PL, UK. Correspondence and requests for materials should be addressed to G.W.E. (email: geoffrey.evatt@manchester.ac.uk) or to K.H.J. (email: katherine.joy@manchester.ac.uk).

When meteorites fall onto a large area of inland Antarctica, the subsequent ice flow dynamics direct many of them to localized surface regions called meteorite stranding zones (MSZs)[1]. This concentration phenomenon (see Supplementary Note 1) allows for efficient human-led recovery missions[2,3]: up until December 2015, some 34,927 meteorites (official approved names of stones classified by the Meteoritical Nomenclature Committee[4]) were recovered from the surface of Antarctica, representing 66.3% of the world's total number of collected specimens. (NB there are many Antarctic stones recovered by some national collection programs that have never been formally classified and others that have only been given provisional names). However, while meteorite falls should be distributed almost uniformly across the Earth's surface, meteorite collection data (see Table 1) reveals that the proportion of iron-based meteorites (iron meteorites and stony-iron meteorites) recovered from Antarctica, 0.7%, is significantly lower than the proportion recovered after witnessed falls (see Supplementary Note 2) from the rest of the World[4-8], 5.5%—a statistical difference at over the 99.9% confidence level. This comparison suggests that one or more physical mechanisms are resulting in an apparent shortfall of iron-based meteorite falls in Antarctica. Another notable statistic of the collection data comes from Antarctica's moraine-free LaPaz Icefield MSZ (86° 22′ S, 70° 0′ W), which has produced a significantly lower proportion of iron-based meteorites (0.3%) than the rest of Antarctica combined (the latter data includes debris-covered regions, glacial moraines, ice tongues and so on)—a difference at the 94% confidence level. With collection methods across Antarctica by different human searching programmes broadly similar, using visual inspection from snowmobiles (rapidly covering large areas of blue ice) or on foot (a more localized approach often focusing on rock-rich glacial moraine fields), this further suggests that the physical cause for the disparity is more pronounced in regions of debris-free MSZ ice.

It is our hypothesis that the under-representation of iron-based meteorites in Antarctic data is caused by solar energy penetrating the clear ice of the MSZs[9]. To expand: with the intensity of transmitted radiation diminishing with depth[10], a meteorite being transported up toward the surface of a MSZ[2,11] will be exposed to an increasing level of solar warming. If the rising meteorite could reach a shallow-enough depth for the solar energy to enable melting of its surrounding ice, the meteorite would sink relative to the upwelling ice. Furthermore, if the thermal properties of iron meteorites enabled them to sink faster than other meteorite classes, and counter all annual upward glacial transportation, it would provide a physical mechanism that traps iron meteorites within the ice, thus explaining their under-representation on the surface.

Under our hypothesis, sinking (relative to the upwelling ice) of englacial meteorites in an Antarctic MSZ will be seasonally dependent. In the winter months, with little-to-no solar radiation, sinking will not be possible, and so all classes of englacial meteorites can be expected to rise with the speed of the ice (whose surface is still ablating due to scouring by the strong winter katabatic winds[12]). Conversely, in the long daylight hours of the summer months, solar warming will enable susceptible meteorites to have a local rate of melting that offsets the upwelling process, and hence, for certain meteorite classes, sinking relative to the ice surface can (re)commence. The question, thus, becomes: could iron and stony-iron meteorites have a propensity to achieve a summertime-averaged relative velocity that is sufficiently large (and negative) to offset all annual upwelling, thereby permanently trapping them below the ice surface?

To answer this question, we first present the results of a series of controlled laboratory experiments that prove englacial solar warming can cause shallow-enough sub-surface meteorites to move down through the encasing ice, under the action of thawing and freezing. We then present a mathematical model of the energy balance within the system which gives a close fit to the laboratory results, allowing us to confidently apply it to an Antarctic MSZ situation. In so doing, we show how the thawing and freezing process will typically negate all annual upward transportation of a MSZ meteorite with a high-enough thermal conductivity (for example, iron), while allowing meteorites with lower conductivities to emerge from the ice. As a consequence of this filtering mechanism, the model suggests a few tens of centimetres beneath the ice surface of a MSZ, there are sub-layers of ice that potentially contain a (sparse) distribution of meteorites with high-thermal conductivities. With meteorites constantly being englacially transported towards many MSZs[13], these layers (which are hidden from surface-searching methods) could harbour an additional reserve of iron-rich meteorites. If accessed, this layer would lead to a significant increase in our library of iron and stony-iron meteorite types, which will directly help our understanding of early solar system-formation processes and the diversity of planetesimals that were present[14,15].

Results

Laboratory results. Our experiments centred on subjecting a meteorite encased in a block of ice to the radiation from a solar-simulator lamp held directly above the ice surface, and focused onto the meteorite (see the Methods section). Two classes of meteorite were tested: an ordinary chondrite meteorite (North West Africa (NWA) 869 L3.9–L6), and an achondritic iron meteorite (Sikhote-Alin). Both samples were of near-prolate-spheroidal geometry, with the axis of symmetry horizontally aligned with diameter 15 mm, and width 10 mm. Under the laboratory conditions, both classes of meteorite proved able to warm up enough that they could melt their surrounding

Table 1	Meteorite collection statistics.		
Meteorite class	**Iron and stony-iron**	**Others**
LaPaz Icefield MSZ finds (Antarctica)	5 (0.3%)	1,665 (99.7%)
Frontier Mountains MSZ finds (Antarctica)	0 (0.0%)	798 (100.0%)
Total Antarctic Finds (all search programmes)	239 (0.7%)	34,688 (99.3%)
Rest of world falls	60 (5.5%)	1,037 (94.5%)
Rest of world finds (excluding falls)	1,145 (6.9%)	15,505 (93.1%)

MSZ, meteorite stranding zone.
Statistics of classified named meteorite stones including the number and percentage of iron-based meteorite finds from the LaPaz Icefield MSZ Antarctica; Frontier Mountain MSZ Antarctica; the whole of Antarctica; the number of observed (and then collected) meteorite 'falls' from the world excluding Antarctica; and the number of meteorite 'finds' from the world excluding Antarctica. Data is taken from the Meteoritical Society bulletin of classified and named meteorite samples[4], updated as of 18 December 2015. We note that this official data set does not include named meteorites with only provisional or undocumented meteorites names. Iron meteorites include all iron groups. Stony-iron meteorites include pallasites and mesosiderite types.

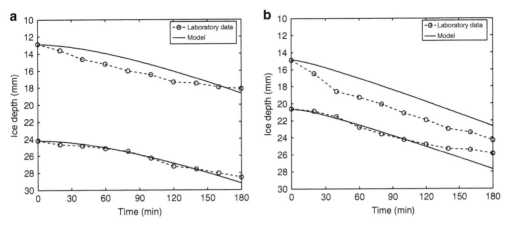

Figure 1 | Laboratory results. Experimental results (circles and dashed line) for the upper-surface depth (relative to the ice surface) of a sinking englacial meteorite as time progresses, where the samples are exposed to solar warming from above the ice. Two sets of results for an (**a**) ordinary chondrite and (**b**) iron meteorite. These data points have a measurement error of under ±1mm. The solid lines represent the corresponding results of our energy balance model, solved using laboratory parameter values (see the Methods section).

ice and sink down, as shown in Fig. 1. The average speed of sinking of the iron meteorite was approximately 2.4 mm per hour, markedly higher than that for the chondrite sample at 1.5 mm per hour (although as shown in Fig. 1, their initial depths were not identical).

To relate these laboratory experiments to the Antarctic setting, we created a mathematical energy balance model that could be applied to each of the physical situations; close alignment between the model and laboratory results would give confidence to the model's Antarctic predictions. Our energy balance model necessarily considers a variety of competing energy fluxes[16]: the atmosphere's turbulent and electromagnetic heat exchanges; heat conduction within the ice, meteorite and surrounding melt-water layers; and energy to and from ice-phase transitions (see the Methods section). To run simulations, the model requires measurements of certain material properties of the meteorites: the albedo of the iron meteorite was measured as 0.159, and the chondrite meteorite as 0.106; an independent estimate of the iron meteorite's thermal conductivity was 25 W m^{-1}K^{-1} (ref. 17) and that for the chondrite meteorite was 1.5 W m^{-1}K^{-1} (ref. 18). Using these, the model's prediction for each of the experiments is overlain on the data points of Fig. 1. The closeness of fit between the experimental results and the model predictions is striking. Thus, it allows us to apply the model with confidence to the Antarctic situation, in anticipation that over the longer Antarctic time scales a more-pronounced divergence between the specimens' behaviours would be revealed.

Antarctic results. We parameterised our energy balance model to Antarctica's (much studied) Frontier Mountain meteorite trap[19] (72° S, 160° W, see collection statistics in Table 1) and allowed the model's shortwave energy flux, longwave energy flux and sensible heat flux to vary seasonally (see the Methods section). To compare the progress between a chondrite and an iron meteorite, we used a meteorite thermal conductivity of $k_m = 1.5$ W m^{-1}K^{-1} for the chondrite meteorite numerical experiments[17,18], and $k_m = 25$ W m^{-1}K^{-1} for the iron meteorite calculations[17]. A meteorite width of 3 cm and, initially, an averaged meteorite broadband (exterior) surface albedo α_m of 0.13 were used for both samples. Using Antarctic climatic parameters (see Table 2), results for the englacial progress of a chondrite and an iron meteorite are shown in Fig. 2a. The results clearly demonstrate the anticipated divergence between meteorite classes; over the longer Antarctic time scales a meteorite with a

high-enough thermal conductivity (for example, iron) can potentially remain trapped below the ice surface (in this instance at a depth of around 35 cm), whereas a meteorite with a lower thermal conductivity (that is, iron-poor chondritic and achondritic types) will emerge onto the surface. For our particular parameter values, we found that meteorites with thermal conductivities higher than ~4 W m^{-1}K^{-1} will remain trapped within the ice. However, given the inevitable stochastic fluctuations around our mean values, meteorites predicted to be trapped at a shallow-enough depth can sometimes still be expected to emerge from the surface. This low propensity for Frontier Mountain iron/stony-iron meteorites to reach the surface is consistent with the collection data of Table 1. The results in Fig. 2 also demonstrate that chondrites can undergo an annual freeze–thaw cycle while trapped within the uppermost half-metre of ice, which is consistent with in-ice weathering observations from englacial meteorites[20].

The energy balance model is able to identify further meteorite parameters that might affect englacial trapping or release, for example, its broadband surface albedo and mass density. It is found that variations in meteorite surface albedo alter the predicted meteorite depths by only a modest amount: a significant ±50% change in meteorite surface albedo was found to alter the iron meteorite depth of Fig. 2a by around only ±2 cm. This small depth variation is not unexpected, as these large albedo deviations correspond to only a ±7.5% variation in absorbed solar energy; and with meteorite surface albedo unlikely to differ widely (including our measured values) and consistently between iron and chondritic meteorites (assuming the dark fusion crust is reasonably intact), we conclude that meteorite surface albedo is not the cause for the under-representation of iron-based meteorites. Likewise, meteorite mass density is not found to be a viable differentiator for englacial sinking between meteorite classes, as mass density can be shown to have a negligible impact on slow heat-transfer processes and englacial gravitational separation[6], and any alteration in the melting point of ice due to the pressure of the overlying meteorite is minute[21]. As such we can use the model to conclude that thermal conductivity is the dominant parameter governing the divergent englacial behaviour between meteorite classes.

A notable parameter that uniformly affects the dynamics within our modelling is the ice broadband albedo α_i (which also acts as a direct proxy for variations in the solar flux scattered back from ice in the vicinity of the surface). So far we have used a default mean value of 0.62 based on field measurements of blue

Table 2 | Energy balance model parameter values.

Parameter	Description	Laboratory value	Antarctic value
k_i	Thermal conductivity of ice $(W\,K^{-1}m^{-1})$	2.22	2.07
k_w	Thermal conductivity of water $(W\,K^{-1}m^{-1})$	0.58	0.58
k_m	Thermal conductivity of meteorite $(W\,K^{-1}m^{-1})$	1.5, 25	1.5, 25
c_a	Heat capacity of air $(J\,kg^{-1}K^{-1})$	1,005	1,005
ρ_a	Air density $(J\,kg^{-1}K^{-1})$	1.29	0.95
ρ_i	Density of ice $(kg\,m^{-3})$	916.2	916.2
L_m	Latent heat of melting ice $(J\,kg^{-1})$	3.34×10^5	3.34×10^5
L_v	Latent heat vapourization, water $(J\,kg^{-1})$	22.6×10^5	22.6×10^5
γ_i	Attenuation coefficient of blue ice (m^{-1})	2.5	2.5
γ_w	Attenuation coefficient of water (m^{-1})	0.001	0.001
V	Ice sheet heave velocity (metres per year)	—	0.065
v	Ice sheet sublimation rate (metres per year)	Negligible	$V/2$
α_i	Blue-ice albedo $(-)$	0.62	0.62
α_m	Meteorite exterior-surface albedo $(-)$	0.106-0.159	0.13
σ	Stefan–Boltzmann's constant $(W\,m^{-2}K^{-2})$	5.667×10^{-8}	5.667×10^{-8}
ϵ	Emmissivity of ice $(-)$	0.94	0.94
θ	Solar elevation angle $(°)$	90	Computed
T_a	Air temperature $(°C)$	-1	$\bar{T}_a(1 - \hat{S}(t))$
$\bar{T}_a$	Lowest air temperature $(°C)$	—	-40
S_{net}	Incoming shortwave energy $(W\,m^{-2})$	1,440	Fig. 4
Q_{long}	Incident longwave radiation $(W\,m^{-2})$	300	$\bar{Q}_1 + \bar{Q}_2\hat{S}(t)$
$\bar{Q}_1$	Longwave energy parameter $(W\,m^{-2})$	—	93
$\bar{Q}_2$	Longwave energy parameter $(W\,m^{-2})$	—	47.5
$\bar{u}$	Average wind speed $(m\,s^{-1})$	2	11
u_*	Friction velocity $(m\,s^{-1})$	0.1	0.1
T_∞	Ice temperature at bottom $(°C)$	-4	—
z_∞	Ice depth (m)	0.05	—
ϕ	Heat flux in region 4 $(W\,m^{-2})$	—	0
s_h	Solar shading	0	7.5%
w	Meteorite width (m)	0.01	0.03

Parameter values used in our energy balance model, for both the laboratory study and the Antarctic analogy (based on the Frontier Mountain meteorite trap area; see the Methods section).

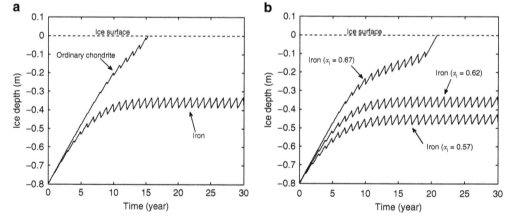

Figure 2 | Antarctic results. Energy balance model results for the Antarctic situation, where parameter values are chosen in accordance with conditions at the Frontier Mountain Meteorite Trap area[19]. The results in **a** show the progress as time progresses of two meteorites, with thermal conductivities $k_m = 1.5\,W\,m^{-1}K^{-1}$ (a typical value for an ordinary chondritic meteorite) and $k_m = 25\,W\,m^{-1}K^{-1}$ (a typical value for an iron meteorite). In **b** the thermal conductivity is held fixed at $25\,W\,m^{-1}K^{-1}$, but the ice surface albedo α_i is varied by $\pm 7.5\%$, highlighting the sensitivity of the meteorite's progress to the reflectivity of the ice surface (and thus also the downwelling shortwave energy flux S_{net}).

ice[12]. Yet this parameter varies depending on the optical quality of the local ice (for example, blue ice versus white ice versus snow-covered ice). In Fig. 2b we show how our results for the iron meteorite are altered when α_i is varied by $\pm 7.5\%$ (equivalent to a $\pm 12\%$ variation in downwelling solar radiation): the higher albedo is now sufficient for the iron meteorite to emerge, whereas the lower albedo deepens the meteorite's resting depth by some 10 cm. As this result suggests, a light covering of snow (or shading

by local topography) on a MSZ could allow iron-based meteorites to emerge, helping explain why a small number of iron-based meteorites are still found on the surface of Antarctica's MSZs, and trapped in glacial moraine fields. Conversely, it is possible to adjust the model parameters (for example, to simulate periods of higher average summer insolation) so as to have the chondritic meteorites also permanently englacially trapped. Interestingly, blue-ice fields in Greenland of similar latitude and elevation, but with a higher average temperature profile than Antarctica, have not yielded any surface meteorites, consistent with our work and aligning with previous suggestions for their absence[22].

Finally, we found the intuitive result that larger meteorites are transported to the ice surface more readily then smaller ones—this being due to the ability of smaller meteorites to more efficiently transmit heat internally to the underlying ice for melting. We observed a near-linear dependence of the englacial resting depth against the meteorite width w, where iron meteorites of widths $> 22\,cm$ emerged at the ice surface (all other parameters as given in Table 2). This result, of course, assumes the meteorite does not break up during the englacial freeze/thaw process. This is an important caveat to make, for the high number of freeze/thaw cycles an iron meteorite can be expected to go through (see Fig. 2) may explain why smaller masses of iron meteorites are recovered from Antarctica as compared with the rest of the world[2].

Discussion

The plausible implication of these results is the existence of a sparsely distributed layer of iron-based meteorites underneath the surface of Antarctica's MSZs. With collection methodologies from MSZs currently based on visual recognition at the ice surface[2], any attempt to determine if such a layer exists would require a significant change in detection and collection strategies. Furthermore, any such collection approach must be able to easily cover relatively large areas of ice. To highlight why, one can use existing collection data to infer a rough number density of iron-based meteorites within a particular MSZ. For the debris-free LaPaz Icefield (Table 1, where we assume that the meteorite fall data to accurately represent the proportion of meteorites recovered: see Supplementary Note 2), some 92 iron-based meteorites appear 'missing' from the searched area of roughly $100\,km^2$. This allows us to crudely estimate the LaPaz number density of distributed englacial iron meteorites at around one per square kilometre. Yet the near absence of terrestrial rock from debris-free MSZs, means the number of false positives that would be detected in a meteorite recovery mission would be negligible. So even though the density of missing iron meteorites is low, suitably focused meteorite collection programmes from debris-free areas of MSZs, may well be feasible.

Along with the value of answering an outstanding scientific question, the motivations for accessing such a layer are clear: every new iron or stony-iron meteorite sample recovered has the potential to have originated from the core[23] (or core–mantle boundary) of its own unique parent asteroid body providing insights into the number, diversity, evolution[24] and destruction[15,25,26] of protoplanets that existed in the early solar system. This knowledge would fill critical gaps in our understanding of both how different meteorite groups are related to one another[27], and the chemical heterogeneity of the solar nebula[24,28,29], from which these bodies were accreted and differentiated.

Methods

Energy balance model. Our energy balance model is designed to capture the essential physics underpinning the sinking process of an englacial meteorite while avoiding extraneous details. The model does not attempt to offer highly accurate Antarctic site-specific results; it aims to address the proof-of-concept question of englacial sinking through the inclusion of dominant (or aggregated) terms in the various energy balance equations and so should be taken to offer generalized quantitative predictive results at this stage.

We consider three distinct stages within our modelling. Chronologically, the evolution starts with the meteorite heating due to absorption of downwelling solar radiation through the ice; however, it remains fully encased in ice while temperatures remain below freezing throughout. Then, once the top surface of the meteorite is sufficiently heated it is able, through conduction, to induce melting of the overlying ice, allowing an upper water layer to form; this water layer can itself

evolve through its interaction with both the overlying ice and the underlying meteorite. Next, if the meteorite continues to warm sufficiently, melting at its lower surface will also commence. When it does, we allow the sub-meteorite melt water to be squeezed upwards by the weight of the meteorite adding to the overlying water layer, thereby enabling the meteorite to sink downwards and fill the displaced volume. These three stages can stop at any time, and freezing reoccur, as the meteorological inputs vary, thus bringing the sinking process to a halt (this being the case for the Antarctic situation during the winter months).

To model this process in the simplest and most elucidating manner, we consider a one-dimensional representation to the physical problem, as sketched in Fig. 3. Thus, the model has four distinct regions at increasing depth, z: (i) an upper ice layer that is exposed to the atmosphere; (ii) a water layer (which need not always exist); (iii) the meteorite; and (iv) a lower ice layer. To model the Antarctic situation, one must incorporate the relative motion of the meteorite to the ice surface. With the MSZ surface in equilibrium at $z = 0$, say, the upwelling ice velocity V must be matched by the ablation rate (which is the sum of the energetic sublimation rate v, and the non-energetic rate that ice is scoured off the surface by the katabatic winds, $V - v$).

To frame the problem mathematically, we need to consider the energy balance at the five boundaries/interfaces of the four regions. It is of note that two of these interfaces are 'free-boundaries', whose locations are variables that require solving as part of the problem: the upper ice/water interface ($z = a$), and the water/meteorite interface ($z = b$). The time-varying solution to these two variables thus determines the dynamics of the meteorite within the ice. In solving for them, we must also solve for the other variables involved within the model, namely the temperatures T_j in regions $j = 1 - 4$ (note that the air temperature is a model input).

At the atmosphere/ice interface ($z = 0$), the energy balance is given by[16]:

$$S_{net} + Q_{long} - \epsilon\sigma\left(273^4 + 4(273^3)T_1\right) + \hat{H}(T_a - T_1) - \rho_i(L_m + L_v)v = -k_i\frac{\partial T_1}{\partial z}, \quad (1)$$

where, sequentially, the terms on the left hand side represent: the contribution of the shortwave solar flux; the incoming longwave flux; the linearized outgoing longwave radiation (Stefan–Boltzmann's law); the sensible heat flux; and sublimation, respectively. The right hand side is the heat flux into the ice, which is given by Fourier's heat-transfer law. All model parameters are defined in Table 2.

At the ice/water interface ($z = a$; once it exists) the temperature must be at $0\,°C$. Further, the energy balance tells us that the energy flux for melting the ice (or its refreezing) plus the heat flux into the ice must equate with the heat flux from the water layer:

$$(\dot{a} + V)\rho_i L_m - k_i\frac{\partial T_1}{\partial z} = -k_w\frac{\partial T_2}{\partial z}, \quad (2)$$

where the overdot denotes differentiation with respect to time. Note that the inclusion of $+ V$ on the left hand side is due to the frame being fixed with respect to the upwelling ice. When the water layer does not exist, we neglect the phase change term in (2) and k_w becomes k_i.

At the water/meteorite interface ($z = b$), we must consider the shortwave solar energy that is reaching the meteorite surface (the longwave flux having been absorbed at the surface), and how this flux decays with depth below the surface. To achieve this we make use of the Beer–Lambert's exponential decay law with extinction coefficient γ_i. The balance at this interface is given by the solar energy flux plus the heat flux from water layer equating with the heat flux into the

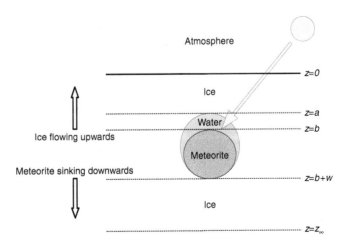

Atmosphere

Ice

Water

Meteorite

Ice

Ice flowing upwards

Meteorite sinking downwards

$z = 0$
$z = a$
$z = b$
$z = b + w$
$z = z_\infty$

Figure 3 | Model geometry. A (not to scale) schematic diagram highlighting the boundaries and geometry of the mathematical model for the Antarctic situation, in which an englacial meteorite is exposed to solar radiation.

meteorite:

$$Q_r e^{-\gamma_i b'} - k_w \frac{\partial T_2}{\partial z} = -k_m \frac{\partial T_3}{\partial z}. \tag{3}$$

Here Q_r may be thought of as the shortwave energy that would be absorbed by the meteorite in the absence of absorption/scattering by the ice, and thus $Q_r = (1 - \alpha_i)(1 - \alpha_m)(1 - s_h)S$, where S is the total incoming solar flux, α_i is the ice surface broadband albedo, α_m is the broadband meteorite albedo and s_h is the percentage of incoming solar radiation that is shaded (thus capturing the effects of local topography). Due to the inclination of the sun to the horizontal (solar elevation angle θ), we make use of an effective distance b' that represents the distance travelled while being attenuated within the ice and melt water (where we neglect the tiny effect of refraction within the very-thin melt-water layer):

$$b' = \frac{1}{\sin(\theta)} \left(a + (b - a) \frac{\gamma_w}{\gamma_i} \right). \tag{4}$$

At the lower meteorite/ice interface ($z = b + w$), where w is the meteorite width, one needs to compute whether the temperature is high enough to melt the ice. When the interface temperature is at $0\,°C$, melting is permitted, and the energy balance is given by the heat flux from the meteorite layer equating with the energy flux for melting plus the heat flux into the lower ice layer:

$$-k_m \frac{\partial T_3}{\partial z} = \rho_i L_m (\dot{b} + V) - k_i \frac{\partial T_4}{\partial z}. \tag{5}$$

Again, the $+V$ term is added due to the moving frame of reference. When the temperature is below melting point, we neglect the phase change term in (5).

At the bottom of the lower ice region we shall prescribe a temperature-gradient condition of the form

$$\frac{\partial T_4}{\partial z} = \phi \quad \text{at} \quad z = z_\infty. \tag{6}$$

In the laboratory, where the ice block is relatively warm and thin (5 cm), and the incoming heat fluxes held constant, we use a measured value of the ice temperature T_∞ at z_∞. Thus, the 'far-field' temperature gradient is

$$\phi = \frac{T_\infty - T_1(0)}{z_\infty} \quad \text{for laboratory conditions.} \tag{7}$$

However, in the Antarctic setting we are unable to prescribe such a measured temperature at z_∞, as the temperature profile within the ice varies with time (albeit slowly). As an alternative, we note that the ice underneath the meteorite must tend to the temperature profile of the surrounding ice, and so we must prescribe a condition drawn from the meteorite-free situation, that is, we need to match the temprature flux at z_∞ with that of the ice thermal boundary layer. To achieve this, we can utilize the fact that the magnitude of the (meteorite-free) annual ice-temperature variation in the boundary layer diminishes rapidly with depth, which means that in practice there is only a small average annual temperature gradient at z_∞ (ref. 30). To the order of accuracy of other assumptions made in this paper, it is reasonable to assume that z_∞ is sufficiently deep so that the temperature gradient can be taken as zero there; hence we take

$$\phi = 0 \quad \text{for Antarctic conditions.} \tag{8}$$

As a robustness check, we computed results for small variations from $\phi = 0$, and found only minor quantitative differences between them, thus showing that this is a reasonable condition to impose.

To prescribe the underlying equations within each region, we need to consider the conservation of heat energy. While one could use the full heat equation ($\rho c T_t = k T_{zz}$, where c is the specific heat capacity of the medium in question), we are able to simplify matters by considering the time scales involved. By noting that the critical depth scale in the model is the annual ice uplift height H ($H = V \times 1\,\text{yr}$), it allows us to compute the associated time scale of sinking, Λ_1, as ~ 10 days (for the Antarctic parameters given in Table 2). The latter is found by comparing the heat flux required for sinking the meteorite $\left(\rho_i L_m \dot{b} \right)$ with the solar forcing felt on it $\left(Q_r e^{-\gamma_i b'} \right)$, and so $\Lambda_1 = \rho_i L_m H / Q_r e^{-\gamma_i H}$. In contrast, the time scale Λ_2 for which the full heat equation will relax to its steady-state version ($T_{zz} \approx 0$), can be shown to be the order of 1 h ($\Lambda_2 = \rho_i c H^2 / k_i$). With this large relative difference between time scales ($\Lambda_1 \gg \Lambda_2$), we need only consider the steady-state version within each region $j = 1$–4 (where we are implicitly assuming z_∞ is suitably shallow for the steady-state heat equation to be used, and simultaneously deep enough for the annual temperature variation to be minor), namely:

$$\frac{\partial^2 T_j}{\partial z^2} \approx 0. \tag{9}$$

It is of note that time dynamics are still present within this model, via the annual variation in incoming solar radiation, thereby making our model quasi-steady. In addition, the steady-state heat equation coupled with the prescription of $\phi = 0$ for the Antarctic situation, sets the ice-temperature within region 4 as that of the meteorite base. This removes any dominating heat flux within region 4, making results computable from regions 1–3 only, and so an exact position for z_∞ is not required in that instance.

Rearrangement of the above equations yields a numerically tractable set of non-linear differential equations for a and b in terms of the input parameters. The particular form of their solutions depends on which stage of the sinking process is currently in effect. During the first stage, when the meteorite is encased in ice and moving with the upwelling ice, the speed of the interfaces a and b will be $-V$ and the surface temperature of the meteorite will be below zero. In this case, calculation of the linear temperature profile is straightforward from equations (1–7), with no time evolution of a and b to solve for.

During the second stage, where melting of the lower ice has not yet commenced but the upper water layer is in existence ($b - a > 0$), one can compute the location of the ice/water interface, a, from the dynamic equation

$$\dot{a} = \frac{1}{\rho_i L_m} \left(-Q_r e^{-\gamma_i b'} - k_i \phi - \frac{k_i \Sigma}{k_i + ha} \right) - V, \tag{10}$$

where

$$\Sigma = S_{\text{net}} + Q_{\text{long}} - \epsilon\sigma(273^4) - \hat{H}T_a - \rho_i(L_m + L_v)v, \tag{11}$$

$$h = \hat{H} + \epsilon\sigma4(273^3), \tag{12}$$

$$\hat{H} = \rho_a c_a \frac{u_*^2}{\bar{u}}, \tag{13}$$

$$S_{\text{net}} = S(1 - \alpha_i)(1 - s_h) - Q_r e^{-\gamma b'}. \tag{14}$$

This last term in S_{net} is necessary to ensure conservation of total solar radiation. Within this relation we are (in effect) assuming that any scattered shortwave energy within the ice only affects the ice at the atmospheric interface, that is, the shortwave energy directly warms only the ice surface and the meteorite (which heats up the remaining ice by conduction). This assumption, which greatly simplifies the analysis, is expected to lower very slightly the temperature of the ice near the meteorite compared with reality. This consequently reduces the rate of sinking, and thus is, by design, a slightly conservative estimate (note that the numerical experiment confirms that, even when the last term in S_{net} is removed completely, there is still little quantitative change to the results).

Once melting of the lower meteorite/ice surface ($z = b + w$) has commenced, and thus the meteorite starts to sink, one must switch to computing the coupled pair of equations,

$$\dot{a} = \frac{1}{\rho_i L_m} \left(-\frac{wk_w Q_r e^{-\gamma_i b'}}{wk_w + (b - a)k_m} - \frac{k_i \Sigma}{k_i + ah} \right) - V, \tag{15}$$

$$\dot{b} = \frac{1}{\rho_i L_m} \left(\frac{(b - a)k_m Q_r e^{-\gamma_i b'}}{wk_w + (b - a)k_m} + k_i \phi \right) - V. \tag{16}$$

To determine whether or not melting of the lower surface has begun, one can simply check whether or not the velocity of b in (14) is positive: while it is positive, we solve for a and b from (13) and (14), and if its numerical value ever becomes negative (no melting), one takes b as fixed and reverts to determining the evolution of a from equation (8).

The final aspect to note in the Antarctic representation of the model is that the atmospheric energy fluxes are seasonally dependent. To account for this we calculated and used the six-hourly average shortwave flux $S(t)$ from the libRadtran

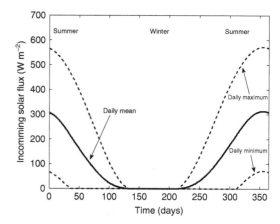

Figure 4 | Ice surface solar energy. The computed incoming shortwave energy flux $S(t)$ reaching the Frontier Mountain meteorite trap ice surface, showing the daily mean (solid line), maximum daily value (upper dashed line) and minimum daily value (lower dashed line). These were calculated using the libRadtran atmospheric radiative-transfer model[31] (as detailed in the Methods section).

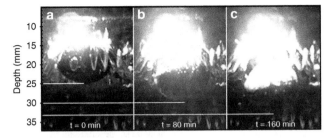

Figure 5 | Laboratory stills. Laboratory measurements of an iron meteorite sinking through ice as discussed in the Methods section. Depth (z-axis) in mm is taken relative to the upper ice surface. Shown are three side-view images of the meteorite sinking downwards, taken at 80-min intervals. The images relate to the first data point (0 min), fifth data point (80 min) and ninth data point (160 min) of Fig. 1 (right panel), of the meteorite starting from 14.9 mm encapsulation depth in the ice. To align these images with the results of Fig. 1 (which show the depth of the upper meteorite surface) one must subtract the meteorite width of 10 mm from the depth of the meteorite base, which are indicated by the white lines (the location of the base is easier to observe, due to the reduced amount of glare from the light source).

atmospheric radiative-transfer model[31], which incorporated a pseudo-spherical approximation with parameter inputs made relevant to the (relatively) well-parameterised Frontier Mountain Meteorite Trap area climatology[32-36] (see Table 1 for details of meteorite collection in this locality). The period of this time average was chosen so as to maintain as much diurnal granularity as possible, without violating our assumption of a quasi-steady heat model, that is, $6 \text{ h} \gg \Lambda_2 \approx 1 \text{ h}$ (although results for durations between 1 h and 12 h showed only a minor quantitative difference). The computed maximum, minimum and mean daily values for the incoming shortwave solar flux S are shown in Fig. 4, where the diurnal variations are sinusoidal (and so the six-hourly averages lie within the shown range). To reflect the fact that the seasonality in shortwave energy has a direct effect on air temperature T_a and thus the incoming longwave energy flux Q_{long}, we allow these parameters to also be seasonally dependent[37]. To achieve this in a straightforward manner, we use a normalized version of the six-hourly shortwave profile, denoted by $\hat{S}(t)$ ($\hat{S}(t) \in [0, 1]$), to approximate their dynamics, namely:

$$Q_{\text{long}} = \bar{Q}_1 + \bar{Q}_2 \hat{S}(t), \qquad (17)$$

$$T_a = \bar{T}_a \left(1 - \hat{S}(t)\right), \qquad (18)$$

where $\bar{T}_a$ is a negative constant measured in °C (thus the maximum air temperature is attained during summer and the minimum temperature during winter), and $\bar{Q}_1$ and $\bar{Q}_2$ are both positive constants. With this model formulation, we are able to solve for the Antarctic situation, as well as the laboratory case (where $\theta = 90°$ and the incoming energy fluxes are held constant).

The parameter values used in the laboratory simulation and the Antarctic simulation are all stated in Table 2. The *in situ* Antarctic parameter values are all in relation to the Frontier Mountain Meteorite Trap area[19]. Suitable proxies were taken for the blue-ice thermal conductivity k_i (ref. 38), the surface roughness estimate u^* ref. 30 and the solar attenuation parameter through water γ_w (ref. 39) and ice γ_I (refs 12,39). When a range of parameters were provided, we used an averaged value. The percentage incoming solar radiation that is shaded, s_h, is a time-averaged value inferred from the neighbouring mountain elevations[17]. The six-hourly solar elevation angles θ were computed[40], while in the laboratory the solar-simulator lamp was held directly overhead (that is, 90°). The meteorite width w used in the Antarctic situation was chosen so as to be indicative of typical collected Antarctic meteorite specimens[2] (although these smaller collected sizes/masses may be the consequence of larger material that broke up on the surface through repeated freeze-thaw cycles and wind action[2]). The scaling of the Antarctic longwave radiation was taken to be consistent with seasonal observations[37]. The thermal conductivity of the IIAB iron (Sikhote-Alin) was inferred from the IAB iron meteorite Campo del Cielo[17]; ± 25% variations to this reasonable estimate still yielded results close to the data points (which was to be expected over the shorter laboratory time scales). The meteorite surface albedos (fusion crust cover) were independently measured for this study (see the Acknowledgements section).

The results were computed using a code written in Matlab, which the authors can supply on request. Potential extensions to our modelling approach are discussed in Supplementary Note 3.

Laboratory experiments. Our experiments centred around subjecting a meteorite encased in a block of ice ($400 \times 400 \times 50$ mm) to the radiation from an Oriel Solar

Simulator arc lamp (irradiance $1{,}440 \text{ W m}^{-2}$) that was held at a constant 10 mm above the ice surface and focused onto the englacial meteorite. This was conducted in a (otherwise dark, non-reflecting) temperature-controlled room, with ambient temperature $-1 °C$. The meteorite movement was recorded over a 3-h period using an HD time-lapse camera positioned at the side of the ice block, from which the meteorite's progress could be measured (with an error of under ± 1 mm), see Fig. 5, and to confirm that the outward-facing ice surfaces were not melting. With this experimental set-up, we were able to successfully conduct four controlled laboratory experiments, as shown in Fig. 1.

To create the blocks of ice with a meteorite included, we first slowly froze the lower half the block in a container. We then placed a meteorite on top and carefully added an upper layer of cold water, which we then slowly froze. Once frozen, we removed the container. This whole process could take 2–3 days, so as to reduce the number of air bubbles within the ice. It also helped fully bond the meteorite to its surrounding ice.

Further to our four main experiments, controlled experiments were performed to show that the meteorite only sank in the presence of the solar simulator. We also recorded temperatures within the ice to confirm that when the meteorite was sinking its base was at 0 °C, thus confirming the assumptions within our energy balance model.

References

1. Righter, K. *et al.* (eds) *35 Seasons of U.S. Antarctic Meteorites (1976-2010): A Pictorial Guide to the Collection* (John Wiley & Sons, 2014).
2. Harvey, R. The origin and significance of Antarctic meteorites. *Chem. der Erde-Geo.* **63**, 93–147 (2003).
3. Delisle, G. *et al.* Meteorite finds by EUROMET near Frontier Mountain, North Victoria Land, Antarctica. *Meteorit. Planet. Sci.* **28**, 126–129 (1993).
4. Meteoritical Society bulletin of classified and named meteorite samples. Available at http://www.lpi.usra.edu/meteor/ (accessed on 18th December 2015).
5. Corrigan, C. *et al.* A Statistical look at the US Antarctica Meteorite Collection. *35 Seasons of U.S. Antarctic Meteorites (1976-2010): A Pictorial Guide to the Collection* (eds Righter, K. *et al.*) Chapter 10, 173–187 (John Wiley & Sons, 2014).
6. Nagata, T. A possible mechanism of concentration of meteorites within the meteorite ice field in Antarctica. Proceeding of the second symposium on Yamato meteorites. *Mem. Natl Inst. Polar Res. Spec. Issue* **8**, 70–92 (1978).
7. Koeberl, C. & Cassidy, W. A. Differences between Antarctic and non-Antarctic meteorites: An assessment. *Geochim. Cosmochim. Acta* **55**, 3–18 (1991).
8. Cassidy, W. A. Meteorites, ice and antarctica: a personal account. *Antarct. Sci.* **16**, 87–88 (2004).
9. Bintanja, R. On the glaciological, meteorological, and climatological significance of Antarctic blue ice areas. *Rev. Geophys.* **37**, 337–359 (1999).
10. Warren, S. G., Brandt, R. E. & Grenfell, T. C. Visible and near-ultravioloet absorption spectrum of ice from transmission of solar radiation into snow. *Appl. Op.* **45**, 5320–5334 (2006).
11. Cassidy, W. A. *et al.* The meteorite collection sites of Antarctica. *Meteoritics* **27**, 490–525 (1992).
12. Bintanja, R. Surface heat budget of Antarctic snow and blue ice: interpretation of spatial and temporal variability. *J. Geophys. Res. (Atmos)* **105**, 24387–24407 (2000).
13. Corti, G. *et al.* Physical modeling of the influence of bedrock topography and ablation on ice flow and meteorite concentration in Antarctica. *J. Geophys. Res.* **113**, 1–18 (2008).
14. Bryson, J. F. J. *et al.* Long-lived magnetism from solidification-driven convection on the pallasite parent body. *Nature* **517**, 472–475 (2015).
15. Yang, J., Goldstein, J. I. & Scott, E. R. D. Iron meteorite evidence for early formation and catastrophic disruption of protoplanets. *Nature* **466**, 888–891 (2007).
16. Evatt, G. W. *et al.* Glacial melt under a porous debris layer. *J. Glaciol.* **61**, 825–836 (2015).
17. Opeil, C. P., Conslmagno, G. J. & Britt, D. T. The thermal conductivity of meteorites: New measurements and analysis. *Icarus* **208**, 449–454 (2010).
18. Opeil, C. P. *et al.* Stony meteorite thermal properties and their relationship with meteorite chemical and physical states. *Meteorit. Planet. Sci.* **47**, 319–329 (2012).
19. Folco, L. *et al.* The Frontier Mountain meteorite trap (Antarctica). *Meteorit. Planet. Sci.* **37**, 209–228 (2002).
20. Harvey, R. & Score, R. Direct evidence of in-ice or pre-ice weathering of Antarctic meteorites. *Meteoritics* **26**, 343 (1991).
21. Wettlaufer, J. S. & Worster, M. G. Premelting dynamics. *Ann. Rev. Fluid Mech.* **38**, 427–452 (2006).
22. Haack, H. *et al.* Results from the Greenland Search for meteorites expedition. *Meteorit. Planet. Sci.* **42**, 17271733 (2007).
23. Benedix, G. K., Kaak, H. & McCoy, T. J. in *Meteorites and Cosmochemical Processes* Vol 1, 267–285 (Elsevier, 2014).
24. Kruijer, T. S. *et al.* Protracted core formation and rapid accretion of protoplanets. *Science* **244**, 1150–1154 (2014).

25. Tarduno, J. A. *et al.* Evidence for a dynamo in the main group pallasite parent body. *Science* **338**, 939–942 (2012).
26. Bottke, W. F. *et al.* Iron meteorites as remnants of planetesimals formed in the terrestrial planet region. *Nature* **439**, 821–824 (2006).
27. Burbine, T. H. *et al. Meteoritic parent bodies: their number and identification. Asteroids III* 653–667 (University of Arizona Press, 2002).
28. Smoliar, M. I., Walker, R. J. & Morgan, J. W. Re-Os Ages of Group IIA, IIIA, IVA, and IVB Iron Meteorites. *Science* **271**, 1099–1102 (1996).
29. Antonelli, M. A. Early inner solar system origin for anomalous sulfur isotopes in differentiated protoplanets. *PNAS* **111**, 17749–17754 (2014).
30. Cuffey, K. M. & Paterson, W. S. B. *The Physics of Glaciers* 4th edn (Butterworth-Heinemann, 2010).
31. Mayer, B. & Kylling, A. Technical note: the libRadtran software package for radiative transfer calculations description and examples of use. *Atmos. Chem. Phys. Discuss.* **5**, 13191381 (2005).
32. Kurucz, R. L. in *Infrared Solar Physics* (ed. Rabin, D. M. & Jefferies, J. T.) 154 (Springer, 1992).
33. Diaz, S. B. *et al.* Climatologies of ozone and UV-B irradiances over Antarctica in the last decades. *Gayana* **68**, 157–160 (2004).
34. Tomasi, C. *et al.* Mean vertical profiles of temperature and absolute humidity from a 12-year radiosounding data set at Terra Nova Bay (Antarctica). *Atmos. Res.* **71**, 139169 (2004).
35. Tomasi, C. *et al.* Aerosols in polar regions: a historical overview based on optical depth and *in situ* observations. *J. Geophys. Res.* **112**, D16205 (2007).
36. Adhikari, L., Wang, Z. & Deng, M. Seasonal variations of Antarctic clouds observed by CloudSat and CALIPSO satellites. *J. Geophys. Res.* **117**, D04202 (2012).
37. Town, M., Walden, Von, P. & Warren, S. G. Spectral and broadband longwave downwelling radiative fluxes, cloud radiative forcing, and fractional cloud cover over the south pole. *J. Climate* **18**, 4235–4252 (2005).
38. Favier, V. *et al.* Modeling the mass and surface heat budgets in a coastal blue ice area of Adelie Land, Antarctica. *J. Geophys. Res.* **116**, F03017 (2011).
39. Hecht, E. *Optics* 4th edn (Addison Wiley, 2002).
40. Reda, I. & Andreas, A. Solar position algorithm for solar radiation applications. *Sol. Energy* **76**, 577589 (2004).

Acknowledgements

We are grateful to Dr Andrew Fearnside and Dr Phil Masding for their meteorite albedo measurements, and to Dr Jonathan Kingslake for useful comments. We are also grateful for the time and illuminating insights provided by the four annonymous reviewers of our paper. I.D.A. acknowledges the support of a Royal Society Wolfson Research Merit Award, and K.H.J. acknowledges the help of the Leverhulme Trust (#2011-569), Royal Society (RS/UF140190) and the Science and Technology Facilities Council grant ST/M001253/1. Funding for this project was received from the Engineering and Physical Sciences Research Council, UK, via the MAPLE platform grant EP/I01912X/1.

Author contributions

G.W.E. supervised the work, including posing the problem, describing the mathematics, writing the paper and raising the necessary funds. M.J.C. performed the bulk of the experiments, developed the mathematical model and statistical analysis. K.H.J. led aspects involved with the physical meteorites, as well as the Antarctic meteorite collection missions and their geophysical setting. A.R.D.S. and P.C. guided and arranged the laboratory experiments and derived the Antarctic solar energy input files. I.D.A. helped develop and solve the underlying mathematical model, advised on the experimentations and helped raise the funding. All authors contributed significantly to the editing of the paper.

Additional information

19

Formation of lunar swirls by magnetic field standoff of the solar wind

Timothy D. Glotch[1], Joshua L. Bandfield[2], Paul G. Lucey[3], Paul O. Hayne[4], Benjamin T. Greenhagen[4], Jessica A. Arnold[1], Rebecca R. Ghent[5] & David A. Paige[6]

Lunar swirls are high-albedo markings on the Moon that occur in both mare and highland terrains; their origin remains a point of contention. Here, we use data from the Lunar Reconnaissance Orbiter Diviner Lunar Radiometer to support the hypothesis that the swirls are formed as a result of deflection of the solar wind by local magnetic fields. Thermal infrared data from this instrument display an anomaly in the position of the silicate Christiansen Feature consistent with reduced space weathering. These data also show that swirl regions are not thermophysically anomalous, which strongly constrains their formation mechanism. The results of this study indicate that either solar wind sputtering and implantation are more important than micrometeoroid bombardment in the space-weathering process, or that micrometeoroid bombardment is a necessary but not sufficient process in space weathering, which occurs on airless bodies throughout the solar system.

[1] Department of Geosciences, Stony Brook University, Stony Brook, New York 11794-2100, USA. [2] Space Science Institute, 4750 Walnut St #205, Boulder, Colorado 80301, USA. [3] Hawaii Institute of Geophysics and Planetology, University of Hawaii, Honolulu, Hawaii 96822, USA. [4] Jet Propulsion Laboratory, M/S 183-301, 4800 Oak Grove Drive, Pasadena, California 91109, USA. [5] Department of Earth Sciences, University of Toronto, Toronto, Ontario, Canada M5S 3B1. [6] University of California Los Angeles, Box 951567, Los Angeles, California 90095-1567, USA. Correspondence and requests for materials should be addressed to T.D.G. (email: timothy.glotch@stonybrook.edu).

Lunar swirls have been documented since the Apollo era[1], and since that time, a variety of swirl morphologies has been identified, ranging from the complex structures present in the Reiner Gamma formation, Mare Ingenii and Mare Marginis swirls, to simple bright patches with diffuse edges, as at Descartes. Intermediate morphologies occur near Airy and Gerasimovich craters[2]. Several swirl formation mechanisms have been proposed, including (1) solar wind standoff due to the presence of local magnetic fields, preventing solar wind sputtering and implantation, nanophase iron formation, and the resulting surface darkening associated with space weathering[3], (2) recent comet impacts or micrometeoroid swarms that scoured the lunar surface, leaving a fine-gained, unweathered material and possibly imparting a remnant magnetization[4-6], and (3) electrostatic levitation and deposition of high-albedo, fine-grained, feldspar-enriched dust[7]. The association of all swirls with magnetic field anomalies of varying strength[8-10] has driven the development of each of these hypotheses. In this work, we use the unique mid-infrared data set from the Diviner Lunar Radiometer to distinguish between these hypotheses. We show that Diviner data support the solar wind standoff model for lunar swirl formation and disqualify the impact swarm and dust transport hypotheses.

Previous near-infrared observations of lunar swirls show them to be optically immature compared with the surrounding terrains[10]. Space weathering leads to optical maturation of the surfaces of airless bodies and is thought to be caused by two main processes: (1) solar wind sputtering and/or implantation of hydrogen atoms, leading to the formation of nanophase metallic iron blebs and (2) micrometeoroid bombardment that leads to the formation of agglutinitic glass and a reduced vapour-deposited coating[11]. Under the solar wind standoff model, horizontal magnetic fields at the lunar swirl sites deflect the solar wind, preventing most sputtering and implantation of solar wind ions[12,13]. If this is the case, micrometeoroid bombardment should be the major relevant space-weathering process[14] at swirl sites. A relative lack of solar wind interaction with the lunar surface at swirl sites is supported by recent observations from the Moon Mineralogy Mapper (M^3) instrument that confirm the Clementine observations of a substantial reduction in optical maturity and also show a reduction of the presence of surface hydroxyls at the swirl sites[15]. Recent observations in the ultraviolet by the Lunar Reconnaissance Orbiter Camera Wide Angle Camera subsystem also suggest that micrometeoroid bombardment, and not solar wind interaction, plays the dominant role in space weathering at the lunar swirls[16].

Diviner has three narrow band infrared channels near 8 μm, from which the silicate Christiansen Feature (CF) can be calculated[17]. The position of the CF is related to the degree of mineral silicate polymerization and provides an indicator of composition[18]. However, analysis of the global observations by Diviner has revealed that the CF position is also influenced by optical maturity. Optically mature surfaces are shifted to longer wavelengths by ~0.1 μm (or about ~20% of the full range of lunar CF values that have been measured) compared with immature surfaces with the same composition[19]. A global CF map[19] shows substantial anomalies related to young, fresh craters such as Tycho and Jackson, which have CF positions at shorter wavelengths than the surrounding mature terrain.

Here we find that Diviner data show the CF anomalies are due to abnormal space weathering at both the mare and highland swirl sites. This result, in addition to the relative lack of thermophysical anomalies at the swirl sites, strongly supports the solar wind standoff model and disqualifies the micrometeoroid/comet swarm and dust levitation models for swirl formation. Diviner data can be used to determine the composition and degree of space weathering of the lunar surface[17,19,20] and to determine thermophysical properties using daytime and night-time thermal infrared measurements[21,22]. The ability to characterize both the compositional and thermophysical properties of the lunar regolith make Diviner well suited to examine the swirls and differentiate between the three proposed formation mechanisms.

Results

Diviner CF anomaly at swirl sites. We analysed Diviner data covering 12 swirl regions including both the mare and highlands sites (Figs 1 and 2; Table 1, Supplementary Figs 1–10). At both the Reiner Gamma and Van de Graaff crater sites, the distributions of CF values on and off the swirls are clearly separated. Garrick-Bethell et al.[7] demonstrated that only a small portion of the Reiner Gamma 'dark lanes' (small curvilinear low-albedo regions that occur between some high-albedo swirls) are truly darker than the surrounding terrain at visible wavelengths. We found that the CF distribution of this portion of the dark lanes within Reiner Gamma is nearly identical to the off-swirl distributions (Fig. 1). The off-swirl locations were defined to be well outside the visible light albedo boundaries of the swirls identified in Lunar Reconnaissance Orbiter Camera wide angle camera or Clementine data. The average off-swirl CF values at Reiner Gamma (8.31 μm) and Van de Graaff (8.20 μm) are close to the previously determined global average mare and highland CF values[17]. The average on-swirl values for Reiner Gamma and Van de Graaff are shifted to shorter wavelengths by 0.08 and 0.09 μm, respectively. In addition, three-point Diviner spectra acquired both on and off the swirls at both the sites show clear differences in the spectral shape, which is reflected in the differences in modelled CF values for these sites. Swirl spectra were acquired from the regions outlined by boxes in Fig. 1 and the average spectra and resulting modelled CF values are shown in Fig. 2.

The low CF position values of each of the swirls are clearly associated with lower optical maturity compared with the surrounding terrains (Table 1). Figure 3 shows the relationship between CF position and the optical maturity parameter (OMAT)[23] derived from Clementine multispectral data at Reiner Gamma. Compared with the scene average, swirl pixels have both lower overall CF and higher OMAT (lower maturity) values.

At Reiner Gamma, Airy, Firsov and Rima Sirsalis, the complex swirls have easily identifiable dark lanes between the high-albedo regions. At Airy, the background terrain albedo varies substantially, so it is difficult to determine if the dark lanes are, in fact, lower albedo than the surrounding terrain. However, at Firsov and Rima Sirsalis, the dark lanes (Fig. 4a,b) have demonstrably lower visible albedos than the surrounding terrain. In Fig. 4c,d, we plot the Clementine 950-nm/750-nm reflectance ratio against the 750-nm ratio for regions on the swirl, in the dark lanes, and on the local surrounding terrain. The dark lanes at these sites are the darkest sampled regions in the scene. At these sites, the averages and distributions of the CF values just off the swirls and within the dark lanes of each swirl are within 0.01 μm, and dark lane OMAT values are nearly identical to the surrounding terrain (Table 1). These data are consistent with the hypothesis that magnetic field lines over the dark lanes have a vertical orientation, allowing the solar wind to interact with the surface[24]. It has previously been suggested that the dark lanes and low-albedo off-swirl regions associated with some swirls result from an increased rate of space weathering as the charged solar wind particles deflected by local magnetic fields are funneled to these regions[14]. However, maturity enhancements in the dark lanes are not

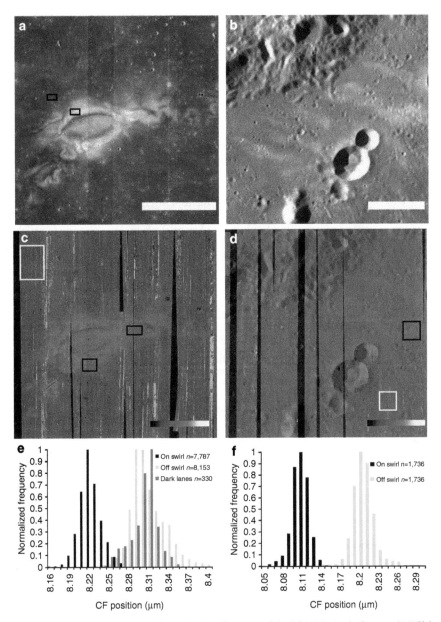

Figure 1 | Diviner CF position data for the Reiner Gamma and Van de Graaff crater swirls. (**a**) Wide Angle Camera (WAC) base map for Reiner Gamma. Black boxes indicate the regions used for surface roughness and thermal model analysis in Figs 5 and 8. Scale bar, 50 km. (**b**) WAC base map for Van de Graaff crater. Scale bar, 20 km. (**c**) CF position map of the Reiner Gamma swirl stretched from 8.15 to 8.45 μm overlayed on Lunar Reconnaissance Orbiter Camera (LROC) WAC mosaic. (**d**) CF position map of the Van de Graaff crater swirls stretched from 8.0 to 8.4 μm overlayed on LROC WAC mosaic. (**e**) Normalized histogram of CF values on the Reiner Gamma swirl (black boxes in **c**), off the swirl (white box in **1c**), and in the dark lanes (yellow boxes in **c**). (**f**) Normalized histogram of CF values on the Van de Graaff crater swirls (black box in **b**) and off the swirl (white box in **1b**).

detected in Diviner data. Rather, the nearly identical CF distributions within the dark lanes and the local terrains near the swirls (for example, Fig. 1) suggest that the main effects of space weathering at mid-infrared wavelengths (shift in CF position to longer wavelengths) reach a maximum early in the space-weathering process and are not enhanced on increased flux of solar wind particles.

The dust levitation model of lunar swirl formation[7] suggests that the spectral character of the swirls should be affected by a minor addition of feldspar, which is slightly enriched in the finest size fraction of the lunar soil[25]. Detailed analyses of the modal abundances of major minerals in sieved fractions of lunar soils show that feldspar is typically enriched by ~2–4 wt. % compared with coarser size fractions[26]. Therefore, the dust levitation model suggests an enrichment of feldspar in the optically active surface

layer of mare swirls by a comparable amount, assuming that all of the levitated dust is < 10 μm in diameter. Laboratory spectroscopic measurements of olivine-feldspar mixtures in a simulated lunar environment[27] (Fig. 5) show that ~16 wt.% feldspar would need to be added to account for a typical mare swirl CF shift from ~8.3 to ~8.2 μm. At the highlands sites, ~20 wt.% feldspar would need to be added to the swirl sites to account for a CF shift from ~8.2 to ~8.1 μm. While it is a simplification to model the swirl sites as olivine-feldspar mixtures, the laboratory data do suggest that the required feldspar contributions to the swirl CF shifts are ~5 to 10 times higher than the levels predicted by the dust levitation model. Swirls in both the highlands and mare consistently show CF shifts of ~0.04 to 0.09 μm to shorter wavelengths. Rather than an addition of a small amount of feldspar, this shift is comparable to CF shifts attributed to the optically immature surfaces. The Diviner data, therefore, suggest that a layer of fine particulate feldspar-enriched material is not responsible for the lunar swirl albedo anomalies and spectral properties. This interpretation supports visible and near-infrared measurements from the M[3] instrument[15].

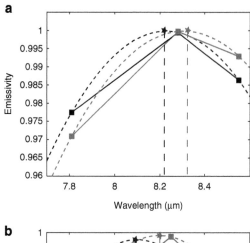

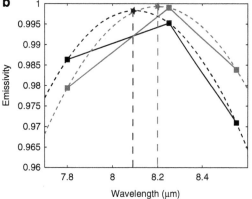

Figure 2 | Average Diviner spectra. Spectra are averaged using data from the white and black boxes outlined in Fig. 1. (**a**) Reiner Gamma swirl. (**b**) Van de Graaff crater swirl. Black spectra are average on-swirl spectra. Red spectra are average off-swirl spectra from the boxes outlined in Fig. 1. The parabolic traces represent the fitting function used to determine the CF position (stars) from the spectra.

Surface roughness at swirl sites. In addition to compositional analysis, we can use the daytime Diviner data to address the physical state of the lunar surface. We employ a roughness model[28] (Fig. 6) and Diviner daytime measurements in channels 3 and 6 (12.5–25 μm) to determine the spectral aniosthermality caused by surface roughness at Reiner Gamma. Surface roughness varies with length scale, and becomes greater with decreasing scale[29]. Previous lunar surface roughness estimates have included Lunar Orbiter Laser Altimeter data at metre to decametre scales[30], visible photometric studies typically at sub-mm scales[31,32] and additional sub-mm surface roughness estimates from *in situ* high-resolution imagery[29]. At sub-mm to cm scales, typical lunar surface roughness values, defined as a root mean square (RMS) slope are 16–25° (ref. 29). Previous thermal infrared observations of the Moon and Mars have shown substantial effects due to surface roughness at ∼cm scales[33–37]. Diviner data offer the ability to characterize cm-scale surface roughness at the high spatial resolution (128–256 ppd (pixels per degree)) typical of the instrument. Figure 6 shows the modelled change in the channel 3 − channel 6 brightness temperature differences (ΔBT) on and off the Reiner Gamma swirl as a function of time of day for different RMS surface slope along with ΔBT data from 6 am to 6 pm local time. Within a standard

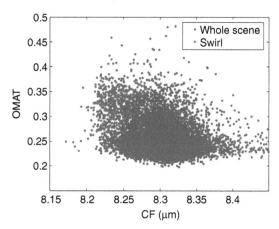

Figure 3 | Comparison of CF position and OMAT maturity index for Reiner Gamma. On-swirl pixels (taken from the regions outlined by white boxes in Fig. 1) clearly have lower CF and higher OMAT (lower maturity) values than the median scene.

Table 1 | CF and OMAT parameter values for each swirl examined in this study.

Swirl	Centre Location (lon., lat.)	On-swirl CF (μm)	Off-swirl CF (μm)	Dark lane CF (μm)	On-swirl OMAT	Off-swirl OMAT	Dark lane OMAT
Airy	3.5E, −18.5N	8.12 ± 0.02	8.16 ± 0.03	8.15 ± 0.03	0.30 ± 0.02	0.25 ± 0.03	0.26 ± 0.01
Descartes	15.9E, −10.6N	8.09 ± 0.01	8.18 ± 0.01	NA	0.31 ± 0.03	0.22 ± 0.01	NA
Firsov	113.5E, 0.5N	8.11 ± 0.02	8.19 ± 0.01	8.18 ± 0.02	0.31 ± 0.02	0.24 ± 0.01	0.24 ± 0.01
Gerasimovich	236.7E, −22.9N	8.09 ± 0.01	8.17 ± 0.01	NA	0.31 ± 0.02	0.22 ± 0.01	NA
Hopmann	159.4E, −50.8N	8.18 ± 0.01	8.23 ± 0.02	NA	0.30 ± 0.02	0.25 ± 0.01	NA
Ingenii mare	163.0E, −35.9N	8.25 ± 0.02	8.30 ± 0.03	NA	0.28 ± 0.02	0.22 ± 0.01	NA
Marginis highlands	96.0E, 20.4N	8.12 ± 0.02	8.20 ± 0.02	NA	0.26 ± 0.02	0.22 ± 0.01	NA
Marginis mare	84.9E, 13.4N	8.25 ± 0.01	8.31 ± 0.02	NA	0.22 ± 0.02	0.19 ± 0.01	NA
Moscoviense	148.9E, 25.4N	8.26 ± 0.02	8.30 ± 0.03	NA	0.21 ± 0.01	0.19 ± 0.01	NA
Reiner Gamma	301.0E, 7.4N	8.23 ± 0.02	8.31 ± 0.03	8.31 ± 0.03	0.35 ± 0.02	0.24 ± 0.01	0.26 ± 0.01
Rima Sirsalis	306.0E, −7.2N	8.29 ± 0.01	8.31 ± 0.01	8.31 ± 0.01	0.23 ± 0.01	0.22 ± 0.02	0.21 ± 0.01
Van de Graaff	171.1E, −27.4N	8.11 ± 0.01	8.20 ± 0.02	NA	0.30 ± 0.02	0.23 ± 0.01	NA

CF, Christiansen feature; E, East; lat, latitude; lon, longitude; N, North; NA, not available; OMAT, optical maturity.

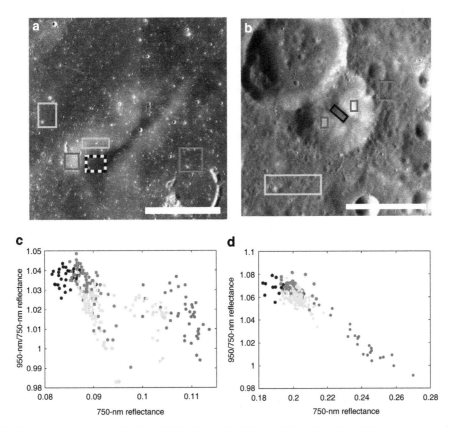

Figure 4 | Clementine albedo variations associated with swirl dark lanes. Dark lanes at Rima Sirsalis and Firsov crater have lower visible albedos than surrounding terrain. (**a**) Rima Sirsalis 750-nm reflectance. Scale bar, 25 km. (**b**) Firsov crater 750-nm reflectance. Scale bar, 25 km. (**c**) Rima Sirsalis 950/750-nm reflectance ratio versus 750-nm reflectance. Coloured points correspond to the coloured boxes in **a**. Black dots correspond to the white dotted box in **a**. (**d**) Firsov crater 950/750-nm reflectance ratio versus 750-nm reflectance. Coloured points correspond to the coloured boxes in **b**.

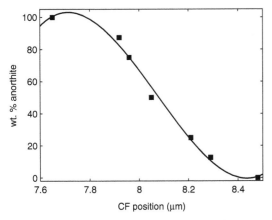

Figure 5 | CF positions of olivine-plagioclase mixture spectra acquired in a simulated lunar environment. Regression of the data using a degree 3 polynomial shows that to go from a CF of 8.3 to 8.2 μm, ~16 wt.% additional feldspar needs to be added to the mixture. To go from 8.2 to 8.1 μm, ~20 wt. % additional feldspar must be added.

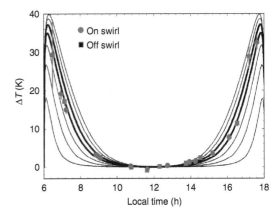

Figure 6 | Reiner Gamma roughness model. Modelled brightness temperature deviations between the channels 3 and 6 as function of the local time are plotted for RMS surface slopes varying in 5° increments from 10° (bottom line) to 40° (top line). Measured data from the Reiner Gamma swirl are displayed as magenta circles. Data from a nearby off-swirl surface are displayed as blue squares. Error bars are 1 s.d. Curves representing 25–30° roughness are bold. Diviner data from Reiner Gamma swirl indicate surface roughness in this range.

deviation of the temperature measurements, Diviner data indicate nearly identical surface roughnesses of ~25–30° both on and off the swirl (bold curves in Fig. 6). This surface roughness reflects the finely structured lunar regolith at cm scales and places tight constraints on geologic processes related to the formation of the swirls. That is, the viability of the dust levitation and comet swarm impact models hinges on the ability to demonstrate that the roughness characteristics of the swirls would not be altered by (1) piling of fine-grained dust, resulting in smoother surfaces at

cm scales or (2) regolith disturbance and impact scouring resulting in a layer of fine particulates at cm scales.

Thermophysical characteristics of swirls. Diviner night-time data and thermal models can also help to differentiate between the competing formation mechanisms for the lunar swirls. Cooling of the lunar surface throughout the night provides information on the physical nature and degree of processing of the lunar regolith, including rock abundance and vertical structure and particle size of the regolith fines[21,22]. We constructed 128-ppd night-time temperature maps using Diviner channel 8 (50–100 μm) brightness temperatures acquired between local times of 19:30 and 05:30 (Fig. 7). Reported values are positive or negative deviations from the scene average normalized for local time variations. On-swirl and off-swirl temperature averages were determined using the same regions of interest shown in the boxes in Fig. 1. The average normalized temperature on the sampled portion of the Reiner Gamma swirl is -0.8 ± 1.5 K as opposed to the average off-swirl temperature of -0.3 ± 1.8 K (1-σ s.d.). The sampled portion of the swirl, on average, is only ~ 0.5 K colder than the off-swirl surface, but it is visible in the stretched night-time temperature image (white arrows in Fig. 7a). At the Van de Graaff crater site, the average on-swirl normalized night-time temperature is 0.4 K, while the off-swirl average is 0.0 K.

The measured temperature differences, though small, can be tied directly to the physical properties of the swirl surfaces. To do this, we use a thermal model to constrain the differences in regolith properties between the swirl and non-swirl surfaces. This standard lunar regolith model[22,38] uses Diviner night-time temperature data to constrain the upper regolith density profile. Using the on- and off-swirl surfaces delineated in Fig. 1a, and the Clementine 750-nm albedos of these regions, we show that the temperature difference between the on- and off-swirl surfaces can be accounted for completely by the albedo difference between the swirls (Fig. 8). In addition, we added a 2-mm low thermal inertia (~ 30 J m^{-2} K^{-1} s$^{-1/2}$, $\sim 50\%$ of the standard model) layer on top of the standard swirl model. Such a low thermal inertia layer would be expected from the admixture of additional fine particulates with typical lunar regolith[39]. It is possible, however, that this model does not capture the effects of the minor admixtures of fine particulates with the bulk lunar regolith. Nevertheless, our results show that even this 2-mm thick layer would produce night-time temperatures much lower than those that are observed. Put another way, the swirls are surficial features that, thermophysically, are nearly indistinguishable from the surrounding terrain. The surficial nature of the swirls as determined by Diviner is consistent with Mini-RF radar observations of these features[40].

The lunar regolith has been highly processed by impact gardening that produces a delicate vertical structure with unique thermophysical properties. Any deviation from this structure shows clear night-time temperature differences that are easily identified in Diviner data[21,22,28] and are absent from the swirl sites. For example, recent impacts by swarms of meteors or comets,[4–6] although they might not produce a 2-mm thick low thermal inertia layer as modelled in Fig. 8, would almost certainly disturb the regolith vertical structure and lead to clearly observable temperature and roughness phenomena in the Diviner data.

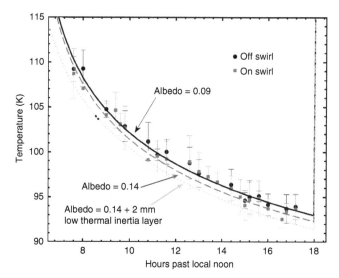

Figure 8 | Reiner Gamma thermal model. Diviner night-time temperature observations on (blue dots) and off (black dots) the Reiner Gamma swirl are plotted and compared with two thermal models. Error bars indicate standard 1 s.d. in temperature measurements within the spatially binned measurement regions (Fig. 1a). The black solid line and blue dashed line show the predicted temperatures using the standard regolith model[21,37]. Predicted night-time temperatures for a 2-mm thick low thermal inertia layer on top of the standard swirl thermal model (green dotted line) are too low to explain temperatures observed by Diviner.

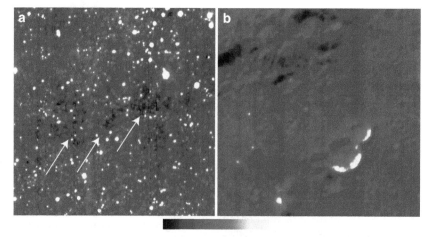

Figure 7 | Diviner night-time temperature data for the Reiner Gamma and Van de Graaff crater swirls. Area of each image is same as Wide Angle Camera base maps in Fig. 1a,b. (**a**) Average normalized night-time temperature for Reiner Gamma, stretched from -3.6 to 7.0 ΔK. Arrows show the colder temperatures associated with the swirl. (**b**) Average normalized night-time temperature for Van de Graaff crater, stretched from -7.0 to 15 ΔK.

Discussion

Previous spectroscopic observations of the swirls at visible and near-infrared wavelengths have been used to argue for the solar wind standoff model[3,14,24], the micrometeoroid/comet swarm model[4-6] or the feldspathic dust pile model. More recently, M^3 spectra have been used to propose a hybrid model for swirl formation that invokes both regolith disturbance and dust lofting[41]. The mid-infrared Diviner data presented here place tight constraints on the swirl formation process and must be accounted for by any proposed swirl formation mechanism. The proposed feldspathic dust lofting[7], meteoroid/comet swarm[4-6] and hybrid[41] swirl formation mechanisms would likely result in thermophysical and spectral properties that are inconsistent with the Diviner measurements. The proposed dust lofting and hybrid processes cannot account for the observed CF shift in Diviner data while the meteoroid/comet swarm and hybrid processes cause regolith disturbance that would result in a substantial thermal anomaly in the night-time Diviner data. On the other hand, the solar wind standoff mechanism for swirl formation would lead to a slight shift of the CF to shorter wavelengths compared with the surrounding regolith due to a reduction in space weathering, but show no difference in the unique lunar regolith thermophysical properties because impact gardening is otherwise unaffected. These properties are exactly what are observed, and the Diviner observations, in addition to M^3 and Mini-RF[14,15,40] observations and laboratory experiments[42], strongly support the solar wind standoff hypothesis for lunar swirl formation. Because they are clearly associated with magnetic field anomalies, lunar swirls present an opportunity to study and isolate the effects of solar wind sputtering/implantation and micrometeoroid bombardment processes on optical maturity. The immature nature of the swirls suggests that solar wind sputtering and implantation, which is prevented by local magnetic fields at the swirl sites, is the primary process for space weathering and optical maturation of airless body surfaces that do not have a global magnetic field, while micrometeoroid bombardment and resulting vapour deposition plays a subordinate role. The results of this study should motivate additional space-weathering experiments and work on the mature lunar regolith samples to test this hypothesis.

Methods

Diviner CF calculations and average spectra. The Diviner Lunar Radiometer Experiment on board the Lunar Reconnaissance Orbiter is an infrared radiometer with two solar reflectance channels and seven channels in the thermal infrared wavelength range, including three narrow band channels that span the wavelength ranges 7.55–8.05 μm, 8.10–8.40 μm and 8.38–8.68 μm. These '8 μm' channels can be used to calculate the position of the silicate CF[17] that is directly sensitive to the degree of silica tetrahedral polymerization and bulk SiO_2 content[18]. By viewing lunar surface targets multiple times under different viewing conditions, the precision of individual Diviner CF values, which are known to be systematically affected by illumination and viewing geometry, are estimated at <0.02 μm (refs 17,43).

For this study, we used data collected between 8:30 and 16:30 local solar time, with emission angles <5° from nadir. These restricted ranges minimize the effects of the observation conditions on Diviner CF estimates. All data used in the current study were reduced using the most recent available corrections[44].

CF values were modelled following the methods of Greenhagen et al.[17] In brief, the shapes of the CFs for the regolith surfaces of airless bodies can be approximated as parabolas. Each three-point Diviner spectrum is fit as a parabola by solving the following system of three equations to find the wavelength of the parabola maximum, λ_{Max}:

$$y_3 = A\lambda_3^2 + B\lambda_3 + C \tag{1}$$

$$y_4 = A\lambda_4^2 + B\lambda_4 + C \tag{2}$$

$$y_5 = A\lambda_5^2 + B\lambda_5 + C \tag{3}$$

$$\lambda_{Max} = -B/(2A) \tag{4}$$

In these equations, y_3, y_4 and y_5 are the emissivities for Diviner channels 3, 4 and 5, respectively, and A, B and C are constants. The calculated CF value is λ_{Max}.

To generate the representative three-point spectra, we averaged groups of 7,897 and 8,181 Diviner spectra to produce representative on-swirl and off-swirl spectra, respectively, at Reiner Gamma and model the CFs for these regions. At the Van de Graaff crater site, we averaged 1,736 spectra each for the on-swirl and off-swirl regions.

Thermal model. The one-dimensional thermal model employs standard forward difference approximations for spatial and temporal derivatives and incorporates important constraints and improvements based on fits to the Diviner data[22,38]. Most important of these are depth-dependent and temperature-dependent thermal conductivity, as well as spatially variable regolith porosity and rock abundance. Model layer thicknesses are typically <1 mm, and the bottom boundary at 2 m is well below the ~10 cm skin depth of the diurnal thermal wave. We allow the model to fully equilibrate over at least 10 years before reporting the results.

Roughness model. The effects of roughness on thermal emission are simulated using a modified statistical shadowing model similar to several that have previously been described in the literature[29,36,37]. Using this model, we predicted the surface temperatures for slopes from 0 to 90° and azimuth orientations from 0 to 360°. We then convolved each calculated temperature with its expected fractional surface area for the given RMS slope, based on the assumed Gaussian height distribution[30]. Infrared emission at the given measurement wavelength is then calculated as a weighted sum of the contributions of all the surface elements (both illuminated and shadowed) within the field of view[28]. The model also incorporates a shadow approximation[31,33] that accounts for the statistical likelihood of a surface to be within a shadow cast by another surface.

Night-time data. Night-time (19:30 to 05:30 local time) temperature maps were constructed from the Diviner Level 2 Gridded Data Products available at the Planetary Data System. A total of 46 separate Diviner channel 8 (50–100 μm) brightness temperature maps were assembled from a single Lunar Reconnaissance Orbiter mapping cycle collected over the course of a lunar day. The average brightness temperature was subtracted from each map to normalize the temperature for the local time variations. The normalized maps were averaged to produce the final temperature deviation maps. This methodology ensures that any temperature differences between on swirl and off swirl are preserved.

References

1. El-Baz, F. The Alhazen to Abdul Wafa Swirl Belt: an extensive field of light-colored sinuous markings. *NASA Spec. Pub.* **315**, 29–93 (1972).
2. Blewett, D. T., Hawke, B. R. & Richmond, N. C. A magnetic anomaly associated with an albedo feature near Airy crater in the lunar nearside highlands. *Geophys. Res. Lett.* **34**, L24206 (2007).
3. Hood, L. L. & Schubert, G. Lunar magnetic anomalies and surface optical properties. *Science* **208**, 49–51 (1980).
4. Schultz, P. H. & Srnka, L. J. Cometary collisions on the Moon and Mercury. *Nature* **284**, 22–26 (1980).
5. Pinet, P. C., Shevchenko, V. V., Chevrel, S. D., Daydou, Y. & Rosemberg, C. Local and regional lunar regolith characteristics at Reiner Gamma formation: optical and spectroscopic properties from Clementine and Earth-based data. *J. Geophys. Res.* **105**, 9457–9476.
6. Starukhina, L. V. & Shkuratov, Y. G. Swirls on the Moon and Mercury: meteoroid swarm encounters as a formation mechanism. *Icarus* **167**, 136–147 (2004).
7. Garrick-Bethell, I., Head, III J. W. & Pieters, C. M. Spectral properties, magnetic fields, and dust transport at lunar swirls. *Icarus* **212**, 480–492 (2011).
8. Hood, L. L. & Williams, C. R. The lunar swirls—distribution and possible origins. *Proc. Lunar Planet. Sci. Conf.* **19Th**, 99–113 (1989).
9. Richmond, N. C. et al. Correlations between magnetic anomalies and surface geology antipodal to lunar impact basins. *J. Geophys. Res.* **110**, E05011 (2005).
10. Blewett, D. T. et al. Lunar swirls: Examining crustal magnetic anomalies and space weathering trends. *J. Geophys. Res.* **116**, E02002 (2011).
11. Hapke, B. Space weathering from Mercury to the asteroid belt. *J. Geophys. Res.* **106**, 10,039–10,074 (2001).
12. Lin, R. P. et al. Lunar surface magnetic fields and their interaction with the solar wind: results from lunar prospector. *Science* **281**, 1480–1484 (1998).
13. Kurata, M. et al. Mini-magnetosphere over the Reiner Gamma magnetic anomaly region on the Moon. *Geophys. Res. Lett.* **32**, L24205 (2005).
14. Kramer, G. Y. et al. Characterization of lunar swirls at Mare Ingenii: a model for space weathering at magnetic anomalies. *J. Geophys. Res.* **116**, E04008 (2011).
15. Kramer, G. Y. et al. M3 spectral analysis of lunar swirls and the link between optical maturation and surface hydroxyl formation at magnetic anomalies. *J. Geophys. Res.* **116**, E00G18 (2011).
16. Denevi, B. W. et al. Characterization of space weathering from lunar reconnaissance orbiter camera ultraviolet observations of the Moon. *J. Geophys. Res.* **119**, 976–997 (2014).

17. Greenhagen, B. T. *et al.* Global silicate mineralogy of the Moon from the Diviner lunar radiometer. *Science* **329**, 1507–1509 (2010).

18. Logan, L. M., Hunt, G. R., Salisbury, J. W. & Balsamo, S. R. Compositional implications of Christiansen Frequency maximums for infrared remote sensing applications. *J. Geophys. Res.* **78**, 4983–5003 (1973).

19. Lucey, P. G., Paige, D. A., Greenhagen, B. T., Bandfield, J. L. & Glotch, T. D. Comparison of Diviner Christiansen feature position and visible albedo: composition and space weathering implications. *Lunar Planet Sci. XLI,* abstract 1600 (2010).

20. Glotch, T. D. *et al.* Highly silicic compositions on the Moon. *Science* **329**, 1510–1513 (2010).

21. Bandfield, J. L. *et al.* Lunar surface rock abundance and regolith fines temperatures derived from LRO Diviner Radiometer data. *J. Geophys. Res.* **116**, E00H02 (2011).

22. Vasavada, A. R. *et al.* Lunar equatorial surface temperatures and regolith properties from the Diviner lunar radiometer experiment. *J. Geophys. Res.* **117**, E00H18 (2012).

23. Lucey, P. G., Blewett, D. T. & Hawke, B. R. Imaging of lunar surface maturity. *J. Geophys. Res.* **105**, 20,377–20,386 (2000).

24. Hemingway, D. & Garrick-Bethell, I. Magnetic field direction and lunar swirl morphology: insights from Airy and Reiner Gamma. *J. Geophys. Res.* **117**, E10012 (2012).

25. Pieters, C. M., Fischer, E. M., Rode, O. & Basu, A. Optical effects of space weathering: the role of the finest fraction. *J. Geophys. Res.* **98**, 20817–20824 (1993).

26. Taylor, L. A., Pieters, C. M., Keller, L. P., Morris, R. V. & McKay, D. S. Lunar mare soils: space weathering and the major effects of surface-correlated nanophase Fe. *J. Geophys. Res.* **106**, 27,985–27,999 (2001).

27. Arnold, J. A., Glotch, T. D., Thomas, I. R. & Bowles, N. E. Plagioclase-olivine mixtures in a simulated lunar environment. *Lunar Planet. Sci. XLIV,* abstract 2972 (2013).

28. Hayne, P. *et al.* Thermophyscial properties of the lunar surface from Diviner observations. *EGU Gen. Assembly,* abstract EGU2013-10871-1 (2013).

29. Helfenstein, P. & Shepard, M. K. Submillimeter-scale topography of the lunar regolith. *Icarus* **72**, 342–357 (1999).

30. Rosenburg, M. A. *et al.* Global surface slopes and roughness of the Moon from the lunar orbiter laser altimeter. *J. Geophys. Res.* **116**, E02001 (2011).

31. Hapke, B. Bidirectional reflectance spectroscopy III—correction for macroscopic roughness. *Icarus* **59**, 41–59 (1984).

32. Shkuratov, Y. G. *et al.* Interpreting photometry of regolith-like surfaces with different topographies: shadowing and multiple scattering. *Icarus* **173**, 3–15 (2005).

33. Smith, B. G. Lunar surface roughness: shadowing and thermal emission. *J. Geophys. Res.* **72**, 4059–4067 (1967).

34. Johnson, P. E., Vogler, K. J. & Gardner, J. P. The effect of surface roughness on lunar thermal emission spectra. *J. Geophys. Res.* **98 (E11)**, 20825–20829 (1993).

35. Danilina, I. *et al.* Roughness effects on sub-pixel radiative temperature dispersion in a kinetically isothermal surface. In *Proceedings of Second Recent Advances in Quantitative Remote Sensing* (ed. Sobrino, J.) 13–18 (Publicacions de la Universitat de Valencia, 2006).

36. Bandfield, J. L. & Edwards, C. S. Derivation of martian surface slope characteristics from directional thermal infrared radiometry. *Icarus* **193**, 139–157 (2008).

37. Bandfield, J. L. Effects of surface roughness and graybody emissivity on martian thermal infrared spectra. *Icarus* **202**, 414–428 (2009).

38. Hayne, P. O. *et al.* Diviner lunar radiometer observations of the LCROSS impact. *Science* **330**, 477–479 (2010).

39. Presley, M. & Christensen, P. R. Thermal conductivity measurements of particulate materials. 2. Results. *J. Geophys. Res.* **102**, 6551–6566.

40. Neish, C. D. *et al.* The surficial nature of lunar swirls as revealed by the Mini-RF instrument. *Icarus* **215**, 186–196 (2011).

41. Pieters, C. M., Moriarty, D. P. & Garrick-Bethell, I. Atypical regolith processes hold the key to enigmatic lunar swirls. *Lunar Planet Sci. XLV,* abstract 1408 (2014).

42. Bamford, R. A. *et al.* Minimagnetospheres above the lunar surface and the formation of lunar swirls. *Phys. Rev. Lett.* **109**, 081101 (2012).

43. Allen, C. C., Greenhagen, B. T., Donaldson Hanna, K. L. & Paige, D. A. Analysis of lunar pyroclastic deposit FeO abundances by LRO Diviner. *J. Geophys. Res.* **117**, E00H28 (2012).

44. Greenhagen, B. T. *et al.* The Diviner lunar radiometer compositional data products: description and examples. *Lunar Planet Sci. XLII,* abstract 2679 (2011).

Acknowledgements

Funding for this work was provided by the Diviner Lunar Radiometer Experiment extended mission science investigation and the Remote, In Situ, and Synchrotron Studies (RIS⁴E) team of NASA's Solar System Research Virtual Institute (SSERVI). This is SSERVI publication number SSERVI-2014-158.

Author contributions

T.D.G. conceived of the study and performed the Diviner CF and temperature data analysis with the assistance of J.L.B., P.G.L. and B.T.G.. P.O.H. supplied the roughness and thermal models discussed in the paper and R.R.G. and D.A.P. assisted with their interpretation. J.A.A. acquired the simulated lunar environment spectra of plagioclase/olivine mixtures and provided interpretations. T.D.G. wrote the paper with the input of all the co-authors.

Additional information

20

Extensive volatile loss during formation and differentiation of the Moon

Chizu Kato[1,2], Frederic Moynier[1,3], Maria C. Valdes[2], Jasmeet K. Dhaliwal[4] & James M.D. Day[4]

Low estimated lunar volatile contents, compared with Earth, are a fundamental observation for Earth–Moon system formation and lunar evolution. Here we present zinc isotope and abundance data for lunar crustal rocks to constrain the abundance of volatiles during the final stages of lunar differentiation. We find that ferroan anorthosites are isotopically heterogeneous, with some samples exhibiting high $\delta^{66}Zn$, along with alkali and magnesian suite samples. Since the plutonic samples were formed in the lunar crust, they were not subjected to degassing into vacuum. Instead, their compositions are consistent with enrichment of the silicate portions of the Moon in the heavier Zn isotopes. Because of the difference in $\delta^{66}Zn$ between bulk silicate Earth and lunar basalts and crustal rocks, the volatile loss likely occurred in two stages: during the proto-lunar disk stage, where a fraction of lunar volatiles accreted onto Earth, and from degassing of a differentiating lunar magma ocean, implying the possibility of isolated, volatile-rich regions in the Moon's interior.

[1] Institut de Physique du Globe de Paris, Université Paris Diderot, CNRS UMR 7154, Paris 75005, France. [2] Department of Earth and Planetary Sciences and McDonnell Center for Space Sciences, Washington University in St Louis, St Louis, Missouri 63130, USA. [3] Institut Universitaire de France, Paris 75005, France. [4] Geosciences Research Division, Scripps Institution of Oceanography, La Jolla, California 92093-0244, USA. Correspondence and requests for materials should be addressed to F.M. (email: moynier@ipgp.fr).

The present-day inventory of volatile elements in inner Solar System terrestrial bodies is a function of initial delivery and subsequent volatile depletion histories. Unravelling this history can, therefore, elucidate mechanisms of planetary growth and differentiation. In the case of the Earth–Moon system, the contrasting fate of volatiles, between a relatively volatile-poor Moon and a water-rich Earth, provides important evidence for their formation and evolution. Despite their clear importance, however, the fate of the volatile elements and the origin of the volatile depletion of the Moon is not well understood[1,2]. An important observation of Apollo lunar samples was their strong depletion in volatile elements compared with terrestrial samples[3]. More recent studies of the volatile content in the Moon have shown that isotopes of moderately volatile elements, such as Cl and Zn, are highly depleted, compared with bulk silicate Earth[4–6]. In addition, some lunar rocks are enriched in the heavier isotopes of Fe compared with bulk silicate Earth[7]. This observation was originally interpreted in terms of Fe volatilization during the giant impact[7]; however, igneous processes provide a more robust interpretation for these results[8,9]. In contrast, studies of OH contents and D/H in apatites from mare basalts, as well as Apollo pyroclastic glass beads, have argued for elevated volatile contents in portions of the lunar interior[10–13]. Reconciling these seemingly contrasting observations remains critical to examining models of lunar formation (for example, ref. 14) and of lunar mantle differentiation[15], since a volatile-rich Moon is not a logical consequence of large-scale collisional events, or magma ocean differentiation[4]. To examine the origin of lunar volatiles, we employ high-precision Zn isotopes and abundances. Zinc is a moderately volatile element with a 50% condensation temperature of 726 K under solar nebula conditions[16]. Isotopic fractionation during magmatic processes has a negligible effect ($<0.1‰$ for $\delta^{66}Zn$; per mil deviation of $^{66}Zn/^{64}Zn$ from the JMC-Lyon standard)[17]. However, Zn isotopes are highly fractionated during high-temperature volatilization in planetary rocks and variations in $\delta^{66}Zn$ larger than 1‰ in terrestrial igneous rocks have only been observed in association with evaporation events[18,19]. The absence of isotopic fractionation during igneous processes makes Zn isotopes powerful tracers of high-temperature evaporation during volatile depletion episodes in planets. Recent work on lunar samples has focused largely on using Zn isotopes to study mare basalts, regolith and pyroclastic glass beads[6,18,20]. These studies have shown that mare basalts are consistently enriched in the heavier Zn isotopes compared with terrestrial basalts, which has been interpreted as a consequence of a whole-scale evaporation event on the Moon[6], or during magma ocean differentiation[4]. It has also been suggested that similar isotopic fractionation could possibly have occurred during eruptive degassing of volatiles from mare basalt lavas into vacuum[5].

To obtain the most comprehensive understanding of the Zn isotopic composition of lunar samples, we investigated the Zn isotopic compositions and abundances of lunar crustal rocks, which include magnesian and alkali suite samples (collectively termed here as MGS) and ferroan anorthosite (FAN) plutonic rocks. MGS are interpreted to have formed within the FAN-dominated lunar crust[21] and, therefore, have not experienced degassing in a vacuum. FAN samples were formed by crystallization of a plagioclase floatation crust above a magma ocean after $>80\%$ differentiation[21]. FANs have highly variable volatile element abundances, with Zn ranging from ~0.1 to >50 p.p.m. (for example, ref. 22), possibly consistent with transport and redistribution of Zn at the lunar surface. New data are also presented for the pyroclastic green glass 15426 to further examine whether pyroclastic glasses are enriched in the light isotopes of Zn due to recondensation of isotopically light vapour[18]. Additional data are also reported for three lunar regolith samples and several mare basalt samples to confirm

previously observed heavy Zn isotope enrichment in these lithologies[6,18,20].

Results

Lunar basalts. New data (Supplementary Table 1) for four high-Ti basalts ($\delta^{66}Zn = +1.5 \pm 0.6‰$, 2 s.d.) and three low-Ti basalts ($\delta^{66}Zn = +1.5 \pm 0.4‰$, 2 s.d.) (Supplementary Note 1) fall within the range observed previously for mare basalts[6,20], and confirm that there is a consistent enrichment in the heavier isotopes of Zn compared with terrestrial basaltic rocks ($\delta^{66}Zn = +0.28 \pm 0.05‰$ for the bulk silicate Earth[17]) (Fig. 1). Combining our new data together with literature data[6,20] (Supplementary Table 2), we calculate a new average value for lunar basalts of $\delta^{66}Zn = +1.4 \pm 0.5‰$ (2 s.d., $n = 26$). A few exceptions (five samples out of 30) are isotopically lighter and are excluded from this average due to secondary condensation effects (Supplementary Discussion).

Lunar crust and interior. In contrast to the relative homogeneity of mare basalts, FAN samples show large Zn isotopic variability ($\delta^{66}Zn$ between -11.4 and $+4.2‰$) and the light isotope enrichment is correlated with the reciprocal Zn abundance (Fig. 2). In comparison, the two MGS rocks that we measured are both isotopically heavy (Supplementary Table 1).

Pyroclastic glass and regolith. The pyroclastic green glass (15426) is enriched in Zn compared with mare basalts (~50 p.p.m. versus ~2 p.p.m., which is in good agreement with previous studies[22]) and is isotopically lighter than the mare basalts ($\delta^{66}Zn = -0.98‰$). This result confirms consistent light isotopic enrichment of Zn in both high- and low-Ti pyroclastic glass beads[18,20], with $\delta^{66}Zn$ for the pyroclastic glasses ranging from -1 to $-4.1‰$. Three regolith samples are enriched in the heavier isotopes of Zn, with $\delta^{66}Zn$ up to $+6.4‰$, confirming previous results[18,20] that the lunar regolith is isotopically heavier than lunar basalts through impact gardening processes (Supplementary Discussion). The Zn abundances of the different samples fall within the range of previous measurements: mare basalts: $\sim1–5$ p.p.m.; FAN: 0.6–75 p.p.m.; MGS: ~3 p.p.m.; and pyroclastic green glass: ~50 p.p.m. (Supplementary Table 2).

Discussion

Enrichment in the lighter isotopes of Zn in Apollo 17 glasses was originally interpreted as the consequence of degassing of basaltic

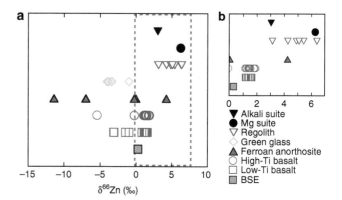

Figure 1 | $\delta^{66}Zn$ values for lunar rocks. The grey dashed box in **a** represents the magnified area shown in **b**. Lunar mare basalts, alkali and magnesium suite samples are enriched in the heavier isotopes of Zn in comparison with the terrestrial mantle composition (orange square). Ferroan anorthosites exhibit significant isotopic variability, to light and heavy Zn isotopic compositions. Isotopic data for lunar samples are from this study and refs 6,18,20. The bulk silicate Earth value is from ref. 17.

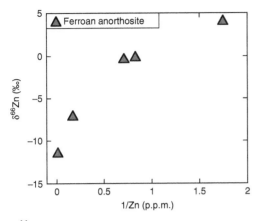

Figure 2 | δ^{66}Zn versus reciprocal Zn concentration in ferroan anorthosites. The correlation between δ^{66}Zn and 1/Zn suggests redistribution and mixing of Zn between a surface reservoir rich in Zn that is isotopically light and a Zn poor, isotopically heavy primary magmatic reservoir.

magma during a lava-fountaining event[18]. During magmatic ascent, the volatile elements are thought to have vapourized, leading to light isotope enrichment through kinetic isotopic fractionation. This vapour is then considered to have recondensed and coated the bead surfaces[18]. The new results on the Apollo 15 green glass (15426) offer the same interpretation for this pyroclastic deposit. The green glass 15426 is coarser (with a mean radius of the grains of $\sim 100\,\mu m$) than the two Apollo 17 glasses (mean radii of $\sim 50\,\mu m$)[23]. Therefore, a prediction, based on surface/volume, is for more limited light isotope enrichment of Apollo 15 glass beads due to surface/volume effects, as our results show (Apollo 15426 δ^{66}Zn $\sim -1‰$, [Zn] ~ 53 p.p.m. versus Apollo 74220 δ^{66}Zn -3.4 to $-4‰$ and [Zn] ~ 230 p.p.m.).

Enrichment in the heavier isotopes of Zn in lunar mare basalts confirms that most lunar mare basalts have a homogeneous Zn isotopic composition, with 25 samples (out of 30) enriched by $>1‰$ (δ^{66}Zn $= +1.42 \pm 0.5‰$; 2 s.d., $n = 26$). It has previously been argued that the heavy isotope enrichment in mare basalts reflects a global phenomenon, where basalts representing four different Apollo mission sampling locations (Apollo 10, 12, 15 and 17), as well as unknown regions sampled by lunar meteorite LaPaz Icefield 02205, all show nearly identical enrichments[6]. The heavy Zn isotopic composition in the source of the mare basalts together, with the depletion in absolute abundances of moderately volatile elements, suggests that the lunar material suffered a whole-scale high-temperature evaporation process, with the preferential partitioning of the lighter isotopes of Zn in the vapour. This evaporation process most probably occurred during large-scale melting of the Moon, such as during the giant impact[6].

Highland rocks represent the primary crust of the Moon and are the oldest known lunar samples (for example, ref. 24). The large range of Zn abundances in FAN samples suggests two possible scenarios: (1) the source of the FAN had a highly variable Zn content, which is reflected in the FAN samples, or (2) the Zn abundance represents secondary redistribution of Zn at the surface of the Moon during cataclasis of the FAN, as well as mixing with some chondritic components. Partial melting does not seem to extensively fractionate Fe/Zn due to the fact that these two elements share similar ionic radii[25]. The five different FAN samples analysed here have variable Fe/Zn (Supplementary Table 3), which, therefore, suggests that either the source of the anorthosites had variable Fe/Zn, that Fe and Zn were fractionated during plagioclase crystallization, or that the Zn elemental variations are a result of secondary evaporative processes. Since Zn isotopes are not fractionated by $>0.2‰$ during igneous

processes[17], the large Zn isotopic variations, which are correlated with the Zn abundance (Fig. 2), indicate that the Zn systematics of FAN are due to high-temperature evaporation effects. Thus, Zn-rich FAN contains a large fraction of an isotopically light vapour, while the most Zn-poor FAN (15415) retains primary magmatic characteristics.

Apollo sample 15415 is unbrecciated, pristine[26] and is enriched in the heavier isotopes of Zn (δ^{66}Zn $= 4.2‰$). On the other hand, 65315, which is enriched in Zn (75 p.p.m.) is a brecciated cataclastic anorthosite with exogenously derived (meteoritic) highly siderophile element additions[27]. Similarly, 67955 is also a brecciated sample, formed from an impact melt, with a significant component of meteoritic siderophile elements[28] and has disturbed Zn isotope systematics. These lines of evidence indicate that the brecciated samples incorporated isotopically light Zn-rich vapour during cataclasis or residence on the lunar surface. Transport of Zn at the surface of the Moon, during later impact bombardment of the crust and during the assimilation of exogenous Zn, implies that the Zn isotopic composition of cataclastic FAN samples do not represent the composition of their source reservoirs.

Significant enrichments in the heavy isotopes of Zn observed in the MGS, which solidified within the lunar crust, point towards a higher volatile depletion in their source than mare basalts. The origin of the MGS is generally interpreted as magmatic intrusions within the FAN crust (for example, ref. 21), formed by either a very high degree of partial melting[29] or by assimilation of FAN crust[30]. Evaporation and condensation in a vacuum at the surface of the Moon has, therefore, played no role in the composition of these samples. Enriched heavier isotopes of Zn observed in the MGS points towards a source already depleted in volatile elements. In turn, this provides strong support to loss of Zn and moderately volatile elements during a whole-scale evaporation of the Moon early in its history either following the giant impact or during a lunar magma ocean phase[4,6], and not locally during degassing, as has been proposed for some of the mare basalts (ref. 5). A third scenario would be that the impactor was already depleted in volatile elements before the giant impact (for example, refs 31,32). The new data cannot constrain this scenario; however, a second process of volatile loss is required to explain the differences in Zn isotopic composition between mare basalts and pristine highland samples. A more likely scenario is, therefore, that volatile element of the Moon occurred during lunar formation and differentiation.

An outstanding question is how Zn and other moderately volatile elements were lost from the Moon, while water, which is far more volatile (for example, ref. 32), seems to have been retained in some samples (for example, refs 10–13). It has been suggested that hydrodynamic escape after the giant impact was not a viable mechanism to explain the loss of water (or potassium) from the Earth–Moon system and that the volatile element content of the Moon was inherited from the precursor material[33]. However, this model fails to explain the extreme volatile depletion of the mare basalt source rocks, which are the most representative derivative rock type from the lunar mantle, or the later evolution of volatiles in the Moon revealed from crustal rocks. In a two-phase vapour-melt numerical model of the dynamical and thermodynamical evolution of the proto-lunar disk[34], where the vapour phase is viscous due to magnetorotational instability, the volatile elements are predicted to accrete to Earth, leaving the material that formed the Moon enriched in refractory elements and correspondingly depleted in volatiles. Calculations of the chemical composition of the vapour of the lunar protoplanetary disk[35] indicates that Zn enters the volatile phase of the disk at temperatures $<3,000$ K, leaving the Moon depleted in Zn and more volatile elements. Our new results support this mode of origin, but also suggest that continued volatile depletion in a lunar magma ocean is required[4] to explain differences in $\delta^{66}Zn$ between mare basalts and pristine lunar crustal rocks.

Methods

Sample preparation. One-gram chips of samples (Supplementary Note 1) were powdered with an agate mortar and pestle. Samples were then dissolved in a mixture of HF/HNO_3 for several days in Teflon beakers at $\sim 140\,°C$. Zinc purification was achieved in a 0.25-ml column using AG-1X8 (200–400 mesh) anion-exchange resin. After applying the sample solution in HBr onto the resin, further HBr was added to remove elements causing matrix effects. Purified Zn was collected in a HNO_3 solution following the procedure of ref. 18. The resulting Zn fraction was further purified in smaller 100-μl columns following the same ion-exchange chromatography procedure.

Sample analysis. Samples were analysed using a Thermo-Fisher Neptune Plus multicollector inductively coupled plasma mass spectrometer at the Department of Earth and Planetary Sciences, Washington University in St Louis. Masses 62, 63, 64, 65, 66, 67 and 68 were measured by Faraday cups positioned, respectively, and the intensity of ^{62}Ni was measured to control and correct the isobaric interference of ^{64}Ni (ref. 6). All samples fell on the mass-dependent fractionation line (Supplementary Fig. 1), and the typical analytical uncertainty (2 s.d.), is 0.04‰ for $\delta^{66}Zn$ and 0.05‰ for $\delta^{67}Zn$ and $\delta^{68}Zn$. Instrumental mass bias was corrected by bracketing of the standard and the sample and is further discussed in ref. 17. The Zn concentrations were also obtained from multicollector inductively coupled plasma mass spectrometer data by comparing the intensity of the sample signal with that of the standard, with an estimated precision of $\pm 10\%$.

References

1. Taylor, S. R. & McLennan, S. M. *Planetary Crusts: Their Composition, Origin and Evolution* (Cambridge Univ. Press, 2009).
2. Taylor, S. R. The Moon re-examined. *Geochim. Cosmochim. Acta* **141**, 670–676 (2014).
3. Wolf, R. & Anders, E. Moon and Earth: compositional differences inferred from siderophiles, volatiles, and alkalis in basalts. *Geochim. Cosmochim. Acta* **44**, 2111–2124 (1980).
4. Day, J. M. D. & Moynier, F. Evaporative fractionation of volatile stable isotopes and their bearing on the origin of the Moon. *Philos. Trans. A Math. Phys. Eng. Sci.* **372**, 20130259 (2014).
5. Sharp, Z. D., Shearer, C. K., McKeegan, K. D., Barnes, J. D. & Wang, Y. Q. The chlorine isotope composition of the Moon and implications for an anhydrous mantle. *Science* **329**, 1050–1053 (2010).
6. Paniello, R. C., Day, J. M. D. & Moynier, F. Zinc isotopic evidence for the origin of the Moon. *Nature* **490**, 376–379 (2012).
7. Poitrasson, F., Halliday, A. N., Lee, D. C., Levasseur, S. & Teutsch, N. Iron isotope differences between Earth, Moon, Mars and Vesta as possible records of contrasted accretion mechanisms. *Earth Planet. Sci. Lett.* **223**, 253–266 (2004).
8. Weyer, S. *et al.* Iron isotope fractionation during planetary differentiation. *Earth Planet. Sci. Lett.* **240**, 251–264 (2005).
9. Wang, K., Jacobsen, S. B., Sedaghatpour, F., Chen, H. & Korotev, R. L. Lunar dunite reveals the same iron isotopic composition of the bulk silicate Earth and Moon. *Lunar Planet. Sci. Conf.* **46**, 1980 (2015).
10. Saal, A. E., Hauri, E. H., Van Orman, J. A. & Rutherford, M. J. Hydrogen isotopes in lunar volcanic glasses and melt inclusions reveal a carbonaceous chondrite heritage. *Science* **340**, 1317–1320 (2013).
11. McCubbin, F. M. *et al.* Nominally hydrous magmatism on the Moon. *Proc. Natl Acad. Sci. USA* **107**, 11223–11228 (2010).
12. Saal, A. E. *et al.* Volatile content of lunar volcanic glasses and the presence of water in the Moon's interior. *Nature* **454**, 192–195 (2008).
13. Hauri, E. H., Saal, A. E., Rutherford, M. J. & Van Orman, J. A. Water in the Moon's interior: Truth and consequences. *Earth Planet. Sci. Lett.* **409**, 252–264 (2015).
14. Cameron, A. The origin of the Moon and the single impact hypothesis V. *Icarus* **126**, 126–137 (1997).
15. Wood, J. A., Dicky, J. S. Jr., Marvin, U. B. & Powell, B. N. Lunar anorthosites and a geophysical model of the moon. *Proc. Apollo 11 Lunar Sci. Conf.* **1**, 965–988 (1970).
16. Lodders, K. Solar System abundances and condensation temperatures of the elements. *Astrophys. J.* **591**, 1220–1247 (2003).
17. Chen, H., Savage, P. S., Teng, F.-Z., Helz, R. T. & Moynier, F. Zinc isotopic fractionation during magmatic differentiation and the isotopic composition of the bulk Earth. *Earth Planet. Sci. Lett.* **369-370**, 34–42 (2013).
18. Moynier, F., Albarède, F. & Herzog, G. F. Isotopic composition of zinc, copper, and iron in lunar samples. *Geochim. Cosmochim. Acta* **70**, 6103–6117 (2006).
19. Moynier, F. *et al.* Nature of volatile depletion and genetic relationships in enstatite chondrites and aubrites inferred from Zn isotopes. *Geochim. Cosmochim. Acta* **75**, 297–307 (2011).
20. Herzog, G. F., Moynier, F., Albarède, F. & Berezhnoy, A. A. Isotopic and elemental abundances of copper and zinc in lunar samples, Zagami, Pele's hairs, and a terrestrial basalt. *Geochim. Cosmochim. Acta* **73**, 5884–5904 (2009).
21. Shearer, C. K. *et al.* Thermal and magmatic evolution of the Moon. *Rev. Mineral. Geochem.* **60**, 365–518 (2006).
22. Ebihara, M., Wolf, R., Warren, P. H. & Anders, E. Trace elements in 59 mostly highland Moon rocks. *Proc. Lunar Planet. Sci. Conf.* **22**, 417–426 (1992).
23. Morgan, J. W. & Wandless, G. A. Surface-correlated trace elements in 15426 lunar glasses. *Proc. Lunar Planet. Sci. Conf.* **15**, 562–563 (1984).
24. Carlson, R. W. & Lugmair, G. W. The age of ferroan anorthosite 60025: oldest crust on a young Moon. *Earth Planet. Sci. Lett.* **90**, 119–130 (1988).
25. Lee, C-T. A. *et al.* The redox state of arc mantle using Zn/Fe systematics. *Nature* **468**, 681–685 (2010).
26. Ganapathy, R., Morgan, J. W., Krähenbühl, U. & Anders, E. Ancient meteoritic components in lunar highland rocks: clues from trace elements in Apollo 15 and 16 samples. *Proc. Lunar Sci. Conf.* **4**, 1239–1261 (1973).
27. Day, J. M. D., Walker, R. J., James, O. B. & Puchtel, I. S. Osmium isotope and highly siderophile element systematics of the lunar crust. *Earth Planet. Sci. Lett.* **289**, 595–605 (2010).
28. Norman, M. D. & Nemchin, A. A. A 4.2 billion year old impact basin on the Moon: U-Pb dating of zirconolite and apatite in lunar melt rock 67955. *Earth Planet. Sci. Lett.* **388**, 387–398 (2014).
29. Longhi, J. Preliminary modeling of high pressure partial melting: Implications for early lunar differentiation. *Proc. Lunar Planet. Sci. Conf.* **12**, 1001–1018 (1981).
30. Warren, P. H. Anorthosite assimilation and the origin of the Mg/Fe-related bimodality of pristine Moon rocks: Support for the magmasphere hypothesis. *J. Geophys. Res.* **91**, D331–D343 (1986).
31. Albarède, F., Albalat, E. & Lee, C.-T. A. An intrinsic volatility scale relevant to the Earth and Moon and the status of water in the Moon. *Meteorit. Planet. Sci.* **50**, 568–577 (2015).
32. Albarède, F. Volatile accretion history of the terrestrial planets and dynamic implications. *Nature* **461**, 1227–1233 (2009).
33. Nakajima, M. & Stevenson, D. J. Hydrodynamic escape does not prevent the "wet" Moon formation. *Lunar Planet. Sci. Conf.* **45**, 2770 (2014).
34. Charnoz, S. & Michaut, C. Evolution of the protolunar disk: dynamics, cooling timescale and implantation of volatiles into the Earth. *Workshop on the Early Solar System Impact Bombardment* III, 3002 (2015).
35. Visscher, C. & Fegley, B. Jr. Chemistry of impact-generated silicate melt-vapor debris disks. *Astrophys. J. Lett.* **767**, L12 (2013).

Acknowledgements

We are grateful to CAPTEM and the NASA Johnson Space Center curatorial staff for provision of samples. This work was supported by funding from NASA (NNX12AH70G to F.M. and NNX11AG34G to J.M.D.D.). F.M. thanks the European Research Council for funding under the H2020 framework program/ERC grant agreement #637503 (Pristine), as well as the financial support of the UnivEarthS Labex program at Sorbonne Paris Cité (ANR-10-LABX-0023 and ANR-11-IDEX-0005-02), the ANR through a chaire d'excellence Sorbonne Paris Cité and the INSU through the PNP programme.

Author contributions

C.K. collected the data and drafted the paper; F.M. and J.M.D.D. conceived the project and M.C.V. and J.K.D. assisted in the data collection. C.K., F.M. and J.M.D.D. wrote the paper and all authors contributed to discussion and interpretation of results.

Additional information

21

Correlated compositional and mineralogical investigations at the Chang'e-3 landing site

Zongcheng Ling[1,2,3], Bradley L. Jolliff[2], Alian Wang[2], Chunlai Li[3], Jianzhong Liu[4], Jiang Zhang[1], Bo Li[1], Lingzhi Sun[1], Jian Chen[1], Long Xiao[5], Jianjun Liu[3], Xin Ren[3], Wenxi Peng[6], Huanyu Wang[6], Xingzhu Cui[6], Zhiping He[7] & Jianyu Wang[7]

The chemical compositions of relatively young mare lava flows have implications for the late volcanism on the Moon. Here we report the composition of soil along the rim of a 450-m diameter fresh crater at the Chang'e-3 (CE-3) landing site, investigated by the Yutu rover with *in situ* APXS (Active Particle-induced X-ray Spectrometer) and VNIS (Visible and Near-infrared Imaging Spectrometer) measurements. Results indicate that this region's composition differs from other mare sample-return sites and is a new type of mare basalt not previously sampled, but consistent with remote sensing. The CE-3 regolith derived from olivine-normative basaltic rocks with high $FeO/(FeO + MgO)$. Deconvolution of the VNIS data indicates abundant high-Ca ferropyroxene (augite and pigeonite) plus Fe-rich olivine. We infer from the regolith composition that the basaltic source rocks formed during late-stage magma-ocean differentiation when dense ferropyroxene-ilmenite cumulates sank and mixed with deeper, relatively ferroan olivine and orthopyroxene in a hybridized mantle source.

[1] Shandong Provincial Key Laboratory of Optical Astronomy and Solar-Terrestrial Environment, Institute of Space Sciences, Shandong University, Weihai 264209, China. [2] Department of Earth & Planetary Sciences and McDonnell Center for the Space Sciences, Washington University, St Louis, Missouri 63130, USA. [3] Key Laboratory of Lunar and Deep Space Exploration, National Astronomical Observatories, Chinese Academy of Sciences, Beijing 100012, China. [4] Institute of Geochemistry, Chinese Academy of Sciences, Guiyang 550002, China. [5] Planetary Science Institute, School of Earth Sciences, China University of Geosciences, Wuhan 430074, China. [6] Institute of High Energy Physics, Chinese Academy of Sciences, Beijing 100049, China. [7] Key Laboratory of Space Active Opto-Electronics Technology, Shanghai Institute of Technical Physics, Chinese Academy of Science, Shanghai 200083, China. Correspondence and requests for materials should be addressed to Z.L. (email: zcling@sdu.edu.cn).

The combination of precise chemical and physical properties of Apollo and Luna samples with known locations provides 'ground truth' for lunar remote sensing studies[1,2]. After some 40 years since the Apollo and Luna missions, China's Chang'e-3 (CE-3) landing and Yutu rover mission in December, 2013, provided the next robotic *in situ* measurements on the Moon[3-6]. The CE-3 landing site is in the northern part of the Imbrium basin (340.49°E, 44.12°N). Serial eruptions flooded the Imbrium basin from ~3.5 Ga to ~2 Ga (ref. 7), making this region of great scientific value for the detection of lava variations during extended volcanic activity[6-8]. The CE-3 landing site represents a relatively young (~2.96 Ga) medium-Ti lunar basalt exposure[7-10]. The top Eratosthenian lava flows at the Mare Imbrium surface on which CE-3 landed are interpreted to be 10–60 m thick and up to 1,200 km in length[7-11].

Here we combine the data from two payload elements of the Yutu rover, the Active Particle-induced X-ray Spectrometer (APXS) and Visible and Near-infrared Imaging Spectrometer (VNIS; see Methods section for the instrument descriptions), and report the composition and mineralogy of the region explored by the rover. Our analysis indicates that this young lunar mare region has unique compositional characteristics, and represents a new type of mare basalt that has not been sampled by previous Apollo and Luna missions and lunar meteorite collections. APXS data suggest that the regolith is extremely rich in FeO, high in CaO, intermediate in TiO_2, modest in Al_2O_3 and poor in SiO_2. We present the mineralogical information derived from APXS chemical data and VNIS spectral data, showing self-consistent and well-correlated results in mineral modes (for example, the proportions of high-Ca pyroxene, low-Ca pyroxene and olivine) and mineral chemistries (both data sets indicate an abundance of high-Ca ferropyroxene plus Fe-rich olivine). The more accurate *in situ* chemical and mineralogical measurements of the new basalt type provide ground-truth validation of remote sensing studies that also indicated the olivine-rich mineralogy of the basalt unit where CE-3 landed. Because the CE-3 landing site was on the ejecta of a fresh impact crater, we take the local regolith composition to be essentially that of the basalt excavated by the impact crater, and then consider the implications of the measured composition. Results suggest an origin from late-stage magma-ocean cumulates that crystallized after ilmenite saturation and hybridized with an olivine-rich cumulate. The CE-3 landing site and *in situ* analyses of the rocks and soils derived from the fresh crater near the landing site provide key new ground truth for some of the youngest volcanism on the Moon.

Results

Landing site description and Yutu rover operations. The location of CE-3 from a global to a ground view is shown in Fig. 1. CE-3 landed on the rim of a young crater (~27–80 Myr old[6]), initially informally named Purple Palace[4] and now formally named Zi Wei (Fig. 1c). The diameter of the crater is ~450 m, which would have excavated ~40–50 m beneath the surface. The ejecta of the impact should cover the entire CE-3 landing site and the region explored by the Yutu rover, evidenced by the blocky surface seen by the landing camera (Fig. 1d) and big boulders encountered by Yutu during its traverse (Fig. 1d,e). The Panoramic Camera imaged two types of rocks; one is a mainly light-toned and coarse-grained rock and the other is a relatively darker, fine-grained rock (Supplementary Figs 1 and 2, Supplementary Note 1). During 32 days of surface operations, Yutu travelled 114 m in this region and made four sets of *in situ* and stand-off measurements at four locations (Fig. 1d)[3,4,6]. The APXS and VNIS aboard the Yutu rover acquired compositional and spectral measurements at four locations (CE3-0005, -0006, -0007 and -0008), as shown in Fig. 1d. A detailed description of

the instruments, measurements and data processing are given in the Methods section and refs 12–16.

Chemical compositions and normative mineralogies from APXS. APXS spectra show peaks of Mg, Al, Si, K, Ca, Ti, Cr, Fe, Ni, Sr, Zr and Y from the CE-3 soils (Fig. 2a). We used the peak-area ratio of measured samples and the calibration target (Supplementary Table 1) to derive the chemical compositions of three measured soils (CE3-0006_2, -0006_3 and -0008, Table 1). In general, the major element concentrations of the three soils at the CE-3 landing site are similar to each other and represent a distinctive composition (Table 1, Fig. 2b–d). They are characterized by low SiO_2 (~41.2 wt.%), very high FeO (~22.8 wt.%), high CaO (~12.1 wt.%), intermediate TiO_2 (~5.0 wt.%) and modest Al_2O_3 (~9.7 wt.%).

When compared with Apollo and Luna soils and basaltic rocks (Fig. 2b–d)[2,17], the TiO_2 versus FeO relation of the CE-3 soils bears some similarity with Apollo 12 ilmenite basalts, but CE-3 soils have higher FeO and TiO_2 (Fig. 2b). The CE-3 soils have MgO concentrations in the range 6.3–11.0 wt.% (Table 1) with derived Mg# ($=$ Mg/(Mg + Fe) × 100) < 50 (Fig. 2c) at the low end, but a higher CaO compared with other mare samples (Fig. 2d), and deviating from the KREEP—feldspathic highlands—mare compositional triangle based on the returned lunar samples. These compositional features suggest that the CE-3 soils differ significantly from other known lunar basaltic materials.

On the basis of chemical composition (Table 1), we calculated the abundances of normative minerals of CE-3 soils using a CIPW (Cross, Iddings, Pirsson and Washington) norm. The major CIPW norm results are summarized in Table 1 and the detailed results are shown in Supplementary Tables 2,3. For CE3-0006_2 and CE3-0006_3 (sampling sites ~10 cm apart), we calculated the mean value as 'Mean_0006.' For the norms summarized in Table 1, we combined the high-Ca pyroxene components as diopside (Di) and the Ca-poor components as hypersthene (Hy). Given the analytical uncertainties associated with the APXS data (Supplementary Table 4), the main difference is in MgO, which is significantly higher in CE3-0008 (11 wt.%) compared with CE3-0006 (6.7 wt.%). The difference in MgO translates to a difference in the relative abundance of olivine and pyroxene and in the ratio of Di to Hy (Table 1). The higher MgO concentration of the CE3-0008 soil results in a higher Mg# (46) compared with CE3-0006 (34) and a higher Fo ($=$ Mg/(Mg + Fe) × 100) in olivine, that is, ~Fo_{51} for the CE3-0008 soil and ~Fo_{40} for the CE3-0006 soil. The CE3-0006 soil is also richer in the high-Fe endmember for both Di and Hy (Table 1) as a result of the difference in MgO. Considering analytical uncertainties for Al, Si, Ca, Fe and Ti, normative abundances of plagioclase and ilmenite are the same in -0006 and -0008, within analytical uncertainties.

In Table 1, the 'Means_all' column shows the average chemical composition and normative mineralogy summary of the CE-3 landing site soils. The soils have a high percentage of normative pyroxene (~42 wt.%), with most being high-Ca pyroxene, Di (29 wt.%), that is, about two times the Hy (13 wt.%). The normative feldspar content (27 wt.%) is within the range of many lunar basaltic samples. The normative olivine content (20 wt.%, corresponding to 17 vol.%) of CE-3 is at the high end of the range for known lunar basalts (for example, Apollo 12 olivine basalt has ~20 vol.% olivine[2]). In the CE-3 soils, olivine is Fe-rich with relatively low average Fo content (~43). The normative ilmenite contents of the three CE-3 soils are similar, averaging ~9 wt.%. The average Mg# of the soils is ~38, indicating the exceptionally ferroan character of source rocks that make up the local surface soils.

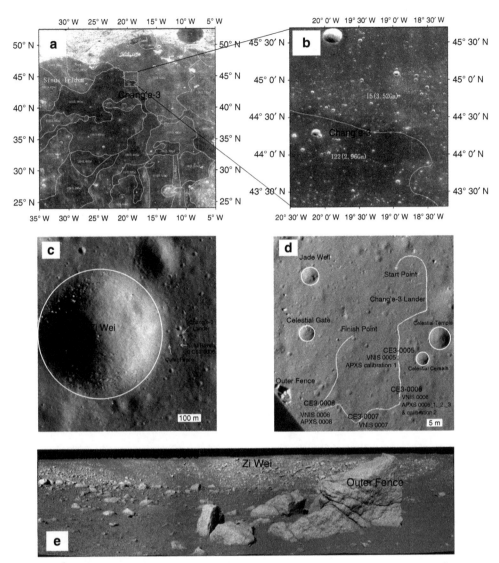

Figure 1 | Location of the Chang'e-3 landing site. (**a**) Chang'e-1 CCD image with boundaries of typical mare basalt units[7]. (**b**) Chang'e-2 CCD image and (**c**) LROC NAC image (LROC NAC M1142582775R). (**d**) The traverse map of the Yutu rover and the locations of APXS and VNIS measurements. (**e**) Panoramic view of the 'Zi Wei' crater by the Panoramic Camera on the Yutu rover at the CE3-0008 site.

Mineral chemistry and mineral modes based on VNIR spectra. The visible-NIR (near-infrared) spectra (Fig. 3a,b) of four VNIS observations show characteristic 1 and 2 μm absorption features owing to the electron transfer of Fe^{2+} in the M1 and M2 sites of lunar mafic silicates[18]. The spectra (Fig. 3a) have obvious absorption features and relatively flat profiles, indicating a low degree of space weathering, consistent with the fact that the CE-3 landing site sits on a relatively young Eratosthenian lava flow and the fresh ejecta of the young and fresh Zi Wei crater.

To estimate the average composition of minerals contributing to the spectra, we apply the modified Gaussian model (MGM)[19,20] to deconvolve the spectral bands. We find that the spectra from CE3-0005 and CE3-0008 sites have wide and strong 1 μm absorption bands but shallow 2 μm band depths (Fig. 3a,b), thus they should have a higher 1–2 μm band area ratio (BAR), which implies the presence of a significant amount of olivine in the soils of these two sites[15,21–23]. The absorption components of all four continuum-removed spectra (Fig. 3b) are calculated using MGM, as mixtures of three endmembers, high-Ca pyroxene (HCP), low-Ca pyroxene (LCP) and olivine (Supplementary Note 5, Table 2). This combination is the most complicated for this type of spectral deconvolution[24–26]. The results of the MGM deconvolution are shown in Table 2.

Extensive laboratory studies of terrestrial and synthetic pyroxenes provide the basis to correlate the 1- and 2-μm band positions with their chemical compositions[27–30]. We plot the central positions of deconvolved 1 and 2 μm bands from HCP and LCP components based on the data of Adams[27] and Cloutis and Gaffey[28] (Fig. 3c). Here we define the HCP as wollastonite (Wo) > 30 and LCP as Wo < 30, keeping with previous work by Sunshine et al.[31] and Klima et al.[30]. By comparison, the compositional features of LCP of the four soils are similar (Fig. 3c) and very Fe-rich, relative to orthopyroxene examined by Adams[27] and Cloutis and Gaffey[28]. However, the HCP compositions of CE-3 soils occur in two groups; CE3-0006 and CE3-0007 are slightly richer in Ca and in Fe than CE3-0005 and CE3-0008, consistent with APXS data (Table 1). The pyroxene chemistry of the CE-3 soils derived from VNIS data thus supports their general Fe-rich character, with CE3-0006 and CE3-0007 having even higher Fe contents, consistent with APXS results.

The volume percentage ratio of HCP and LCP (HCP/LCP) can be estimated using the band-strength ratios of 1 and 2 μm bands from MGM deconvolution of VNIS spectra[19,24,29,30,32]. The HCP/LCP vol.% ratios for four CE-3 soils were calculated using both the 1- and 2-μm band-strength ratios. The results for each

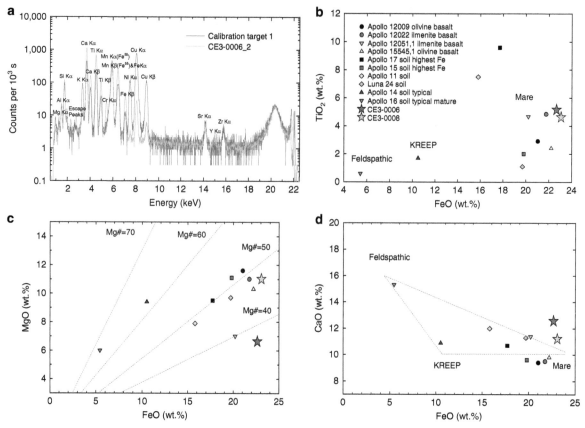

Figure 2 | X-ray spectrum and chemical compositions of Chang'e-3 soils from APXS. (**a**) APXS spectrum CE3-0006_2 overlain on the calibration spectrum. Comparison of Chang'e-3 site surface soil samples with Apollo and Luna samples[2,17] in (**b**) FeO versus TiO_2, (**c**) FeO versus MgO and (**d**) FeO versus CaO.

soil using two ratios are consistent (Table 2), indicating an equivalent compositional effect on both 1 and 2 μm bands. Overall, the HCP/LCP ratios in four CE-3 soils are similar, with HCP about two times LCP in abundance.

The MGM-derived band positions of olivine shift as a function of Fo content, thus they can be used to estimate olivine chemistry[20,25]. We plot central positions of two deconvolved M1 component bands of CE-3 olivine (Fig. 3d) with trend lines determined on terrestrial samples by Sunshine and Pieters[20]. The central positions of the two olivine M1 bands occur at 870–884 nm and 1234–1261 nm for the four CE-3 soils, suggesting they are Fe-rich (30 < Fo < 55, Fig. 3d). Specifically, spectra indicate that CE3-0005 and CE3-0008 soils have higher olivine Fo values than the other two soils.

The precise location of the M2 band ($\sim$1050 nm) of olivine in the VNIS spectrum is difficult to determine via MGM deconvolution[20,33] (Supplementary Note 5). However, we plotted the central positions of olivine M2 bands of four CE-3 soils derived from MGM deconvolution in Fig. 3d, which also shows a trend along the trend line determined by Sunshine and Pieters[20]. Therefore, olivine chemistry of the CE-3 soils derived from VNIS data supports their general Fe-rich character, consistent with normative analysis of the APXS results (Fo $\sim$43 on average, Table 1). The band-strength ratios of the HCP 1-μm band to the olivine M1 band near 1.25 μm could also be used to estimate their volume percentage[24]. The four CE-3 soils can be divided into two groups (Table 2): the CE3-0005 and -0008 soils are richer in olivine on the basis of VNIS analysis (HCP/OL = 2.0 and 2.3) than CE3-0006 and CE3-0007 (HCP/OL = 3.0 and 3.3, respectively).

Refinement of mineral mode by correlated APXS and VNIS. A key result from both the APXS and VNIS data is the inferred abundance of olivine. The APXS data indicate relatively low SiO_2 and high FeO + MgO, resulting in a significant proportion of olivine in the norm (10 vol.% in CE3-0006 and 30 vol.% in CE3-0008, Table 1). Our MGM analysis of the VNIS spectra also reflects high olivine contents (the HCP/OL ratio is 3.0 for CE3-0006 and 2.3 for CE3-0008, Table 2). A high olivine content coupled with intermediate to high TiO_2 makes the CE-3 soil and the basalt from which it derives unique among the known lunar samples, similar to a basalt type that has been inferred from orbital data[34], but not until the CE-3 mission verified by *in situ* or sample analysis.

Second, olivine chemistry derived from the norm analysis based on the APXS composition (Table 1) and from the central positions of the olivine M1 bands in VNIS spectra (Fig. 3d) both support the Fe-rich character of olivine in CE-3 soils. Fe-rich olivine was predicted on the basis of remote sensing of this area[21], thus the Fe-rich olivine found by CE-3 indicates a relatively evolved magma from late-stage volcanic activity in the Imbrium basin[35].

We also find similarity in pyroxene features inferred from normative analyses of the APXS composition (Table 1) and deconvolved VNIS spectra (Table 2). For example, both sets of analyses suggest CE-3 soils are rich in HCP. A good match was found between HCP/LCP ratios in CE3-0006 soil derived from VNIS (2.0, Table 2) and the Di/Hy components derived from APXS analysis (2.0, Table 1). The CIPW norm analysis based on APXS composition indicates a much higher Di/Hy ratio for CE3-0008 (8.7, Table 1) than the HCP/LCP derived from VNIS

Table 1 | Compositional data in weight percent and results of CIPW norm of Chang'e-3 soils from Yutu APXS after calibration.

	CE3-0006_1[†]	CE3-0006_2	CE3-0006_3	CE3-0008	Mean_0006[†]	Mean_all[†]
SiO_2	41.8 ± 3.1	42.3 ± 3.2	41.6 ± 3.1	39.6 ± 3.0	42.0 ± 3.1	41.2 ± 3.1
TiO_2	5.0 ± 0.1	5.2 ± 0.1	5.1 ± 0.1	4.6 ± 0.1	5.2 ± 0.1	5.0 ± 0.1
Al_2O_3	10.0 ± 0.7	9.8 ± 0.6	10.0 ± 0.7	9.3 ± 0.6	9.9 ± 0.7	9.7 ± 0.6
Cr_2O_3*	0.3	0.3	0.3	0.3	0.3	0.3
FeO	21.7 ± 2.0	22.7 ± 2.1	22.6 ± 2.1	23.0 ± 2.1	22.6 ± 2.1	22.8 ± 2.1
MnO*	0.3	0.3	0.3	0.3	0.3	0.3
MgO	8.1 ± 0.1	6.3 ± 0.1	7.0 ± 0.1	11.0 ± 0.1	6.7 ± 0.1	8.1 ± 0.1
CaO	12.3 ± 0.4	12.6 ± 0.4	12.6 ± 0.4	11.2 ± 0.3	12.6 ± 0.4	12.1 ± 0.4
Na_2O*	0.3	0.3	0.3	0.3	0.3	0.3
K_2O	0.11 ± 0.01	0.12 ± 0.01	0.11 ± 0.01	0.11 ± 0.01	0.12 ± 0.01	0.11 ± 0.01
P_2O_5*	0.1	0.1	0.1	0.1	0.1	0.1
Plag (wt.%)	28	27	28	26	28	27
Di (wt.%)[‡]	29	31	31	26	31	29
Hy (wt.%)[‡]	15	20	15	3	18	13
$\sum$ Px (wt.%)	45	52	46	29	49	42
Ol (wt.%)	16	10	15	34	12	20
Ilm (wt.%)	10	10	10	9	10	9
Plag (vol.%)	34	33	34	32	33	33
Di (vol.%)	31	33	32	28	33	31
Hy (vol.%)	14	18	14	3	16	12
$\sum$ Px (vol.%)	45	51	46	31	49	43
Ol (vol.%)	14	8	12	30	10	17
Ilm (vol.%)	7	7	7	6	7	7
Plag An	88	88	88	88	88	88
Oliv Fo	46	38	41	51	40	43
Mg#	40	33	36	46	34	38

Among the four locations, only two (CE3-0006 and CE3-0008) were analysed by APXS.
*The compositional data are normalized to 99% to allow for missing elements. (Missing elements Cr_2O_3, MnO, Na_2O and P_2O_5 are set to 0.3, 0.3, 0.3 and 0.1 wt.% for more reasonable calculations of CIPW norms. Fe^{3+}/Fe_{Total} are assumed to be 0.0 for lunar minerals for the CIPW calculations.
†CE3-0006_1 is for reference in the test mode with a detection distance of ~5 cm above the surface soils, and thus is not included in the calculation of Mean_0006. Mean_0006 is the average value of CE3-0006_2 and CE3-0006_3. Mean_all is the average value of CE3-0006_2 and CE3-0006_3 and CE3-0008. Among the four locations, only two (CE3-0006 and CE3-0008) were analysed by APXS.
‡In the CIPW calculation, we refer to the sum of HCP components (Di Wo, Di En and Di Fs) as Di, whereas the Hy is the sum of Hy En and Hy Fs. Mg#, 100Mg/(Mg + Fe) atomic.

(2.3, Table 2). This large difference results in part from the effect of the normative pyroxene components in which the Hy component has no Ca. In reality the LCP pyroxene (in this case, pigeonite) does contain Ca. The effect is greatest in CE3-0008 because it contains so much olivine that there is little normative Hy and abundant Di. Moreover, the VNIS and APXS target areas were within a short distance of each other (<1 m), so we do not expect such a large variation in this less (space-) weathered basaltic regolith.

To provide a better comparison between APXS data and VNIS data, we use typical compositions of lunar mare minerals (that is, olivine, augite, pigeonite and plagioclase) as endmembers in a mixing-model calculation instead of the normative calculation results (Supplementary Table 7). The results of this mixing analysis are shown in Table 3.

From the mixing analysis, the refined mineral mode of CE3-0008 yields an Aug/Pig ratio of 2.4 (Table 3), which matches well with the VNIS HCP/LCP ratio (2.3, Table 2) based on VNIS spectral deconvolution. For CE3-0006, the Aug/Pig value in the refined mineral mode (2.0) and the HCP/LCP value in the VNIS-derived mode (2.0) are essentially the same. Considering these results, the APXS and VNIS mineral modal data are consistent.

The CE3-0008 soil may have a greater abundance of material contributed from deeper levels of the nearby Zi Wei crater (Fig. 1e), with a composition similar to the nearby light-toned rocks (Supplementary Figs 2 and 3) such as the 'Outer Fence' boulder. From an image-based mineral modal estimation of Outer Fence, we infer ~20 vol.% in plagioclase phenocrysts (Supplementary Fig. 3, Supplementary Table 5, Supplementary Note 2), which is in general agreement with our estimations

(~33 vol.%) of nearby CE3-0008 soil. Moreover, the regolith reflectance differences observed by the Lunar Reconnaissance Orbiter Camera Narrow Angle Cameras (LROC NAC) are rather limited around the landing site (Supplementary Fig. 4, Supplementary Note 3). The Al_2O_3 content of CE-3 landing site based on experience of lunar samples is most likely in the range of 7–10 wt.% (Supplementary Fig. 5, Supplementary Note 4). The reflectance of the nearby rocks at the CE-3 landing site may result from texture-related human-eye brightness exaggeration (that is, some workers[36] initially interpreted the rock as an aluminous basalt with a plagioclase content possibly exceeding 40 vol.% (see Supplementary Notes 2–4 for additional discussion).

Correspondence between the landing site and remote sensing. Lunar Prospector (LP) Gamma-Ray Spectrometer (GRS) results suggest that soils developed on north-central Imbrium mare basalts, including the CE-3 landing site, have high FeO (>20 wt.%, half-degree per pixel binning) and TiO_2 (5.20 wt.%, 2° per pixel binning)[6,37]. Clementine data for areas near the CE-3 site indicate ~19 wt.% FeO and 5–7 wt.% TiO_2. The LP-GRS TiO_2 data in this part of Imbrium are variable, however. Considering both the LP-GRS data and the Clementine ultraviolet–visible data, the values are broadly similar to those of some widespread western Procellarum mare regions (for example, centred ~18° N and 303° E) where the surface is spectrally similar and where LP-GRS TiO_2 values are similar (4–6 wt.%). TiO_2 is normally used as the primary compositional criterion to classify remote sensing data for lunar basalts and indicates that many of the lunar basalts are actually intermediate in TiO_2 content (for example, 4.5–7.5 wt.%)[38]. The CE-3 composition falls in this range, thus we regard them as intermediate Ti basalt (similar to Neal et al.[36]). The

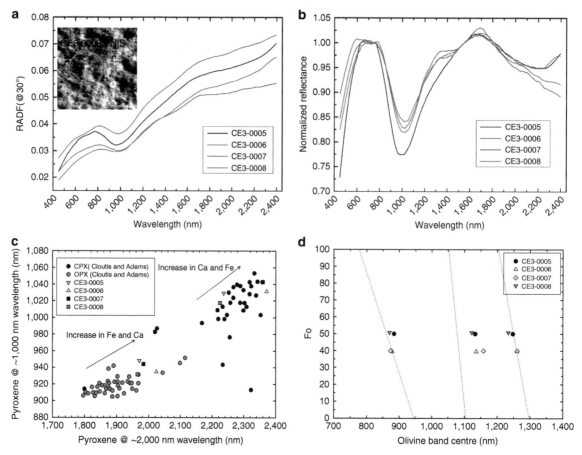

Figure 3 | Visible-NIR spectral properties and mineral chemistry of Chang'e-3 soils from VNIS. (**a**) Combined VNIS spectra (450–2,400 nm) from sites 0005, 0006, 0007 and 0008. The inset image is from site CE3-0006 of the VNIS (450–950 nm) image mode at 750 nm. The dashed circle indicates the region measured by the VNIS-point spectral mode (900–2,400 nm). (**b**) VNIS spectra after continuum removal. (**c**) Pyroxene VNIS peak positions of the CE-3 soils overlain on experimental results from Adams[27] and Cloutis and Gaffey[28]. (**d**) Fo values of olivine in four CE-3 soils derived from VNIS spectra, overlain on calibration lines (Sunshine and Pieters[20]).

intermediate Ti content of CE-3 provides an important calibration point for the TiO_2 estimation by spectral data (for example, Clementine UVVIS)[38,39], which has been problematic (for example, Gillis et al.[40] suggested about 50% uncertainty in the Imbrium basalts). The CE-3 TiO_2 data provide ground truth for further revision or evaluation of TiO_2 estimation methods for remote sensing studies. The high CaO content of the CE-3 soils (compared with other mare basalts, Fig. 2d) and the abundance of HCP inferred from the visible-NIR spectra support a high Ca content in the area of this flow unit. This result is consistent with the LP-GRS results (CaO of ~11.1 wt.%, same region in north-central Imbrium at 2° per pixel[37]) and Clementine UVVIS mineral mapping (rich in clinopyroxene and olivine, but poor in orthopyroxene)[2,41]. LRO Diviner data[42] show a Christiansen feature value near 8.55 µm, indicating a low SiO_2 content in this region, which qualitatively agrees with the low SiO_2 (41.2 wt.% average, Table 1) of the CE-3 APXS measurements. Moreover, the high olivine content and Ti enrichment in the late-stage volcanism of the Procellarum and Imbrium regions have been interpreted by many authors[21–23,34,43,44] according to spectral data from telescopic observations, the UVVIS camera on Clementine, and the Moon Mineralogy Mapper (M^3) on Chandrayaan-1, and these observations are supported in the Imbrium region by CE-3 results.

Discussion

Given a basalt that is rich in TiO_2 (5%) yet also rich in olivine (>10%), we consider its possible origin. The composition of CE-3 basalt, inferred from APXS analyses of immature soils at the site derived from the uppermost young basalt flow, is FeO-rich, with an Mg/(Mg + Fe) value of ~0.4. This composition suggests derivation from late-stage magma-ocean cumulates that crystallized after ilmenite saturation and thus were rich in ferropyroxene and ilmenite, possibly hybridized with an intermediate to late-stage olivine-orthopyroxene cumulate. Such an origin may be common among late-stage basalts in the Procellarum-Imbrium region (Procellarum KREEP Terrane or 'PKT')[45,46], producing olivine-bearing, ilmenite-rich ferrobasalts such as those of the flow sampled by CE-3 in north-central Imbrium as well as the extensive Ti-rich, olivine-bearing basalts of Western Procellarum[21,34]. Extended magmatic activity in the PKT may have been driven by radioactive decay and heating of late-stage, incompatible-element-rich cumulates that were also rich in Ti. Olivine enrichment would not be expected with such cumulates because lunar magma ocean (LMO) residual melts at the stage of ilmenite saturation (>95% LMO crystallization) would most likely be saturated with pigeonite and augite[47,48], thus requiring localized mixing or hybridization with an olivine-rich cumulate. In that case, the olivine-rich cumulate would, itself, be fairly iron-rich, for example, as might have been produced from the LMO after some 75% solidification. These would be late-stage LMO cumulates, likely enriched in incompatible trace elements (ITEs), including U and Th, consistent with prolonged mare basaltic activity in this part of the Moon[49]. An origin of such hybridized magmas, as formed by partial melting of upper mantle cumulates, is similar to the origin inferred by Snyder et al.[47] for high-Ti basalts.

Table 2 | MGM results of the four VNIS spectra.

Spectrum	CE3-0005			CE3-0006		
Mineral	Centre	FWHM	Strength	Centre	FWHM	Strength
Olivine	884	138	− 0.103	878	135	− 0.026
	1,134	139	− 0.076	1,136	122	− 0.073
	1,248	291	− 0.099	1,260	237	− 0.059
HCP‡	1,029	144	− 0.196	1,031	145	− 0.179
	2,236	337	− 0.070	2,371	404	− 0.099
LCP‡	948	110	− 0.106	935	122	− 0.091
	1,971	240	− 0.035	2,024	373	− 0.051
HCP/LCP* (1 μm ratio)	1.8 (LCP/(LCP + HCP) ≈ 0.35)			2.0 (LCP/(LCP + HCP) ≈ 0.34)		
HCP/LCP* (2 μm ratio)	2.0 (LCP/(LCP + HCP) ≈ 0.33)			1.9 (LCP/(LCP + HCP) ≈ 0.34)		
HCP/OL† (1 μm/1.25 μm)	2.0 (OL/(OL + HCP) ≈ 0.34)			3.0 (OL/(OL + HCP) ≈ 0.25)		

Spectrum	CE3-0007			CE3-0008		
Mineral	Centre	FWHM	Strength	Centre	FWHM	Strength
Olivine	874	148	− 0.038	870	136	− 0.037
	1,158	121	− 0.045	1,123	146	− 0.068
	1,261	251	− 0.053	1,234	268	− 0.063
HCP‡	1,043	166	− 0.175	1,017	142	− 0.144
	2,359	429	− 0.145	2,224	395	− 0.069
LCP‡	944	130	− 0.076	932	111	− 0.060
	1,983	309	− 0.065	1,965	271	− 0.030
HCP/LCP* (1 μm ratio)	2.3 (LCP/(LCP + HCP) ≈ 0.30)			2.4 (LCP/(LCP + HCP) ≈ 0.29)		
HCP/LCP* (2 μm ratio)	2.2(LCP/(LCP + HCP) ≈ 0.31)			2.3 (LCP/(LCP + HCP) ≈ 0.30)		
HCP/OL† (1 μm/1.25 μm)	3.3 (OL/(OL + HCP) ≈ 0.23)			2.3 (OL/(OL + HCP) ≈ 0.30)		

*The HCP/LCP is the volume ratio of HCP to LCP, determined by the 1- and 2-μm band-strength ratio for the two phases. The normalized band-strength ratios, LCP/(HCP + LCP), indicating the fractions of LCP in the mixtures, are also given for references in brackets. The consistencies in LCP/(LCP + HCP)s for the 1- and 2-μm bands are important in showing that olivine and other mineral absorptions are not skewing the MGM results[32].
†The HCP/OL is the volume ratio of HCP to olivine, determined by the peak strength ratio for the two phases (1 μm peak of HCP versus ~1.25 μm peak of olivine). The normalized band-strength ratios, OL/(HCP + OL), are also indicated in brackets.
‡Spectrally, HCP has the maximum absorption of the 1-μm feature located beyond 980 nm and the 2-μm feature, beyond 2,200 nm, whereas LCP has these features located at <980 nm and 2,200 nm, respectively.

Table 3 | Mineralogy of Chang'e-3 soils derived from APXS data using mixture modelling of the chemical composition.

APXS Mixing components	CE3-0006_1	CE3-0006_2	CE3-0006_3	CE3-0008	Mean_0006‡	Mean_all‡
Olivine	11.9	9.2	10.5	23.4	9.8	14.4
Augite	36.4	37.1	36.2	31.6	36.6	35.0
Pigeonite	16.9	18.4	17.9	13.2	18.2	16.5
Plagioclase	27.7	27.4	27.8	25.8	27.6	27.0
Ilmenite	8.0	8.3	8.5	7.7	8.4	8.2
*Apatite**	0.2	0.2	0.2	0.2	0.2	0.2
*Cr-spinel**	− 0.1	0.1	0.0	-0.2	0.0	0.0
Sum	101.2	100.7	101.1	101.8	100.9	101.2
Olivine_chem	Fo38.2	Fo34.7	Fo36.1	Fo49.6	Fo35.4	Fo40.1
Augite_chem	En30.1Wo33.7Fs36.2	En24.8Wo34.9Fs40.6	En30.5Wo33.7Fs35.8	En39.5Wo32.2Fs28.4	En27.7Wo34.3Fs38.2	En31.6Wo33.6Fs34.9
Pigeonite_chem	En35.7Wo20.1Fs44.2	En28.1Wo21.0Fs50.9	En29.6Wo20.7Fs49.7	En35.7Wo20.1Fs44.2	En28.9Wo20.9Fs50.3	En31.1Wo20.6Fs48.3
χ^2/ν†	13.8	12.9	12.9	19.6	12.9	15.1

Mineral proportion values in wt.%.
*Minerals shown in italics (apatite and Cr-spinel) are based on assumed input values for P_2O_5 and Cr_2O_3 in similar mare basalts (Papike et al.[17]) to more accurately model Ca and Fe. Model mineral component compositions are given in Supplementary Table 7.
†χ^2/ν is a measure of the goodness of fit of the mixing model and is the error-weighted sum of squares of differences between model and actual compositions divided by the number of chemical parameters (oxides) minus the number of mixing components (minerals)[53].
‡Mean_0006 is the average value of CE3-0006_2 and CE3-0006_3. Mean_all is the average value of CE3-0006_2 and CE3-0006_3 and CE3-0008.

In conclusion, from a correlated analysis of the regolith derived from rocks at the CE-3 landing site, freshly excavated by Zi Wei crater, we recognize a new type of lunar basalt with a distinctive mineral assemblage compared with the samples from Apollo and Luna, and the lunar meteorites. The chemical and mineralogical information of the CE-3 landing site provides new ground truth for some of the youngest volcanism on the Moon.

Methods

Instruments and data descriptions. The APXS is designed to conduct *in situ* elemental measurements of lunar regolith, using ^{55}Fe and ^{109}Cd as the excitation sources. The APXS is installed on the arm of the Yutu rover with an effective detection area of ~50 mm in diameter[13]. The VNIS employs the Acousto-Optic Tunable Filter (AOTF) technique to provide hyperspectral images in the 450–950-nm region and point spectral measurements in the 900–2,400-nm region[14–16]. The VNIS is installed on the front of the rover at a height of 0.69 m, observing the lunar surface at a fixed 45° view angle. The targeting area of the VNIS imager is ~16 × 21 cm. The targeting area of the VNIS-point spectrometer is a circle of ~7 cm diameter, inside of the targeting area (Fig. 3a) of the VNIS imager. The distance between the sampling areas of the APXS and VNIS at one location is within 1 m. The detailed descriptions of these two instruments and their calibration procedures can be found in refs 12–16.

This study is based on level 2C (APXS) and 2B (VNIS) data released by the Ground and Research Application System (GRAS) of the Chang'e-3 Project. The APXS conducted two calibration target measurements and four sets of lunar soil measurements at two locations (CE3-0006 and -0008). Measurement 0006_1 was a test measurement at ~5 cm distance from the surface of the regolith, 0006_2 was at the same location as 0006_1 but at a distance of ~ 2.5 cm above the surface and 0006_3 was at a lateral distance of about 10 cm from 0006_2. The VNIS acquired four hyperspectral images (450–950 nm) and four point NIR spectral measurements (900–2,400 nm) at four locations (CE3-0005, -0006, -0007 and -0008). The locations of the APXS and VNIS measurements are shown in Fig. 1d. The APXS data at CE3-0006_1-3 and CE3-0008 correspond to the VNIS spectra from CE3-0006 and CE3-0008.

APXS data processing and analysis. The APXS has a calibration target with known chemical composition (Fig. 2a). We employed the peak-area ratio of measured samples and calibration target (proportional to the elemental ratios of corresponding elements) to derive the chemical compositions of the CE-3 soils. Our APXS data processing began with level 2C data, which has undergone energy, dead-time and temperature corrections[13]. The four raw spectra were accumulated for 2657 s, 2778 s, 2050 s and 3627 s. We first normalized the spectral counts of all four APXS spectra to 10^3 seconds (Fig. 2a). The background was then removed from raw spectra to eliminate the effects of multiple scattering. The peak areas of individual elements were derived by spectral curve fitting using GRAMS software (Galactic Industries Corporation, NH, USA). A Gaussian function was used to fit the peak shape. Multiple iterations in curve fitting were conducted until convergence was reached. The resulting peak area for each element was then divided by that from the calibration target, and multiplied by the known concentration of that element in the calibration target, to obtain the nominal concentrations of elements. The concentrations of seven major elements (Mg, Al, Si, K, Ca, Ti and Fe) as oxides were then normalized to 99% to obtain the final concentration of each of these elements in a CE-3 surface sample, with the remaining 1% set for elements not measured but known to be present, including Cr_2O_3, MnO, Na_2O and P_2O_5, to 0.3, 0.3, 0.3 and 0.1 wt.%, respectively (Table 1). This method has the advantage of normalization of different observation geometries and instrument effects (for example, the CE3-0006_1 test measurement also shows a relatively reasonable result compared with CE3-0006_2, although it was taken at a distance of ~ 2.5 cm higher than the other three measurements). Neal et al.[36] also reported derived chemical compositions of the CE-3 soils, but using instead a Fundamental Parameters Method. Their results are in general agreement with our derived composition, but differ in detail, especially for Al. A detailed analysis of the differences is in the Discussion section of this paper.

VNIS data processing and analysis. Our VNIS spectral data processing began with level 2B data, which has undergone dark current, flat-field, temperature corrections and radiometric and geometric calibration[15]. For each of the four locations, we averaged ~ 9182 pixels spectra (450–950 nm) of the sampling area from VNIS in the image mode, then connected the averaged spectrum with a point spectrum (900–2400 nm) measured from a circular area within the imaged area (Fig. 3a). To correct the step artifacts between the two spectral ranges due to the response differences of the two detectors, a factor obtained from 900 nm of the point spectrum was used to multiply the averaged spectrum in the 450–900 nm region. We derived the single scattering albedo of the combined spectra, which is independent of the illumination geometry of the visible-NIR measurements. Based on the Hapke radiative transfer model[50,51], the radiance factor was converted at the standard illumination geometry ($i = 30°$, $e = 0°$ and $\alpha = 30°$) to facilitate comparison. The spectra were then smoothed using the Savitzky–Golay smoothing method[52], which is a spectral smoothing algorithm that essentially performs a polynomial regression to the data points in a moving window (Fig. 3b). The spectral background (continuum) was removed using a straight line that connects the spectral points at 750 nm and 1700 nm. Mineral modes of four CE-3 soils were extracted by spectral deconvolution using the MGM (refs 19,20,24–26,32; Supplementary Figs 6 and 7, Supplementary Table 6, Supplementary Note 5). The software was downloaded from the Reflectance Experiment Laboratory (http://www.planetary.brown.edu/mgm/).

References

1. Heiken, G. H., Vaniman, D. T. & French, B. M. (eds). *Lunar Sourcebook: A User's Guide to the Moon* (Cambridge University Press, 1991).
2. Lucey, P. G. et al. in *New Views of the Moon* Vol. 60 (eds Jolliff, B. L., Wieczorek, M. A., Shearer, C. K. & Neal, C. R.) 83–219 (Mineralogical Society of America, 2006).
3. Ip, W. H., Yan, J., Li, C. L. & Ouyang, Z. Y. Preface: the Chang'e-3 lander and rover mission to the Moon. *Res. Astron. Astrophys.* **14,** 1511 (2014).
4. Li, C. L. et al. Analysis of the geomorphology surrounding the Chang'e-3 landing site. *Res. Astron. Astrophys.* **14,** 1514–1529 (2014).
5. Xiao, L. China's touch on the Moon. *Nat. Geosci.* **7,** 391–392 (2014).
6. Xiao, L. et al. A young multilayered terrane of the northern Mare Imbrium revealed by Chang'E-3 mission. *Science* **347,** 1226–1229 (2015).
7. Hiesinger, H., Jaumann, R., Neukum, G. & Head, J. W. Ages of mare basalts on the lunar nearside. *J. Geophys. Res.* **105,** 29239–29275 (2000).
8. Schaber, G. G. Lava flows in Mare Imbrium: Geologic evaluation from Apollo orbital photography. *Proc. Lunar Planet Sci. Conf.* **4,** 73–92 (1973).
9. Bugiolacchi, R. & Guest, J. E. Compositional and temporal investigation of exposed lunar basalts in the Mare Imbrium region. *Icarus* **197,** 1–18 (2008).
10. Zhao, J. et al. Geologic characteristics of the Chang'E-3 exploration region. *Sci. China-Phys. Mech. Astron.* **57,** 569–576 (2014).
11. Hiesinger, H., Head, J. W., Wolf, U., Jaumann, R. & Neukum, G. Lunar mare basalt flow units: thicknesses determined from crater size-frequency distributions. *Geophys. Res. Lett.* **29,** 89 (2002).
12. Peng, W. X. et al. Active particle-induced X-ray Spectrometer for CHANG'E-3 YuTu Rover Mission and its first results. In *45th Lunar and Planetary Science Conference.* Abstract no. 1699 (2014).
13. Fu, X. H. et al. Data processing for the Active Particle-induced X-ray Spectrometer and initial scientific results from Chang'e-3 mission. *Res. Astron. Astrophys.* **14,** 1595–1606 (2014).
14. He, Z. P. et al. Operating principles and detection characteristics of the Visible and Near-Infrared Imaging Spectrometer in the Chang'e-3. *Res. Astron. Astrophys.* **14,** 1567–1577 (2014).
15. Liu, B. et al. Data processing and preliminary results of the Chang'e-3 VIS/NIR Imaging Spectrometer in-situ analysis. *Res. Astron. Astrophys.* **14,** 1578–1594 (2014).
16. Liu, B. et al. Reflectance conversion methods for the VIS/NIR imaging spectrometer aboard the Chang'E-3 lunar rover: based on ground validation experiment data. *Res. Astron. Astrophys.* **13,** 862–874 (2013).
17. Papike, J. J., Ryder, G. & Shearer, C. K. in *Planetary Materials* (ed. Papike, J. J.) 1–234 (Mineralogical Society of America, 1998).
18. Burns, R. G. *Mineralogical Applications of Crystal Field Theory,* 2nd ed. (Cambridge University Press, 1993).
19. Sunshine, J. M. & Pieters, C. M. Estimating modal abundances from the spectra of natural and laboratory pyroxene mixtures using the modified Gaussian model. *J. Geophys. Res.* **98,** 9075–9087 (1993).
20. Sunshine, J. M. & Pieters, C. M. Determining the composition of olivine from reflectance spectroscopy. *J. Geophys. Res.* **103,** 13675–13688 (1998).
21. Staid, M. I. et al. The mineralogy of late stage lunar volcanism as observed by the Moon Mineralogy Mapper on Chandrayaan-1. *J. Geophys. Res.* **116,** E00G10 (2011).
22. Besse, S. et al. Compositional variability of the Marius Hills volcanic complex from the Moon Mineralogy Mapper (M^3). *J. Geophys. Res.* **116,** E00G13 (2011).
23. Varatharajan, I., Srivastava, N. & Murty, S. V. Mineralogy of young lunar mare basalts: assessment of temporal and spatial heterogeneity using M^3 data from Chandrayaan-1. *Icarus* **236,** 56–71 (2014).
24. Clénet, H. et al. A new systematic approach using the Modified Gaussian Model: Insight for the characterization of chemical composition of olivines, pyroxenes and olivine–pyroxene mixtures. *Icarus* **213,** 404–422 (2011).
25. Isaacson, P. J. & Pieters, C. M. Deconvolution of lunar olivine reflectance spectra: Implications for remote compositional assessment. *Icarus* **210,** 8–13 (2010).
26. Isaacson, P. J. et al. Remote compositional analysis of lunar olivine-rich lithologies with Moon Mineralogy Mapper (M^3) spectra. *J. Geophys. Res.* **116,** E00G11 (2011).
27. Adams, J. B. Visible and near-infrared diffuse reflectance spectra of pyroxenes as applied to remote sensing of solid objects in the solar system. *J. Geophys. Res.* **79,** 4829–4836 (1974).
28. Cloutis, E. A. & Gaffey, M. J. Pyroxene spectroscopy revisited: Spectral-compositional correlations and relationship to geothermometry. *J. Geophys. Res.* **96,** 22809–22826 (1991).
29. Klima, R. L., Pieters, C. M. & Dyar, M. D. Spectroscopy of synthetic Mg-Fe pyroxenes I: spin-allowed and spin-forbidden crystal field bands in the visible and near-infrared. *Meteorit. Planet. Sci.* **42,** 235–253 (2007).
30. Klima, R. L., Dyar, M. D. & Pieters, C. M. Near-infrared spectra of clinopyroxenes: effects of calcium content and crystal structure. *Meteorit. Planet. Sci.* **46,** 379–395 (2011).
31. Sunshine, J. M. et al. High-calcium pyroxene as an indicator of igneous differentiation in asteroids and meteorites. *Meteorit. Planet. Sci.* **39,** 1343–1357 (2004).
32. Kanner, L. C., Mustard, J. F. & Gendrin, A. Assessing the limits of the Modified Gaussian Model for remote spectroscopic studies of pyroxenes on Mars. *Icarus* **187,** 442–456 (2007).
33. Cloutis, E. A., Gaffey, M. J., Jackowski, T. L. & Reed, K. L. Calibrations of phase abundance, composition, and particle size distribution for olivine-orthopyroxene mixtures from reflectance spectra. *J. Geophys. Res.* **91,** 641–11,653 (1986).
34. Staid, M. I. & Pieters, C. M. Mineralogy of the last lunar basalts: results from Clementine. *J. Geophys. Res.* **106,** 27887–27900 (2001).
35. Basaltic Volcanism Study Project. *Basaltic Volcanism on the Terrestrial Planets* (Pergamon Press, New York, 1981).
36. Neal, C. R., Wu, Y. Z., Cui, X. Z., Peng, W. X. & Ping, J. S. Regolith at the Chang'e-3 Landing Site: a new type of Mare Basalt Composition. In *46th Lunar and Planetary Science Conference.* Abstract no. 1641 (2015).
37. Prettyman, T. H. et al. Elemental composition of the lunar surface: analysis of gamma ray spectroscopy data from Lunar Prospector. *J. Geophys. Res.* **111,** E12007 (2006).

38. Giguere, T. A., Taylor, G. J., Hawke, B. & Lucey, P. G. The titanium contents of lunar mare basalts. *Meteorit. Planet. Sci.* **35,** 193–200 (2000).

39. Lucey, P. G., Blewett, D. T. & Jolliff, B. L. Lunar iron and titanium abundance algorithms based on final processing of Clementine ultraviolet-visible images. *J. Geophys. Res.* **105,** 20297–20305 (2000).

40. Gillis-Davis, J. J., Lucey, P. G. & Hawke, B. R. Testing the relation between UV–VIS color and TiO_2 content of the lunar maria. *Geochim. Cosmochim. Acta* **70,** 6079–6102 (2006).

41. Lucey, P. G. Mineral maps of the Moon. *Geophys. Res. Lett.* **31,** L08701 (2004).

42. Greenhagen, B. T. *et al.* Global silicate mineralogy of the Moon from the Diviner Lunar Radiometer. *Science* **329,** 1507–1509 (2010).

43. Pieters, C. M. *et al.* Late high-titanium basalts of the western maria: geology of the Flamsteed region of Oceanus Procellarum. *J. Geophys. Res.* **85,** 3913–3938 (1980).

44. Thiessen, F., Besse, S., Staid, M. I. & Hiesinger, H. Mapping lunar mare basalt units in mare Imbrium as observed with the Moon Mineralogy Mapper (M^3). *Planet. Space Sci.* **104,** 244–252 (2014).

45. Jolliff, B. L., Gillis, J. J., Haskin, L. A., Korotev, R. L. & Wieczorek, M. A. Major lunar crustal terranes: surface expressions and crust-mantle origins. *J. Geophys. Res.* **105,** 4197–4216 (2000).

46. Haskin, L. A., Gillis, J. J., Korotev, R. L. & Jolliff, B. L. The materials of the lunar Procellarum KREEP Terrane: a synthesis of data from geomorphological mapping, remote sensing, and sample analyses. *J. Geophys. Res.* **105,** 20403–20415 (2000).

47. Snyder, G. A., Taylor, L. A. & Neal, C. R. A chemical model for generating the sources of mare basalts: combined equilibrium and fractional crystallization of the lunar magmasphere. *Geochim. Cosmochim. Acta* **56,** 3809–3823 (1992).

48. Elardo, S. M. *et al.* The origin of young mare basalts inferred from lunar meteorites Northwest Africa 4734, 032, and LaPaz Icefield 02205. *Meteorit. Planet. Sci.* **49,** 261–291 (2014).

49. Wieczorek, M. A. & Phillips, R. J. The 'Procellarum KREEP Terrane': Implications for mare volcanism and lunar evolution. *J. Geophys. Res.* **105,** 20417–20430 (2000).

50. Hapke, B. *Theory of Reflectance and Emittance Spectroscopy* (Cambridge University Press, 2005).

51. Li, S. & Li, L. Radiative transfer modeling for quantifying lunar surface minerals, particle size, and submicroscopic metallic Fe. *J. Geophys. Res.* **116,** E09001 (2011).

52. Savitzky, A. & Golay, M. J. Smoothing and differentiation of data by simplified least squares procedures. *Anal. Chem.* **36,** 1627–1639 (1964).

53. Korotev, R. L., Haskin, L. A. & Jolliff, B. L. A simulated geochemical rover mission to the Taurus Littrow valley of the Moon. *J. Geophys. Res.* **100,** 14403–14420 (1995).

Acknowledgements

This research was partly funded by the National Natural Science Foundation of China (U1231103, 41473065, 41373068 and 41490634), National Science and Technology Infrastructure Work Projects (2015FY210500), Natural Science Foundation of Shandong Province (JQ201511 and ZR2015DQ001) and Key Research Program of the Chinese Academy of Sciences (KGZD-EW-603). Support from the Department of Earth & Planetary Sciences and the McDonnell Center for the Space Sciences at Washington University in St Louis for B.L.J., A.W. and Z.L. is greatly appreciated. We thank Ryan Clegg-Watkins for assistance in processing the LROC NAC images, and Randy Korotev, Carle Pieters, Bin Liu and Xiaohui Fu for helpful discussions. The authors thank Mariek Schmidt and two anonymous reviewers for constructive comments, which led to significant improvements in the manuscript.

Author contributions

Z.L. wrote the manuscript and performed calculations. B.L.J. contributed to the CIPW and mixing modelling as well as science interpretations. A.W. assisted with the APXS and VNIS data processing and interpretations. C.L. and J.Z.L. contributed to the geological interpretations of the data. J.Z., B.L., L.S and J.C helped to process the APXS and VNIS data. L.X., J.J.L. and X.R. helped to produce the camera data and geological analysis. W.P., H.W. and X.C. are team members of the APXS instrument and helped with the APXS data preprocessing. Z.H. and J.W. are team members of the VNIS instrument and helped with VNIS data preprocessing. We thank the team of the CE-3 Project for their successful work, especially the GRAS of Lunar Exploration, for their valuable and efficient assistance with providing the data and data calibration.

Additional information

22

Upward electrical discharges observed above Tropical Depression Dorian

Ningyu Liu[1], Nicholas Spiva[1,‡], Joseph R. Dwyer[1,†], Hamid K. Rassoul[1], Dwayne Free[2] & Steven A. Cummer[3]

Observation of upward electrical discharges from thunderstorms has been sporadically reported in the scientific literature. According to their terminal altitudes, they are classified as starters (20-30 km), jets (40-50 km) and gigantic jets (70-90 km). They not only have a significant impact on the occupied atmospheric volumes but also electrically couple different atmospheric regions. However, as they are rare and unpredictable, our knowledge of them has been built on observations that typically record only one type of such discharges. Here we report a close-distance observation of seven upward discharges including one starter, two jets and four gigantic jets above Tropical Depression Dorian. Our optical and electromagnetic data indicate that all events are of negative polarity, suggesting they are initiated in the same thundercloud charge region. The data also indicate that the lightning-like discharge channel can extend above thunderclouds by about 30 km, but the discharge does not emit low-frequency electromagnetic radiation as normal lightning.

[1] Department of Physics and Space Sciences, Florida Institute of Technology, 150 West University Boulevard, Melbourne, Florida 32901, USA. [2] Space Coast Intelligent Solutions, Melbourne, Florida 32934, USA. [3] Department of Electrical and Computer Engineering, Duke University, Durham, North Carolina 27708, USA. † Present address: Department of Physics, University of New Hampshire, Durham, New Hampshire 03824, USA. ‡ Deceased. Correspondence and requests for materials should be addressed to N.L. (email: nliu@fit.edu).

Upward electrical discharges from thunderstorms known as starters[1-5], jets[4-8] and gigantic jets[4,9-16] belong to a larger group of electrical discharge phenomena in the middle and upper atmosphere caused by thunderstorm/lightning activities, which are termed transient luminous events[17-19]. Past observations indicated that starters and jets appear as a cone of blue light shooting upward from thunderstorms with a dimmer fan near their tops[1,3-6], while gigantic jets display a tree-like structure and more complex dynamics[9,10,14], and they bridge thunderstorms and the ionosphere, allowing a rapid transfer of a large amount of charge between the lower and upper atmosphere[12,16]. Most of the starters and jets reported to date occurred above land storms[1,3-8], but gigantic jets predominately occur above tropical storms over oceans and coasts[20,21].

The upward electrical discharges can be produced by thunderstorms through two principal mechanisms[2,22]. A standard, simple model of the charge structure of thunderstorms consists of two cloud charge layers of opposite polarities centred at different cloud altitudes and a screening charge layer around the cloud top that has the same polarity as the lower cloud charge. The upward electrical discharges can be developed from electrical breakdown, beginning either between the two cloud charge layers or between the upper cloud charge and the screening charge, where electric field is typically strongest. If a proper charge imbalance condition is created by electrical or meteorological processes, the initiated upward electrical discharge can penetrate through the charge layer it is directed to, and escape from the cloud top[2,22]. As the directions of the electric field are opposite at those two regions, the resulting upward electrical discharges have different polarities. This theory has been verified by observations reported later, indicating that the upward discharges beginning between the upper cloud charge and the screening charge tend to develop into starters or jets[3], while those beginning between two cloud charge layers evolve into gigantic jets[12,16].

The underlying electrical discharge process driving the development of starters, jets and gigantic jets is known as leaders[2,23-29], the same as normal lightning. Leader discharges are responsible for electrically breaking down air to form a hot ($>$5,000 K), highly conductive channel, and their initiation and propagation mechanism is not well understood at present[30]. Metre-long leaders can be generated and studied in laboratory experiments. However, the kilometre-long leaders of natural electrical discharges possess significantly different characteristics, because the involved spatial and temporal scales are much larger and there are no well-defined counterparts of electrodes and discharge gaps as laboratory experiments. Observing various electromagnetic emissions from natural leaders using optical and radio instruments is the primary experimental means to study their discharge characteristics inside or outside thunderstorms[31-35]. From a theoretical perspective, the similarity laws of a particular electrical discharge or a particular stage of a discharge can be formulated and used if the same basic discharge processes dominate at different air pressures or densities[18,19,36]. Recently, theoretical studies have predicted that the leaders of the upward discharges propagate at a similar speed as lightning leaders, but require a significantly longer timescale to create a new section of the leader channel, which is found to be inversely proportional to the square of air density[26-29].

Compared with normal lightning that frequently occurs during thunderstorms, starters, jets and gigantic jets are rare. Only one recent study reported an observation of all three types of the upward electrical discharges above a single storm, but the polarities of the events could not be unambiguously determined, because the storm was far away ($\sim$400 km) and electromagnetic measurements of the discharges were unavailable[4]. Here we present a close-distance ($\sim$80 km) observation of one starter, two jets and four gigantic jets above Tropical Depression Dorian. Our optical images and electromagnetic data indicate all of them are driven by negative leaders, suggesting that they originate between the two cloud charge regions. Our data also indicate that the leader channel above the cloud is charged similarly to a lightning leader channel, but it does not radiate low-frequency electromagnetic radiation as the lightning leader at lower altitudes. In addition, the upward leader can transfer a large amount of charge to the middle and upper atmosphere, even if it never reaches the ionosphere.

Results

The parent storm and lightning activities. The seven upward electrical discharge events occurred above Tropical Depression Dorian over the Atlantic Ocean between 3:45 Coordinated Universal Time (UTC) and 4:12 UTC on 3 August 2013. Tropical Depression Dorian formed from the remnants of the Tropical Storm Dorian, which started as a strong tropical wave off the west African coast on 22 July and evolved into Tropical Storm Dorian on 24 July when it was located about 300 km west of Cape Verde Islands en route to the southeast coast of United States[37]. Three days later, the storm weakened into a tropical wave, and on 2 August, when Dorian almost reached the coast of southeastern Florida, its remnants regenerated into Tropical Depression Dorian. In the early morning of 3 August, an hour before the first event, the ASCAT measurements from EUMETSAT Metrop-B Satellite indicated that the average wind speed of Dorian was 55 km h^{-1} with a maximum of 65-67 km h^{-1} in a localized area. Meanwhile, ground radar data from Melbourne, Florida, intermittently indicated that the wind speed was as high as 83 km h^{-1} at 1.5-1.7 km altitude on the south side of the circulation.

The GOES satellite infrared images from 2:45 UTC to 4:45 UTC show that several isolated small convection cells existed initially and they rapidly intensified, expanded and merged together. The entire storm also expanded rapidly, with its west edge reaching the east coast of Florida around 4:30 UTC. There were two active convection cores, when the events occurred. The one at the northwest corner of the storm, which was also closer to the observation site, was the parent cell producing the upward discharges. Figure 1a shows the GOES infrared image at 4:01 UTC, on which are plotted the locations of the lightning events recorded by the National Lightning Detection Network (NLDN[38]) from 3:30 UTC to 4:30 UTC. The cloud top temperature of the coldest area of that cell was 190-200 K and the NLDN lightning events clustered around that region. The open red circles denote the locations of the upward discharge events that overlap with the dense area of the NLDN lightning. Their distances to the observation site vary from 75 to 79 km.

Figure 1b shows a time scatter plot of the peak currents of the NLDN lightning events in the rectangular area ($0.7° \times 0.7°$), covering the core of the parent cell, in Fig. 1a. There are a total of 266 NLDN events between 3:30 UTC and 4:30 UTC, including 12 positive cloud-to-ground (CG), 110 negative CG, 133 positive intracloud (IC) and 11 negative IC events. The polarity of a lightning flash is defined by the polarity of the charge that effectively moved downward. Most of the lightning events are $-$CGs or $+$ICs, indicating that the storm cell is normally electrified, that is, the main positive charge layer of the storm resides over its main negative charge layer[39]. The average NLDN lightning rate is $\sim$4.5 events per min and the maximum average rate over a 5-min interval is 18.6 events per min, both of them falling in the normal range of thunderstorm cells[40], (p. 25). As shown by the figure, the first event occurred in the early electrification stage of the cell, and 8 min later 5 events occurred in a 4-min interval. The last event occurred after the most

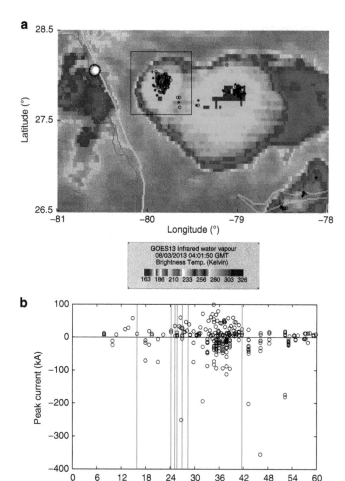

Figure 1 | The parent storm and lightning activities. (a) The GOES infrared image of Tropical Depression Dorian that produced seven upward electrical discharges on 3 August 2013, on which the locations of the NLDN lightning events and the upward discharges are superimposed. Dark symbols represent the locations of the lightning events: ' + '—positive CG lightning; ' O '—negative CG; ' * '—positive IC lightning; and ' □ '—negative IC. Red circles denote the locations of the upward discharges and the solid white dot represents the observation site. **(b)** A time scatter plot of the peak currents of the NLDN lightning events located within the black box in **a**. Black and blue circles represent ICs and CGs, respectively. The lightning activities are dominated by − CGs and + ICs, indicating that the parent cell was normally electrified. Vertical red lines show the occurrence times of the upward discharge events.

electrically active stage of the cell, about 13 min apart from the previous event.

A more detailed analysis indicates that NLDN lightning events occurred in very close temporal and spatial proximity to each event, except the sixth event (a gigantic jet). However, it should be noted that NLDN lightning detection efficiency is not perfect[38] and the video of the sixth event does show that lightning flashes occurred before and during this event. For the other six events, no NLDN events were found within a 20-s time period centred around each event, except a short time interval of 1–2 s containing the event. The fifth event (a gigantic jet) is the only one immediately before which CG activity was detected by NLDN. The sudden increase of the NLDN flash rate before each event is consistent with early studies of jets[7].

Video images of the events. The upward discharge events were recorded by a low-light-level Watec camera and an all-sky camera

installed on the campus of Florida Institute of Technology, and another all-sky camera about 10 km northwest. The recorded videos are available as Supplementary Movies 1–3. Figure 2 shows a few selected image fields (16.7 ms exposure time, Global Positioning System (GPS) time stamped) extracted from interlaced videos of 2–3 s for individual events recorded by the Watec camera. The seven events at their full extents are shown in Fig. 2a. The altitude labels to the right of each image vary from one event to another, because the distance of the jet to the camera is different. According to the videos, lightning flash(es) preceded every event and accompanied its development. Events 1 and 3 terminate at 51–55 km altitude, the tops of events 2, 5, 6 and 7 are outside the field of view of the camera, resulting in a > 77–82 km terminal altitude, and event 4 terminates at ~ 26 km. On the basis of their significantly different terminal altitudes and temporal dynamics (as shown by the videos), we classify event 4 as a starter, events 1 and 3 as jets and the rest of the events as gigantic jets.

The video images show that all the events have a tree-like structure. For the starter and jets, they vanished in 50–60 ms after they reached their full extents, but the gigantic jets lasted much longer after their final jump to the ionosphere (that is, the sudden establishment of a discharge channel between the upward discharge and the ionosphere). For the two jets, after their main branches reached 42 and 47 km altitude, respectively, several branches were generated sequentially or simultaneously near their tops. The gigantic jets 2 and 5 initially propagated upward similarly as the jets. When they reached 39 and 48 km altitudes, respectively, multiple branches were produced at their tops similar to the jets, and then in the next video field one of those branches (event 2) or a branch below the top (event 5) made the final jump. Both events were followed by an intense lightning flash, which seems to fuel the short bases of the upward discharges to emit extremely bright light. The final jump was made at a lower altitude of 35 km from the tops of the upward discharges for events 6 and 7. After the final jump, the temporal dynamics were very similar for all gigantic jets, except that no visible lightning flashes followed events 6 and 7, and re-brightening of the discharge volume occurred for event 7. Compared with previously reported positive jets and starters[3,6,8], the starter and jets have more branches and lack a diffuse fan top. This suggests that the upward leaders are of negative polarity, which is verified by the associated electromagnetic signatures (discussed below). The morphology and temporal dynamics of the gigantic jets are generally similar to the negative gigantic jets observed previously[9,10,14]. Figure 2b shows the detailed development of the starter, which lasted about 260 ms and had multiple branches connecting to a common, bright base (the multiple branches of the starter are more clearly shown by Supplementary Fig. 1 that presents a composite image of this event).

Figure 2c shows selected fields of event 1 that started around 3:45:51 UTC, following lightning flashes that began 150 ms earlier. The upward leader exited the cloud top at about 15.6 km altitude with a single main channel tilting from the vertical with an angle of 21°. For the next ~ 270 ms, the leader continued moving in that direction, while constantly spawning dimmer channels in a narrow cone of about 30°. Its vertical speed fluctuated between 4.5×10^4 and 1.2×10^5 m s^{-1} until its top reached 42 km altitude (fields 17 and 18), the uncertainty of which is ± 4 km given that the leader channel might be tilted towards or away from the camera with an angle of 21° as well. The leader then appears unable to continue its steady propagation and dimmer channels originated from its top simultaneously and sequentially, as shown in the fields from 19 to 25. In field 19, a short, hardly visible vertical channel extended upward from the leader tip; it disappeared in the next field and then a small

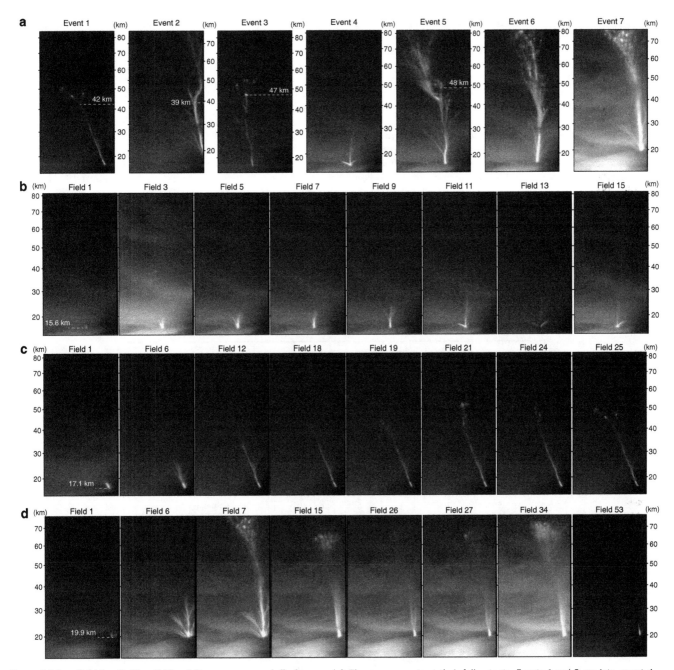

Figure 2 | Low-light-level video fields of the seven upward discharges. (**a**) The seven events at their full extents. Events 1 and 3 are jets, event 4 is a starter and the rest of the events are gigantic jets. Selected video fields of (**b**) the starter (event 4), (**c**) a jet (event 1) and (**d**) a gigantic jet (event 7).

tree-like structure with a relatively larger vertical and horizontal extent suddenly appeared in field 21, resulting in a speed of $6.6 \times 10^5\,\mathrm{m\,s^{-1}}$ if it was formed by a discharge wave started from the leader tip. After field 25, the luminosity of the entire leader channel decreased rapidly and completely vanished in four fields. Given the different spatial structures and temporal dynamics of the discharges at the tip of the leader after it reached 42 km altitude, it is reasonable to speculate that the subsequent discharge activities near the leader tip show the streamer zone[27–29] preceding the leader tip. If this is true, the vertical extent of the streamer zone is about 11 km for this particular leader tip at 42 km altitude. Following the same argument, the leader for event 3 reached an altitude as high as 47 km, as shown in Fig. 2a, which means that the leader channel extended a distance of more than 30 km above the thundercloud top.

Figure 2d shows the development of event 7, the most impulsive upward discharge event in our data set. The leader emerged from the cloud top with several distinct branches around 4:11:38 UTC. The centre branch had the highest top. Its vertical speed was initially $6.8 \times 10^4\,\mathrm{m\,s^{-1}}$ and then increased from 1.6×10^5 to $2.1 \times 10^5\,\mathrm{m\,s^{-1}}$. It reached 34.8 km altitude in field 6, and then jumped to >77.1 km altitude in the next video field, resulting in a speed of $>2.5 \times 10^6\,\mathrm{m\,s^{-1}}$. The speed and its variation with altitude are consistent with previously reported negative gigantic jets[9,14,27,29]. After the jump, relatively stationary bright beads and dimmer glows appeared at the top of the discharge. The luminosity of the top gradually decayed afterwards, while bead-like structures with short trails moved upward from about 50 km altitude along the pre-existing channels, as shown in field 15. The luminosity continued to decrease until field 26, when the top of the gigantic jet as well as the scattered light from cloud lightning activity started to re-brighten. The re-brightening reached its strongest stage in fields 34 and 35, which lasted 7 fields, and upward motion of the beads at the top as well as horizontal displacement of the entire discharge volume is visible. After the main body of the gigantic jet vanished, a short bright column base above the cloud, as shown in field 53, persisted for a while, and the entire duration of the discharge was as long as 1.2 s. To the best of our knowledge, this is the longest duration of the upward cloud discharges that have ever been reported.

Magnetic field measurements and source waveforms. The electromagnetic radiation from the upward discharges and accompanied cloud discharges was measured by a low frequency (LF) magnetic field detector at Florida Institute of Technology

and by a ULF (ultralow frequency) magnetic field sensor at Duke University[12,16]. Figure 3 shows the magnetic field measurements of a jet (event 1 in Fig. 2a,c), a gigantic jet (event 7 in Fig. 2a,d) and a starter (event 4 in Fig. 2a,b). For all seven events, strong LF pulses, as shown in Fig. 3a,c,e, started to appear about 0.2–1 s before the upward discharges emerged above the cloud. Such LF pulses are known to be produced by in-cloud discharges[12,16]. For some events, the LF pulses started with a large narrow bipolar pulse with a width of 10–20 μs, similar to the first pulse in Fig. 3c, which has recently been shown to be associated with the initiation of the in-cloud discharges that led to two gigantic jets[16]. Additional electromagnetic radiation data collected by Kennedy

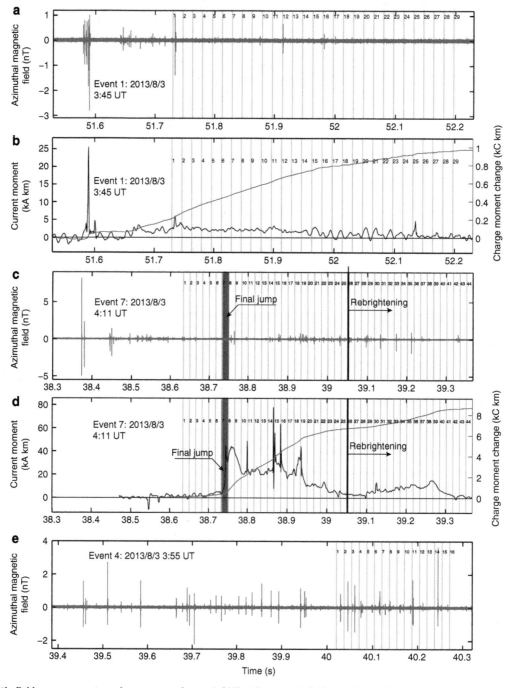

Figure 3 | Magnetic-field measurements and source waveforms. (**a,b**) The LF magnetic-field waveform and the source waveforms (current moment and charge moment change), respectively, for the jet; (**c,d**) the same waveforms for the gigantic jet; and (**e**) the LF waveform for the starter. The current moments and charge moment changes are derived from the ULF measurements. Each vertical strip bounded by two green lines corresponds to an video field, with its field number given between the two boundary lines. The grey strips in **c** and **d** correspond to the video field showing the final jump.

Space Center indicate that the onset altitudes of the in-cloud discharges resulting in all the upward discharge events varied between 12–14 km altitudes.

Figure 3a,c show that there are no persistent LF activities during the propagation of the upward leader above the cloud. This indicates that the negative leader above the cloud did not radiate strong-enough LF pulses to be detected by the LF sensor ~80 km away, and that the cloud discharges were not as active as the early initiating stages of the upward discharges. However, the discharge activity, as indicated by the LF pulses, associated with the starter was continual before and throughout the event, suggesting the starter occurred during the active stage of the cloud flash. Figure 3a also shows that there are no strong LF signatures associated with the sequence of the discharge events occurring at the negative leader tip from the video field 18 to 25. It is known that the propagation of negative leaders consists of discrete steps and LF pulses are produced by the stepping process of the negative leaders of IC and CG lightning. The absence of LF activities suggests that the stepping of negative leaders above thunderclouds occurs on a longer timescale than the sensitive range of the LF sensor, possibly resulting from a larger spatial scale of the discharge at higher altitudes as suggested by the scaling laws of electrical discharges in air[28,29,41]. Figure 3c also shows that several distinct LF pulses appear right after the final jump of the upward discharge, which are probably produced by the rejuvenation of the cloud discharges due to the established electrical connection between the thundercloud and the ionosphere[16].

Previous studies have demonstrated that gigantic jets can transfer a significant amount of charge from thunderclouds to the ionosphere[10,12,13,16]. All four gigantic jets observed carried strong-enough currents so that the associated current moment waveforms can be unambiguously determined from the ULF data. The resulting total charge moment change varies from 3.1 to 8.7 kC km and the total deposited charge in the middle and upper atmosphere varies from 48 to 134 C if a channel length of 65 km is assumed. The wide range of the total charge moment change agrees with an ensemble built from previous, separate observations[10,12,13,16]. The charge moment change can also be unambiguously extracted for a jet (event 1) to be 0.98 kC km, corresponding to 56 C charge transfer if the channel is assumed to be uniformly charged and have a length of 35 km, but not for the other jet (event 3) and the starter (event 4). Overall, the amount of charge moved upward from the thunderstorm by all the discharges in about 30 min is at least 383 C.

Figure 3b,d show the waveforms of the current moments and charge moment changes of event 1 and event 7. Figure 3b shows that the current carried by the upward leader of the jet begins to exceed the detectable level more than 100 ms before the first video field of the jet. The current moment stays at an approximately constant value of 1.5–2.5 kA km from 51.75 to 51.95 s during the upward propagation of the leader, which means that the current decreases as the leader propagates upward. Given the measured total charge moment change is 0.98 kC km for this event and there is only a single leader channel, the linear charge density of the channel can be estimated to be about 1.5 mC m^{-1}, assuming that the charge is uniformly distributed along the channel and the channel length is 35 km, which is consistent with the linear charge density of a lightning leader[40], (p. 123–126).

For the gigantic jet's waveforms shown in Fig. 3d, the initial current moment up to field 5 is relatively small and the resulting charge moment change increases slowly to about 0.4 kC km, comparable to the final value of event 1. During the video field showing the final jump, the current moment of the gigantic jet rapidly increases to about 40 kA km, consistent with previous work[11,12,16]. The current moment maintains at such a high level

for 30 ms, and then decreases to 20 kA km and stays there for the next 160 ms. About 65% (85 C) of the total amount of charge transferred between the thunderstorm and the ionosphere by this event occurs during this ~200 ms period. The re-brightening is accompanied by an increase in the current flowing in the discharge channel, resulting in a charge moment change of 1.8 kC km (21% of the total charge moment change of the event). The other gigantic jets (without re-brightening) have similar current moment and charge moment waveforms up to the moment of re-brightening, with the charge moment change before the final jump varying in the range of 0.3–1 kC km.

The negative polarity of the jet and the gigantic jet is unambiguously shown by the current moments and charge moment changes derived from the ULF magnetic field measurements. Although a reliable current moment waveform cannot be derived from the ULF data for the starter event, the conclusion of its negative polarity can be drawn based on the following reasons. As discussed above, its morphology is different from previously reported positive starters. Supplementary Fig. 2 shows that there are no signatures of downward negative leaders that would be expected to accompany a starter of positive polarity. The interpretation of negative polarity is consistent with fractal modelling results, as discussed in Supplementary Note. Finally, previous studies on the downward 'attempted leader' between thunderstorms and ground[42] indicated it is possible for a negative leader that is rooted in positive leaders in the main negative cloud charge region not to propagate far after escaping from thunderstorms.

Discussion

Event 1 in our data set is quite unique, because there is only a single leader discharge channel. The agreement between its linear charge density with typical values of lightning leaders provides an important new piece of evidence for the theory of jets and gigantic jets being escaped leaders from thunderstorms[2,22,24,25]. The two escaping mechanisms for the upward leaders[2,22] have been verified recently[3,12,16]. Our observation further indicates that starters, jets and gigantic jets can be developed from leaders initiated from the same region in the thundercloud charge structure, that is, between the two main thundercloud charge reservoirs. Below, we offer an explanation why an upward leader develops into a starter/jet or a gigantic jet through examining the characteristics of the leaders, resulting in events 1 and 7. From the leader theory, the electric potential difference between the leader tip and the ionosphere can be determined if its altitude and streamer zone size are known[24,25,28,29], assuming that the electric field in the streamer zone is the critical field for streamer propagation. If this field is assumed to be the critical field for negative streamer propagation, which is about two to three times larger than that field for positive streamers, the current derived from a simple leader model[30], (p. 62) with the known potential and speed is about three times larger than the value found from the measured current moment with an assumption of the lower end of the leader at 13 km altitude. In addition, for event 1 the leader reaches 42 km altitude and the streamer zone extends to 51 km. If the electric field at the streamer zone boundary is the critical field for negative streamers, the electric field of the leader will probably exceed the breakdown field at 75–85 km altitudes, because it decreases slower than the electrical breakdown threshold field as altitude increases. As a result, high-altitude electrical discharges known as sprites should have occurred, but they did not. We conclude that the electric field in the streamer zone is smaller, and we assume it is the critical field for positive streamers.

With this assumption, the leader tip potential and current of event 1 are estimated to be 10 MV and 35 A, respectively, when it reaches 42 km altitude. The current agrees reasonably well (about 40% larger) with the value derived from the current moment. The two quantities for the event 7 right before the final jump are 28 MV and 180 A, respectively. However, when the leaders just exit from the cloud, their potential and current could be significantly different from those values. According to the binary leader theory of lightning development[30], (p. 153), a leader acquires an average potential of the thundercloud volume occupied by itself, and as the leader develops its potential may undergo substantial changes. Assuming the lower end of the leader is at 13 km altitude, the leader current for event 1 is ~ 340 A and the derived potential is 100 MV, when the leader just exits from the thunderstorm. For event 7, they are 270 A and 70 MV, respectively. Surprisingly, the leader of the jet event initially has a larger current and potential than those of the gigantic jet. However, the current moment for the jet leader decreases as the leader propagates upward, while that for the gigantic jet increases. The decreasing current moment while the channel length is increasing means that the current and potential of the leader decrease. With this information, an explanation why the two leaders evolve into the jet or gigantic jet may be formulated as follows. For the jet, the supporting in-cloud positive leader probably stops propagating or extends very slowly; hence, it does not explore an extensive cloud charge region to compensate the decrease in the leader potential due to the extension of the upward negative leader in the low-potential region above the cloud. As a result, the overall potential of the leader decreases rapidly as the upward leader develops. On the other hand, the positive leader for the gigantic jet probably moves through an extensive cloud region and effectively slows down the decrease in the potential due to the extension of the upward leader, allowing it to make the final jump to the ionosphere. Therefore, the dynamic development of both the upward leader and accompanied in-cloud discharges is critical for the formation of starters, jets and gigantic jets. At present, the formation of the upward electrical discharges is best studied by fractal models, but the altitude dependence of the leader channel size is not taken into account by a typical fractal model. Consequently, an escaped upward leader simulated by the fractal model tends to reach high altitudes. An improvement to the fractal model would be introducing the altitude-dependent spatial scale of the leader in the model, so that the potential of the dynamic binary leader system can be more accurately calculated.

Methods

Video acquisition and analysis. Supplementary Movie 1 was recorded by a Watec 120N + camera and Supplementary Movies 2 and 3 were recorded by two Sandia Allsky cameras. All the cameras are coupled to separate triggering systems to record a few-second video when specified trigger criteria are met. The NTSC video from the Watec camera is time stamped by a GPS-synchronized video time inserter. However, a comparison of the timings of the NLDN lightning events and the cloud flashes visible in the videos of the events indicates that the GPS time stamped on the video probably delays by 33 ms (two video fields). This delay has been corrected in Fig. 3. The images shown in Fig. 2 are cropped video fields obtained by deinterlacing the Watec video with FFmpeg software. The raw video data in avi format recorded by the video acquisition system are converted to MP4 format to reduce the data size also by using FFmpeg. The locations and heights of the events were obtained by a combined analysis of the Watec images using information from the star field and NLDN lightning locations. The Allsky system uses a HB-710E Star Light B/W charge-coupled device camera with an auto iris, fisheye lens and the recorded video is stamped with the computer clock time synchronized to the network time.

Electromagnetic measurements and analysis. Electromagnetic signals radiated by lightning and other natural electrical discharges can be measured either locally or remotely, to extract the information about the source discharges. The horizontal magnetic field waveforms presented in Fig. 3a,c,e were recorded with an LF magnetic field detector, sensitive to the frequency range of 1–300 kHz, at the Florida Tech observation site. At a field site near Duke University, radio emissions in frequency band of 0.1–400 Hz (ULF) are measured with two pairs of magnetic

induction coils. The ULF system is ideal to measure slowly varying sources, for instance, the current flowing in a discharge channel between a cloud and either ground (lightning) or the ionosphere (gigantic jets); on the other hand, the LF detector with its higher sensitive frequency range can monitor more rapidly varying sources with a timescale down to a few microseconds, for example, cloud discharges. The current moment (the integral of the current along the entire channel) is extracted from the ULF magnetic field measurements[12]. The cumulative charge moment change can then be obtained by integrating the current moment over the discharge period.

References

1. Wescott, E. M. *et al.* Blue starters: brief upward discharges from an intense Arkansas thunderstorm. *Geophys. Res. Lett.* **23**, 2153–2156 (1996).
2. Krehbiel, P. R. *et al.* Upward electrical discharges from thunderstorms. *Nat. Geosci.* **1**, 233–237 (2008).
3. Edens, H. E. Photographic and lightning mapping observations of a blue starter over a New Mexico thunderstorm. *Geophys. Res. Lett.* **38**, L17804 (2011).
4. Chou, J. K. *et al.* Optical emissions and behaviors of the blue starters, blue jets, and gigantic jets observed in the Taiwan transient luminous event ground campaign. *J. Geophys. Res.* **116**, A07301 (2011).
5. Suzuki, T., Hayakawa, M., Hobara, Y. & Kusunoki, K. First detection of summer blue jets and starters over Northern Kanto area of Japan: lightning activity. *J. Geophys. Res.* **117**, A07307 (2012).
6. Wescott, E. M., Sentman, D., Osborne, D., Hampton, D. & Heavner, M. Preliminary results from the Sprites94 aircraft campaign: 2. Blue jets. *Geophys. Res. Lett.* **22**, 1209–1212 (1995).
7. Wescott, E. M., Sentman, D. D., Heavner, M. J., Hampton, D. L. & Vaughan, Jr. O. H. Blue jets: their relationship to lightning and very large hailfall, and their physical mechanisms for their production. *J. Atmos. Solar Terr. Phys.* **60**, 713–724 (1998).
8. Wescott, E. M. *et al.* New evidence for the brightness and ionization of blue starters and blue jets. *J. Geophys. Res.* **106**, 21549 21554 (2001).
9. Pasko, V. P., Stanley, M. A., Matthews, J. D., Inan, U. S. & Wood, T. G. Electrical discharge from a thundercloud top to the lower ionosphere. *Nature* **416**, 152–154 (2002).
10. Su, H. T. *et al.* Gigantic jets between a thundercloud and the ionosphere. *Nature* **423**, 974–976 (2003).
11. Kuo, C. L. *et al.* Discharge processes, electric field, and electron energy in ISUAL-recorded gigantic jets. *J. Geophys. Res.* **114**, A04314 (2009).
12. Cummer, S. A. *et al.* Quantification of the troposphere-to-ionosphere charge transfer in a gigantic jet. *Nat. Geosci.* **2**, 617–620 (2009).
13. van der Velde, O. A. *et al.* Multi-instrumental observations of a positive gigantic jet produced by a winter thunderstorm in Europe. *J. Geophys. Res.* **115**, D24301 (2010).
14. Soula, S. *et al.* Gigantic jets produced by an isolated tropical thunderstorm near Réunion Island. *J. Geophys. Res.* **116**, D19103 (2011).
15. Yang, J. & Feng, G. L. A gigantic jet event observed over a thunderstorm in mainland China. *Chin. Sci. Bull.* **57**, 4791–4800 (2012).
16. Lu, G. *et al.* Lightning development associated with two negative gigantic jets. *Geophys. Res. Lett.* **38**, L12801 (2011).
17. Pasko, V. P. Recent advances in theory of transient luminous events. *J. Geophys. Res.* **115**, A00E35 (2010).
18. Ebert, U. *et al.* Review of recent results on streamer discharges and discussion of their relevance for sprites and lightning. *J. Geophys. Res.* **115**, A00E43 (2010).
19. Liu, N. Y. In *The Lightning Flash* 2nd edn (ed. Cooray, V.) 725–786 (The Institution of Engineering and Technology, 2014).
20. Chen, A. B. *et al.* Global distributions and occurrence rates of transient luminous events. *J. Geophys. Res.* **113**, A08306 (2008).
21. van der Velde, O. A. *et al.* Analysis of the first gigantic jet recorded over continental North America. *J. Geophys. Res.* **112**, D20104 (2007).
22. Riousset, J. A., Pasko, V. P., Krehbiel, P. R., Rison, W. & Stanley, M. A. Modeling of thundercloud screening charges: Implications for blue and gigantic jets. *J. Geophys. Res.* **115**, A00E10 (2010).
23. Petrov, N. I. & Petrova, G. N. Physical mechanisms for the development of lightning discharges between a thundercloud and the ionosphere. *Tech. Phys.* **44**, 472–475 (1999).
24. Raizer, Y. P., Milikh, G. M. & Shneider, M. N. On the mechanism of blue jet formation and propagation. *Geophys. Res. Lett.* **33**, L23801 (2006).
25. Raizer, Y. P., Milikh, G. M. & Shneider, M. N. Leader streamers nature of blue jets. *J. Atmos. Solar Terr. Phys.* **69**, 925–938 (2007).
26. Riousset, J. A., Pasko, V. P. & Bourdon, A. Air-density-dependent model for analysis of air heating associated with streamers, leaders, and transient luminous events. *J. Geophys. Res.* **115**, A12321 (2010).
27. da Silva, C. L. & Pasko, V. P. Simulation of leader speeds at gigantic jet altitudes. *Geophys. Res. Lett.* **39**, L13805 (2012).
28. da Silva, C. L. & Pasko, V. P. Vertical structuring of gigantic jets. *Geophys. Res. Lett.* **40**, 3315–3319 (2013).

29. da Silva, C. L. & Pasko, V. P. Dynamics of streamer-to-leader transition at reduced air densities and its implications for propagation of lightning leaders and gigantic jets. *J. Geophys. Res.* **118**, 13561 13590 (2013).

30. Bazelyan, E. M. & Raizer, Y. P. *Lightning Physics and Lightning Protection* (IoP Publishing Ltd, 2000).

31. Rison, W., Thomas, R. J., Krehbiel, P. R., Hamlin, T. & Harlin, J. A GPS-based three-dimensional lightning mapping system: Initial observations in central New Mexico. *Geophys. Res. Lett.* **26**, 3573–3576 (1999).

32. Winn, W. P. *et al.* Lightning leader stepping, K changes, and other observations near an intracloud flash. *J. Geophys. Res.* **116**, D23115 (2011).

33. Hill, J. D., Uman, M. A. & Jordan, D. M. High-speed video observations of a lightning stepped leader. *J. Geophys. Res.* **116**, D16117 (2011).

34. Stolzenburg, M. *et al.* Luminosity of initial breakdown in lightning. *J. Geophys. Res.* **118**, 2918–2937 (2013).

35. Marshall, T. *et al.* Initial breakdown pulses in intracloud lightning flashes and their relation to terrestrial gamma ray flashes. *J. Geophys. Res.* **118**, 10907 10925 (2013).

36. Pasko, V. P. Theoretical modeling of sprites and jets. In *Sprites, Elves and Intense Lightning Discharges of NATO Science Series II: Mathematics, Physics and Chemistry* vol. 225 (eds Füllekrug, M., Mareev, E. A. & Rycroft, M. J.) 253–311 (Springer, 2006).

37. Brown, D. P. Tropical storm Dorian. Tech. rep. http://www.nhc.noaa.gov/data/tcr/AL042013_Dorian.pdfNOAA (2013).

38. Cummins, K. & Murphy, M. An overview of lightning locating systems: history, techniques, and data uses, with an in-depth look at the U.S. NLDN. *IEEE Trans. Electromagn. Compat.* **51**, 499–518 (2009).

39. Krehbiel, P. R. In *The Earth's Electrical Environment* 90–113 (National Academy Press, 1986).

40. Rakov, V. A. & Uman, M. A. *Lightning: Physics and Effects* (Cambridge Univ. Press, 2003).

41. Liu, N. Y. & Pasko, V. P. Effects of photoionization on propagation and branching of positive and negative streamers in sprites. *J. Geophys. Res.* **109**, A04301 (2004).

42. Shao, X. M., Krehbiel, P. R., Thomas, R. J. & Rison, W. Radio interferometric observations of cloud-to-ground lightning phenomena in Florida. *J. Geophys. Res.* **100**, 2749–2783 (1995).

Acknowledgements

We thank Jérémy Riousset for various discussions on the formation mechanisms of the upward electrical discharges and fractal modelling. This research was supported in part by NSF grants AGS-0955379 and AGS-1348046, and a DARPA grant HR0011-10-1-0061 to the Florida Institute of Technology, and by an NSF grant AGS-1047588 and a DARPA grant HR0011-10-1-0059 to Duke University.

Author contributions

N.Y.L. drafted the manuscript, supervised the project and analysed the images, storm/lightning data and electromagnetic signals. N.S. operated the Watec camera system, analysed the images and performed the star field analysis. J.R.D. and H.K.R. analysed the images. D.F. operated the two Sandia Allsky cameras and analysed the images. S.A.C collected and analysed the electromagnetic signals. All authors contributed to the discussion of the results and the preparation of the manuscript.

Additional information

23

Hot super-Earths stripped by their host stars

M.S. Lundkvist[1,2], H. Kjeldsen[1], S. Albrecht[1], G.R. Davies[1,3], S. Basu[4], D. Huber[1,5], A.B. Justesen[1], C. Karoff[1,6], V. Silva Aguirre[1], V. Van Eylen[1], C. Vang[1], T. Arentoft[1], T. Barclay[7,8], T.R. Bedding[1,5], T.L. Campante[1,3], W.J. Chaplin[1,3], J. Christensen-Dalsgaard[1], Y.P. Elsworth[1,3], R.L. Gilliland[9], R. Handberg[1], S. Hekker[1,10], S.D. Kawaler[11], M.N. Lund[1,3], T.S. Metcalfe[1,12], A. Miglio[1,3], J.F. Rowe[7,13], D. Stello[1,5], B. Tingley[1] & T.R. White[1,14]

Simulations predict that hot super-Earth sized exoplanets can have their envelopes stripped by photoevaporation, which would present itself as a lack of these exoplanets. However, this absence in the exoplanet population has escaped a firm detection. Here we demonstrate, using asteroseismology on a sample of exoplanets and exoplanet candidates observed during the Kepler mission that, while there is an abundance of super-Earth sized exoplanets with low incident fluxes, none are found with high incident fluxes. We do not find any exoplanets with radii between 2.2 and 3.8 Earth radii with incident flux above 650 times the incident flux on Earth. This gap in the population of exoplanets is explained by evaporation of volatile elements and thus supports the predictions. The confirmation of a hot-super-Earth desert caused by evaporation will add an important constraint on simulations of planetary systems, since they must be able to reproduce the dearth of close-in super-Earths.

[1] Stellar Astrophysics Centre (SAC), Department of Physics and Astronomy, Aarhus University, Ny Munkegade 120, DK-8000 Aarhus C, Denmark. [2] Zentrum für Astronomie der Universität Heidelberg, Landessternwarte, Königstuhl 12, 69117 Heidelberg, Germany. [3] School of Physics and Astronomy, University of Birmingham, Birmingham B15 2TT, UK. [4] Department of Astronomy, Yale University, New Haven, Connecticut 06511, USA. [5] Sydney Institute for Astronomy (SIfA), School of Physics, University of Sydney, New South Wales 2006, Australia. [6] Department of Geoscience, Aarhus University, Høegh-Guldbergs Gade 2, DK-8000 Aarhus C, Denmark. [7] NASA Ames Research Center, Moffett Field, California 94035, USA. [8] Bay Area Environmental Research Institute, 596 1st Street West, Sonoma, California 95476, USA. [9] Center for Exoplanets and Habitable Worlds, The Pennsylvania State University, 525 Davey Lab, University Park, Pennsylvania 16802, USA. [10] Max Planck Institute for Solar System Research, D-37077 Göttingen, Germany. [11] Department of Physics and Astronomy, Iowa State University, Ames, Iowa 50011, USA. [12] Space Science Institute, Boulder, Colorado 80301, USA. [13] SETI Institute, Mountain View, California 94043, USA. [14] Institut für Astrophysik, Georg-August-Universität Göttingen, Friedrich-Hund-Platz 1, D-37077 Göttingen, Germany. Correspondence and requests for materials should be addressed to M.S.L. (email: lundkvist@phys.au.dk).

Models predict that the envelopes of exoplanets orbiting close to their host stars are stripped by photoevaporation, which should be evident as an absence of very hot super-Earth sized exoplanets. The simulations by ref. 1 show a deficit in the number of exoplanets with radii between 1.8 and $4\,R_\oplus$, and that these exoplanets should become comparatively rare for fluxes exceeding $100\,F_\oplus$ due to photoevaporation. In addition, the simulations reveal a corresponding increase in the number of rocky planets with $R < 1.8\,R_\oplus$ caused by the presence of the stripped cores. The existence of a paucity in the radius distribution of close-in exoplanets caused by evaporation is also supported by other theoretical works[2-4].

Previous studies have detected a deficit of exoplanets in the radius-period (or semi major axis) diagram[4-6]. This so-called sub-Jovian pampas or sub-Jovian desert extends from 3 to $10\,R_\oplus$ for periods shorter than 2.5 days[5,6]. However, the absence in the distribution of exoplanets caused by evaporation has escaped a secure confirmation[2-8] primarily due to uncertain host star parameters. This can now be changed with asteroseismology. Asteroseismology studies the stellar pulsations, and it allows us to determine the properties of many exoplanet host stars to high accuracy[9-12], which in turn markedly improves the planetary properties.

NASA's Kepler mission has provided high-quality data for thousands of potential exoplanets and their host stars[7,13-15]. Here we exploit these data, using asteroseismology, to make a robust detection of the hot-super-Earth desert, a region in the radius-flux diagram completely void of exoplanets. We find that the hot-super-Earth desert is statistically significant and not caused by selection effects or false positives. The detection of the existence of a hot-super-Earth desert confirms that photoevaporation does play a role in shaping the exoplanet population that we see today. This imposes an important constraint on simulations of the formation and evolution of exoplanetary systems since this effect needs to be taken into account.

Results

The seismic sample of exoplanets. Using asteroseismology, we obtained accurate stellar mean densities and radii for 102 exoplanet host stars (both confirmed and candidate exoplanets). These are shown in an asteroseismic Hertzsprung–Russell Diagram in Fig. 1 (the methods used to determine the parameters are discussed in Methods, while Supplementary Table 1 contains the data). The asteroseismic mean densities and radii, combined with precise periods and transit depths as well as the stellar effective temperature, allowed us to calculate very precise planetary radii and incident fluxes for the subset of Kepler exoplanets that orbit the 102 host stars (typically more precise than 10%, see Fig. 5).

We determined the flux that the exoplanet receives from its host star, using the following expression for the time-averaged incident flux in units of the Earth value (assuming circular orbits):

$$\frac{F}{F_\oplus} = \left(\frac{\rho_*}{\rho_\odot}\right)^{-2/3} \left(\frac{P}{1\,\mathrm{yr}}\right)^{-4/3} \left(\frac{T_{\mathrm{eff},*}}{T_{\mathrm{eff},\odot}}\right)^4. \quad (1)$$

Here ρ_* is the stellar mean density obtained from asteroseismology, P is the orbital period, and T_{eff} is the effective temperature, with $T_{\mathrm{eff},\odot} = 5{,}778\,\mathrm{K}$ being the effective temperature of the Sun. To find the radius we used the planet–star radius ratio (R_p/R_*), which can be obtained from the transit depth ($\delta F/F$) and the stellar radius from grid-modelling:

$$R_p = \sqrt{\frac{\delta F}{F}} R_* = \left(\frac{R_p}{R_*}\right) R_* . \quad (2)$$

The periods and the planet–star radius ratios have been

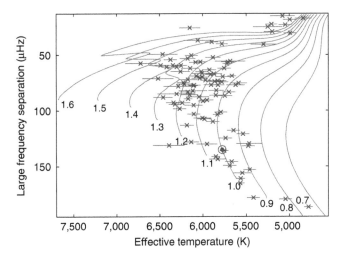

Figure 1 | The seismic host stars. Asteroseismic Hertzsprung–Russell diagram showing the large frequency separation as a function of stellar effective temperature (with 1σ uncertainties) for the 102 exoplanet host stars in the asteroseismic sample. The grey lines show evolution tracks of different masses (in solar units) for solar composition (adapted from ref. 41). The location of the Sun is indicated by the grey solar symbol (circle with a dot).

obtained from ref. 16 (62 exoplanets) or the NASA Exoplanet Archive's cumulative KOI (Kepler Object of Interest) list (http://exoplanetarchive.ipac.caltech.edu/cgi-bin/TblView/nph-tblView?app=ExoTbls&config=cumulative, accessed on 1 July 2015) with preference given to the former. The uncertainties were estimated using propagation of (Gaussian) uncertainties, where the dominant contribution to the uncertainty on the incident flux stems from the temperature uncertainty.

The hot-super-Earth desert. In Fig. 2, we show the exoplanet radius as a function of the incident flux for 157 of the 162 exoplanets. Five exoplanets were removed from the subsample because their radius estimates had an uncertainty in excess of 20% (in order to not have our sample polluted by bad data points, see Methods for details). For illustration we also show in Fig. 2 all Kepler KOIs with apparent sizes below $30\,R_\oplus$ determined to better than 20%, that have a calculated flux and are not in our seismic sample (the non-seismic sample, the incident fluxes and radii for these KOIs have been taken from the NASA Exoplanet Archive).

Figure 2 clearly displays a complete absence of exoplanets with sizes between 2.2 and $3.8\,R_\oplus$ and an incident flux above 650 times the Earth value (shaded area in Fig. 2). We constrained the size of the empty region by bootstrapping with 1 million iterations of the exoplanets present in the seismic subsample and using the boundaries that left the region empty in 95.45% (2σ) of the iterations (see Methods for further information). This empty region in the radius-flux diagram is the hot-super-Earth desert, and its location agrees with the theoretical prediction[1]. We note that some data points from the non-seismic sample will fall in the region of the hot-super-Earth desert, if no cut is made to weed-out uncertain data points (see for example Fig. 7 of ref. 7 and Methods).

As the boundaries and Fig. 2 suggest, we opted to model the hot-super-Earth desert as a simple box region. While such a simplistic model may not capture the full effects of evaporation on the planet population, we do believe it to encompass the main features. A more sophisticated model taking into account how the amount of evaporation scales with incident flux and planet mass could be a next step.

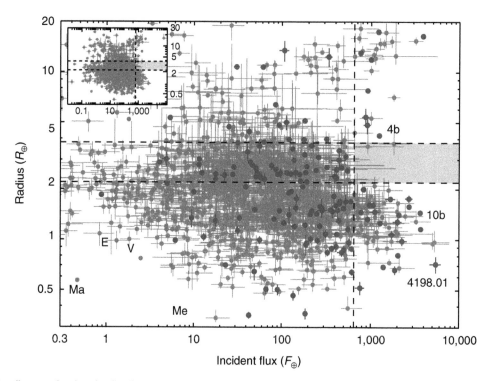

Figure 2 | Radius-flux diagram showing the distribution of exoplanets. The 157 exoplanets in the seismic sample are plotted with 1σ errorbars (in blue), while the non-seismic sample is shown in grey (the inset shows the full non-seismic sample). The exoplanets Kepler-4b, Kepler-10b and KOI-4198.01 are identified in the diagram. The location of the four rocky solar system planets; Mercury (Me), Venus (V), Earth (E) and Mars (M) is indicated with the green writing (no points). The vertical dashed line marks an incident flux of $650\,F_{\oplus}$, while the horizontal dashed lines indicate radii of 2.2 and $3.8\,R_{\oplus}$, respectively. The location of the hot-super-Earth desert has been shaded.

As an aside, it should be pointed out that in the seismic subsample shown in the radius-flux diagram in Fig. 2, KOI-4198.01 (which we call Zenta) appears somewhat isolated. If Zenta is a bona fide exoplanet, then it could potentially be a very interesting object, since it has the highest incident flux of the exoplanets in the seismic subsample, and it is below $1\,R_{\oplus}$ in size. We have inspected the light curve of Zenta to make sure the transits look like genuine exoplanet transits, and we have obtained a few spectra of the host star with the Nordic Optical Telescope (on La Palma). These spectra will be the subject of a subsequent analysis.

Significance. We employed different techniques to assess the significance of the hot-super-Earth desert. First, under the assumption that the period-, or equivalently the incident flux distribution does not change with planet radius[17] (the null hypothesis), we tested whether the hot-super-Earth desert could occur by chance. This was done by drawing exoplanets randomly from the planet population below $650\,F_{\oplus}$ and counting how many exoplanets fell in the radius range of the desert (see Methods for details). We find that only 8 of our 10 million simulations returned zero exoplanets in the desert. Thus, it is very unlikely to observe the desert if the incident flux is not a function of radius, which is in agreement with our observation of a gap in the distribution.

Second, we used a Gaussian mixture model to represent the seismic subsample as it would look with no desert. Here the underlying assumption is that the radius-flux distribution can be described by a sum of log-normal distributions[18]. From the model we created a histogram (Fig. 3), and we found that fewer than 0.4% of our simulations return the observed number of planets (zero) in the region of the hot-super-Earth desert (see also Methods). This shows that the gap in the radius-flux diagram is significant. It is worth noting that from the non-seismic sample alone, this inference cannot be made (it gives a p value of 8%). We

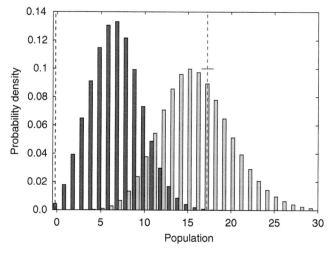

Figure 3 | Simulated number of exoplanets in and below the hot-super-Earth desert. Histograms of the number of exoplanets in (blue) and below the hot-super-Earth desert (in radius, green) using a Gaussian mixture model of the seismic subsample under the assumption of no gap. The two dashed lines show the observed number of planets in (dark blue) and below the desert (dark green); 0 ± 0.04 and 17 ± 0.7, respectively. The histograms are based on 5 million realizations.

do not believe that this could be due to selection effects between the seismic and the non-seismic sample, since any detected hot super-Earth planet would have been a high-priority target to the Kepler mission. From the Gaussian mixture model treatment of the seismic sample, we also found a slight, although not statistically significant, overdensity below the desert (see Fig. 3), similar to that expected if the rocky cores are left over from evaporation[1].

Selection effects and false positives. It should be emphasized that there are some selection biases in the sample. The limitations in the detection sensitivity of Kepler are the reason for the lack of small exoplanets with low incident flux. Also, for the asteroseismic subsample, the selected stars were on the Kepler short-cadence target list, which is the reason for the low number of large exoplanets with low incident flux (short-cadence slots were prioritized for multi-planet systems over single-planet systems, which favours small planets[19], and exoplanets showing many transits were discovered early in the mission and kept). While the completeness of the sample is hard to quantify[14,20], no known selection effects[21] would produce the paucity that we observe and the sample is complete down to $2R_{\oplus}$ for short-period exoplanets[8] (additionally, any missing small planets from below the gap, would only make the desert more pronounced). We also attempted to account for detection biases in our sample by imposing a signal-to-noise ratio (SNR) criterion and assuming that no exoplanets meeting that criterion with a radius above $1.4R_{\oplus}$ would have have been missed[9]. We found this not to affect the presence of the hot-super-Earth desert (see Fig. 7 and Methods for further details).

Despite our basic vetting, the seismic subsample of exoplanets will contain some false positives (FPs). The overall FP-rate for the sample is found to be low[22] (in particular for the multi-planet systems[23]), but it does vary over the sample. For example, the FP-rate is lower for exoplanets with radii $2-4R_{\oplus}$ than for those with smaller or larger radii[24]. However, clearly no FPs have filled the hot-super-Earth desert, and our simulations show we would not be significantly affected by the presence in the sample of the percentage of FPs suggested by ref. 24 (see Methods).

The sub-Jovian pampas. A trend, agreeing with our results, has been seen in the radius-period (or semi major axis) diagram by previous studies[4–6]. They detected a deficit of exoplanets in the radius-range $3-10R_{\oplus}$ with periods shorter than around 2.5 days (the sub-Jovian pampas or sub-Jovian desert)[5,6]. While both our hot-super-Earth desert and the sub-Jovian pampas lie at high temperatures (be it high incident flux or short periods), the radius-range is somewhat different, since we find the hot-super-Earth desert to extend only up to $3.8R_{\oplus}$. Therefore, we investigated the radius-range of the hot-super-Earth desert further to determine whether it could be an extension of the sub-Jovian pampas.

Of the exoplanets in the seismic subsample above our flux boundary of $650F_{\oplus}$, four exoplanets are present above $10R_{\oplus}$. These are all confirmed exoplanets (Kepler-1b, 2b, 7b and 14b), and thus agree with the upper limit set by the previous studies. In the radius-range between 4 and $10R_{\oplus}$, another four exoplanets are present in our seismic sample. Of these, two are confirmed exoplanets (Kepler-4b and 56b) and a third (KOI-5.01) is a candidate in a multi-planet system (where the FP-rate is lower[23]). Most important to the location of the upper boundary of the hot-super-Earth desert is Kepler-4b, which is located at $R = 4.2R_{\oplus}$ and $F = 1243F_{\oplus}$ (see Fig. 2), and thus effectively sets the upper boundary. Kepler-4b has a density of around $1.9\,\text{g\,cm}^{-3}$ (similar to the density of Neptune), and is consequently volatile rich[25], which agrees with its location above the desert. Similarly, Kepler-56b also has a density estimate consistent with a volatile-rich composition[26], thus also agreeing with its location above the desert.

We have examined the transits of Kepler-4b for evidence that it could be evaporating, but we failed to find any asymmetry in the transits or any transit-to-transit depth variations, which could both indicate atmospheric loss[27,28]. Still, it cannot be ruled out that evaporation of Kepler-4b could be ongoing at a level, which

we cannot detect and possibly at a level that does not influence the radius evolution of the planet.

To assess the significance of an extended gap scenario, we tested two additional sets of boundaries, allowing for the presence of exoplanets from the seismic sample within the gap. The first scenario had the same flux boundary as the hot-super-Earth desert ($650F_{\oplus}$), but spanned the radius-range from 2.2 to $10R_{\oplus}$ in agreement with the upper limit stated for the sub-Jovian pampas. This meant that four seismic exoplanets were present in the tested region (Kepler-4b, Kepler-56b, KOI-5.01 and KOI-1314.01). However, since the sub-Jovian pampas was defined in orbital period rather than incident flux, we cannot replicate the exact limit in the radius-flux diagram found by for instance ref. 6. Thus, we also considered the possibility that we should move the boundary to higher incident flux. Therefore, we tested a region with the aforementioned limits in radius, but bounded by an incident flux of $1,000F_{\oplus}$ instead of $650F_{\oplus}$, which only leaves Kepler-4b in the region of the gap (even though a flux limit this high does not seem to agree with the sharp cutoff in the seismic sample in the $2.2-3.8R_{\oplus}$ region). We find that both of the tested scenarios are less significant than the hot-super-Earth desert, with the $650F_{\oplus}$ radius-extended scenario being by far the least significant one.

It can be noted in connection to the sub-Jovian pampas that some exoplanets seem to occupy that gap[6,29], and that they do not all appear to be FPs[29]. Reference 29 investigates three planet candidates located in the sub-Jovian pampas, and they find that two of the three are likely true planets. While these two planets fall comfortably within the sub-Jovian pampas, one of them is too large to fall in the hot-super-Earth desert, and the other one has uncertainties large enough that it could as well be outside the hot-super-Earth desert (it sits $<1\sigma$ from the upper radius-limit[29]).

Discussion

For exoplanets in the radius range in question, radius is thought to be a good proxy for composition[21,30]. This allows for the transition from a predominantly rocky to a volatile-rich make-up to be expressed in terms of radius, and this transition has been found to occur around $1.6-1.8R_{\oplus}$ by different studies[21,30,31]. Thus, the majority of exoplanets in the $2.2-3.8R_{\oplus}$ range are expected to be volatile rich, though some of them could be water worlds[21] (for comparison, the radius of Neptune is $\sim3.8R_{\oplus}$). This agrees with the theory that these exoplanets could be stripped of their envelopes when they are too close to their host star. Thus, we can infer from our hot-super-Earth desert that hot exoplanets below $\sim2.2R_{\oplus}$ most likely have a predominantly rocky composition.

Dynamical interactions may in principle also be responsible for shaping the gap in the radius-flux diagram, for example, due to orbital decay or inward migration of planets at late evolutionary stages. However, it seems unlikely that orbital decay played a major part in clearing out the particular part of parameter space associated with the hot-super-Earth desert since the planets would either need to be more massive or on shorter orbits[32]. Other migration channels such as a combination of planet-planet scattering, tidal circularisation and the Kozai mechanism could have played a role in shaping the location of the hot-super-Earth desert through migration of exoplanets that were initially part of a triple (or larger) system[33,34]. These effects have not been considered in our work, but they could be responsible for later migration of some of the planets that sit above the hot-super-Earth desert (and inside the sub-Jovian pampas, such as Kepler-4b). In addition, the flux boundary is likely a function of the planet mass with heavier planets being able to better withstand the evaporation. Therefore, while we find that the hot-super-Earth desert is more significant than the other regions we tested, we are not in a position to unambiguously decide whether the hot-super-Earth desert is an extension of the sub-Jovian pampas or a separate feature in the radius-flux diagram.

We have established the existence of a hot-super-Earth desert in the radius-flux diagram. Its presence confirms that photoevaporation plays an important role in planetary evolution, with the mass-loss history depending on the incident stellar flux. This represents a mechanism not seen in our own solar system, by which some volatile-rich exoplanets are stripped of their atmospheres by their host stars. Consequently, our detection of a hot-super-Earth desert will add an important constraint for simulations of the evolution of planetary systems.

Methods

Preparation of the power spectra. Asteroseismology is the study of stellar oscillations. In the case of solar-like stars, the frequencies of the oscillations are almost regularly spaced in a Fourier transform of the time series (a power spectrum, see the inset in Fig. 4). The dominant regular structure yields the large frequency separation, which carries information about the stellar mean density[35].

We have searched all 275 exoplanet host stars with a Kepler magnitude brighter than 13.5 and with short cadence Kepler data (sampled every 58.85s) for an asteroseismic signal. A magnitude limit of 13.5 was chosen since we have essentially no chance of detecting oscillations in a solar-like star fainter than this[36]. To be able to search for the large frequency separation (Δv) for each of the stars, we first made weighted power spectra. The power spectrum for each star was calculated in the following manner: (1) The time series for each quarter (data from Kepler are divided into quarters of ~3 months duration due to the roll of the spacecraft) was cleaned for bad data points using sigma-clipping (with 4σ) of a high-pass filtered time series (high-pass filter was 7 min) to take out the effect of all slow variations. (2) Using a high-pass filter, the long-term variation of the noise per data point was estimated and taken as the scatter (σ). (3) Using $1/\sigma^2$ as the statistical weight per data point, we calculated the power spectrum following ref. 37. (4) For each quarter we calculated a separate power spectrum, and subsequently we combined the power spectra for all quarters into one single spectrum using a weighted mean. The weights were given by $1/(\text{median}(\text{power}))^2$, where the median of the power between 2 and 4 mHz was used. This serves the purpose of down-weighting power spectra for quarters with higher noise levels with respect to the others. Also, when combining several power spectra this way, we change the statistics of the power spectrum from being described by a χ_2^2 to approaching a normal distribution (as stated by the central limit theorem)[38]. An example of a part of a power spectrum can be seen in the inset in Fig. 4.

Extraction of large frequency separations.

Extraction of large frequency separations. A clear asteroseismic signature was found in 102 of the host stars using a matched filter response function (MFR)[39] to search for the large frequency separation. The method takes advantage of the

near-regular spacing of the high-order, low-degree p-modes in the power spectrum of solar-like stars. It does this by summing the smoothed power at specific frequencies, which have been calculated from the asymptotic relation[40] in the version:

$$v_{n,\ell} \approx \Delta v\left(n + \frac{\ell}{2} + \varepsilon\right) - \ell(\ell+1)D_0 . \tag{3}$$

Here n is the radial order of the mode (related to the number of nodes in the radial direction), ℓ is the degree of the mode (the number of surface nodes), ε is a parameter sensitive to the near-surface layers of the star, while D_0 is sensitive to the conditions near the core.

When summing the power at frequencies given by different values of Δv (collapsing over different values of the other parameters in expression (3)), the result is the MFR giving the summed power as a function of Δv (see ref. 39 for details). An example for the host star KIC 9414417 can be seen in Fig. 4. The large frequency separation corresponding to the most prominent peak in the MFR is then the large frequency separation of the star. The uncertainty on the large frequency separation is determined as the full width at half maximum of the peak.

Grid-modelling of the host stars. We used four pipelines to determine the stellar parameters for the 102 exoplanet host stars. These were Asteroseismology Made Easy (AME)[41], SEEK[42], BAyesian STellar Algorithm (BASTA)[12] and the Yale-Birmingham (YB)[43–45] pipeline. The YB pipeline derived the properties from five different grids of stellar models, which brings us to a total of eight different grids of stellar models. These pipelines have been used extensively for asteroseismology[9–11], and further description of the pipelines can be found in the literature.

As inputs to the grid-modelling we used for each star its large frequency separation (Δv) found from asteroseismology and two spectroscopic inputs; the effective temperature (T_{eff}) and the metallicity ([Fe/H]). The values that were used for the 102 host stars can be found in Supplementary Table 1.

We chose to use the mean density and radius returned by AME and then determined the uncertainty by adding in quadrature the uncertainty returned by AME and the scatter over the values returned by the other seven grids. Three stars were too massive for the AME grid, so for these we used the median parameters from the other seven pipelines and estimated the uncertainties by adding in quadrature the median formal uncertainty and the scatter over all seven grids. We note that the parameters returned by the various pipelines were consistent.

Many of the host stars in our seismic sample are present in other large host star samples[9,12], and we have compared the densities and radii obtained for our sample with these other results. We find our parameters to be fully consistent (within 1σ) with the results from ref. 9 with the exception of Kepler-22. However, this is due to the fact that we are using a very different large frequency separation, since the signal originally found[46] is no longer thought to be the correct one (H. Kjeldsen et al. (manuscript in preparation)). When comparing the densities and radii for the host stars that we have in common with ref. 12 (32 stars), we find that all densities and 29 of the 32 radii are consistent within 1σ with the remaining three radii differing by just above 1σ, leading us to conclude that our densities and radii are in agreement with those previously determined.

Vetting of the seismic subsample of exoplanets. To do some basic vetting of our seismic subsample, we chose to limit our sample to exoplanets that had an uncertainty in radius of <20%. A large uncertainty on radius was primarily due to large uncertainties on R_p/R_*, which can be caused by grazing transits where the planet only partly covers the star. This removed five exoplanets from the sample: KOI-371.01, KOIs 2612.01 and 2612.02, KOI-3194.01 (which in addition has an impact parameter (b (ref. 47)) larger than unity) and KOI-5086.01 (also $b > 1$). A radius cut of 30% would remove three of these targets (it would leave KOIs 2612.01 and 2612.02 in the sample). It should be noted that none of these exoplanets were situated in the hot-super-Earth desert. Instead of limiting our sample by using the uncertainty in radius, we also tried using the impact parameter (with the criterion $b < 1$), which would remove some of the grazing transits. This removed the two exoplanets mentioned above from the asteroseismic subsample, but we opted for the stricter 20% limit on the radius uncertainty.

We also tried to vet the subsample by using asterodensity profiling[16,48–50]. Here the ratio of stellar mean densities derived from the orbit and, in our case, grid-modelling ($\rho_{*,\text{tr}}/\rho_{*,\text{seis}}$) is considered, and a value very different from unity points to either very eccentric orbits or a blend scenario (these are the two largest effects). However, it was difficult to put meaningful constraints on the density ratio since a conservative value did not eliminate any candidates and a more aggressive value would risk throwing away high-eccentricity exoplanets. Thus, we did not pursue this further.

If the cut in radius uncertainty is made at a higher value than 20%, then exoplanets from the non-seismic sample will appear in the desert. We have examined the points that appear if the cut is instead made at 30 or 40%. Using the information from the NASA Exoplanet Archive (from 1 July 2015), when we make the cut at 20% one exoplanet from the non-seismic sample is present in the top of the desert (with its 1σ errorbars easily placing it outside the desert). If we increase this value to 30%, then two additional planets enter the hot-super-Earth desert, one very close to the lower flux boundary, and one which, since we downloaded the data, has been flagged as a FP.

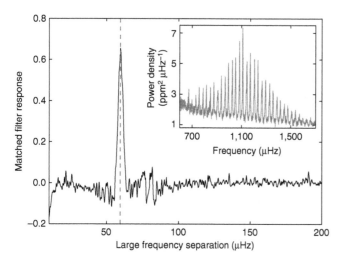

Figure 4 | The large frequency separation. Output from the matched filter response function for the host star KIC 9414417. The large peak seen at ~53 µHz (and marked by the dashed line) indicates the large frequency separation for the star. The inset shows a section of the power spectrum of KIC 9414417 (smoothed with a Gaussian filter with a width of 1 µHz), where the regular spacing between the oscillation frequencies is clear.

If the cut is instead made at 40%, then a total of 13 exoplanets occupy the region of the hot-super-Earth desert including those discussed above. Of these 13 planets, two are FPs and one is the confirmed exoplanet Kepler-319b. However, on checking the radius and flux for Kepler-319b listed in the discovery paper[51], it is clear that this planet is in fact situated far from the desert (with $R = 1.63\ R_\oplus$ and $F = 261.6\ F_\oplus$ (ref. 51)), which brings us down to 10 exoplanets in the desert.

We have manually inspected these 10 remaining exoplanets situated in the hot-super-Earth desert. They all orbit stars of spectral type F or G, and we find that the reason for the very uncertain exoplanet parameters is very uncertain parameters for the host stars. We find that all of them have uncertainties consistent with a location outside the desert, and that two of them are likely FPs judged on inconsistency between the stellar density derived from the transit and that derived from the stellar mass and radius (these planets are on short orbits and are thus unlikely to have large eccentricities). It is noteworthy that excluding data points with high uncertainties does not exclude a specific stellar spectral type, for instance, it simply limits the number of bad data points in the sample.

We have plotted histograms of the relative uncertainties on the radius and incident flux, which can be seen in Fig. 5. We note that there is a clear bimodal distribution in both histograms, and that the uncertainties for the seismic subsample are lower than the typical uncertainties in the non-seismic sample. This emphasizes our point that the properties of the seismic sample are determined to a high accuracy. The bimodal distributions in relative uncertainty in flux and radius show that the non-seismic sample is divided into a 'low' uncertainty and a 'high' uncertainty population, and the division between the two populations lie at $\sim 30\%$ in radius. Thus, making a cut at 20% should ensure that we are only plotting the best data points from the non-seismic sample, and we have verified that we are not cutting away a population of planets around M-dwarfs (which would have high uncertainties in radius) by doing so.

Determining the boundaries of the hot-super-Earth desert. We constrained the size of the hot-super-Earth desert by doing a bootstrap with 1 million iterations of the exoplanets present in the seismic subsample and using the boundaries that make 95.45% (2σ) of the iterations return an empty desert. To be specific, we first randomly drew 157 exoplanets with replacement from the seismic subsample. Then we assigned each of these a radius and a flux randomly selected from Gaussians centred on the parameters for the drawn exoplanet with a standard deviation equal to the uncertainty. Subsequently, we determined how many of these exoplanets that were situated in the hot-super-Earth desert. This was repeated 1 million times, after which we calculated the percentage of iterations without planets in the hot-super-Earth desert (which is the observed number). We used this information to change the boundaries of the desert, and we repeated the above procedure until we had obtained the 2σ limits. This procedure does not yield unique boundaries, although they are well constrained due to the small uncertainties on the exoplanets in the sample. However, to determine the exact extent of the hot-super-Earth desert is beyond the scope of this work, and it will in addition depend on whether or not one will allow any exoplanets in the desert.

The Gaussian mixture model. We have used a Gaussian mixture model (GMM), which is a probabilistic model that is the sum of a finite number of Gaussian

distributions (we used the Python Scikit-Learn Gaussian Mixture Model[52]). We used the GMM to describe the planet population in log–log radius-flux space and then applied tests to the model to assess the probability that we had detected the hot-super-Earth desert. The distribution of planets in flux and radius is expected to form a correlated log-normal distribution as an outcome of a stochastic planet formation process that produced many correlated, fractional changes in planet sizes and orbits[18]. Thus, it is justified to use the GMM, which fits a sum of bivariate Gaussians to the data.

The two different hypotheses that we tested using the GMM and the data are the null hypothesis and the irradiated hypothesis. The null hypothesis states that the radius-flux distribution is smooth, thus that there is no hot-super-Earth desert present in the data. The irradiated hypothesis states that there is a gap in the population density and that there is an overdensity at radii lower than the gap.

We leave the number of summed normal distributions as a parameter to be determined by the data in order to allow for different formation processes, selection effects and other biases. The number of Gaussian components is determined by selecting the model with the lowest Bayesian Information Criterion (BIC). We apply the fit to three different samples; the seismic subsample of exoplanets, the non-seismic subsample and the combined sample. For each sample we use the minimum BIC to determine the number of components used in the GMM. For the seismic subsample, the typical number of components selected by the BIC is one.

The fit applied by the GMM does not treat statistical uncertainties on the data points. To ensure our tests are robust we have used a Monte Carlo approach to draw each data point from its statistical uncertainties. We generate 1,000 draws from the uncertainties and for each draw we fit the GMM, and each time we determine the number of components by selecting the lowest BIC. From each of these 1,000 models, we draw 5,000 populations and record the number of planets that occupy the gap for each. This then provides the probability distribution of planets in the gap under the null hypothesis, since we fit our model to the data under this assumption.

Figure 6 shows examples of the real data together with simulated samples drawn from the fit. We artificially injected a hot-super-Earth desert ($2.2 \leq R_{\mathrm{p}}/R_\oplus \leq 3.8$ and $F \geq 650\ F_\oplus$) into the drawn samples by subtracting $2.7\ R_\oplus$ from the planetary radius if the planet fell within the desert. While somewhat crude, this introduces the gap and an overdensity below the gap.

In the seismic sample, no planets are observed in the hot-super-Earth desert. Figure 3 shows the probability distribution of planets expected in the hot-super-Earth desert under the null hypothesis together with the observed value (0 ± 0.04). The uncertainty comes from the small chance that a system actually occupies the gap due to the uncertainties on the planetary radius and flux. Furthermore, we show the expected population distribution below the desert ($0.4 \leq R_{\mathrm{p}}/R_\oplus \leq 2.2$ and $F \geq 650\ F_\oplus$) also with the observed value (17 ± 0.7). The probability of observing no planets in the hot-super-Earth desert in the seismic sample given the null hypothesis is $p = 0.4\%$, which is sufficiently small that we reject the null hypothesis. We observe a slight overdensity in the planet population below the desert, but this is very weak and not statistically significant.

We repeat the analysis with the non-seismic and the combined samples. For the non-seismic sample we find the probability of observing the data in the desert under the null hypothesis is $p = 8\%$, which supports the rejection of the null hypothesis but is not significant under the typical requirements of either $p < 5\%$ or $p < 1\%$. For the combined data we find a small improvement with $p = 0.3\%$, which is clearly dominated by the seismic sample.

We checked our method using the simulated data with and without a gap. We found results that were consistent with those reported here for the real data. It should be noted, specifically, that in the simulated-gap seismic sample, we

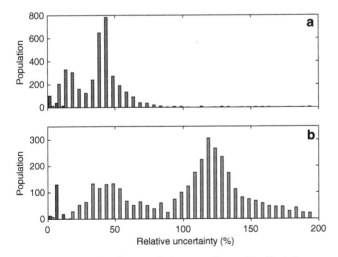

Figure 5 | Distribution of uncertainties on radius and incident flux.
Histogram showing the distribution of the relative uncertainty in radius (**a**) and incident flux (**b**) for both the seismic and the non-seismic sample. The seismic sample is shown in blue and the non-seismic sample is shown in grey. Note the bimodal distribution in the non-seismic sample and the low uncertainties in the seismic sample compared with the non-seismic one.

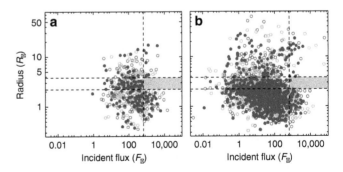

Figure 6 | Radius-flux diagrams showing the real and simulated data.
(**a**) One draw of the 157 exoplanets in the seismic subsample (filled blue cicles) as well as a model of the subsample made from the Gaussian mixture model with (grey open circles) and without an artificial gap (green open circles). The vertical dashed line shows where the incident flux is equal to $650\ F_\oplus$, while the horizontal dashed lines indicate radii of 2.2 and $3.8\ R_\oplus$, respectively. The location of the hot-super-Earth desert has been shaded. (**b**) Same, but for the non-seismic sample.

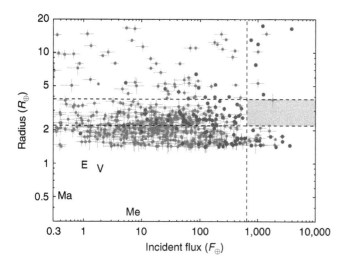

Figure 7 | Radius-flux diagram showing the debiased seismic subsample. The debiased sample of exoplanets with a threshold SNR of 10 and a radius fulfilling $R_{min} < R_x < R_p$ with $R_x = 1.4\,R_\oplus$ is shown (in blue) with 1σ errorbars. The grey points are all the exoplanets in the non-seismic sample fulfilling $R_{min} < R_x < R_p$, also with 1σ errorbars. The location of the four rocky solar system planets; Mercury (Me), Venus (V), Earth (E) and Mars (Ma) is shown with the green writing for reference. The vertical dashed line shows where the incident flux is equal to $650\,F_\oplus$, while the horizontal dashed lines indicate radii of 2.2 and $3.8\,R_\oplus$ respectively. The location of the hot-super-Earth desert has been shaded.

consistently found we could reject the null hypothesis of no gap, while we typically did not confirm an overdensity below the gap.

Debiasing the seismic subsample. In an attempt to account for detection biases in our seismic subsample, we debiased the sample following the approach described by ref. 9. First, we determined the minimum planetary radius that should be detectable for a given host star[8]:

$$R_{min} = R_*\left(\sigma_{CDPP} \cdot SNR_{lim}\right)^{0.5}\left(\frac{n_{tr}t_{dur}}{6\,h}\right)^{-0.25}. \qquad (4)$$

Here σ_{CDPP} is the 6 h Combined Differential Photometry Precision[53], SNR_{lim} the required signal-to-noise ratio (SNR), n_{tr} the number of observed transits and t_{dur} the duration of a transit. We chose a SNR threshold of 10 (ref. 8), and for each exoplanet in the sample we estimated R_{min} by using the median 6 h σ_{CDPP} over all observed quarters (obtained from the Mikulski Archive for Space Telescopes, MAST, https://archive.stsci.edu/kepler/, accessed on 8 July 2015), the transit durations from NASA's Exoplanet Archive's cumulative KOI list (http://exoplanetarchive.ipac.caltech.edu/cgi-bin/TblView/nph-tblView?app=ExoTbls config=cumulative, accessed on 8 July 2015), and crudely estimating the number of observed transits by dividing the total lifetime of Kepler (around 1470 days) by the period of the exoplanet.

After having calculated R_{min} for all exoplanets in the seismic sample, we found for a range of exoplanet radii (R_x) the number of exoplanets fulfilling the inequality $R_{min} < R_x < R_p$. Finally, our debiased sample consists of the exoplanets that fulfil the inequality $R_{min} < 1.4\,R_\oplus < R_p$, where $1.4\,R_\oplus$ was the value of R_x that returned the maximum number of exoplanets. The debiased sample can be seen in Fig. 7 along with the debiased non-seismic sample (for illustration) using the R_x determined from the seismic subsample. It can be seen that the desert is still evident (also, see below).

Further tests. As well as using the Gaussian mixture model to assess the hot-super-Earth desert, we also took another approach to verify the significance of the missing data points. We did this by first dividing the seismic subsample of exoplanets in two groups, with $F/F_\oplus > 650$ and $F/F_\oplus < 650$. We then used the exoplanet radius distribution for the $F/F_\oplus < 650$ sample to generate a sample with randomly selected exoplanet radii for a number of exoplanets corresponding to the number of exoplanets with $F/F_\oplus > 650$. Afterwards we determined how many of the selected exoplanets had a radius between 2.2 and $3.8\,R_\oplus$. This simulation was repeated 10^7 times, and we determined the number of times we found the same number of exoplanets as we observe in the hot-super-Earth desert (zero), in this radius range. We found this to happen in only 8 of the 10 million simulations.

This test was repeated to measure the effect of FPs on the detection. This was done by randomly removing points from the seismic subsample according to the percentages given by ref. 24 and then repeating the above analysis. Of course this

approach does not take into consideration any non-uniformity of the FP-rate with flux (for instance there are more eclipsing binaries at higher incident flux[23]), but on the other hand we do not compensate for the FP-rate for multi-planet systems being lower than for single-planet systems[23], or that we have many confirmed exoplanets in the sample (for which the FP-rate should be essentially zero). Thus, this approach should give a fairly conservative estimate of the effect of FPs on the hot-super-Earth desert. We find that 39 of our 10 million simulations return zero exoplanets in the radius region of the desert, meaning that the presence of FPs in our sample does not significantly affect the detection of the hot-super-Earth desert.

To assess the importance of potential systematic errors on the detection of the desert, we investigated the impact on the incident flux of a 100 K temperature offset and also of a non-zero eccentricity of $e = 0.5$. We find that the effect of both of these changes is of the same magnitude, and using this test we determined that they have no impact on the detection of the hot-super-Earth desert.

We also used the test on the debiased sample (seen in Fig. 7), and here 54 of the 10 million simulations returned the same number of exoplanets in the desert as we observe; thus the detection of the hot-super-Earth desert is not greatly changed by using the debiased sample instead. We performed a bootstrap on the debiased sample similarly to what was done for the full sample. Here we found that the boundaries of the hot-super-Earth desert that we determined from bootstrapping the full sample ($2.2 < R_p/R_\oplus < 3.8$ and $F > 650\,F_\oplus$) are stronger when considering the debiased sample. They change from being 2σ limits to being just above 3.5σ, meaning that $> 99.95\%$ of the 1 million simulations left the hot-super-Earth desert empty (380 yielded one planet).

References

1. Lopez, E. D. & Fortney, J. J. The role of core mass in controlling evaporation: the Kepler radius distribution and the Kepler-36 density dichotomy. *Astrophys. J.* **776,** 2 (2013).
2. Owen, J. E. & Wu, Y. Kepler planets: a tale of evaporation. *Astrophys. J.* **775,** 105 (2013).
3. Owen, J. E. & Wu, Y. Atmospheres of low-mass planets: the "Boil-off". *Astrophys. J.* **817,** 107 (2016).
4. Kurokawa, H. & Nakamoto, T. Mass-loss evolution of close-in exoplanets: evaporation of hot jupiters and the effect on population. *Astrophys. J.* **783,** 54 (2014).
5. Szabó, G. M. & Kiss, L. L. A short-period censor of sub-jupiter mass exoplanets with low density. *Astrophys. J.* **727,** L44 (2011).
6. Beaugé, C. & Nesvorný, D. Emerging trends in a period-radius distribution of close-in planets. *Astrophys. J.* **763,** 12 (2013).
7. Rowe, J. F. *et al.* Planetary candidates observed by Kepler. V. Planet sample from Q1-Q12 (36 months). *Astrophys. J. Suppl.* **217,** 16 (2015).
8. Howard, A. W. *et al.* Planet occurrence within 0.25AU of solar-type stars from Kepler. *Astrophys. J. Suppl.* **201,** 15 (2012).
9. Huber, D. *et al.* Fundamental properties of Kepler planet-candidate host stars using asteroseismology. *Astrophys. J.* **767,** 127 (2013).
10. Chaplin, W. J. *et al.* Asteroseismic fundamental properties of solar-type stars observed by the NASA Kepler mission. *Astrophys. J. Suppl.* **210,** 1 (2014).
11. Campante, T. L. *et al.* An ancient extrasolar system with five sub-Earth-size planets. *Astrophys. J.* **799,** 170 (2015).
12. Silva Aguirre, V. *et al.* Ages and fundamental properties of Kepler exoplanet host stars from asteroseismology. *Mon. Not. R. Astron. Soc.* **452,** 2127–2148 (2015).
13. Borucki, W. J. *et al.* Characteristics of planetary candidates observed by Kepler. II. Analysis of the first four months of data. *Astrophys. J.* **736,** 19 (2011).
14. Batalha, N. M. *et al.* Planetary candidates observed by Kepler. III. Analysis of the first 16 months of data. *Astrophys. J. Suppl.* **204,** 24 (2013).
15. Mullally, F. *et al.* Planetary candidates observed by Kepler. VI. Planet sample from Q1-Q16 (47 months). *Astrophys. J. Suppl.* **217,** 31 (2015).
16. Van Eylen, V. & Albrecht, S. Eccentricity from transit photometry: small planets in Kepler multi-planet systems have low eccentricities. *Astrophys. J.* **808,** 126 (2015).
17. Morton, T. D. & Swift, J. The radius distribution of planets around cool stars. *Astrophys. J.* **791,** 10 (2014).
18. Farr, W. M., Mandel, I., Aldridge, C. & Stroud, K. The occurrence of earth-like planets around other stars. Preprint at http://arxiv.org/abs/1412.4849 (2014).
19. Latham, D. W. *et al.* A first comparison of Kepler Planet candidates in single and multiple systems. *Astrophys. J.* **732,** L24 (2011).
20. Petigura, E. A., Marcy, G. W. & Howard, A. W. A plateau in the planet population below twice the size of Earth. *Astrophys. J.* **770,** 69 (2013).
21. Wolfgang, A. & Lopez, E. How rocky are they? The composition distribution of Kepler's sub-neptune planet candidates within 0.15AU. *Astrophys. J.* **806,** 183 (2015).
22. Désert, J.-M. *et al.* Low false positive rate of Kepler candidates estimated from a combination of spitzer and follow-up observations. *Astrophys. J.* **804,** 59 (2015).

23. Lissauer, J. J. *et al.* Validation of Kepler's multiple planet candidates. II. Refined statistical framework and descriptions of systems of special interest. *Astrophys. J.* **784,** 44 (2014).

24. Fressin, F. *et al.* The false positive rate of Kepler and the occurrence of planets. *Astrophys. J.* **766,** 81 (2013).

25. Borucki, W. J. *et al.* Kepler-4b: a hot neptune-like planet of a G0 star near main-sequence turnoff. *Astrophys. J.* **713,** L126–L130 (2010).

26. Huber, D. *et al.* Stellar spin-orbit misalignment in a multiplanet system. *Science* **342,** 331–334 (2013).

27. Rappaport, S. *et al.* Possible disintegrating short-period super-mercury orbiting KIC 12557548. *Astrophys. J.* **752,** 1 (2012).

28. Rappaport, S. *et al.* KOI-2700b—a planet candidate with dusty effluents on a 22 hr orbit. *Astrophys. J.* **784,** 40 (2014).

29. Colón, K. D., Morehead, R. C. & Ford, E. B. Vetting Kepler planet candidates in the sub-Jovian desert with multiband photometry. *Mon. Not. R. Astron. Soc.* **452,** 3001–3009 (2015).

30. Lopez, E. D. & Fortney, J. J. Understanding the mass-radius relation for sub-neptunes: radius as a proxy for composition. *Astrophys. J.* **792,** 1 (2014).

31. Rogers, L. A. Most 1.6 Earth-radius planets are not rocky. *Astrophys. J.* **801,** 41 (2015).

32. Essick, R. & Weinberg, N. N. Orbital decay of hot jupiters due to nonlinear tidal dissipation within solar-type hosts. *Astrophys. J.* **816,** 18 (2016).

33. Fabrycky, D. & Tremaine, S. Shrinking binary and planetary orbits by Kozai cycles with tidal friction. *Astrophys. J.* **669,** 1298–1315 (2007).

34. Nagasawa, M., Ida, S. & Bessho, T. Formation of hot planets by a combination of planet scattering, tidal circularization, and the Kozai mechanism. *Astrophys. J.* **678,** 498–508 (2008).

35. Chaplin, W. J. & Miglio, A. Asteroseismology of solar-type and red-giant stars. *Annu. Rev. Astron. Astrophys.* **51,** 353–392 (2013).

36. Chaplin, W. J. *et al.* Predicting the detectability of oscillations in solar-type stars observed by Kepler. *Astrophys. J.* **732,** 54 (2011).

37. Frandsen, S. *et al.* CCD photometry of the δ-Scuti star κ^2 bootis. *Astron. Astrophys.* **301,** 123 (1995).

38. Appourchaux, T. O. n. Maximum likelihood estimation of averaged power spectra. *Astron. Astrophys.* **412,** 903–904 (2003).

39. Gilliland, R. L. *et al.* Asteroseismology of the transiting exoplanet host HD 17156 with hubble space telescope fine guidance sensor. *Astrophys. J.* **726,** 2 (2011).

40. Tassoul, M. Asymptotic approximations for stellar nonradial pulsations. *Astrophys. J. Suppl.* **43,** 469–490 (1980).

41. Lundkvist, M., Kjeldsen, H. & Silva Aguirre, V. AME—Asteroseismology made easy. Estimating stellar properties by using scaled models. *Astron. Astrophys.* **566,** A82 (2014).

42. Quirion, P.-O., Christensen-Dalsgaard, J. & Arentoft, T. Automatic determination of stellar parameters via asteroseismology of stochastically oscillating stars: comparison with direct measurements. *Astrophys. J.* **725,** 2176–2189 (2010).

43. Basu, S., Chaplin, W. J. & Elsworth, Y. Determination of stellar radii from asteroseismic data. *Astrophys. J.* **710,** 1596–1609 (2010).

44. Gai, N., Basu, S., Chaplin, W. J. & Elsworth, Y. An in-depth study of grid-based asteroseismic analysis. *Astrophys. J.* **730,** 63 (2011).

45. Basu, S., Verner, G. A., Chaplin, W. J. & Elsworth, Y. Effect of uncertainties in stellar model parameters on estimated masses and radii of single stars. *Astrophys. J.* **746,** 76 (2012).

46. Borucki, W. J. *et al.* Kepler-22b: A 2.4 Earth-radius planet in the habitable zone of a sun-like star. *Astrophys. J.* **745,** 120 (2012).

47. Seager, S. & Mallén-Ornelas, G. A unique solution of planet and star parameters from an extrasolar planet transit light curve. *Astrophys. J.* **585,** 1038–1055 (2003).

48. Tingley, B., Bonomo, A. S. & Deeg, H. J. Using stellar densities to evaluate transiting exoplanetary candidates. *Astrophys. J.* **726,** 112 (2011).

49. Kipping, D. M. Characterizing distant worlds with asterodensity profiling. *Mon. Not. R. Astron. Soc.* **440,** 2164–2184 (2014).

50. Sliski, D. H. & Kipping, D. M. A high false positive rate for Kepler planetary candidates of giant stars using asterodensity profiling. *Astrophys. J.* **788,** 148 (2014).

51. Rowe, J. F. *et al.* Validation of Kepler's multiple planet candidates. III. Light curve analysis and announcement of hundreds of new multi-planet systems. *Astrophys. J.* **784,** 45 (2014).

52. Pedregosa, F. *et al.* Scikit-learn: machine learning in python. *J. Mach. Learn. Res.* **12,** 2825–2830 (2011).

53. Christiansen, J. L. *et al.* The derivation, properties, and value of Kepler"s combined differential photometric precision. *Publ. Astron. Soc. Pac.* **124,** 1279–1287 (2012).

Acknowledgements

Funding for the Stellar Astrophysics Centre is provided by The Danish National Research Foundation (Grant agreement no.: DNRF106). The research is supported by the ASTERISK project (ASTERoseismic Investigations with SONG and Kepler) funded by the European Research Council (Grant agreement no.: 267864). This research has made use of NASA's Astrophysics Data System and the NASA Exoplanet Archive, which is operated by the California Institute of Technology, under contract with the National Aeronautics and Space Administration under the Exoplanet Exploration Program.

Author contributions

M.S.L. led the work, did grid modelling, computed the exoplanet parameters, did the bootstrapping, made the debiased sample and wrote the manuscript. H.K. determined the large frequency separations and performed simulations to assess the significance of the hot-super-Earth desert. M.S.L. and H.K. also designed the project and inspected the exoplanets in the non-seismic sample that fell in the hot-super-Earth desert. S.A. provided feedback on the radius-flux diagram and helped with the structure of the manuscript. G.R.D. ran the Gaussian mixture model and helped with the manuscript. V.V.E. gave feedback on the radius-flux diagram and checked the light curve of KOI 4198.01. D.H. helped with the structure of the manuscript. S.B., C.K. and V.S.A. did grid-modelling. C.V. did an independent analysis of some of the Kepler data. A.B.J. investigated the light curve of Kepler-4b. All authors participated in the interpretation of the results and commented on the manuscript.

Additional information

24

Observations of discrete harmonics emerging from equatorial noise

Michael A. Balikhin[1,*], Yuri Y. Shprits[2,3,*], Simon N. Walker[1], Lunjin Chen[4], Nicole Cornilleau-Wehrlin[5,6], Iannis Dandouras[7,8], Ondrej Santolik[9,10], Christopher Carr[11], Keith H. Yearby[1] & Benjamin Weiss[3]

A number of modes of oscillations of particles and fields can exist in space plasmas. Since the early 1970s, space missions have observed noise-like plasma waves near the geomagnetic equator known as 'equatorial noise'. Several theories were suggested, but clear observational evidence supported by realistic modelling has not been provided. Here we report on observations by the Cluster mission that clearly show the highly structured and periodic pattern of these waves. Very narrow-banded emissions at frequencies corresponding to exact multiples of the proton gyrofrequency (frequency of gyration around the field line) from the 17th up to the 30th harmonic are observed, indicating that these waves are generated by the proton distributions. Simultaneously with these coherent periodic structures in waves, the Cluster spacecraft observes 'ring' distributions of protons in velocity space that provide the free energy for the waves. Calculated wave growth based on ion distributions shows a very similar pattern to the observations.

[1] Department of Automatic Control and Systems Engineering, University of Sheffield, Mappin Street, Sheffield S1 3JD, UK. [2] Department of Earth Planetary and Space Sciences, UCLA, 595 Charles Young Drive East, Box 951567, Los Angeles, California 90095-1567, USA. [3] Department of Earth Atmospheric and Planetary Sciences, MIT, 77 Massachusetts Avenue, Cambridge, Massachusetts 02139-4307, USA. [4] W.B. Hanson Center for Space Sciences, Department of Physics, The University of Texas at Dallas, 800 West Campbell Road, Richardson, Texas 75080-3021, USA. [5] LPP, CNRS, École Polytechnique, Palaiseau 91128, France. [6] LESIA, Observatoire de Paris, Section de Meudon, 5, Place Jules Janssen, Meudon 92195, France. [7] CNRS, IRAP, 9, Avenue du Colonel Roche, Toulouse BP 44346-31028, France. [8] UPS-OMP, IRAP, 14, Avenue Edouard Belin, Toulouse 31400, France. [9] Department of Space Physics, Institute of Atmospheric Physics ASCR, Bocni II/1401, 14131 Praha 4, Czech Republic. [10] Faculty of Mathematics and Physics, Charles University in Prague, V Holesovickach 2, 18000 Praha 8, Czech Republic. [11] Blackett Laboratory, Imperial College London, South Kensington Campus, London SW7 2AZ, UK. * These authors contributed equally to this work. Correspondence and requests for materials should be addressed to Y.Y.S. (email: yshprits@igpp.ucla.edu).

Oscillations of the electric and magnetic field in plasmas, usually referred to as plasma waves, have been observed in the Earth's magnetosphere, interplanetary space, and most recently, outside of the heliosphere. In space, plasma waves exhibit a wide variety of modes and are classified according to their frequency, polarization characteristics, types of oscillation (longitudinal or transverse) and their dispersion relation, which is the relation between the frequency of the wave and its vector of propagation.

The OGO (Orbiting Geophysical Observatory) 3 space mission detected plasma waves that were very closely confined to the terrestrial magnetic equatorial region[1,2]. These emissions were observed above the proton gyrofrequency—the frequency at which a proton gyrates around the field line. Due to their close confinement to the equator, they were named 'equatorial noise', but are also referred to as fast magnetosonic waves or magnetosonic noise due to their properties. These emissions are one of the most common waves observed in space. While these waves are observed only very close to the geomagnetic equator, they are seen on around 60% of equatorial satellite traversals in the inner magnetosphere[3]. When these waves were discovered[1], it was also noted that they may also be in resonance with harmonics of electron bounce motion (periodic motion of trapped electrons along the field line between the mirror points) and thus may be potentially generated by electrons in the plasma.

Observations by Interplanetary Monitoring Platform 6 Satellite, Hawkeye 1 (Explorer 52) Satellite and the Geostationary Operational Environmental Satellite[2,4] showed cursory evidence for discrete frequency bands, suggesting that these waves may interact with protons, alpha particles and heavy ions trapped near the equator. However, the width and spacing of these bands in frequency appeared to be non-uniform and could not be accurately measured, except at the frequencies of the lowest harmonics. The spectral frequencies of these bands were in some cases approximately at harmonics of the proton gyrofrequency but did not match them exactly (see Supplementary Note 2). Clear observational and analytical evidence for this type of frequency spectrum has so far remained elusive. A number of very detailed follow-up studies, including a recent detailed statistical study using measurements from the Polar mission[5], either failed to find the discrete waves, or found spectral structures at frequencies different from harmonics of the local proton gyrofrequency[6]. The suggested explanation for the discrepancy between theory and observations was that the waves may be generated at different locations (near the equator) and propagate to the point of observation. However, since observations showing a clear harmonic structure were not available, the theory remained unverified by observations.

Multi-point Cluster observations presented in this study show remarkable observations of very distinct harmonic emissions coinciding with multiples of gyrofrequency on two Cluster spacecraft. The waves are observed exactly in the source region. Using the observed distributions of rings, we calculated the growth rates of magnetosonic waves and show that the results of the calculations are consistent with the observed harmonics between the 17th and 30th harmonic resonances. The presented observations of distinct periodic emissions exactly at the harmonics of the gyrofrequency together with the simulations of wave growth that are based on the observed ion distributions, definitively show that magnetosonic emissions are generated by unstable ion ring distributions.

Results

Cluster mission. The European Space Agency (ESA) Cluster mission[7] consists of four identically instrumented spacecraft in a polar, eccentric orbit (apogee 18.6, perigee 3 Earth radii) with a period of 57 h. Launched in August 2000, the mission has been operating since February 2001. During its lifetime, the inter-satellite separation has varied from less than a few hundred kilometres to over 20,000 km, to explore processes occurring within the magnetosphere at different spatial scales (see Supplementary note 3).

To resolve the long-standing scientific question of the generation and propagation of the equatorial noise, ESA's Cluster mission conducted a special Inner Magnetosphere Campaign aimed at studying the structure of these waves in their source region. Figure 1 shows that on 6 July 2013, all four spacecraft were close to the geomagnetic equator. Clusters 3 and 4 were very close to each other, within 60 km, while Cluster 1 was ~800 km from Clusters 3 and 4, and Cluster 2 was around 4,400 km in the earthward direction from the trio. Supplementary Fig. 6 and Supplementary Note 8 show the expected wavelength of the magnetosonic waves. Satellites 3 and 4 are separated by a maximum of 3–5 wavelengths, depending upon the propagation direction with respect to the separation vector.

The observations of waves made by the Cluster Spatio-Temporal Analysis of Field Fluctuations (STAFF) instrument on 6 July 2013, between 18:40 and 18:55 UT, not only present observational evidence for their generation, but also show the most remarkable example of their banded structure ever observed in space. Despite being commonly referred to as magnetosonic noise, the emissions observed by the Clusters 3 and 4 spacecraft separated by 60 km have a remarkably clear discrete structure between the 17th and 30th harmonics of the proton gyrofrequency (Fig. 2) in the frequency range in which equatorial noise is usually observed. This previously unobserved, well organized and periodic structure provides definitive evidence that these waves are generated by protons. The exact match between the harmonics and observed emissions lines shows that these observations are made right in the wave source region.

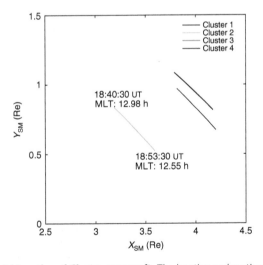

Figure 1 | Location of Cluster spacecraft. The location and motion of the four Cluster spacecraft during the period 18:40:30 to 18:53:30 UT on 6 July 2013 in which the emissions were observed (Clusters 1, 2, 3 and 4 are shown in black, green, red and blue respectively). The coordinate system used (known as Solar Magnetic (SM)) is aligned with the Earth's magnetic field, (the z direction is aligned with the magnetic dipole, and the Sun direction lies in the xz plane) in units of Earth radii (Re). Since spacecraft Cluster 3 and Cluster 4 (C3,4) are separated by only 60 km, their traces lie virtually on top of each other. Cluster 1 is around 1,000 km from C3,4, while C2 is around 4,000 km distant.

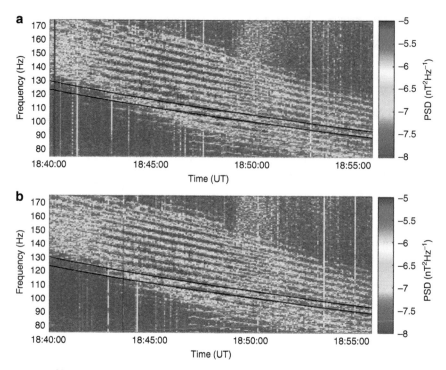

Figure 2 | Dynamical spectrograms. Observations by the STAFF instrument on (**a**) Cluster 3 and (**b**) Cluster 4 of the harmonic structure of magnetosonic waves near the equator. The figure shows the colour coded magnetic field Power Spectral Density (PSD) as a function of time and frequency. The 20th and 21st harmonics of the proton gyrofrequency are marked by solid black lines. Harmonics up to 30th are clearly seen, and one can outline the traces of the 31st and higher harmonics.

The Cluster measurements enabled not only the observation of the fine structure of the wave spectrum but also provided multi-satellite measurements of this emission at very short separation distances. The periodic pattern of emissions between the 17th and 30th harmonics observed on Cluster 4 is almost an exact replication of that observed by Cluster 3. The similarity of the signals has been analysed with the use of the coherency function (Supplementary Fig. 1 and Supplementary Note 1). The high coherence (>0.8) between the signals at harmonic frequencies of the gyrofrequency show that their separation is less than the wave coherency length and that this remarkably organized periodic structure is at least 60 km in scale which encompasses several wavelength (Supplementary Fig. 1, Supplementary Note 6).

Supplementary Fig. 2 and Supplementary Note 4 shows a comparison of the wave observations made by all four Cluster spacecrafts. While Cluster 1 observes similar discrete pattern of waves, the waves are not coherent with the Cluster 3 and Cluster 4 observations. Cluster 2 is more distant from Clusters 3 and 4, and did not observe similar type emissions.

Cluster measurements also allow to determine the polarization properties of waves to confirm that the observed emissions are the same type as the usually observed magnetosonic noise waves (Fig. 3). The fluctuating wave magnetic field on Cluster 4 is orientated parallel/antiparallel to the background magnetic field, the wave propagates at highly oblique wave normal angles, and shows linear polarization, confirming that these are typical equatorial magnetosonic waves[8]. For comparison, the polarization properties resulting from spacecrafts 1 and 3 are shown in Supplementary Figs 3 and 4.

The Cluster spacecraft also provided an opportunity to observe the source of free energy for this wave. It has been suggested[4,9] that ring-like particle distributions in velocity space may lead to wave generation through the development of instabilities. Figure 4 shows the momentum space distribution of protons near Alfvén speed (the characteristic speed at which low

frequency waves propagate within a plasma) observed by the Cluster Ion Spectrometry Composition Distribution function (CIS CODIF) analyser instrument. Particle distributions at all CIS measured energies are shown in Supplementary Figs 5 and 6 and Supplementary Note 5. The observed 'ring' distribution is unstable and results in the generation of waves[9].

The unique observations by the multiple Cluster spacecraft in the vicinity of the geomagnetic equator clearly show the fine periodic structure of magnetosonic waves generated in their source region and the simultaneous occurrence with 'ring-type' ion distributions.

Excitation of waves and growth rates. The linear growth rate can be expressed as a sum of different harmonics of an integral over perpendicular velocity that depends on the gradients of ion phase space density in the velocity space, and can be expressed as[10]

$$\sum_{n=-\infty}^{+\infty} \int_0^\infty dv_\perp \left(W_{n\perp} \frac{\partial f}{\partial v_\perp} + W_{n\parallel} \frac{\partial f}{\partial v_\parallel} \right)_{v_\parallel = v_{res\parallel}} \quad (1)$$

where n is the harmonic number, $v_\perp$ and $v_\parallel$ are the perpendicular and parallel velocities with respect to background magnetic field, and $W_{n\perp}$ and $W_{n\parallel}$ are weighting functions. $v_\parallel$ in equation (1) is taken at resonance velocities corresponding to different order resonances of harmonic number n. Since the waves are highly oblique, the resonance occurs with the ions of $v_\parallel \sim 0$ only when the wave frequency is approximately equal to multiples of the ion gyrofrequency, while there are few resonant ions when the wave frequency is not in the vicinity of multiples of the ion gyrofrequency. The injection of protons will create a ring-type distribution, where phase space density has a positive df/dv along the v direction. This ion distribution may be unstable and provide the free energy for the wave excitation with growth rate maximizing at multiples of the ion gyrofrequency.

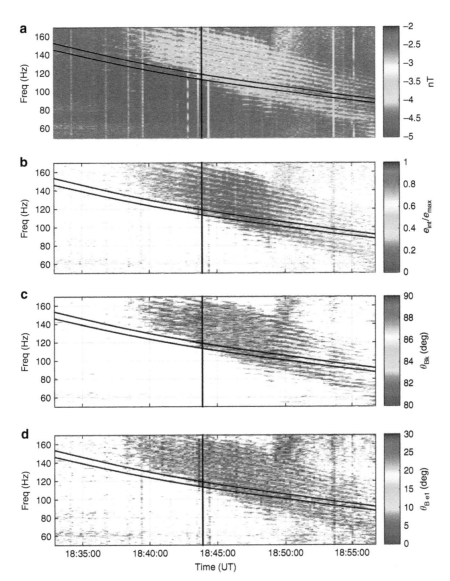

Figure 3 | Polarization properties of the magnetosonic waves observed by Cluster 4 on 6 July 2013. (**a**) The spectrum of the waveform STAFF-SC Bz component. (**b**) The ellipticity of the waves representing the polarization of the emissions. Values close to unity indicate circular polarization while those in the region of zero are indicative of linear polarization. (**c**) The wave normal angle with respect to the external magnetic field. (**d**) The angle between the external magnetic field and the oscillating magnetic field of the wave. The horizontal black lines represent the 20th and 21st harmonics of the proton gyrofrequency. Cluster 4 crossed the geomagnetic equator at the time marked by the vertical black line.

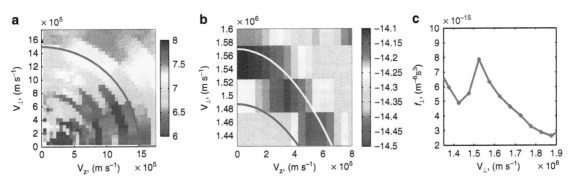

Figure 4 | Observations of the ion distribution in velocity space. (**a**) Distributions of proton fluxes in velocity space at 18:54:33 ᴜᴛ. (**b**) Distribution of phase space density for quasi-perpendicular ions at 18:54:33 ᴜᴛ. The white line denotes velocity contour of $1.57 \times 10^6 \, m^{-3}$ while the blue line denotes a velocity contour of value equal to Alfven speed $1.487 \times 10^6 \, m^{-3}$. (**c**) A plot of the ring distribution for the phase space density of protons gyrating near the equatorial plane (particles bouncing very near the equator). The blue line connecting the data points outlines the shape of the distribution. y axis is the density in phase space, and x axis is the velocity of particles.

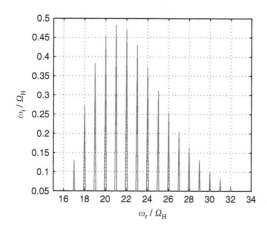

Figure 5 | Theoretical linear growth rates based on the measured ion distributions. Growth rates ω_i normalized to the proton gyrofrequency are given as a function of wave frequency ω_r which is also normalized to the proton gyrofrequency (Ω_H). Growth rate of magnetosonic waves is calculated for nearly perpendicularly directed wave vector (89.5° angle between wave vector direction and the background magnetic field).

Observations of ion distributions enable the calculation of the wave growth rates. Linear growth rates[11] are calculated using the measured ion distributions, magnetic field and plasma density measurements inferred from wave observations. For the growth rate calculation, we use a background magnetic field $B_0 = 305\,\text{nT}$, electron number density of $20 \times 10^6\,\text{m}^{-3}$ and corresponding Alfvén speed $1.487 \times 10^6\,\text{m s}^{-1}$. To represent the observed ring, we assume a Gaussian ring distribution with number density $0.008 \times 10^6\,\text{m}^{-3}$, peak velocity $1.57 \times 10^6\,\text{m s}^{-1}$ and a width of $0.2 \times 10^6\,\text{m s}^{-1}$ (as shown in Supplementary Fig. 7) and evaluate the gradient of the proton distribution in velocity space to calculate the growth rate.

The growth rates show the frequencies at which waves should be theoretically observed. Figure 5 shows the linear growth rate as a function of frequency normalized to the proton gyrofrequency. The general structure of peaks is very similar to those observed (as shown in Fig. 2), with maximum growth rates occurring between the 17th and 30th harmonics. Traces of the higher harmonics can be also seen in Fig. 2

Discussion

Fast magnetosonic waves have recently attracted much attention because they are capable of accelerating particles to high energies or providing a mechanism that results in the loss of these particles into the atmosphere[5,12,13] and may be important for space weather. The observed discrete structure of magnetosonic waves may play an important role in the acceleration and scattering of electrons and ions by these waves. The discrete nature of these waves may change how these waves interact with electrons during both gyro and bounce resonance interactions and may determine acceleration and loss rates for electrons in the radiation belts. The presence of such highly structured waves may be also used in the future as a tell-tale sign of ion ring distributions.

Similar wave generation mechanisms may also operate in the magnetospheres of the outer planets, close to the Sun and in distant corners of the universe. Understanding the mechanisms behind the generation of waves is most important for laboratory plasma and for finding new ways to remotely heat plasma.

Methods

Wave propagation analysis. The data sets obtained by the STAFF-SC instrument consist of a time series of vector measurements of the magnetic field. During the period of observation, the data are sampled at 450 Hz. The spectral information, shown in Fig. 2, is obtained in this particular case with the use of the fast Fourier transform technique. This results in a frequency representation of the time series data. The polarization parameters are obtained as follows. For each frequency resulting from the fast Fourier transform process, the three spectral components corresponding to the three component measurements are combined to form the spectral matrix. By analysing the complete spectral matrix using singular value decomposition (SVD) techniques[14], it is possible to obtain the wave vector direction (without distinguishing between parallel and antiparallel directions), the orientation and the size of the polarization ellipse, and the planarity of the polarization (not shown).

SVD is a general method used to factorize a real or complex matrix, and a corresponding detailed implementation is discussed in a study[14]. This factorization process is similar to performing a least-squares fit to the data, but without actually solving the minimization problem.

Wave growth calculation. Wave growth is calculated by solving the kinetic dispersion relation in a uniformly magnetized plasma[11,15] consisting of three components: a cold electron component, a cold proton component and a hot proton ring. Several assumptions are made to facilitate the calculation using a non-Maxwellian ion distribution. First, cold plasma is dominant over the hot proton component, which allows the cold dispersion relation to be used to approximate the real part of the kinetic dispersion relation. This assumption is valid because the measured proton ring density is much smaller than the cold plasma density. Second, the growth rate is small compared with the wave frequency, which is also verified by our calculation results. With these two assumptions, we can obtain the temporal growth rate in terms of proton phase space density gradients in velocity space (as shown in the equation (1)), evaluated at resonant protons satisfying $\omega - k_{\parallel}\, v_{\parallel} = n\Omega$.

Instrumentation. The data presented in this paper were collected by the Fluxgate Magnetometer (FGM)[16], the STAFF-SC search coil magnetometer[17] and the CIS CODIF mass-resolving ion spectrometer[18]. During the period of observation presented here, the satellites were operating in science burst mode 1. In this mode, FGM sampled the DC magnetic field at 67 Hz whilst STAFF-SC sampled the AC magnetic field at 450 Hz through a 180 Hz filter.

References

1. Russell, C. T., Holzer, R. E. & Smith, E. J. OGO 3 observations of ELF noise in the magnetosphere 2. The nature of the equatorial noise. *J. Geophys. Res.* **75**, 755–768 (1970).
2. Gurnett, D. A. Plasma wave interactions with energetic ions near the magnetic equator. *J. Geophys. Res.* **81**, 2765–2770 (1976).
3. Santolik, O. *et al.* Systematic analysis of equatorial noise below the lower hybrid frequency. *Ann. Geophys.* **22**, 2587–2595 (2004).
4. Perraut, S. *et al.* A systematic study of ULF waves above FH+ from GEOS 1 and 2 measurements and their relationship with proton ring distributions. *J. Geophys. Res.* **A87**, 6219–6236 (1982).
5. Tsurutani, B. T. *et al.* Extremely intense ELF magnetosonic waves: a survey of polar observations. *J. Geophys. Res.* **119**, 964–977 (2014).
6. Santolik, O., Pickett, J. S., Gurnett, D. A., Maksimovic, M. & Cornilleau-Wehrlin, N. Spatiotemporal variability and propagation of equatorial noise observed by Cluster. *J. Geophys. Res.* **107**, 1495 (2002).
7. Escoubet, C. P., Schmidt, R. & Goldstein, M. L. Cluster—science and mission overview. *Space Sci. Rev.* **79**, 11–32 (1997).
8. Boardsen, S. A. *et al.* Van Allen Probe observations of periodic rising frequencies of the fast magnetosonic mode. *Geophys. Res. Lett.* **41**, 8161–8168 (2014).
9. Horne, R. B., Wheeler, G. V. & Alleyne, H. S. C. K. Proton and electron heating by radially propagating fast magnetosonic waves. *J. Geophys. Res.* **105**, 27597–27610 (2000).
10. Kennel, C. F. Low Frequency Whistler Mode. *Phys. Fluids* **9**, 2190 (1966).
11. Chen, L., Thorne, R. M., Jordanova, V. K. & Horne, R. B. Global simulation of magnetosonic wave instability in the storm time magnetosphere. *J. Geophys. Res.* **115**, A11222 (2010).
12. Horne, R. B. *et al.* Electron acceleration by fast magnetosonic waves. *Geophys. Res. Lett.* **34**, L17107 (2007).
13. Shprits, Y. Y. Potential waves for pitch-angle scattering of near-equatorially mirroring energetic electrons due to the violation of the second adiabatic invariant. *Geophys. Res. Lett.* **36**, L12106 (2009).
14. Santolik, O., Parrot, M. & Lefeuvre, F. Singular value decomposition methods for wave propagation analysis. *Radio Sci.* **38**, 10–11 (2003).
15. Boardsen, S. A., Gallagher, D. L., Gurnett, D. A., Peterson, W. R. & Green, S. L. Funnel-shaped, low-frequency equatorial waves. *J. Geophys. Res.* **97**, 967–14,976 (1992).

16. Balogh, A. *et al.* The Cluster magnetic field investigation. *Space Sci. Rev.* **79,** 65–91 (1997).

17. Cornilleau-Wehrlin, N. *et al.* The Cluster Spatio-Temporal Analysis of Field Fluctuations (STAFF) experiment. *Space Sci. Rev.* **79,** 107–136 (1997).

18. Rème, H. *et al.* First multispacecraft ion measurements in and near the Earth's magnetosphere with the identical Cluster ion spectrometry (CIS) experiment. *Ann. Geophys.* **19,** 1303–1354 (2001).

Acknowledgements

This research was supported by the Cluster mission. Y.Y.S. would like to acknowledge the support of The Presidential Early Career Award for Scientists and Engineers through NASA grant NNX10AK99G, NSF GEM AGS-1203747 and UC Lab Fee grant 12-LR-235337. O.S. acknowledges support from grants Praemium Academiae, LH12231, and LH14010. M.A.B. acknowledges a Royal Society International Collaboration Grant and EPSRC grant EP/H00453X/1, project PROGRESS funded from the European Union's Horizon 2020 research and an innovation programme under the grant agreement No 637302, and International Space Science Institute (Bern). L.C. acknowledges the support of NSF grant AGS-1405041.

Author contributions

Y.Y.S. and M.A.B. led the study and coordinated the efforts among different institutions. The manuscript was largely written by Y.Y.S. with contributions from M.A.B. and all other coauthors. N.C.W., C.C., I.D., B.W. and K.H.Y. provided the data sets and advice on their usage. S.N.W., M.A.B. and O.S. performed the wave analysis, Y.Y.S. and L.C. performed the particle analysis, and L.C. performed growth rate calculations. All authors commented on the manuscript and participated in the analysis of the results.

Additional information

Competing financial interests: The authors declare no competing financial interests.

25

Carbon-depleted outer core revealed by sound velocity measurements of liquid iron–carbon alloy

Yoichi Nakajima[1], Saori Imada[2,3], Kei Hirose[3,4], Tetsuya Komabayashi[2,5], Haruka Ozawa[4,6], Shigehiko Tateno[3,6], Satoshi Tsutsui[7], Yasuhiro Kuwayama[8] & Alfred Q.R. Baron[1,7]

The relative abundance of light elements in the Earth's core has long been controversial. Recently, the presence of carbon in the core has been emphasized, because the density and sound velocities of the inner core may be consistent with solid Fe_7C_3. Here we report the longitudinal wave velocity of liquid $Fe_{84}C_{16}$ up to 70 GPa based on inelastic X-ray scattering measurements. We find the velocity to be substantially slower than that of solid iron and Fe_3C and to be faster than that of liquid iron. The thermodynamic equation of state for liquid $Fe_{84}C_{16}$ is also obtained from the velocity data combined with previous density measurements at 1 bar. The longitudinal velocity of the outer core, about 4% faster than that of liquid iron, is consistent with the presence of 4–5 at.% carbon. However, that amount of carbon is too small to account for the outer core density deficit, suggesting that carbon cannot be a predominant light element in the core.

[1] Materials Dynamics Laboratory, RIKEN SPring-8 Center, RIKEN, Hyogo 679-5148, Japan. [2] Department of Earth and Planetary Sciences, Tokyo Institute of Technology, Tokyo 152-8551, Japan. [3] Earth-Life Science Institute, Tokyo Institute of Technology, Tokyo 152-8550, Japan. [4] Laboratory of Ocean-Earth Life Evolution Research, Japan Agency for Marine-Earth Science and Technology, Kanagawa 237-0061, Japan. [5] School of GeoSciences and Centre for Science at Extreme Conditions, University of Edinburgh, Edinburgh EH9 3FE, UK. [6] Institute for Study of the Earth's Interior, Okayama University, Tottori 682-0193, Japan. [7] Research and Utilization Division, SPring-8, Japan Synchrotron Radiation Research Institute, Hyogo 679-5198, Japan. [8] Geodynamics Research Center, Ehime University, Ehime 790-8577, Japan. Correspondence and requests for materials should be addressed to Y.N. (email: yoichi.nakajima@spring8.or.jp).

Sound velocity and density are important observational constraints on the chemical composition of the Earth's core. While properties of solid iron alloys have been extensively examined by laboratory studies to core pressures ($>136\,\text{GPa}$)[1–3], little is known for liquid alloys because of experimental difficulties. The core is predominantly molten, and the longitudinal wave (P-wave) velocity of liquid iron alloy is the key to constraining its composition. However, previous static high-pressure and -temperature (P–T) measurements of liquid iron alloys were performed only below 10 GPa using large-volume presses[4–6]. Shock wave experiments have been carried out at much higher pressures but only along a specific Hugoniot P–T path[7,8].

Carbon is one of the possible light alloying components in the core because of its high cosmic abundance and strong chemical affinity with liquid iron[9]. Its high metal/silicate partition coefficients indicate that thousands of parts per million to several weight percent of carbon could have been incorporated into the core during its formation[9–11]. In addition, recent experimental and theoretical studies[12,13] have suggested that solid Fe_7C_3 may explain the properties of the inner core, in particular its high Poisson's ratio[14,15], supporting the presence of carbon in the core.

In this study, we determine the P-wave velocity (V_P) (equivalent to bulk sound velocity, V_Φ, in a liquid) of liquid $Fe_{84}C_{16}$ at high P–T based on inelastic X-ray scattering (IXS) measurements. Combined with its density data at 1 bar (ref. 16) both velocity and density (ρ) profiles of liquid $Fe_{84}C_{16}$ along adiabatic compression are obtained. They are compared with seismological observations, indicating that both V_P and ρ in the Earth's outer core are not explained simultaneously by liquid Fe-C.

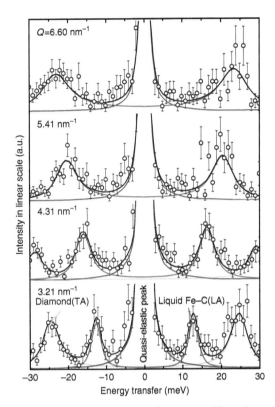

Figure 1 | Typical inelastic X-ray scattering spectra. These data were collected at 26 GPa and 2,530 K at momentum transfers Q, as indicated. The spectra include three components: a quasi-elastic peak near zero energy transfer (blue), longitudinal acoustic (LA) phonon mode of liquid $Fe_{84}C_{16}$ (red), and transverse acoustic (TA) phonon mode of diamond (turquoise).

Results

Longitudinal wave velocity measurements. We collected the high-resolution IXS spectra from liquid $Fe_{84}C_{16}$ (4.0 ± 0.3 wt.% carbon) at static high P–T using both resistance- and laser-heated diamond-anvil cells (Methods; Fig. 1). The starting material was synthesized beforehand as a mixture of fine-grained Fe and Fe_3C at 5 GPa and 1,623 K in a multi-anvil apparatus. Experimental P–T conditions were well above the eutectic temperature in the Fe–Fe_3C binary system (Supplementary Fig. 1). The carbon concentration in the eutectic liquid is known to be 3.8–4.3 wt.% at 1 bar to 20 GPa (ref. 17), almost identical to the composition of our sample. Above 20 GPa, we heated the sample to temperatures comparable or higher than the melting temperature of Fe_3C, a liquidus phase in the pressure range explored, assuring a fully molten sample. The molten state of the specimen was carefully confirmed, before and after the IXS measurements, by the absence of diffraction peaks from the sample (Fig. 2). We sometimes, depending on a sample volume, were also able to observe the diffuse diffraction signal typical of a liquid.

The V_P of liquid $Fe_{84}C_{16}$ was determined between 7.6 and 70 GPa (Fig. 3 and Supplementary Table 1) from dispersion curves for a range of momentum transfer (Fig. 4). It was found to be 15–30% smaller than that of solid Fe (refs 3,18–20) and Fe_3C (refs 21–23; note that a starting material in the present experiments was a mixture of these solid phases) (Fig. 5), confirming that we measured a liquid sample. The velocities of a fictive solid $Fe_{84}C_{16}$ alloy are also estimated assuming a linear velocity change between Fe (ref. 24) and Fe_3C (ref. 23) indicating that V_P drops by 13% upon melting at 2,300 K, a eutectic temperature at 45 GPa (ref. 17). Such a velocity change is comparable to that expected for pure Fe. The difference in V_Φ between solid and liquid $Fe_{84}C_{16}$ is very small (1.8%). On the

other hand, the V_P of our liquid $Fe_{84}C_{16}$ sample is 3–14% faster at 8–70 GPa than that of liquid Fe determined by shock-wave study[8] (Fig. 3).

Earlier ultrasonic measurements performed below 10 GPa reported a change in V_P by <2–3% per 1,000 K for liquid Fe–S alloys[4,5]. Theoretical calculations[25–27] and shock compression data[8] on liquid Fe and Fe–S alloy demonstrated even smaller effects above 100 GPa (<0.5% by 1,000 K). It is therefore very likely that the V_P of liquid $Fe_{84}C_{16}$ is also not sensitive to temperature with the temperature effect much smaller than the uncertainty in the present velocity determinations (±3%).

Thermodynamical equation of state. V_P of a liquid can be described using the Murnaghan equation of state[4] (Methods) as;

$$V_P = \sqrt{\frac{K_{S0}}{\rho_0}\left(1 + \frac{K'_S}{K_{S0}}P\right)^{\frac{1}{2} - \frac{1}{2K'_S}}}, \qquad (1)$$

where K_S and K'_S are adiabatic bulk modulus and its pressure derivative, respectively (zero subscripts denote values at 1 bar and $T = T_0$). Here, consistent with the discussion above, we neglect the temperature dependence of our V_P data, while ρ_0 is taken to be temperature dependent[16] (Methods). We fit equation (1) to our P–V_P data for liquid $Fe_{84}C_{16}$ and find $K_{S0} = 110 \pm 9$ GPa and $K'_S = 5.14 \pm 0.30$ when $T_0 = 2,500$ K (Supplementary Table 2 and Supplementary Fig. 2). The choice of T_0 and, accordingly, the variation in ρ_0 practically changed K_{S0} and K'_{S0} as $(\partial K_{S0}/\partial T) = -9.4 \times 10^{-3}$ GPa K^{-1} and $(\partial K'_{S0}/\partial T) = -2.7 \times 10^{-4}$ K^{-1}. Our value for K_{S0} is similar to that for liquid iron[8] but for K'_S is higher than that for pure iron, $K'_S = 4.7$.

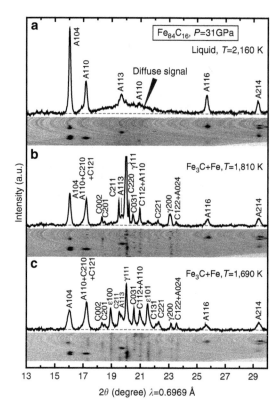

Figure 2 | X-ray diffraction spectra before and after melting. They were collected at 2,160 K (**a**), 1,810 K (**b**) and 1,690 K (**c**) during heating at 31 GPa. The starting material was composed of Fe (ε or γ) and Fe_3C (**c**), and the peaks of Al_2O_3 (**a**) were from a thermal insulator. The coexistence of ε- and γ-Fe phases at 1,610 K was due to a sluggish solid–solid phase transition[49] and the peaks from the ε-phase were lost at 1,810 K. All sample peaks disappeared between 1,810 and 2,160 K. In addition, the background was enhanced slightly, indicating a diffuse scattering signal from a liquid sample.

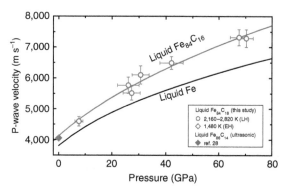

Figure 3 | Compressional wave velocity of liquid $Fe_{84}C_{16}$. Open circles, obtained by laser-heated DAC; open diamond, by external-resistance-heated DAC. The data at 1 bar is from ultrasonic measurements[28] (closed diamond). The red curve represents a thermodynamical fitting result for liquid $Fe_{84}C_{16}$, compared with the velocity of liquid Fe (black curve)[8].

This suggests that liquid $Fe_{84}C_{16}$ becomes progressively stiffer than liquid Fe with increasing pressure. We also found $V_{P0} = 4,121 \pm 177\ \mathrm{m\,s^{-1}}$ for liquid $Fe_{84}C_{16}$ from K_{S0} and ρ_0, in good agreement with a previous study[28] of liquid $Fe_{86}C_{14}$ at 1 bar (4,050 m s^{-1}) and faster than $V_{P0} = 3,860\ \mathrm{m\,s^{-1}}$ for liquid Fe (ref. 8).

To compare the present results with earlier density measurements of liquid Fe–C alloys at high pressure[29,30] the isothermal

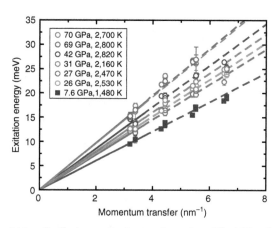

Figure 4 | Longitudinal acoustic phonon dispersion of liquid $Fe_{84}C_{16}$. The dispersion data were obtained at pressures from 7.6 to 70 GPa. Only data collected with the low momentum transfer ($<3.5\,\mathrm{nm^{-1}}$) were used to determine the velocity to avoid possible anomalous dispersion for liquid (see Methods).

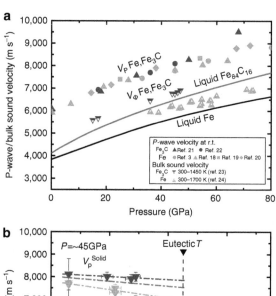

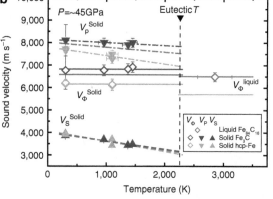

Figure 5 | Comparison of velocities between liquid and solid Fe–C alloys. (**a**), P-wave velocities (V_P) and bulk sound velocities (V_Φ) of solid Fe (turquoise)[3,18–20,24] and Fe_3C (blue)[21–23] were determined by previous IXS and nuclear inelastic scattering (NIS) measurements. The V_P for liquid Fe is from shock-wave study[8]. (**b**), Temperature effects on the sound velocities of Fe–C alloys at $\sim$45 GPa. The V_P ($= V_\Phi$) of liquid $Fe_{84}C_{16}$ and liquid Fe is from the present work at 42 GPa and shock-wave data[8], respectively. The V_P, shear velocity (V_S), and V_Φ for solid Fe and Fe_3C were reported by NIS measurements[23,24]. Red lines for fictional solid $Fe_{84}C_{16}$ are estimated from a linear relationship between Fe and Fe_3C.

bulk modulus for liquid $Fe_{84}C_{16}$ is estimated to be $K_{T0} = 100$ (82) GPa at 1,500 K (2,500 K) from our determination of K_S combined with Grüneisen parameter $\gamma_0 = 1.74$ (ref. 8) and

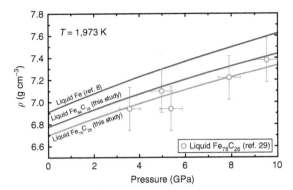

Figure 6 | Comparison with previous density measurements. Blue and red curves demonstrate calculated densities at 1,973 K for liquid Fe (ref. 8) and $Fe_{84}C_{16}$ (present study). The density of liquid $Fe_{75}C_{25}$ (turquoise curve) is estimated assuming linear compositional dependence between pure Fe and $Fe_{75}C_{25}$, which shows good agreement with the previous measurements at 1,973 K (ref. 29).

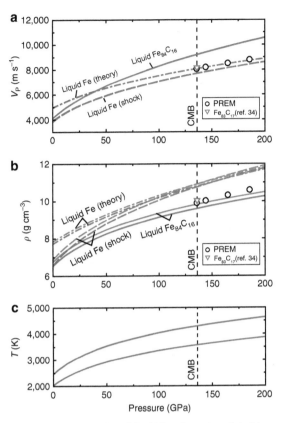

Figure 7 | Velocity and density of liquid $Fe_{84}C_{16}$ extrapolated to core pressures. The P-wave velocity (**a**) and density (**b**) profiles of liquid $Fe_{84}C_{16}$ are calculated along two adiabatic temperature curves (**c**) of 3,600 K (blue) and 4,300 K (red) at the CMB. Those for liquid Fe are from shock-compression experiments[8] and theoretical calculations[26]. The velocity and density for liquid $Fe_{83}C_{17}$ at the CMB (4,300 K) are by theory[34]. PREM denotes seismologically deduced Preliminary Reference Earth model[35].

thermal expansion coefficient[16] (Methods). When applying $C_P/C_V = 1.125$ at 1,820 K for liquid $Fe_{86}C_{14}$ derived from theoretical calculations[31], $K_{T0} = 106–98$ GPa is obtained at the same temperature range. These K_{T0} values for liquid $Fe_{84}C_{16}$ are similar to $K_{T0} = 95–63$ GPa for liquid Fe at 1,500–2,500 K (ref. 8) On the other hand, they are significantly larger than $K_{T0} = 55.4$ GPa for liquid $Fe_{86}C_{14}$ at 1,500 K and $K_{T0} = 50$ GPa for liquid $Fe_{75}C_{25}$ at 1,973 K from previous density measurements[29,30]. However, the calculated density for $Fe_{84}C_{16}$ using the present EoS are in reasonable agreement with the previous density measurements of $Fe_{75}C_{25}$ (ref. 29) (Fig. 6). The disagreement of elastic parameters with such earlier experiments may be attributed either to the limited pressure range of the previous density determinations, or to a different structure or magnetic (or electronic) change in the state of the liquid Fe–C at low pressure, as has been suggested from the change in compressional behaviour of liquid $Fe_{78}C_{22}$ around 5 GPa (ref. 6). Our data were collected above 7.6 GPa, so that the physical properties of liquid Fe–C obtained here should be more applicable to the Earth's core.

ρ of liquid $Fe_{84}C_{16}$ is then given, using the elastic parameters determined above, by;

$$\rho = \rho_0 \left(1 + \frac{K'_S}{K_{S0}} P\right)^{\frac{1}{K'_S}}. \tag{2}$$

Equations (1) and (2) give the V_P and ρ profiles for adiabatic compression (Methods), assuming $\gamma_0 = 1.74$, the same as that of liquid Fe (ref. 8) (Fig. 7). We find $V_P = 9,200$ m s^{-1} and $\rho = 9.82–9.61$ g cm^{-3} at the core-mantle boundary (CMB) for $T_{CMB} = 3,600–4,300$ K (refs 32,33) This indicates that V_P of liquid $Fe_{84}C_{16}$ is 19.6% faster than that of liquid Fe at the CMB[8], implying that the addition of 1 at.% carbon increases the V_P of liquid Fe by 1.2%. The extrapolation of the present experimental data using the Murnaghan equation of state may overestimate the V_P by $2-4\%$ at the CMB (Supplementary Note 1 and Supplementary Fig. 3), but, even if this is the case, 1 at.% carbon enhances the V_P of liquid Fe by as large as 0.8%. Indeed, the effect of carbon is much larger than a recent theoretical prediction of only 0.2% increase in velocity per 1 at.% carbon at 136 GPa (ref. 34). On the other hand, our data show that the incorporation of 1 at.% carbon reduces the density of liquid Fe by 0.6–0.7%, while theory suggested only 0.3% density reduction by 1 at.% carbon[34].

Discussion

We now compare the sound velocity and density of liquid $Fe_{84}C_{16}$ and liquid Fe with the seismologically based PREM model[35] for the outer core (Fig. 8). The V_P and ρ of liquid Fe are 4.6% slower and 10.1–8.6% denser, respectively, than the PREM at the CMB (3,600–4,300 K). To match the PREM values, considering the uncertainty of data extrapolation to higher pressures (Supplementary Note 1), only 5.2–4.0 at.% (1.2–0.9 wt.%) carbon is required to match the velocity, whereas 15.4–12.0 at.% (3.8–2.9 wt.%) carbon is necessary to account for the density. Therefore, carbon cannot be a predominant light element in the outer core.

These results suggest there is <5.2 at.% (1.2 wt.%) carbon in the outer core, consistent with the previous cosmochemical and geochemical arguments. In particular, the silicate portion of the Earth exhibits much higher $^{13}C/^{12}C$ isotopic ratio than that of Mars, Vesta and chondrite meteorites, as may be attributed to a strong enrichment of ^{12}C in core-forming metals[9]. The carbon isotopic fractionation that occurred during continuous core-formation process proposed previously[36,37] will give a reasonable $^{13}C/^{12}C$ ratio in the silicate Earth, and yields 1 wt.% carbon in the core[9]. In addition, Wood et al.[9] demonstrated that carbon strongly affects the chemical activity of Mo and W in liquid metal, so that their abundance in the mantle can be explained by partitioning between silicate melt and core-forming metal with ~0.6 wt.% carbon. It has been repeatedly suggested that the inner core may be composed of Fe_7C_3, which accounts

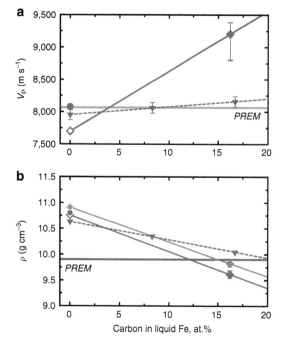

Figure 8 | Effect of carbon on the velocity and density of liquid Fe at 136 GPa. (a) Velocity and (**b**) density for liquid $Fe_{84}C_{16}$ at $T_{CMB} = 4,300\,K$ (red) and $3,600\,K$ (blue). Present results (closed diamonds, solid curves) are compared with theoretical calculations[34] (triangles, broken curve). The data for pure Fe are from shock compression study[8] (open diamonds) and theoretical calculations[26] (closed circles). PREM denotes seismological observations[35] at the CMB.

for high Poisson's ratio observed[14,15]. The crystallization of solid Fe_7C_3 from a liquid outer core with <1.2 wt.% carbon may still be possible if sulfur is also included in the core[38].

Methods

High P–T generation. Molten Fe–C alloy was obtained at high P–T in an external-resistance-heated (EH) or laser-heated (LH) diamond-anvil cell (DAC; Supplementary Table 1) using facilities installed at SPring-8. A disc of pre-synthesized $Fe_{84}C_{16}$ sample, 20–25 μm thick and 100–120 μm in diameter, was loaded into a hole of a rhenium gasket, together with two 12–17 μm thick single-crystal Al_2O_3 sapphire discs that served as both thermal and chemical insulators. The sample was compressed with 300 μm culet diamond anvils to a pressure of interest before heating.

In LH-DAC experiments, the sample was heated at high pressure from both sides by using two 100 W single-mode Yb fibre lasers (YLR-100-AC, IPG Photonics Corp.). The Gaussian-type energy distribution of the laser beam was converted into flat-top one with a refractive beam shaper (GBS-NIR-H3, Newport Corp.). A typical laser spot was 50–70 μm in diameter on the sample, much larger than X-ray beam size (~17 μm). We determined temperature by a spetroradiometric method, and its variations within the area irradiated by X-rays and fluctuations during IXS measurements were < ± 10%. The pressure was obtained from the equation of state for Fe_3C (ref. 39) from the lattice constant observed before melting at 1,800–2,500 K. Its error was derived from uncertainties in both temperature and the volume of Fe_3C. A typical image of a sample recovered after the laser heating experiment at 70 GPa and 2,700 K is given in Supplementary Fig. 4.

Only run #FeC08 was conducted in an EH-DAC. The whole sample was homogeneously heated by a platinum-resistance heater placed around the diamonds. The temperature was obtained with a Pt-Rh (type-R) thermocouple whose junction was in contact with the diamond near a sample chamber. The temperature uncertainty was <20 K. We determined the pressure based on the Raman shift of a diamond anvil[40] before heating at 300 K, whose uncertainty may be as much as ±20%.

IXS measurements. The sound velocity of liquid Fe–C alloy was determined in the DAC by high-resolution IXS spectroscopy at the beamline BL35XU, SPring-8 (ref. 41). Both LH- and EH-DACs were placed into vacuum chambers to minimize background scattering by air. The measurements were carried out with ~2.8 meV energy resolution using Si (999) backscattering geometry at 17.79 keV. The

experimental energy resolutions were determined using scattering from Polymethyl-methacrylate. The incident X-ray beam was focused to about 17 μm size (full width at half maximum) in both horizontal and vertical directions by using Kirkpatrick–Baez mirrors[42]. The X-ray beam size was much smaller than heated area (50–70 μm for LH-DAC). Scattered photons were collected by an array of 12 spherical Si analyzers leading to 12 independent spectra at momentum transfers (Q) between 3.2 and 6.6 nm^{-1} with a resolution $\Delta Q \sim 0.45\,nm^{-1}$ (full width) that was set by slits in front of the analyzer array. The energy transfer range of ± 30 (or − 10 to ± 30) meV was scanned for 1–3 h. Before and after IXS data collections, sample melting was confirmed by X-ray diffraction data (Fig. 2) that was collected, in situ, by switching a detector to a flat panel area detector (C9732DK, Hamamatsu Photonics K.K.)[43].

The IXS spectra included three (sometimes five) peaks (Fig. 1) of Stokes and anti-Stokes components of the longitudinal acoustic (LA) phonon mode from the sample (sometimes also from a diamond), and a quasi-elastic contribution near zero energy transfer. These spectra were fitted with the damped harmonic oscillator (DHO) mode[44] for acoustic phonon modes and with Lorenzian function for quasi-elastic peaks convolved by experimental resolution function. The DHO model function can be described as;

$$S^{DHO}(Q,\omega) = \left[\frac{1}{1 - e^{-\hbar\omega/k_B T}}\right]\frac{A_Q}{\pi}\frac{4\omega\omega_Q\gamma_Q}{(\omega^2 - \Omega_Q^2)^2 + 4\gamma_Q^2\omega^2}, \quad (3)$$

where A_Q, Γ_Q, Ω_Q, k_B and $\hbar$ are the amplitude, width, and energy of inelastic modes, Boltzmann constant and Planck constant, respectively. In the fitting, temperature T was fixed at a sample temperature obtained by a spetroradiometric method or a thermocouple. The excitation energy modes appearing at both Stokes and anti-Stokes sides correspond to the phonon creation and annihilation, respectively. With increasing temperature, as given by the Bose function in equation (3), the intensities of such Stokes and anti-Stokes peaks become similar to each other. A symmetric shape of the present IXS spectra therefore assures that the IXS signals originated from a high-temperature area.

The peak at a finite energy transfer gives the frequency of each mode (Fig. 1). The excitation energies for the LA phonon mode of liquid $Fe_{84}C_{16}$ obtained in a pressure range of 7.6–70 GPa are plotted as a function of momentum transfer (Q) in Fig. 4. The compressional sound wave or P-wave velocity (V_P) corresponds to the long-wavelength LA velocity at $Q \to 0$ limits;

$$V_P = \left(\frac{dE}{dQ}\right)_{Q\to 0}. \quad (4)$$

We made a linear fit to the data obtained at low Q below 3.5 nm^{-1} to determine the P-wave velocity (Supplementary Table 1), because positive dispersion can appear at higher $Q \gg 3\,nm^{-1}$ (ref. 45). For comparison, the results based on a sine-curve fit to all Q-range data, as is usually applied for polycrystalline samples in similar high-pressure IXS measurements[46], are also given in Supplementary Table 1. In general, the error bars of the two determinations of V_P overlap, though the sine fit to large Q does give slightly larger V_P, as would qualitatively be expected from previous measurements on liquid iron[47].

Equation of state for liquid $Fe_{84}C_{16}$. We constructed an equation of state (EoS) for liquid $Fe_{84}C_{16}$ to extrapolate the present V_P data and to estimate its density at the core pressure range. V_P of liquid can be written as;

$$V_P = \sqrt{\frac{K_S}{\rho}}. \quad (5)$$

The pressure dependence of K_S is assumed to be

$$K_S = K_{S0} + K'_S P, \quad (6)$$

where K'_S is the pressure derivative of K_S and pressure and subscript zero indicates a value at 1 bar. The adiabatic Murnaghan EoS can be described as (for example, ref. 4);

$$\rho = \rho_0 \left(1 + \frac{K'_S}{K_{S0}}P\right)^{\frac{1}{K'_S}}. \quad (7)$$

Equation (5) is thus rewritten as;

$$V_P = \sqrt{\frac{K_{S0}}{\rho_0}}\left(1 + \frac{K'_S}{K_{S0}}P\right)^{\frac{1}{2} - \frac{1}{2K'_S}}. \quad (8)$$

The temperature effect on ρ_0 can be expressed by;

$$\rho_0(T) = \rho_0(T_0)/\exp(\int_{T_0}^{T}\alpha dT). \quad (9)$$

The thermal expansion coefficient α is also dependent on temperature as;

$$\alpha(T) = a + bT, \quad (10)$$

where a and b are constants. Previous density measurements[16] of liquid Fe–C alloys at 1 bar give $a = 6.424 \times 10^{-5}\,K^{-1}$ and $b = 0.606 \times 10^{-8}\,K^{-2}$ for liquid

$Fe_{84}C_{16}$ using $\rho_0 = 6.505\ \mathrm{g\,cm^{-3}}$ at $T_0 = 2{,}500$ K as a reference. The result of fitting equation (8) to the present $P - V_P$ data is given in Fig. 3.

Isothermal bulk modulus. We estimate isothermal bulk modulus K_T from isentropic bulk modulus K_S in two ways. The relationship between these two is described as follows;

$$\frac{K_S}{K_T} = \frac{C_P}{C_V} = 1 + \alpha\gamma, \tag{11}$$

where C_P and C_V are heat capacities at constant pressure and volume, respectively. Although γ for liquid Fe-C alloys is not known, $\gamma_0 = 1.74$ has been reported for liquid Fe at 1 bar and 1,811 K (ref. 8) It is close to 1.58 for liquid $Fe_{90}O_8S_2$ estimated from the shock compression data set[48].

Extrapolation of present data to core pressures. With the EoSs determined above (equations (7) and (8)), we extrapolate the P-wave velocity and density of liquid $Fe_{84}C_{16}$ to the core pressure range along adiabatic compression, in which temperature is given by;

$$T = T_0\ exp\left[\int_{\rho_0}^{\rho}(\gamma/\rho)d\rho\right]. \tag{12}$$

Assuming $\gamma = \gamma_0 \times (\rho_0/\rho)$, temperature is simply represented as;

$$T = T_0\ exp\left[\gamma_0\left(1 - \frac{\rho_0}{\rho}\right)\right]. \tag{13}$$

γ_0 is fixed at 1.74 previously obtained for liquid Fe (ref. 8). Using the temperature dependence of K_{S0} and ρ_0 shown above, we calculate density, velocity and temperature profiles along adiabatic compression with various reference temperatures at the CMB. The adiabatic compression profiles of liquid $Fe_{84}C_{16}$ for the low ($T_0 = 2{,}045$ K and $T_{CMB} = 3{,}600$ K)[32] and high ($T_0 = 2{,}457$ K and $T_{CMB} = 4{,}300$ K)[33] temperature cases are calculated in Fig. 7.

References

1. Badro, J. et al. Effect of light elements on the sound velocities in solid iron: implications for the composition of Earth's core. Earth Planet. Sci. Lett. **254**, 233–238 (2007).
2. Sata, N. et al. Compression of FeSi, Fe₃C, Fe₀.₉₅O, and FeS under the core pressures and implication for light element in the Earth's core. J. Geophys. Res. **115**, B09204 (2010).
3. Ohtani, E. et al. Sound velocity of hexagonal close-packed iron up to core pressures. Geophys. Res. Lett. **40**, 5089–5094 (2013).
4. Jing, Z. et al. Sound velocity of Fe-S liquids at high pressure: implications for the Moon's molten outer core. Earth Planet. Sci. Lett. **396**, 78–87 (2014).
5. Nishida, K. et al. Sound velocity measurements in liquid Fe-S at high pressure: Implications for Earth's and lunar cores. Earth Planet. Sci. Lett. **362**, 182–186 (2013).
6. Sanloup, C., van Westrenen, W., Dasgupta, R., Maynard-Casely, H. & Perrillat, J.-P. Compressibility change in iron-rich melt and implications for core formation models. Earth Planet. Sci. Lett. **306**, 118–122 (2011).
7. Huang, H. et al. Evidence for an oxygen-depleted liquid outer core of the Earth. Nature **479**, 513–516 (2011).
8. Anderson, W. W. & Ahrens, T. J. An equation of state for liquid iron and implications for the Earth's core. J. Geophys. Res. **99**, 4273–4284 (1994).
9. Wood, B. J., Li, J. & Shahar, A. Carbon in the core: its influence on the properties of core and mantle. Rev. Miner. Geochem **75**, 231–250 (2013).
10. Chi, H., Dasgupta, R., Duncan, M. S. & Shimizu, N. Partitioning of carbon between Fe-rich alloy melt and silicate melt in a magma ocean – implications for the abundance and origin of volatiles in Earth, Mars, and the Moon. Geochim. Cosmochim. Acta **139**, 447–471 (2014).
11. Nakajima, Y., Takahashi, E., Suzuki, T. & Funakoshi, K. "Carbon in the core" revisited. Phys. Earth Planet. In. **174**, 202–211 (2009).
12. Nakajima, Y. et al. Thermoelastic property and high-pressure stability of Fe₇C₃: implication for iron-carbide in the Earth's core. Am. Mineral. **96**, 1158–1165 (2011).
13. Mookherjee, M. et al. High-pressure behavior of iron carbide (Fe₇C₃) at inner core conditions. J. Geophys. Res. **116**, B04201 (2011).
14. Prescher, C. et al. High Poisson's ratio of Earth's inner core explained by carbon alloying. Nat. Geosci. **8**, 220–223 (2015).
15. Chen, B. et al. Hidden carbon in Earth's inner core revealed by shear softening in dense Fe₇C₃. Proc. Natl Acad. Sci. USA **111**, 17755–17758 (2014).
16. Ogino, K., Nishiwaki, A. & Hosotani, Y. Density of molten Fe-C alloys. J. Jpn Inst. Met. **48**, 1004–1010 (1984).
17. Fei, Y. & Brosh, E. Experimental study and thermodynamic calculations of phase relations in the Fe-C system at high pressure. Earth Planet. Sci. Lett. **408**, 155–162 (2014).
18. Fiquet, G., Badro, J., Guyot, F., Requardt, H. & Krisch, M. Sound velocities in iron to 110 gigapascals. Science **291**, 468–471 (2001).
19. Antonangeli, D. et al. Simultaneous sound velocity and density measurements of hcp iron up to 93 GPa and 1100 K: an experimental test of the Birch's law at high temperature. Earth Planet. Sci. Lett. **331-332**, 210–214 (2012).
20. Mao, Z. et al. Sound velocities of Fe and Fe-Si alloy in the Earth's core. Proc. Natl Acad. Sci. USA **109**, 10239–10244 (2012).
21. Fiquet, G., Badro, J., Gregoryanz, E., Fei, Y. & Occelli, F. Sound velocity in iron carbide (Fe₃C) at high pressure: implications for the carbon content of the Earth's inner core. Phys. Earth Planet. In. **172**, 125–129 (2009).
22. Gao, L. et al. Sound velocities of compressed Fe₃C from simultaneous synchrotron X-ray diffraction and nuclear resonant scattering measurements. J. Synchrotron Radiat. **16**, 714–722 (2009).
23. Gao, L. et al. Effect of temperature on sound velocities of compressed Fe₃C, a candidate component of the Earth's inner core. Earth Planet. Sci. Lett. **309**, 213–220 (2011).
24. Lin, J. et al. Sound velocities of hot dense iron: Birch's law revisited. Science **308**, 1892–1894 (2005).
25. Vočadlo, L., Alfè, D., Gillan, M. J. & Price, G. D. The properties of iron under core conditions from first principles calculations. Phys. Earth Planet. In. **140**, 101–125 (2003).
26. Ichikawa, H., Tsuchiya, T. & Tange, Y. The P-V-T equation of state and thermodynamic properties of liquid iron. J. Geophys. Res. Solid Earth **119**, 240–252 (2014).
27. Umemoto, K. et al. Liquid iron-sulfur alloys at outer core conditions by first-principles calculations. Geophys. Res. Lett. **41**, 6712–6717 (2014).
28. Pronin, L., Kazakov, N. & Filippov, S. Ultrasonic measurements in molten iron. Izv. Vuzov. Chernaya Metall **5**, 12–16 (1964).
29. Terasaki, H. et al. Density measurement of Fe₃C liquid using X-ray absorption image up to 10 GPa and effect of light elements on compressibility of liquid iron. J. Geophys. Res. **115**, B06207 (2010).
30. Shimoyama, Y. et al. Density of Fe-3.5wt% C liquid at high pressure and temperature and the effect of carbon on the density of the molten iron. Phys. Earth Planet. In. **224**, 77–82 (2013).
31. Belashchenko, D. K., Mirzoev, A. & Ostrovski, O. Molecular dynamics modelling of liquid Fe-C alloys. High Temp. Mater. Processess **30**, 297–303 (2011).
32. Nomura, R. et al. Low core-mantle boundary temperature inferred from the solidus of pyrolite. Science **343**, 522–525 (2014).
33. Anzellini, S., Dewaele, A., Mezouar, M., Loubeyre, P. & Morard, G. Melting of iron at Earth's inner core boundary based on fast X-ray diffraction. Science **340**, 464–466 (2013).
34. Badro, J., Côté, A. S. & Brodholt, J. P. A seismologically consistent compositional model of Earth's core. Proc. Natl Acad. Sci. USA **111**, 7542–7545 (2014).
35. Dziewonski, A. M. & Anderson, D. L. Preliminary reference Earth model. Phys. Earth Planet. In. **25**, 297–356 (1981).
36. Wood, B. J., Walter, M. J. & Wade, J. Accretion of the Earth and segregation of its core. Nature **441**, 825–833 (2006).
37. Rubie, D. C. et al. Heterogeneous accretion, composition and core–mantle differentiation of the Earth. Earth Planet. Sci. Lett. **301**, 31–42 (2011).
38. Wood, B. J. Carbon in the core. Earth Planet. Sci. Lett. **117**, 593–607 (1993).
39. Litasov, K. D. et al. Thermal equation of state and thermodynamic properties of iron carbide Fe₃C to 31 GPa and 1473 K. J. Geophys. Res. Solid Earth **118**, 5274–5284 (2013).
40. Akahama, Y. & Kawamura, H. Pressure calibration of diamond anvil Raman gauge to 310 GPa. J. Appl. Phys. **100**, 043516 (2006).
41. Baron, A. Q. R. et al. An X-ray scattering beamline for studying dynamics. J. Phys. Chem. Solids **61**, 461–465 (2000).
42. Ishikawa, D., Uchiyama, H., Tsutsui, S., Fukui, H. & Baron, A. Q. R. Compound focusing for hard x-ray inelastic scattering. Proc. SPIE 8848-88480F (2013).
43. Fukui, H. et al. A compact system for generating extreme pressures and temperatures: an application of laser-heated diamond anvil cell to inelastic X-ray scattering. Rev. Sci. Instrum **84**, 113902 (2013).
44. Fåk, B. & Dorner, B. Phonon line shapes and excitation energies. Phys. B Condens. Matter **234-236**, 1107–1108 (1997).
45. Scopigno, T., Ruocco, G. & Sette, F. Microscopic dynamics in liquid metals: the experimental point of view. Rev. Mod. Phys. **77**, 881–933 (2005).
46. Fiquet, G. et al. Application of inelastic X-ray scattering to the measurements of acoustic wave velocities in geophysical materials at very high pressure. Phys. Earth Planet. In. **143**, 5–18 (2004).
47. Hosokawa, S., Inui, M., Matsuda, K., Ishikawa, D. & Baron, A. Damping of the collective modes in liquid Fe. Phys. Rev. B **77**, 174203 (2008).
48. Huang, H. et al. Melting behavior of Fe-O-S at high pressure: a discussion on the melting depression induced by O and S. J. Geophys. Res. **115**, B05207 (2010).

49. Kubo, A. *et al.* In situ X-ray observation of iron using Kawai-type apparatus equipped with sintered diamond: absence of β phase up to 44 GPa and 2100 K. *Geophys. Res. Lett.* **30,** 1126 (2003).

Acknowledgements

We thank H. Fukui for his advice through IXS measurements. H. Uchiyama, D. Ishikawa, N. Murai and Y. Um are acknowledged for their supports during synchrotron experiments and data analyses, and D. Ishikawa and H. Fukui for implementation of the KB setup. Comments from three anonymous reviewers were helpful. All experiments were performed at BL35XU, SPring-8 (Proposal no. 2012B1356, 2013A1541, 2013B1407, 2014A1368, 2014B1271 and 2014B1536).

Author contributions

Y.N. synthesized a starting material and performed experiments and data analysis. Y.N., S.I., K.H, T.K., S. Tateno, S. Tsutsui, Y.K. and A.B. were involved in IXS measurements. H.O., S. Tateno, S.I. and Y.N. were involved in developing the laser heating system at the beamline. Y.N., K.H. and A.B. wrote the paper. All authors discussed the results and commented on the manuscript.

Additional information

Competing financial interests: The authors declare no competing financial interests.

26

The dune effect on sand-transporting winds on Mars

Derek W.T. Jackson[1], Mary C. Bourke[2] & Thomas A.G. Smyth[3]

Wind on Mars is a significant agent of contemporary surface change, yet the absence of *in situ* meteorological data hampers the understanding of surface–atmospheric interactions. Airflow models at length scales relevant to landform size now enable examination of conditions that might activate even small-scale bedforms (ripples) under certain contemporary wind regimes. Ripples have the potential to be used as modern 'wind vanes' on Mars. Here we use 3D airflow modelling to demonstrate that local dune topography exerts a strong influence on wind speed and direction and that ripple movement likely reflects steered wind direction for certain dune ridge shapes. The poor correlation of dune orientation with effective sand-transporting winds suggests that large dunes may not be mobile under modelled wind scenarios. This work highlights the need to first model winds at high resolution before inferring regional wind patterns from ripple movement or dune orientations on the surface of Mars today.

[1] School of Environmental Sciences, Ulster University, Coleraine BT52 1SA, UK. [2] Department of Geography, Trinity College Dublin, Dublin D2, Ireland. [3] School of the Environment, Earth Sciences Building, Flinders University, GPO Box 2100, Adelaide, South Australia 5001, Australia. Correspondence and requests for materials should be addressed to D.W.T.J. (email: d.jackson@ulster.ac.uk).

The presence of sand dunes on planetary surfaces can be indicative of past and present wind regimes that have sculpted cohesionless material into organized landforms. These landforms provide a unique insight into the evolutionary dynamics of planetary surfaces and are a window into past climatic behaviour. Reconstructing patterns of regional wind behaviour from the orientation of aeolian bedforms has had limited success due to, for example, model limitations and the, as yet, unstudied role that the inheritance of bedform morphology from previous climatic conditions may play. Unlike dunes on Earth, we still do not fully understand airflow dynamics over and around Martian dunes at relevant scales. If bedforms (for example, dunes or ripple movement on dunes) are to be used as a wind direction proxy, then a better understanding of the controls on ripple migration is needed. Here we use three-dimensional (3D) computational fluid dynamic modelling to demonstrate that wind flow dynamics over and around large-scale dune forms are complex on Mars. Model output in this study was validated using ripple displacements measured from High Resolution Imaging Science Experiment (HiRISE) data. This work advocates the use of detailed, high-resolution surface modelling of winds before attempting to understand regional wind patterns from contemporary bedforms on Mars. In the absence of a network of *in situ* instrumentation to measure winds on Mars, our understanding of airflow over the surface of the planet has relied on large-scale, for example, ref. 1, and meso-scale, for example, ref. 2,3, atmospheric circulation models along with the interpretation of landform features from satellite images, for example, refs 4–6. The poor spatial scale of such circulation model data, however, has effectively precluded detailed examination of the forcing mechanisms by which windblown features, such as dunes, move on the surface of Mars. Large- to meso-scale atmospheric circulation models are designed to operate only at scales substantially (2–5 times) larger than the landform feature(s) itself, thereby inhibiting a full understanding of the process response in the system. We must, therefore, adopt a much finer resolution (that is, microscale model) approach to examine the driving mechanisms of any aeolian (windblown) system and its associated landforms. With the availability of high-resolution (0.25 m) HiRISE stereo images of Mars in recent years, high-resolution digital terrain models (DTM) of the surface are now available. This topographic surface can be used to run fine-resolution (sub-landform scale) wind models across complex 3D surface topography such as dune fields.

In the southern hemisphere of Mars, dune fields are contained primarily within large crater basins[7]. The dune sediment is sourced locally from exposed strata in the crater walls and floors[8,9]. In some cases (for example, Proctor Crater) the large diameter of these craters means that some intracrater dune fields may be far enough from the crater rim to experience a largely localized wind regime. Here we examine dunes within Proctor Crater, a 150-km diameter impact crater, located within the southern highlands of Mars (47.041°S; 30.667°E) where the dune field is made up of transverse dune forms. This study has shown that detailed 3D modelling of wind on Mars enables us, for the first time, to see the importance of large dune ridge shapes and their complex modification of localized airflow over the surface of Martian dune sites. Results show that back modelling of wind flow from actual ripple movement patterns can identify those regional winds that are forcing the migration of sediment over the dunes on Mars today. We find that winds travelling from the east southeast (ESE) (110°) are the dominant sand-transporting winds, whereas winds from east northeast (ENE) (75°) and west southwest (WSW) (239°) directions appear to play a more subordinate role. Wind modelling on a microscale (5 m) now provides us with an effective new tool to accompany surface ripple displacement information to help understand dune dynamics on Mars.

Results

Dune characteristics and wind selection. The dunes at the study site (Fig. 1) are large (500 m wavelength; 70–120 m high) and are aligned transverse to an inferred ENE (75°) incident wind[10]. Bright dune forms are located on the floor of the crater and are in the interdune areas. These have a wavelength of 23 m and are considered immobile under current conditions and are most likely transverse granule ripples[10].

A 1 m resolution DTM (Fig. 1b) was used for the 3D computational fluid dynamics (CFD) modelling[11]. We selected three dominant wind directions to model; 239° WSW (primary), 110° ESE (secondary) and 75° ENE (tertiary). These winds were identified in earlier work[10] on Proctor Crater as representing the dominant wind regimes indicated from dune slip-face dip orientations. HiRISE image data that were acquired two Mars years apart were used to measure ripple displacement at 11 sites (Fig. 1c) on three of the largest dunes. Each sample quadrant measured ripple displacement distance and direction over a 10,000 m² area.

3D airflow modelling. Results from CFD modelling allow us to examine the detailed wind flow behaviour and its corresponding association with underlying dune topography at the site within Proctor Crater. High-resolution-modelled winds are seen to vary extensively across the complex dune topography (Fig. 2a), with a general picture of maximum wind (acceleration) along the crestal

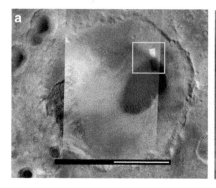
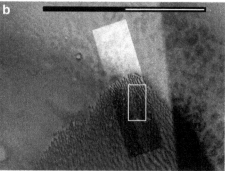

Figure 1 | Study location. Regional position of Proctor Crater, area over which airflow model was run and site locations of ripples studied. (**a**) Location of study site within Proctor Crater, (47°41'26.36'S 29°54'53.54E). The dunefield is the dark area in the image. Note: scale bar length is 100 km. The location of our model location is shown in **b**. (**b**) The eastern section of the dune field containing the study area. Note: scale bar length is 40 km (**c**) Image of the surface over which the simulations were run with site locations marked (area demarked in **b**). Note: scale bar length is 400 m.

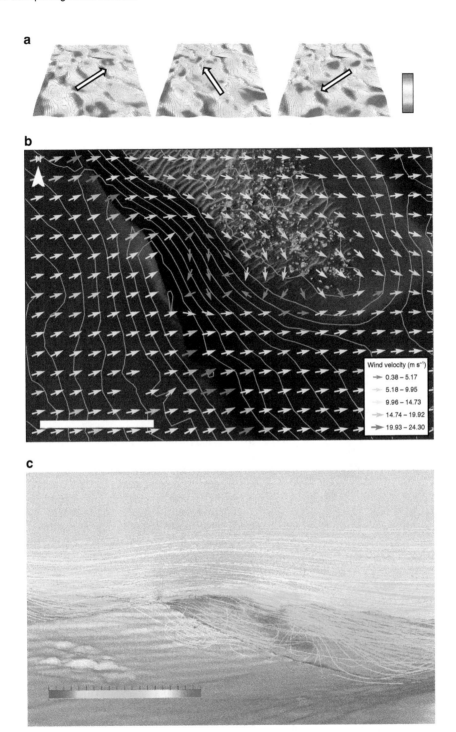

Figure 2 | Airflow modelling results near the dune field surface from three simulated directions. (**a**) Primary, secondary and tertiary wind flow at 3 m above the surface over the entire dune field surface identified in Fig. 1b. Note: scale bar range is 0 m s^{-1} (blue) to 21 m s^{-1} (red). These wind directions were adopted from Fenton et al.[8] who used the slip-face dip azimuths of transverse ridges to derive the primary secondary and tertiary directions. The arrow in the centre of each image represents the incident wind direction at the boundary of the computational domain. Data show topographically accelerated wind velocities at dune crests with maximum wind speeds of 24.3 m s^{-1} and 'dead zones' (velocities ≤1 m s^{-1}) in the interdunes for all three modelled winds. Contour data at 10-m interval derived from a DTM was built using HiRISE stereo image pairs PSP_003800_1325 and PSP_004077_1325. The DTM data were made available by NASA/JPL/University of Arizona. The region shown in **a** is the same as that delineated in Fig. 1c. (**b**) Flow steering by a transverse dune (100 m high) under primary incident wind conditions (239°) south of site 8 (fig. 1c). Vectors indicate wind flow at 3 m above the surface, spaced at 50-m intervals. Contours are plotted at 10-m intervals. Note: scale bar range is 300 m. (**c**) A 3D view of the detached flow for primary winds (239°) at site 8 (location in fig. 1c). The perspective of the image is towards the north-west. Incident wind flow (m s^{-1}) is moving from left to right in the image. Site 8 is situated immediately north of the frame in **c**. Note: scale bar range is 0 m s^{-1} (blue) to 20 m s^{-1} (red).

brinks and minima in the dune troughs. Flows accelerate (relative to incident flow) by up to 150–200% at crest positions, displaying a similar behaviour to dune sites on Earth[12,13]. Deceleration of flow is evident in dune trough locations, which coincides with the brighter dune forms in these locations that show no evidence of recent movement in our data (site 11).

Distinctive steering of flow (Fig. 2b) results from pre-existing dune topography. The large dunes differentially retard the flow from oblique directions, forcing the flow into orientations significantly offset from the upwind direction. The specific crestline orientation relative to the upwind flow direction is of primary importance to the degree of steering.

Overall, steered wind directions mapped over the ripple study sites are an average of only 8 degrees offset from the average ripple migration direction. This demonstrates excellent agreement with CFD-modelled flow and localized surface response.

Ripple movement and modelled localized wind patterns. As the crestal positions of the smaller scale ripples (wavelength 3.4 m) in this study have been shown to move on the order of 0.5–2.0 m over a two Mars–year period, they are considered to be a true response to the modern winds we see on Mars today. Our data show that five of the ripple displacement study sites (2, 4, 6, 8 and 9) have very good correspondence with secondary winds (110°) that have undergone localized steering, displaying only 1–12 degrees of difference between ripple migration direction and steered wind direction. Note that for sites 7, 10 and 11 migration was below detection limits or showed no motion. Study sites 1 and 3 map better onto steered tertiary (75°) incident winds (3 and 19 degree differences, respectfully), while only site 5 mapped onto steered primary (239°) winds.

Our data show decreased flow steering at rounded crests (bottom of Fig. 2b) relative to sharper crests (middle of Fig. 2b). Under circumstances where flow steering has not taken place, the cross-sectional topography dictates if the flow becomes detached, forming eddies that shear off the crestal areas (Fig. 2c). Both flow steering and detachment behaviour modelled for Mars is similar to Earth where crestal sharpness and surface undulations exert a major control[14,15].

Discussion

In terms of the modern surface response to wind forcing, an examination of ripple movement provides some intriguing insights into the dominant incident wind directions that are responsible for active migration of ripple sets[16]. For the ripple sites investigated, we find that all three incident wind directions move ripples and, in general, the use of ripples as a tool for back modelling regional winds can be strengthened from the results presented here. However, the large dune topography exerts a significant control on steering localized winds to produce the ripple migration directions. This has a direct control on the mobility of surface sediments and dictates the fate of surface behaviour, contributing to actively migrating surface ripples. Therefore, the ripple direction may reflect a topographically steered wind direction rather than the regional wind regime. Interestingly, when we plot steered wind speed at ripple sites against observed ripple average migration rates (Fig. 3), we see a good correlation (r^2, 0.7), further substantiating that CFD-modelled results are concurrent with ripple response displaying an exponential increase in migration distance with velocity increase.

Modelling in 3D enables us for the first time to see the importance that large dune ridge shape plays on modifying localized flow to give either steered or detached behaviour. We would expect to see a similar localization of winds with strong topographically steered and altered winds for comparable dune topography and wind regimes of similar magnitude and directions elsewhere on Mars.

The physical cross-sectional dimensions of the main dune ridges appear to play an important role in deflecting and steering

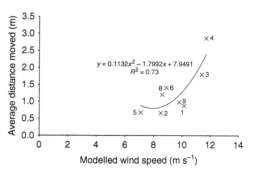

Figure 3 | Ripple migration distance versus locally modelled wind velocity. Relationship between the average observed ripple migration distance over 3.8 Earth years (2.0 Mars years) and local CFD wind velocity.

the localized flow, and seem to dictate which incident flow drives particular ripple set migration directions. Sharply peaked dune crests are likely to steer winds and may also lead to detached flow and large-scale return eddies moving in the opposite direction to incident flow. Rounded dune topographies are less likely to induce this behaviour of detachment and flow maintains itself largely in the same direction as the incident winds albeit steered somewhat in certain cases.

Back modelling–wind flow orthogonally from ripple migration directions examined here, identifies the regional winds that are driving the migration of sediment. Our data show that in the Proctor Crater region, secondary wind travelling from the ESE (110°) is the dominant sand-transporting wind in the dune field. Winds from the tertiary ENE (75°) and primary WSW (239°) directions appear to play a more subordinate role at this site.

The dune slip-face orientations suggest that the formative dune winds are from the ENE[10] (that is, the tertiary winds). The low correlation of tertiary wind direction with active sediment movement in ripples at the Proctor Crater site suggests that the large dune morphology we studied may not be currently maintained by tertiary winds. Rather, the geomorphologically effective winds are from the ESE (that is, the secondary winds). This opens the possibility that the largest dunes at our site in Proctor Crater may be emplaced during former climate conditions.

Other studies have used the brink orientation on large dunes to infer modern regional-scale wind direction. We advocate caution in using this approach and suggest that further study is required before we can confidently infer modern regional wind direction from the brink orientations of large dunes on Mars. Our findings, along with model limitations, may go some way towards explaining the poor correlation between meso-scale model output and mapped inter-crater dune orientations, for example, ref. 17.

Our microscale wind model provides a new tool whereby mapped ripple displacement directions can be used to back-model the incident and formative wind at timescales relevant to modern Martian aeolian dynamics.

Methods

Ripple mapping. For the time-series analysis we used two HiRISE images (PSP_003800_1325 and ESP_021469_1325) that were acquired 3.8 Earth years apart. The images were selected for minimal differences in the orbital parameter and emission angle (3.4°) that reduces parallax distortions. Radiometric distortions are also minimal. The images were acquired at similar times in the season ($L_s = 240.9°$ and 242.7°) and at a similar time of day, minimizing radiometric distortions due to different illumination conditions and effects of potential shadow length.

Eleven sites were selected for ripple analysis and were located on dune flanks and in the interdune of three consecutive transverse ridges. The transverse ridges are spaced ~500 m apart and attain heights of 100 m. Ripple locations were mapped on the two georectified images in ARCGIS and displacement directions and distances were measured.

Wind flow modelling. Near-surface airflow over the study area was simulated using OpenFOAM, a CFD toolbox that integrates the principal equations of fluid of flow over a domain by converting the integral equations to algebraic equations, before solving them iteratively[18]. Flow was calculated using Reynolds-averaged navier-stokes equations, which average the motion of fluid flow over time. Turbulence was modelled using the two-equation re-normalized group k-ε turbulence model. The mesh domain measured $2,900 \times 3,535 \times 600$ m and contained 3.8 million cells. Cell size gradually decreased from the top (maximum size 20×20 m) of the computational domain towards the surface to a minimum resolution of $5 \times 5 \times 2.5$ m.

In terms of the inlet static stability profile at the site, temperature was not calculated in each simulation and, therefore, the vertical wind profile is assumed static at the inlet with no consideration made of diurnal temperature change. Detailed evaluation of patterns of diurnal variability was not the focus of this study and instead topographically induced impacts as a result of local dune topography were highlighted. In each case, wind was defined at the inlet boundary as logarithmic using Richards and Hoxey[19] equations for k-ε turbulence models (equation 1).

$$U(z) = \left(\frac{u_*}{K}\right)\ln\left(\frac{z+z_0}{z_0}\right) \tag{1}$$

Where $U(z)$ is the wind speed at height z, u_* is shear velocity, K is the Von Karman constant (0.4187) and z_0 is the aerodynamic roughness. Note that z_0 was attributed a value of 0.05 m, typical of a semi-arid sparse brush[20] or sparse grass 0.5 m high[21]. With larger roughness lengths of 0.06–0.07 m, near-surface airflow speeds would have been only marginally lower and zones of separation/recirculation marginally greater. These results on steep dunes are, however, negligible[22]. u_* was prescribed a value of 1.45 m s^{-1}, producing a wind speed of 20 m s^{-1} at 30 m above the surface at the inlet. Turbulence kinetic energy (k) and energy dissipation (ε) were simulated at the inlet boundary using conditions prescribed by Richards and Hoxey[19] equations 2 and 3.

$$k = \frac{u_*^2}{\sqrt{C_\mu}} \tag{2}$$

$$\varepsilon = \frac{u_*^3}{K(z+z_0)} \tag{3}$$

Where C_μ is a constant of the k-ε model and equals 0.09. The atmospheric kinematic viscosity was specified as 0.0011 m^2 s^{-1}, this value assumes that the atmosphere is composed of 100% CO_2 at a temperature of 5 °C. Each simulation was considered complete and converged when wind speed, which was probed at 16 equally spaced points within each of the ripple sites, became steady to the third decimal place.

References

1. Anderson, F. S. *et al.* Assessing the Martian surface distribution of aeolian sand using a Mars general circulation model. *J. Geophys. Res. (Planets)* **104**, 18991–19002 (1999).
2. Silvestro, S. *et al.* Pervasive aeolian activity along rover Curiosity's traverse in Gale Crater, Mars. *Geology* **41**, doi:10.1130/G34162.1 (2013).
3. Rafkin, S. C. R. & Michaels, T. I. Meteorological predictions for 2003 Mars Exploration Rover high-priority landing sites. *J. Geophys. Res. (Planets)* **108**, E12, 8091 doi:10.1029/2002JE002027 (2003).
4. Greeley, R., Kraft, M. D., Kuzmin, R. O. & Bridges, N. T. Mars Pathfinder landing site: Evidence for a change in wind regime from lander and orbiter data. *J. Geophys. Res. (Planets)* **105**, 1829–1840 (2000).
5. Cardinale, M., Komatsu, G., Silvestro, S. & Tirsch, D. The influence of local topography for wind direction on Mars: two examples of dune fields in crater basins. *Earth Surf. Proc. Land.* **37**, 1437–1443 (2012).
6. Fenton, L. K., Michaels, T. I., Chojnacki, M. & Beyer, R. A. Inverse maximum gross bedform-normal transport 2: application to a dune field in Ganges Chasma, Mars and comparison with HiRISE repeat imagery and MRAMS. *Icarus* **230**, 47–63 (2014).
7. Hayward, R. K. *et al.* Mars digital dune database and initial science results. *J. Geophys. Res. (Planets)* **112**, E11007 doi:10.1029/2007JE002943 (2007).
8. Fenton, L. K., Toigo, A. D. & Richardson, M. I. Aeolian processes in Proctor Crater on Mars: mesoscale modeling of dune-forming winds. *J. Geophys. Res. (Planets)* **110**, E06005 (2005).
9. Tirsch, D., Jaumann, R., Pacifici, A. & Poulet, F. Dark aeolian sediments in martian craters: composition and sources. *J. Geophys. Res. (Planets)* **116**, E03002 (2011).
10. Fenton, L. K. Potential sand sources for the dune fields in Noachis Terra, Mars. *J. Geophys. Res. (Planets)* **110**, 1–27 (2005).
11. Jackson, D. W. T. *et al.* Investigation of three-dimensional wind flow behaviour over coastal dune morphology under offshore winds using Computational Fluid Dynamics (CFD) and ultrasonic anemometry. *Earth Surf. Proc. Land.* **36**, 1113–1134 (2011).
12. Baddock, M., Livingstone, I. & Wiggs, G. F. S. The geomorphological significance of airflow patterns in interdunes. *Geomorphology* **87**, 322–336 (2007).
13. Smyth, T. A. G., Jackson, D. W. T. & Cooper, J. A. G. Three dimensional airflow patterns within a coastal trough-bowl blowout during fresh breeze to hurricane force winds. *Aeolian Res.* **9**, 111–123 (2013).
14. Delgado-Fernández, I. *et al.* Field characterisation of three-dimensional lee-side airflow patterns under offshore winds at a beach-dune system. *J. Geophys. Res.-Earth* **118**, 706–721 (2013).
15. Jackson, D. W. T. *et al.* Airflow reversal and alternating corkscrew vortices in foredune wake zones during perpendicular and oblique offshore winds. *Geomorphology* **187**, 86–93 (2013).
16. Silvestro, S., Fenton, L. K., Vaz, D. A., Bridges, N. T. & Ori, G. G. Ripple migration and dune activity on Mars: evidence for dynamic wind processes. *Geophys. Res. Lett.* **37**, L20203 (2010).
17. Hayward, R. K. *et al.* Aeolian dunes as ground truth for atmospheric modeling on Mars. *J. Geophys. Res. (Planets)* **114**, E11012 (2009).
18. Versteeg, H. K. & Malalasekera, W. *An Introduction to Computational Fluid Dynamics* 2nd edn (Pearson, 2006).
19. Richards, P. J. & Hoxey, R. P. Appropriate boundary conditions for computational wind engineering models using k-e turbulence model. *J. Wind Eng. Ind. Aerodyn.* **46**, 145–153 (1993).
20. Blackadar, A. K. The vertical distribution of wind and turbulent exchange in a neutral atmosphere. *J. Geophys. Res.* **67**, 3095–3102 (1962).
21. Deacon, E. L. *Vertical Profiles of Mean Wind in the Surface Layers of the Atmosphere.* Geophysical Mem. No. 91 (Meteorological Office, Air Ministry, 1953).
22. Engel, P. Length of flow separation over dunes. *J. Hydraul. Div.* **107**, 1133–1143 (1981).

Acknowledgements

This work was supported by the Natural Environment Research Council (grant number NE/F019483/1), Irish Research Council New Foundations Award and EU FP7 CIG #618892. We thank Dr Meiring Beyers, Klimaat Consulting and Innovation, Inc., Canada, for his inspirational introduction to computational fluid dynamic modelling. The support from the School of the Environment, Flinders University is also acknowledged.

Author contributions

All authors contributed to the main study design and helped carry out the simulations. All authors were involved in writing and editing the manuscript with specific airflow modelling being carried out by T.A.G.S. Results from the airflow modelling were analysed by all authors. Ripple migration analysis was carried out by M.C.B with interpretations of these in relation to airflow results conducted by all authors. The original concept of using CFD airflow modelling on Martian dunes at this scale was initiated by D.W.T.J. and M.C.B.

Additional information

27

Pressure–temperature evolution of primordial solar system solids during impact-induced compaction

P.A. Bland[1], G.S. Collins[2], T.M. Davison[2], N.M. Abreu[3], F.J. Ciesla[4], A.R. Muxworthy[2] & J. Moore[2]

Prior to becoming chondritic meteorites, primordial solids were a poorly consolidated mix of mm-scale igneous inclusions (chondrules) and high-porosity sub-μm dust (matrix). We used high-resolution numerical simulations to track the effect of impact-induced compaction on these materials. Here we show that impact velocities as low as $1.5\,km\,s^{-1}$ were capable of heating the matrix to >1,000 K, with pressure–temperature varying by >10 GPa and >1,000 K over ~100 μm. Chondrules were unaffected, acting as heat-sinks: matrix temperature excursions were brief. As impact-induced compaction was a primary and ubiquitous process, our new understanding of its effects requires that key aspects of the chondrite record be re-evaluated: palaeomagnetism, petrography and variability in shock level across meteorite groups. Our data suggest a lithification mechanism for meteorites, and provide a 'speed limit' constraint on major compressive impacts that is inconsistent with recent models of solar system orbital architecture that require an early, rapid phase of main-belt collisional evolution.

[1] Department of Applied Geology, Curtin University, GPO Box U1987, Perth, Western Australia 6845, Australia. [2] Impacts & Astromaterials Research Centre (IARC), Department of Earth Science & Engineering, Imperial College London, South Kensington Campus, London SW7 2AZ, UK. [3] Earth Science Program, Pennsylvania State University—Du Bois Campus, Du Bois, Pennsylvania 15801, USA. [4] Department of Geophysical Science, University of Chicago, 5734 South Ellis Avenue, Chicago, Illinois 60430, USA. Correspondence and requests for materials should be addressed to P.A.B. (email: p.a.bland@curtin.edu.au).

Chondritic asteroids are among the most primitive objects in the solar system, containing a unique record of solar system formation. Three processes dominated their evolution as geological bodies—thermal metamorphism, aqueous alteration and impacts. Meteorite classification schemes are based on the degree to which a given sample was exposed to these processes. The impact record in chondrites has been studied in detail[1]. Shock metamorphism is not considered a dominant factor in the evolution of the most primitive meteorites, the carbonaceous chondrites (CC), as 85% are ranked S1 ('unshocked'; <4–5 GPa) or S2 ('very weakly shocked'; 5–10 GPa)[1,2]. Associated low-intensity collisions are thought to leave a chondrite essentially unscathed, with no effects from local pressure–temperature (PT) excursions, and minimal post-shock thermal metamorphism[1]. Primitive (that is, unequilibrated) ordinary chondrites rank at higher shock levels: 50% are S3 ('weakly shocked'; 10–20 GPa)[1]. Shock level is calibrated against shock recovery experiments on non-porous or low-porosity crystals and rocks[1], but porous objects respond very differently to impact than non-porous objects[3,4]. When shocking a porous material, extra PdV work is expended to crush out the pore space. After release from the shocked state, that extra work (termed the 'waste heat') heats the material, thus porous material will reach higher temperatures than a non-porous material in a similar impact. Although they are now lithified (often low porosity) rocks, chondrite precursors were highly porous[5–8].

In addition, chondrites are bimodal materials: zero-porosity mm-sized spherical chondrules set in an aggregate composed of sub-μm monomers. But the level of impact processing in chondrites is determined by shock metamorphic textures in large (>50–100 μm) grains (for example, chondrule olivines), accessible via optical microsocpy[1,2]. The assumption has been that if chondrules are unshocked, so is matrix. The possibility that the shock record might be different in matrix has not been considered, and with notable exceptions[9–11], the extent of shock metamorphism in matrix has not been explored.

Alongside petrographic observations and experimental impact studies, computational modelling has been utilized extensively to understand impact shock, generally applied to non-porous targets, but most recently in porous planetesimals[11]. In these models, porosity is parameterized, and derived PT estimates are 'bulk' values, averaged over large (asteroidal) scales, making it difficult to translate model predictions to meteorite observations. No numerical studies have attempted to resolve shock in the bimodal materials that were chondritic precursors. This is unfortunate given that they were arguably the starting point for all inner solar system objects.

An obstacle to understanding impact effects in porous meteorites has been the absence of direct information on the nature of the porosity prior to the shock event[3]. But a methodology that quantitatively relates rock fabric intensity to net compression[7] has overcome this obstacle, allowing us to reconstruct a pre-compaction porosity for matrix and for the parent body as a whole. As outlined in that work[7], it is possible to reconstruct a pre-compaction matrix porosity by deriving the ratio of final (that is, post-compaction) lengths to initial (pre-compaction) lengths of notional lines perpendicular to the matrix uniaxial fabric. This ratio (Z) is a quantitative measure of the degree of compaction. Given the current observed (final) abundance of matrix (A_{mf}), and current (final) bulk porosity (ϕ_{bf}), we can estimate current (final) matrix porosity (ϕ_{mf}) (assuming chondrule porosity $\sim 0\%$). Given Z and ϕ_{mf} the initial (pre-compaction) matrix porosity (ϕ_{mi}) is given by

$\phi_{mf}Z + (1 - Z)$. It follows that initial bulk porosity (ϕ_{bi}) is given by

$$\phi_{bi} = \frac{A_{mf}\phi_{mi}}{[A_{mf}\phi_{mi} + (1 - \phi_{bf})Z]}$$

and initial matrix abundance (A_{mi}) by ϕ_{bi}/ϕ_{mi}.

Thus far the early evolution of the material properties of solar system solids has been explored largely via experiments with analogue materials[5]. Most recently, impact compaction experiments compared final bulk porosities with chondrite values[8]. The saturation limit of the porosity varies for different materials: 0.68–0.42 at a pressure of 1 MPa[8]. The upper bound appears to be a closer approximation to ϕ_{mi} for chondrite parent bodies. In OCs, ϕ_{bf} is typically $\sim 5\%$ (ref. 12), but there are outliers with values up to 21% (ref. 13). For the most porous OCs it has been shown that this is primordial (not fracture) porosity[13]. Similarly, for anhydrous CCs there is a large range: $\phi_{bf} = 7$–28% for CO chondrites, and 4–28% for CV chondrites[14]. Using the relation outlined above, assuming (conservatively), no compaction ($Z = 1$, and $\phi_{mf} = \phi_{mi}$), it is apparent that current matrix porosity (ϕ_{mf}) in the most porous OCs must be 60–70%; 60% in the most porous CVs and 66–70% in the most porous COs. In the CV chondrite Allende, where a detailed rock fabric analysis was performed[7,15], and consistent matrix fabrics observed across the sample, degree of compaction consistent with $Z = \sim 0.5$ was defined[7] (that is, compaction had roughly halved the volume of the matrix aggregate). A ϕ_{mi} value of 70–80% was derived[5]. The data from compaction experiments[8], fabric analysis[7] and an overview of literature porosity, appear consistent with $\phi_{mi} = \sim 70\%$ across most chondrite groups.

If chondrite precursors were initially highly porous, how were they compacted to the (generally) low-porosity rocks that we see today? Recent studies[8] favour impacts as a compaction mechanism, and several other observations support that conclusion. Lithostatic pressure (gravitational compression) was trivial (~ 1 MPa at the centre of a 100-km radius asteroid, equivalent to a depth of 10 m on Earth[16]), indicating that it was not a significant factor in chondrite compaction. In addition, gravitational compaction would generate parent bodies with meteorites from the interior being more compacted (and showing more thermal metamorphism), than meteorites sourced from nearer the surface of the object. This is not observed. There is a lack of positive correlation between compaction and thermal metamorphism in OCs[12], COs[14] or CVs[14]. In the CVs there is a negative correlation: unmetamorphosed CV_rs are highly compacted (3.6% porosity), while more metamorphosed CV_os are not (averaging 19.7%)[14]. Taken together, these observations are consistent with impact-induced compaction of initial high-porosity material occurring on large (parent body?) scales.

Here we use numerical modelling to study the effects of impact-induced compaction on primordial chondritic materials. Our goal was to achieve sufficient resolution in simulations to allow a comparison with meteorite petrography observed in a thin section (100 μm). Although low velocity impacts are expected to produce minor effects in CCs[1,2], resolving at this finer 'mesoscale', we find that this is not the case. Even at 1.5 km s^{-1}, porous matrix can experience peak $T > 1,000$ K as it compacts. PT varies by > 10 GPa and $> 1,000$ K over ~ 100 μm. Chondrules act as heat-sinks, so temperature excursions are brief. These results have implications for key aspects of the chondrite record, and provide a 'speed limit' constraint on the initial collisional evolution of planetesimals.

Results

Mesoscale simulations. We have developed a two-dimensional (2D) 'mesoscale' numerical approach that allows quantification of

Table 1 | Results of mesoscale iSALE simulations of planar shock propagation through a chondrule/matrix mixture.

	CC-like simulations (70% initial matrix)															OC-like simulations (33% initial matrix)					
Impact velocity (km s^{-1})	0.75	1.00	**1.00**	1.00	**1.00**	1.00	1.50	2.00	**2.00**	2.00	**2.00**	2.00	2.50	3.00	**3.00**	0.75	1.00	1.50	2.00	2.50	3.00
Initial matrix fraction (A_{mi})	0.70	0.70	**0.70**	0.70	**0.70**	0.70	0.70	0.70	**0.70**	0.70	**0.70**	0.70	0.70	0.70	**0.70**	0.33	0.33	0.33	0.33	0.33	0.33
Initial matrix porosity (φ_{mi})	0.70	0.60	**0.60**	0.70	**0.70**	0.80	0.70	0.60	**0.60**	0.70	**0.70**	0.80	0.70	0.70	**0.70**	0.70	0.70	0.70	0.70	0.70	0.70
Initial bulk porosity (φ_{bi})	0.49	0.42	**0.41**	0.49	**0.49**	0.56	0.49	0.42	**0.41**	0.49	**0.49**	0.56	0.49	0.49	**0.49**	0.23	0.23	0.23	0.23	0.23	0.23
Bulk shock pressure*	0.64	1.14	**0.87**	0.98	**0.72**	0.70	1.71	3.98	**3.69**	3.02	**2.60**	1.92	5.55	7.62	**6.47**	1.34	2.16	5.12	9.40	14.1	18.9
Bulk shock pressure 1σ (GPa)	0.02	0.06	**0.03**	0.05	**0.03**	0.04	0.18	0.55	**0.64**	0.32	**0.54**	0.19	0.63	0.38	**0.68**	0.02	0.07	0.10	0.20	0.46	0.27
Bulk T(final)†	371	408	**397**	425	**405**	451	539	674	**595**	701	**615**	715	897	1,120	**947**	363	410	530	689	874	1,130
Final bulk porosity (φ_{bf})	0.28	0.09	**0.03**	0.18	**0.10**	0.29	0.05	0.02	**0.00**	0.03	**0.00**	0.03	0.02	0.02	**0.00**	0.08	0.04	0.02	0.02	0.02	0.02
Final matrix porosity (φ_{mf})	0.48	0.17	**0.05**	0.34	**0.22**	0.55	0.11	0.04	**0.00**	0.07	**0.01**	0.09	0.06	0.04	**0.00**	0.39	0.24	0.15	0.13	0.12	0.11
Final matrix fraction (A_{mf})	0.58	0.52	**0.49**	0.54	**0.47**	0.54	0.45	0.51	**0.48**	0.45	**0.42**	0.34	0.43	0.45	**0.42**	0.20	0.16	0.15	0.15	0.15	0.15
Matrix T(final†) mean (K)	424	507	**496**	533	**527**	589	825	934	**814**	1,110	**941**	1,430	1,450	1,790	**1,440**	600	842	1,330	1,810	2,410	3,020
Matrix T(final†) 1σ (K)	29.9	38.3	**40.3**	43.3	**50.2**	71.8	90	108	**92.1**	110	**99.9**	167	185	270	**165**	123	161	235	320	410	436
Matrix P(peak§) mean (GPa)	0.80	1.70	**1.45**	1.28	**1.01**	0.96	3.11	10.2	**8.81**	8.36	**7.68**	4.92	13.0	16.8	**15.4**	1.87	3.22	7.75	13.9	21.1	27.8
Matrix P(peak§) 1σ (GPa)	0.10	0.29	**0.43**	0.17	**0.22**	0.15	0.82	2.51	**1.71**	2.61	**1.96**	1.56	3.29	3.93	**3.01**	0.77	1.22	1.71	2.24	3.08	3.91
Matrix T(peak) mean (K)	436	521	**505**	545	**534**	603	862	1,050	**904**	1,220	**1,020**	1,520	1,720	2,240	**1,690**	640	885	1,460	2,130	2,990	3,880
Matrix T(peak) 1σ (K)	31.6	44.2	**43.6**	47.1	**53.8**	76.9	105	236	**162**	193	**168**	220	448	651	**358**	138	183	278	424	621	758
Chondrule T(final) mean (K)	300	303	**305**	302	**302**	302	310	400	**396**	367	**379**	353	476	565	**590**	307	326	383	448	546	642
Chondrule T(final) 1σ (K)	2.28	7.46	**9.90**	4.78	**3.99**	1.93	18.5	52.8	**46.8**	45.2	**47.1**	41.3	77.0	98	**111**	24.4	32.3	57.1	73.6	82	87
Chondrule P(peak) mean (GPa)	0.71	1.45	**1.26**	1.11	**0.92**	0.85	2.68	8.71	**7.49**	6.53	**6.14**	4.13	10.8	14.8	**13.2**	1.96	3.23	7.18	12.4	18.8	24.5
Chondrule P(peak) 1σ (GPa)	0.19	0.37	**0.47**	0.27	**0.33**	0.23	0.72	2.36	**1.51**	1.79	**1.63**	1.18	2.77	3.39	**2.72**	0.74	0.98	1.33	1.58	2.15	2.77
Chondrule T(peak) mean (K)	304	311	**314**	308	**308**	306	326	438	**431**	396	**410**	380	528	642	**649**	324	349	428	519	625	741
Chondrule T(peak) 1σ (K)	4.19	18.8	**24.3**	10.9	**11.2**	6.15	39.6	76.2	**73.6**	71.0	**69.9**	77.0	110	129	**153**	33.1	53.2	96.7	116	109	106
Matrix melt fraction||	0.00	0.00	**0.00**	0.00	**0.00**	0.00	0.00	0.00	**0.00**	0.01	**0.01**	0.59	0.64	0.94	**0.64**	0.00	0.01	0.29	0.94	1.00	1.00

CC, carbonaceous chondrites; OC, ordinary chondrites.

Initial matrix volume fraction varied from 70 to 33%, producing post-impact matrix volume fraction similar to carbonaceous and ordinary chondrites, respectively. Impact velocities ranged from 0.75 to 3 km s^{-1}. Bold entries show results for simulations with serpentine matrix and forsterite chondrules (approximating CM and CR scenarios). Other entries show results for forsterite matrix and forsterite chondrules.

In all simulations the initial temperature ($T0$) was 300 K. Varying $T0$ does not significantly affect δT (4% lower at $T0 = 600$ K versus $T0 = 300$ K).

*Average pressure in the shock wave as it propagated through the sample region.

†Average temperature in the sample region after release from high pressure.

‡T (final) statistics use temperature of each tracer at time of release from high pressure.

§P (peak) and T (peak) statistics use extremes of pressure and temperature recorded by each tracer during passage of the shock wave. Note that the difference in average peak shock pressure between the chondrules and matrix is a consequence of the large strength difference between the components; the average peak longitudinal stress is the same in the chondrules and matrix (see Methods for more details).

||The fraction of the matrix with T (final) greater than the solidus temperature (1,373 K).

peak shock pressures (P(peak)) and post-shock temperatures (immediately after passage of the shock wave) (T(final)) across a computational mesh composed of a bimodal mixture of chondrules surrounded by a highly porous matrix using the iSALE hydrocode[17–19]. Our approach complements and extends recent experimental work that tracked porosity in analogue materials[8], allowing us to observe how pressure, temperature and porosity vary during an impact within and between chondritic components. Tracer particles recorded the peak- and post-shock

state of the matrix and chondrule material, from which the bulk state was determined. We conducted 16 simulations (Table 1) that spanned a range in impact velocity (0.75–3 km s^{-1}) and initial matrix volume fraction (70 and 33%) to simulate impacts that generated post-compaction matrix volume fractions consistent with CC-like objects and ordinary chondrite (OC)-like objects, respectively. An ANEOS-derived equation of state table for forsterite[20] was used to describe the thermodynamic response of the non-porous disks and the solid component of the

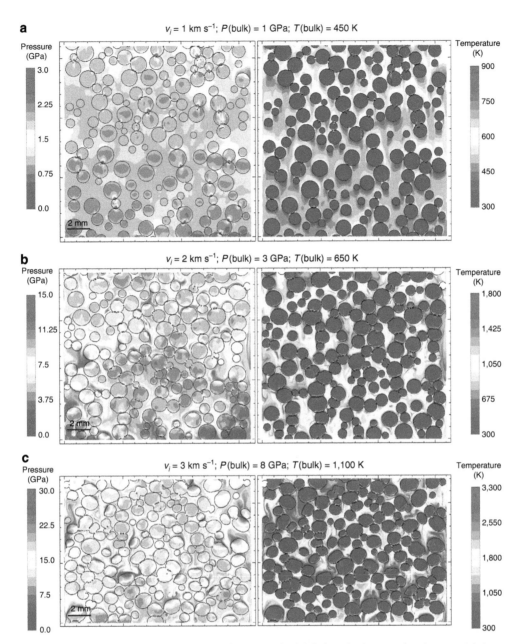

Figure 1 | Variation in peak-shock pressure and peak temperature in mesoscale simulations. In the examples shown, matrix is composed of forsterite, initial matrix volume fraction is 70%, initial matrix porosity is 70% and starting temperature is 300 K. (**a**) 1 km s^{-1} impact results in bulk shock pressure $\sim$1 GPa; (**b**) 2 km s^{-1} impact results in P(bulk) $\sim$3 GPa; (**c**) 3 km s^{-1} impact results in P(bulk) $\sim$7 GPa. The images illustrate the variation in P(peak) and T(peak) recorded by a representative portion of the sample after passage of the shock wave. Note the change in pressure and temperature scale between frames. The extreme variability in matrix PT over short lengthscales is apparent, indicated by swirling contours in matrix. Our simulations reproduce a variety of chondrule/matrix textures observed in the chondrite record (elongate or oriented chondrules; matrix 'flowing' between chondrules; indented chondrules). Chondrule indentation textures become apparent at 2 km s^{-1} (1.5 km s^{-1} at lower matrix vol%, (for example, OCs)); internal chondrule P(peak) $\sim$5 GPa.

matrix. We also conducted five simulations with serpentine matrix and forsterite chondrules (to approximate CM and CR chondrites), using ANEOS input parameters for serpentine[21]. Impact heating will lead to dehydration of hydrous phases, releasing water that may cause textural modifications, but computational limitations meant that we could not model this process explicitly. Compaction of porosity and material strength were modelled using methods described in earlier work[18,19]. We take $\phi_{mi} = 70\%$ as a standard value in our simulations, with selected comparison runs at $\phi_{mi} = 60\%$ and $\phi_{mi} = 80\%$. Initial temperature was 300 K (an argument can be made that this should be higher, given that peak metamorphic temperature in

CV chondrites exceeded 600 K, and compaction post-dated formation of secondary minerals). We explored the effect of varying T(initial). For small (100 K) variations, temperature changes predicted by the models were independent of the starting temperature, for example, 100 K lower T(initial) would reduce final temperatures by that amount.

Our simulations allow us to observe how pressure, temperature and porosity vary during an impact within and between chondritic components, observing detail down to the 100-μm level (Fig. 1). Although low velocity impacts are expected to produce minor effects in CCs[1,2], resolving at this finer 'mesoscale', we find that this is not the case. Predictably, matrix porosity is removed rapidly, decreasing from 40–50% at $0.75 \, km \, s^{-1}$ to $\sim 15\%$ at $1.5 \, km \, s^{-1}$. But matrix and chondrules experience very different conditions, particularly with respect to temperature, in all scenarios (Fig. 1). A 2.0-$km \, s^{-1}$ event into a 70% forsterite matrix precursor (Fig. 1b) with $\phi_{mi} = 70\%$ generated chondrule P(peak) of $6.5 \pm 1.8 \, GPa$: barely detectable in a typical petrographic study[1,2]. However, matrix experiences T(final) of $1,100 \, K$ while chondrules are virtually unheated (ΔT(final) $\sim 70 \, K$). We also observe significant heterogeneity in PT conditions over short ($100 \, \mu m$) lengthscales. Taking the same example ($2.0 \, km \, s^{-1}$ and 70% initial matrix volume fraction: Fig. 1b), we observe a 7–18-GPa range of P(peak) over $100 \, \mu m$ in matrix (average matrix P(peak) is $8.4 \pm 2.6 \, GPa$ in this simulation), and a 700–$2,000$-K variation in matrix T(final) over similar lengthscales (average matrix T(final) is $1,100 \, K$). Chondrule deformation (and flattening) is apparent at velocities of $2 \, km \, s^{-1}$ and higher. The assumed Hugoniot Elastic Limit for the chondrules ($\sim 5 \, GPa$) is exceeded in these higher-velocity simulations, that is, the pressure is so high that they are deformed in a plastic, rather than in a brittle manner, despite the high strain rate. We also note that chondrule edges experience higher shock than chondrule interiors. This occurs where chondrules impact and indent one another (leading to local P(peak) spikes up to $\sim 12 \, GPa$ in the 2-$km \, s^{-1}$ simulation), but it is also a general feature, as matrix compacts around chondrules. Although average chondrule P(peak) is $6.5 \, GPa$ at $2 \, km \, s^{-1}$, interiors only see 4–$5 \, GPa$, while chondrule edges range up to $8 \, GPa$ (it is probable that the real value is higher, but spatial resolution is limited by computational costs). Finally, with regard to the serpentine matrix scenarios, at low shock pressure there is little difference in matrix T(final) between serpentine and forsterite. In simulations $\geq 2 \, km \, s^{-1}$ temperature is buffered by phase changes in serpentine, reducing matrix temperatures in these higher-shock scenarios. However, even here the same general features remain: high matrix heating alongside 'cold' chondrules, and significant PT heterogeneity over short lengthscales.

Previous studies examined shock effects in porous targets, or non-porous targets, but not mixtures. It is known that impacts into porous materials result in much more heating than impacts into non-porous materials, as shock energy is expended in collapsing pores[3,4]. Impacts into a uniformly porous asteroid will compact that material, heating a large volume around the crater that then cools and equilibrates over timescales up to 10–$100 \, Myr$[11]. Primitive chondrites are compacted, but are texturally and compositionally unequilibrated. Our data indicate that impact-induced compaction in chondrites occurred without generating macroscopic thermal metamorphic textures, or chemical equilibration, because of the juxtaposition of two materials with vastly different initial porosities and bulk properties. In the chondrite scenario, chondrules act as a heat sink. Matrix is heated rapidly to 100's of degrees above ambient, but because of the proximity to 'cold' chondrules, it cools rapidly. Rather than slow cooling of a large volume[11], we see sub-mm heterogeneities that equilibrate in $< 10 \, s$. So unless bulk T(final) exceeds temperatures typical of CC peak metamorphism (a condition we observe only in scenarios with impact velocity > 2–$2.5 \, km \, s^{-1}$), macroscopic effects will not be apparent.

Electron microscopy. Although the effect of shock on fine-grained crystalline solids has been investigated[22], there are few detailed transmission electron microscopy (TEM) studies of shock metamorphism in the meteorite matrix[9]. And while petrographic studies of experimentally shocked chondrites are informative[23], the 'unshocked' starting point in this work is a meteorite that has already experienced significant compaction.

Given the limited literature, we conducted our own coordinated field emission (FE)-electron probe microanalysis (EPMA), FE-scanning electron microscope (SEM) and focused ion beam (FIB)/TEM study of a CR2 (GRA 06100), focussing specifically on how impact-induced processes may have affected sub-μm materials in this rock. We devised a novel methodology that used temperature-related crystallographic transitions to quantify matrix temperatures at fine scale. To select regions to extract FIB sections, we first built a map based on FE-SEM back-scattered electron and energy-dispersive spectrometry (EDS). The results of the FIB analysis guided us in interpreting detailed mineralogy from the original map. This combined approach enabled us to draw the first detailed isotherm map for a chondrite, in this case GRA 06100 (Fig. 2) (a follow-up study[10] in which an additional eight FIB sections were extracted found that the mineralogy was consistent with expectations at all the new sites).

No evidence of heating was found in chondrule olivine and pyroxenes. GRA 06100 matrix consists of sub-micron, diffuse masses that have compositions ranging from low-total, Mg, Al-bearing, saponite to ferroan serpentine under a 10-μm EPMA beam. The porosity and mineralogy of GRA 06100 matrix is variable, containing deformed SiO_2 (grains contain nanocrystalline and amorphous regions), Fe-silicide, nanophase fayalitic olivine and ferrosilite, non-stoichiometric ferromagnesiosilica grains, as well as amorphous silicates and phyllosilicates with variable amounts of Al. Although Fe-rich serpentine is the most common matrix phyllosilicate, the smectite aliettite was identified in several regions of the matrix, using a combination of TEM/ EDS and crystallography. Aliettite undergoes four endothermic transitions (150, 220, 605 and $905 \, ^\circ C$)[24], which are of use in determining the minimum temperature reached in a region of matrix. Aliettite recording $220 \, ^\circ C$ was found in close proximity to aliettite recording 605 and $905 \, ^\circ C$.

Near-end member ankerite was found at the edge of a large metal nodule, located $< 20 \, \mu m$ from unheated chondrule olivines. Ankerite of this composition is predicted to occur at $T > 827 \, ^\circ C$ (ref. 25). Also $< 20 \, \mu m$ from unheated chondrule olivines, the high-temperature silicate-sulfate–phosphate scorzalite–lazulite was found. The observed compositions in the scorzalite–lazulite solid solution correspond to $T > 485 \, ^\circ C$ (ref. 26). Finally, pentlandite, the most common sulphide in CR chondrites, is absent from GRA 06100. Where sulphides are present, they consist of pyrrhotite and partly oxidized Fe–sulfides sub-domains. Thermal decomposition of pentlandite occurring in opaque assemblages in GRA 06100 occurs at $T > 600 \, ^\circ C$ (ref. 27).

On the basis of the distribution of different minerals of selected areas of the thin section, we produced a thermal contour map (Fig. 2b). We estimate that $\sim 46\%$ of the section shows no evidence of heating, $\sim 12\%$ is consistent with $493 \, K \leq T \leq 873 \, K$, $\sim 28\%$ $873 \, K \leq T \leq 1,023 \, K$, $\sim 9\%$ $1,023 \, K \leq T \leq 1,178 \, K$ and $\sim 5\%$ records temperatures $\geq 1,178 \, K$. However, the great diversity of mineral assemblages identified in matrix indicates that more FIB sections are needed to fully constrain the temperature regimes recorded by fine-grained material. Note: although phase transition temperatures are defined at low pressure, the comparison with model results is valid as we are comparing to post-shock/pre-equilibration temperatures (that is, low pressure after shock passage but before thermal equilibration between chondrules and matrix).

Macroscale simulations. In addition to the mesoscale simulations, a suite of 'macroscale' planetesimal collision simulations were performed (for example, Fig. 3). Tracer particles were used to track planetesimal material exposed to different bulk post-shock temperatures for comparison with the mesoscale

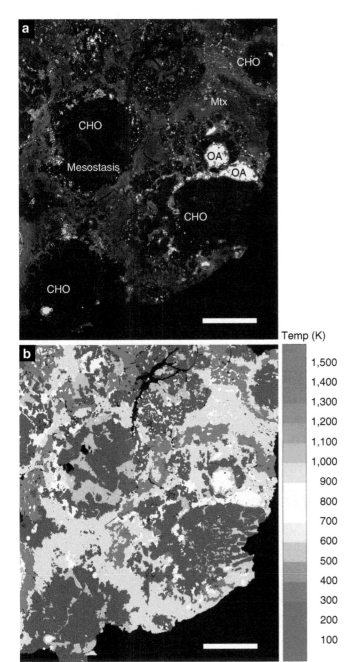

Temp (K)

1,500
1,400
1,300
1,200
1,100
1,000
900
800
700
600
500
400
300
200
100

Figure 2 | Back-scattered electron (BSE) and temperature images of CR2 chondrite GRA 06100. (**a**) BSE image showing chondrules (CHO), mesostasis, opaque assemblages (OA) and matrix (Mtx). (**b**) Temperature map of the same area. Scale bar, 0.5 mm. Chondrule olivines (gray-scale regions) show no evidence of heating or shock. Temperatures were estimated based on the crystallographic transitions of aliettite (423, 493, 878 and 1,178 K)[24]; presence of near-end member ankerite ($T>1,100$ K)[25]; thermal decomposition of pentlandite ($T>873$ K)[27]; scorzalite–lazulite solid solution ($T>758$ K)[26].

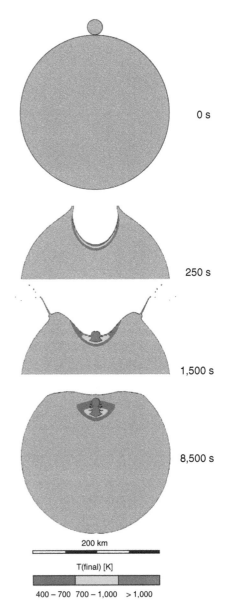

0 s

250 s

1,500 s

8,500 s

200 km

T(final) [K]

400 – 700   700 – 1,000   > 1,000

Figure 3 | Translating mesoscale observations into asteroid-scale impacts. Macroscale hydrocode modelling of impacts between porous planetsimals. In an impact on a real planetesimal, the shock wave attenuates rapidly with distance, but the peak shock PT conditions are similar to the planar impact scenario. Macroscale models consider impact velocities from 1 to 10 km s^{-1} for both OC-like (20% initial bulk porosity) and CC-like (50% initial bulk porosity) objects; assume vertical impact; and with target planetesimals of 200–500 km diameter, and impacting planetesimals of 10–150 km in diameter. This example involves CC-like objects impacting at 4 km s^{-1}. The blue region logs tracers at 400 K < bulk T(final) < 700 K: the range consistent with meteorite data. Of the material that is compacted to meteorite-like porosities, we assume (conservatively) that if > 20% experienced bulk T(final) > 700 K (shown here in green (700–1,000 K) and red (>1,000 K)) it would be observed in the meteorite record. This corresponds to velocities exceeding 3 km s^{-1} in modelled impacts (Fig. 4). Similar velocities are derived from OC-like simulations.

simulation results and meteoritic constraints (Fig. 3). The mass of material in the range 400 K < bulk T(final) < 700 K is shown in blue; material that experienced bulk T(final) > 700 K is shown in green (700–1,000 K) and red (> 1,000 K). Figure 4 shows the total mass of material heated to different final bulk temperatures from an example suite of collisions (1:1,000 impactor-to-target mass ratio; 50% initial porosity in impactor and target) at various impact speeds using the same colour coding. All macroscale simulations assumed head-on collisions between planetesimals,

enforced by the 2D axial symmetry of the numerical model. In reality, the most common impact angle is 45° to the target plane. Numerical simulations of oblique impacts on planar surfaces suggest that the volume of material heated to a given temperature scales with vertical component of impact velocity in the same way as for vertical impacts[28]. Hence, in the context of

a

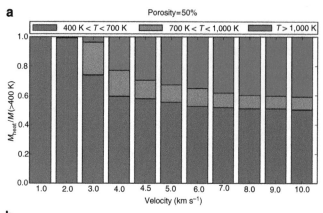

b

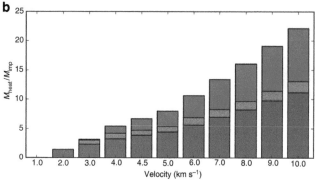

Figure 4 | Fraction of material heated to varying temperature in macroscale collisions. Results are shown for head-on collisions between two planetesimals with mass ratios of 1:1,000 and with initial porosities of 50% (CC-like simulation). Equivalent simulations were performed for OC-like (20% initial bulk porosity) objects, and for a range of other target and impactor diameters. Results are normalized by both the impactor mass (**b**) and the mass of material heated by 100 K, which corresponds to a bulk porosity reduction of ~10% (**a**). The fraction of material heated above 700 K exceeds 0.2 at a vertical impact velocity of ~3 km s^{-1}. Similar velocities are derived from OC-like simulations. The 2D axial symmetry of the numerical model enforces head-on collisions between planetesimals in the macroscale simulations. In reality, the average impact angle is 45° to the target plane. The volume of material heated to a given temperature also scales with a vertical component of impact velocity in the same way as for vertical impacts[28]. Thus, a 3-km s^{-1} vertical impact is approximately equivalent to a 45° impact at $3/\sin(45) = 4.25$ km s^{-1}.

Fig. 4, a 3-km s^{-1} vertical impact is approximately equivalent to a 45° impact at $3/\sin(45) = 4.25$ km s^{-1}.

Discussion

Our observations have a number of implications for understanding the chondrite record: lithification of primordial materials may have been a natural outcome of transient heating of matrices, driven by collisional compaction. Hydrothermal alteration has been suggested as a mechanism for meteorite lithification, but some primitive meteorites have escaped this process and yet are consolidated rocks. Lithostatic pressure was minimal[16], and towards the surface—where primitive chondrites are expected to originate—it would have been even lower. Yet chondrite lithification was efficient even at fine scales: matrices are lithified. Although collisional compression has been proposed as a mechanism[16], impact shock has generally been excluded as few CCs show macroscopic shock indicators. But our data indicate that this is not a valid constraint when considering the response of matrix to shock. In addition, experiments examining

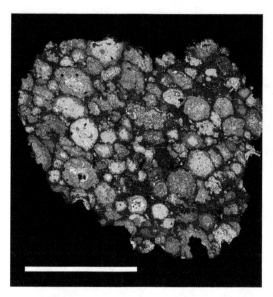

Figure 5 | Chondrule textures in the CR2 chondrite Renazzo. Flattened and indented chondrules highlighted in an element map of Renazzo (green:MgI yellow:Ca; white:AI; red:Fe; blue:Si); textures similar to the 2-km s^{-1} simulation shown in Fig. 1b. Scale bar 6 mm.

the sintering behaviour of olivine have found that it is favoured where a fine-grained (micron-scale) aggregate is rapidly heated to high T^{29}: a situation identical to that encountered by high-porosity matrix during impact-induced compaction.

Our results have implications for chondrite palaeomagnetism. Magnetic fields may be recorded at the grain scale without raising the bulk temperature above the Curie point; and this can occur without generating obvious macroscopic shock metamorphic textures (that is, it may have affected meteorites that have been classified as low shock). It is known that an ambient magnetic field at the time of impact may be amplified or even produced by the impact itself[30]. Large impact-generated fields are possible[31]. But impact-generated fields have not been considered a significant factor in the palaeomagnetic record of primitive chondrites for two principal reasons. It was thought that shock pressures > 40 GPa[30] were required to generate temperatures high enough to impart a thermoremanent magnetization (TRM). In these highly energetic events post-impact cooling times would be long (perhaps 10^5–10^7 Myr). But impact-generated fields are brief ($<$ mins). An additional reason is the absence of macroscopic shock textures in most CCs[30]. Our work indicates that these arguments should be re-examined. We show that it is necessary to model PT evolution at the meteorite scale, in order to constrain laboratory measurements at the meteorite scale. The absence of shock textures in chondrules does not inform our understanding of matrix PT history. The bulk rock will respond as expected in the literature, but on a finer scale, matrix will experience rapid heating and cooling, prior to equilibrating to bulk T(final) in $<$ 10's of seconds. If the principal magnetic carrier phase was hosted in matrix (as is often the case), then the palaeomagnetic record may well have been modified during compaction. Equally, a matrix phase could record an impact-generated field with a duration of 10's of seconds to minutes. A rock could acquire a TRM during rapid cooling of matrix following impact-induced compaction.

Model results showing PT heterogeneity at fine scale and high transient matrix temperatures suggest a unified explanation for anomalous features of chondrite petrography and texture that previously have required complex mechanisms (involving accretion, disruption and dispersal, heating or shock processing in the

Table 2 | Material parameters used in numerical simulations.

Parameter	Chondrule	Matrix (forsterite)	Matrix (serpentine)	Bulk
Initial porosity	0	0.6–0.8	0.6–0.7	0.5
Compaction rate*	NA	0.98	0.98	0.98
Vol. strain at onset of plastic compaction*	NA	$-1E-5$	$-1E-5$	$-1E-5$
Poisson ratio (solid component)[†]	0.23	0.23	0.23	0.23
Intact cohesion[†] (MPa)	1,000	0.1	0.1	10
Intact friction coefficient[†]	1.2	1.2	1.2	1.2
Intact strength limit[†] (GPa)	3.5	0.035	0.035	3.5
Damaged cohesion[†] (MPa)	0.01	0.01	0.01	0.01
Damaged friction coefficient[†]	0.6	0.6	0.6	0.6
Damaged strength limit[†] (GPa)	3.5	0.035	0.035	3.5
Melt temperature[3] (zero pressure) (K)	1,373	1,373	1,098	1,373
Simon approximation constant[‡] (GPa)	1.52	1.52	1.52	1.52
Simon approximation exponent[‡]	4.05	4.05	4.05	4.05
Thermal softening parameter[†]	1.2	1.2	1.2	1.2

NA, not applicable.
*Wunnemann et al.[19]
[†]Collins et al.[18]
[‡]Wünnemann et al.[57]

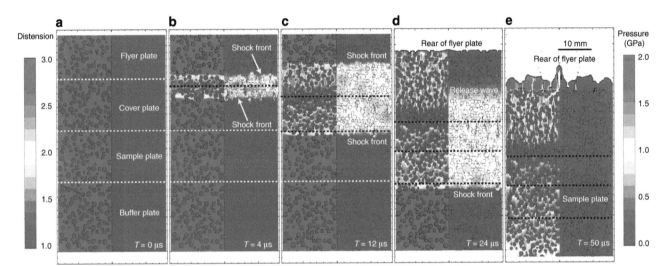

Figure 6 | Time sequence of a typical mesoscale impact simulation. The nominally planar shock wave is shown propagating through a bimodal mixture of explicitly resolved non-porous chondrules surrounded by a high-porosity matrix, with snapshots showing the initial state (**a**), at 4 μs (**b**), 12 μs (**c**), 24 μs (**d**), and ending at 50 μs (**e**). Scale bar, 10 mm. Each time snapshot is divided into two panels, with colour-scales denoting distension (1/(1 − porosity)) (high initially in matrix), and instantaneous longitudinal stress (0 initial throughout the mixture). Upon impact, shockwaves are generated at the flyer-sample interface and propagate both down into the sample and up into the flyer plate, compacting the matrix. In this example, after ~24 μs the shock wave in the sample has reached the sample/buffer plate interface; by this time the shock in the flyer plate has reflected off the rear of the flyer plate as a release wave that propagates back through the flyer, sample and buffer plates. By 50 μs the release wave has left the computational domain and the post-shock state of the sample may be recorded. The variation in peak pressure, peak- and post-shock temperature experienced by both the chondrules and matrix was recorded for subsequent analysis as was the reduction in porosity in the matrix.

nebula, followed by re-accretion and (possibly) brecciation[32,33]. Features such as shock-induced mobilization of sulphides in (apparently) 'unshocked' OCs[32], pyroxene polymorphs in CC matrices that require a high-temperature origin (>1,300 K) followed by very rapid cooling[33,34] and matrix olivines showing dislocations indicative of significant shock deformation with adjacent chondrule olivines unaffected[9,35], are entirely consistent with our observations. In addition, flattened, oriented or indented chondrules are a feature of many primitive meteorites, even those characterized as very weakly shocked (for example, Renazzo, S2; Fig. 5). In our modelling these chondrule textures become apparent at 2 km s^{-1} (Fig. 1b). Chondrule P(peak) in the forsterite simulation is 6.5 GPa (6.1 GPa with serpentine matrix): according to macroscopic shock indicators in chondrules, this is consistent with shock level S2 (very weakly

shocked). CR chondrites are among the most pristine CCs: we might expect them to preserve the most complete record of early compaction. The detailed isotherm map generated from our TEM study of CR chondrite matrix revealed the degree of heterogeneity in GRA 06100. Although (as expected in a C2 chondrite) a large fraction of matrix shows little evidence of heating, mineral indicators in >50% of matrix are consistent with temperatures >900 K, and we are able to define areas that experienced $T > 1,180$ K (ref. 24). We observe 1,000 K heterogeneity on 100 μm lengthscales. In a highly unequilibrated meteorite this required rapid heating and cooling. Chondrules show no evidence of shock metamorphic textures. These observations are all consistent with our model predictions. In the case of GRA 06100, they are a match to a ~1.5-km s^{-1} scenario with 70% initial matrix by volume. Finally, a specific model prediction—

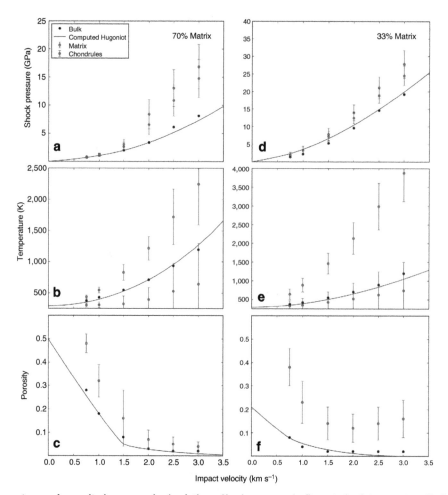

Figure 7 | Pressure, temperature and porosity in mesoscale simulations. Shock pressure (**a,d**), post-shock temperature (**b,e**) and post-shock porosity (**c,f**) as a function of impact velocity for bimodal chondrule/matrix mixtures with matrix volume fraction 70% (**a–c**) and 33% (**d–f**). Black symbols show bulk values of the state quantity, averaged over the entire sample region; red and blue symbols (and error bars) show mean (and s.d.) of the state variable in the matrix and chondrule fraction, respectively. Apart from the peak-shock pressures in the matrix and chondrules, which represent the maximum pressure recorded by the material at any time, all variables are measured at a given instant in time either during the shock passage (bulk pressure) or after release (final temperature and porosity). As a result, while the bulk porosity and final temperature values represent volume-weighted averages of the porosity and temperature of the matrix and chondrules components, the bulk shock pressure values are substantially less than the peak-shock pressures recorded by matrix and chondrules. Solid line shows a Hugoniot curve for the bulk porous material (0.49 for 70% matrix; 0.23 for 33% matrix) computed using the epsilon–alpha porous compaction model.

local melting of matrix in the most compacted primitive chondrites—has also been observed. Leoville is a reduced CV chondrite with porosity of 2% (ref. 14). A suite of observations, including highly variable shock effects on 100-μm scales, and local matrix melting, was explained by a combination of hot accretion and impact[36]. Our modelling suggests a simpler, unified explanation.

The apparent connection between shock level (increasing from CCs, through unequilibrated ordinary chondrites, to equilibrated ordinary chondrites (EOCs)) and asteroid metamorphism[1,2] has prompted some to suggest that asteroid thermal metamorphism was a result of large impacts[37]. Mesoscale simulations showing PT variability with matrix fraction provide an alternative explanation. At an impact velocity of $2 \, \mathrm{km \, s^{-1}}$, chondrule interiors in a CC-like (high matrix fraction) precursor experience P(peak) of $\sim 5 \, \mathrm{GPa}$; an OC-like (low matrix fraction) precursor $\sim 12 \, \mathrm{GPa}$; and a non-porous dunite (EOC approximation) $\sim 24 \, \mathrm{GPa}$. In other words, shock level varies from S1–S4 depending on matrix fraction for the same impact velocity. Shock level is derived from observation of textures in chondrule olivines. There is a widely held view that varying shock level informs us about varying intensity of asteroidal collisions. But our

data suggest that the way we use macroscopic textures to infer shock level can bias our interpretation. It is not required that varying impact energy caused the differences in shock level between the chondrite groups. Rather, varying matrix proportion in chondrite targets controlled the level of shock metamorphism recorded by the chondrules within them.

Both in terms of locating impact-induced compaction within the meteorite record, and exploring its significance with respect to the dynamical evolution of planetesimals, it is useful to constrain the timing of compaction. Chondrule formation defines an upper limit on chondrite accretion times. Chondrites could not have accreted prior to the formation of the youngest chondrules that they contain. In addition, it is apparent from CV[7,15] and CR (this study) data that compaction occurred after aqueous alteration. Recent $^{207}\mathrm{Pb}$–$^{206}\mathrm{Pb}$ ages for CV calcium-aluminium inclusions (CAIs) (the assumed starting point for solar system formation) are $4567.18 \pm 0.50 \, \mathrm{Myr}$[38] and $4567.30 \pm 0.16 \, \mathrm{Myr}$[39]. Using current $^{238}\mathrm{U}/^{235}\mathrm{U}$ values[38] to correct published chondrule ages drives chondrule Pb-isotopic ages 0.95 Myr younger: $4,564.5 \pm 0.50 \, \mathrm{Myr}$ for a multichondrule fraction from Allende[40], $4,563.23 \pm 0.66 \, \mathrm{Myr}$ for a multichondrule fraction from a CR[41] (we can anticipate that individual chondrules within that fraction would have formed later than the 4,563.23 ± 0.66 Myr average). Analysis of two individual Allende chondrules generated ages of 4,567.32 ± 0.42 Myr and 4,566.24 ± 0.63 Myr[39]. Relative [26]Al/[27]Al chondrule ages agree, indicating that CC chondrules were still forming at 4.3 Myr after CAI[42]. Recent [53]Mn–[53]Cr ages for secondary fayalite in CVs are 3.9(+1.0/−0.9) Myr[43]. Together, these data allow us to say that CC compaction did not occur earlier than 4–5 Myr after CAI. The OC data are more variable. Three chondrules from NWA 5697 have ages that range from 4,566.67 ± 0.43 to 4,564.71 ± 0.30 Myr[39]. [26]Al/[27]Al suggests that the youngest OC chondrules date to ∼3.5 Myr after CAI[42]. And [53]Mn–[53]Cr for secondary fayalite from an OC gave 2.4(+1.8/−1.3) Myr[44] after CAI. On balance, these data suggest an upper bound for compaction of OC planetesimals at ∼3.5 Myr after CAI. Considering the chronometry data, our mesoscale simulations allow us to define a 'speed limit' constraint on the major compressive impact event(s) affecting CC planetesimals after 4–5 Myr, and OC planetesimals after ∼3.5 Myr.

A spectrum of meteorite data (porosities, meteorite fabrics, petrography and local PT heterogeneity) are consistent with compaction from high ϕ_{mi} at velocities in the range 0.75–2 km s^{-1} (bulk T(final) ∼400–700 K). In the case of CC-like simulations, velocities >2 km s^{-1} generate bulk $T > 700$ K (evidence for metamorphism at these temperatures is extremely rare in CCs). In the case of OC-like simulations, velocities >1.25 km s^{-1} (corresponding to bulk $T > 600$ K) generate matrix T(final) such that >20% of matrix is heated above the solidus (OC meteorites do not show evidence for widespread matrix melting). To translate our mesoscale observations into a constraint on asteroid-scale impact velocity, we can go to our macroscale numerical simulations of planetesimal collisions (Figs 3 and 4) which account for shock-wave attenuation with distance. Material in the range 400 K < bulk T(final) < 700 K (blue) is consistent with meteorite data, according to the mesoscale simulation predictions. Material that experienced bulk T(final) > 700 K (green and red) is not consistent with data from primitive unequilibrated meteorites. Of the material that is compacted to meteorite-like porosities, we assume that if >20% experienced high bulk T(final), and/or widespread matrix melting (for example, in OC targets), it would be observed in the meteorite record. In the example suite of simulations shown in Fig. 4 it is apparent that if asteroid collisional velocities exceeded 3 km s^{-1} in head-on impacts during the compaction phase (translating to ∼4 km s^{-1} for average 45° impacts) this boundary would be crossed. Put another way, mesoscale simulations showing that velocities >2 km s^{-1} for CC impacts and >1.25 km s^{-1} for OC impacts are incompatible with meteorite data would translate to velocities of 3–4 km s^{-1} in real asteroid collisions.

Although the specific collisional 'speed limit' will depend on a variety of model assumptions, it is apparent (a) that initial impact velocities for OC and CC parent bodies were similar, and (b) that they must have been low. The chondrule age upper limit on accretion time, and chronometry for secondary olivine, indicates that these general constraints applied from ∼4 Myr after solar system formation. In addition to relevance for understanding the early evolution of primitive solar system objects, our 'speed limit' has implications for dynamical models of giant planet migration. According to recent models[45,46], OC-like parent bodies formed at 0.7–3 AU and CC-like bodies formed between the giant planets and from 8.0 to 13.0 AU. The migration of Jupiter resulted in an early (4–5 Myr) depletion and excitation of the asteroid belt region[45,46]. It was then repopulated with material from these two source regions[45,46]. Estimates for asteroid collisional velocities at 4 Myr from models that do not involve early migration of Jupiter[47] are 5–9 km s^{-1}. Grand Tack[45,46] variants would

produce higher mean impact velocities. This is not consistent with our observations. In addition, if OCs and CCs are derived from very different source regions we would expect to see evidence of rather different initial impact velocities. We do not. With respect to the planetary dynamics of the early solar system, our speed limit constraint is consistent with a relatively 'late' phase of main-belt collisional evolution, rather than the early, rapid depletion of a massive main belt[45,46].

Methods

Numerical modelling. The iSALE shock physics code[17–19] was used in two ways to quantify the compaction of porous planetesimal material in an impact. In one suite of models, 2D plane-strain, 'mesoscale' simulations were performed of shock wave propagation through a bimodal mixture of non-porous disks (chondrules) surrounded by a highly porous continuous matrix. The purpose of these simulations was to quantify the heterogeneity in peak pressure, post-shock temperature and post-shock porosity caused by the propagation of a single shock wave through a heterogeneous bimodal 'chondritic' mixture. In another suite of models, 2D axially symmetric, 'macroscale' simulations were performed of head-on collisions between porous planetesimals of different sizes and at speeds of a few km s^{-1}. The purpose of these simulations was to constrain the post-impact distribution of compacted material consistent with meteoritic evidence.

Mesoscale simulations. Chondritic meteorites are comprised of (nominally) zero-porosity spherical chondrules (0.1–1 mm in size) set in a highly porous matrix aggregate composed of sub-μm monomers. The three orders of magnitude difference in lengthscale of these two components allowed us to simulate shock in this material by explicitly resolving the chondrules as disks of non-porous forsterite in the 2D computational grid and surrounding these with a high-porosity continuum of forsterite (in 16 simulations), and serpentine (in 5). The porosity in this continuous matrix is parameterized as it exists on a lengthscale too small to be resolved in the simulation. The epsilon–alpha porous compaction model[19,48] was used to parameterize the effect of pore space compaction during the shock, while an ANEOS equation of state table for dunite/forsterite[20] was used to describe the thermodynamic response of the chondrule material. The non-porous part of the porous matrix material was described by either the dunite/forsterite equation of state, or an ANEOS equation of state table for serpentine[21]. A strength model for geologic materials[18] was used to represent the response of the chondrules and matrix to changes in deviatoric stress: the chondrules were given a high cohesive strength (1 GPa), whereas the porous matrix was assumed to be very weak, with a cohesive strength of 100 kPa. A complete listing of material parameters used for the chondrules and matrix is given in Table 2. To generate a bimodal mixture analogous to the chondrule/matrix mixtures in chondritic meteorites, we randomly placed chondrule analogues (dunite disks) of various sizes ranging from 0.3 to 1 mm in diameter into an otherwise continuous region of the porous matrix analogue until the desired chondrule volume fraction was achieved. All simulations assumed an initial temperature of 300 K and an initial matrix porosity of 70%, apart from selected comparison runs at 60 and 80%. The remaining initial conditions are summarized in Table 2.

In the mesoscale simulations, numerical planar impact experiments were performed in which a flyer plate impacted a target, comprising a sample sandwiched between a cover plate above and a buffer plate below (Fig. 6). The flyer, cover, sample and buffer plates were all comprised of the same bimodal mixture of non-porous chondrule disks, surrounded by a high-porosity matrix. The presence of a cover plate allowed the planar shock wave to achieve a steady form before passing through the sample and then the adjacent buffer plate. The simulation time extended until the sample was released from high pressure by a release wave from the rear of the flyer plate (see Fig. 6).

Figure 6 illustrates the propagation of the shock wave from a representative numerical simulation. Upon impact an upward moving shock wave is generated in the flyer plate and a downward shock wave is generated in the cover plate. As the shock propagates though the cover plate it evolves to a steady wave as the shock front thickness and rise time increase to constant values determined by the mesoscale structure of the material. For the particle size, bulk porosity and shock pressure range in our simulations; the shock front thickness was ∼1–3 chondrule diameters, consistent with front thicknesses determined by mesoscale simulations of granular material compaction[49]. A consequence of the initial increase in shock front thickness is that shock compaction is greatest at the impact plane and decays with distance until a steady wave is achieved. This is evident from the gradient in porosity in the cover and flyer plates near the interface between them. The thickness of the cover plate (a few chondrule diameters) was chosen to ensure that the shock wave was steady when it entered the sample; no gradient in compaction exists within the sample. The mesoscale structure also creates resonant oscillations around the steady wave amplitude, which result in the important heterogeneous temperatures within the sample that are the focus of our study. Such oscillations have been observed in laboratory experiments of layered composites of 'hard' and 'soft' materials[50] and observed and modelled in porous granular materials[51] and porous rocks[52]. The magnitude and duration of the oscillations depends on the impedance mismatch between the components in the system[50], which is very large for the chondrule/matrix system studied here. It is important to note that while the duration of the shock in our numerical simulations is long compared with typical laboratory experiments, it is short compared with shock durations in km-scale impacts on planetesimals (the shock duration is appropriate for cm-scale impactors). However, a systematic verification study that varied the shock duration showed no significant effect of shock duration on either the mean or the variance of the shock temperatures recorded in the simulations, provided the shock duration was longer than the time necessary to achieve a steady wave. Hence, extrapolation of results to longer shock durations is justified. To record changes in temperature, pressure and distension (porosity), passive tracer particles were initially placed one per cell and subsequently followed the particle path of that material. Both peak and instantaneous pressure and temperature were recorded by the tracers, as well as the instantaneous matrix distension (1/(1 – porosity)). Bulk temperature, pressure and distension were computed as volume-weighted averages of all tracers in the shocked sample plate. Note that a consequence of the heterogeneity in shock propagation is that the peak shock pressure recorded by each tracer throughout the duration of shock wave passage can be substantially higher than the instantaneous bulk shock pressure at any time. For example, in the simulation shown in Fig. 1b the instantaneous bulk shock pressure in the shock wave was approximately 3 ± 0.3 GPa and yet the mean peak shock pressure experienced by chondrule and matrix material in the sample was 6.5 ± 1.8 and 8.4 ± 2.6 GPa, respectively. Variations in peak and post-shock state were visualized using contour plots (for example, Fig. 1) and cross-sections through the sample at specific time intervals.

In post-processing, the statistical data from the tracer particles were analysed to obtain the mean and s.d. of each variable in both the chondrule and the matrix analogues (see Table 1). Values of post-shock temperature and porosity were recorded just after the release wave had passed through the sample mixture (for example, ~ 50 µs in Fig. 6). Repeat simulations with the same input parameters but different resolutions (cells across the largest chondrule) and different random distributions of chondrules were performed. In all cases, changes in the mean values were well within the s.d. of that value in a single simulation. To reflect the variability in mean values for simulations with different particle distributions, values in Table 1 are given to three significant figures only.

Figure 7 illustrates the variation in the bulk shock response, as well as in the response of the matrix and chondrule components, with impact velocity. The peak shock pressure experienced by both the matrix and chondrules can be more than twice the bulk shock pressure of the steady shock wave. The massive difference in compressibility between the porous matrix and the non-porous chondrules results in large temperature differences between the matrix and chondrules. Impact velocities $< 4\,\mathrm{km\,s^{-1}}$ are not sufficient to compact all porosity from the matrix/chondrule mixture.

We note that the agreement between the macroscale parameterization of porous compaction (the computed Hugoniot; black solid lines in Fig. 7) and the bulk behaviour observed in the mesoscale simulations (black circles) demonstrates self-consistency between the mesoscale modelling approach and the bulk parameterization used in previous work[53] and the macroscale models of impacts on a planetesimals (see macroscale modelling section). The difference in average peak shock pressure between the chondrules and matrix is a consequence of the large strength difference between the two components. While the average peak longitudinal stress is the same in the chondrule and matrix components, the much higher strength of the chondrules implies that the corresponding peak pressure (the isotropic part of the stress tensor) in the chondrule is lower than in the weaker matrix.

The mesoscale simulations described here provide a significant first step in quantifying the heterogeneous response of chondritic precursor material to shock compaction, which we will refine in future by addressing some of the simplifying assumptions used here. Principal among these is the use of 2D plane-strain geometry, rather than a more realistic three-dimensional (3D) geometry. This assumption was necessary to limit computational cost. In 3D geometry, out-of-plane contacts between chondrules would likely stiffen the bulk response of the mixture, particularly in scenarios where the initial chondrule volume fraction is high. However, based on similar numerical mesoscale studies of pore-space compaction in 2D and 3D[52] we expect qualitatively similar behaviour in both geometries, particularly with regard to the magnitude and lengthscale of the PT heterogeneity and the trends in PT heterogeneity with both impact velocity and initial matrix fraction.

The porous compaction model and the dunite/forsterite and serpentine equation of state tables used to describe both the chondrules and solid-component of the porous matrix are also oversimplified. The compaction model assumes that all of the pressure–volume work deposited by the shock in the porous matrix leads to increase in temperature. In reality, dissipative processes during compaction, such as grain deformation and fracturing lead to an increase in entropy as well as temperature. Neglecting the entropy increase during crushing will result in an overestimate of shock heating, but this is difficult to quantify without experimental measurements of shock heating. The version of ANEOS used to derive the table does not permit both solid–solid and solid–liquid phase transitions to be included at the same time[54]. As in previous work[53] the effect of the latter was regarded as less important than that of the former. Neglecting latent heat of melting implies that temperatures in the table that exceed the solidus are overestimated. In addition, as the real matrix is a multi-component system, the temperature increase in the matrix may be buffered by one of the less refractory components reaching the point of a phase change (for example, melting or vaporization) below the assumed melting point of our single-component chondritic analogue (1,373 K). As a first step to address this issue, we performed a complementary suite of mesoscale simulations that used a serpentine equation of state to represent the solid component of the matrix in place of forsterite (see Table 1). In low-velocity simulations there is little difference in matrix *T*(final) between serpentine and forsterite. In simulations $\geq 2\,\mathrm{km\,s^{-1}}$ temperature is buffered by phase changes in serpentine, reducing matrix *T*(final). However, still evident are high matrix temperatures alongside 'cold' chondrules, and significant PT heterogeneity over short lengthscales.

At higher shock pressures, shock compression experiments of quartz[55] suggest that ANEOS overestimates the temperature increase and underestimates the entropy increase during shock compression, because it assumes a heat capacity in the fluid region that is too low[56]. If this limitation of ANEOS is also important for other silicate rocks, it implies that the shock pressure required to vaporize the matrix is overestimated by ANEOS and that peak- and post-shock temperatures above the liquids are also overestimated. As our primary focus here is relatively low-velocity collisions, which causes matrix heating below and up to the solidus, this limitation of ANEOS is of minor significance to our conclusions.

For the reasons described above, the peak- and post-shock temperatures quoted in Table 1, particularly those above the solidus, can be considered as upper limits for the given impact scenario. On the other hand, peak- and post-shock temperature are a strong function of initial temperature. A less conservative initial temperature in our models could easily compensate for any overestimate in temperature due to inadequacies of the material model. Moreover, the relative trends of increasing temperature with impact velocity and chondrule volume fraction are robust.

Of the assumed model parameters, peak- and post-shock matrix temperatures are most sensitive to the initial porosity of the matrix. Hence, the effect of initial matrix porosity on post-shock temperature provides a good measure of the sensitivity of the temperature estimates to all model assumptions. In simulations that varied the initial matrix porosity between 60 and 80%, we observed increasing post-shock matrix temperature with increasing initial porosity, for the same impact velocity (Table 1, main text). However, as demonstrated in Table 1, the mean post-shock matrix temperature for 60% initial matrix porosity is still 80–70% of the 80% initial matrix porosity values (at $1\,\mathrm{km\,s^{-1}}$ and $2\,\mathrm{km\,s^{-1}}$, respectively). Hence, temperature excursions are unlikely to be overestimated by > 20–30% in our current simulations. Comparing the results of mesoscale simulations for the same 70% initial matrix porosity, but different impact velocities, a matrix temperature increase of 20–30% corresponds to an impact velocity increase of $< 1\,\mathrm{km\,s^{-1}}$. With regard to the 'speed limit' constraint provided by the lack of evidence for high matrix temperatures among meteorites, the uncertainty in matrix temperatures associated with a given mesoscale compaction scenario translates into an uncertainty in impact velocity of $< 1\,\mathrm{km\,s^{-1}}$.

Macroscale simulations. In addition, a suite of 'macroscale' planetesimal collision simulations were performed. In macroscale simulations of planetesimal collisions, practical limits on computational mesh resolution did not allow us to explicitly resolve chondrule-scale heterogeneity in the planetesimals. Instead the material model of the chondrule–matrix mixture was approximated using the same approach as for the matrix material in the mesoscale simulations, but with strength and porosity parameters appropriate for the bulk material (see the 'Bulk' column of Table 2). Target planetesimals 200 and 500 km in diameter were modelled, with a uniform initial temperature (300 K). Two planetesimal bulk porosities were considered: 20%, which is appropriate for an ordinary chondrite (OC) parent body; and 50%, appropriate for a CC parent body. Impacting planetesimals of 10–150 km in diameter and with identical material properties to the target collided with the target at 1–10 $\mathrm{km\,s^{-1}}$.

In an impact on a real planetesimal, the shock wave attenuates rapidly with distance, but the peak-shock PT conditions are similar to the planar impact scenario. Tracer particles tracked planetesimal material exposed to different bulk post-shock temperatures (for example, Fig. 3) for comparison with the mesoscale simulation results and meteoritic constraints. The simplified material model used in our mesoscale simulations may overestimate the temperature excursions by as much as 20–30% (see supporting material), implying that an increase in impact velocity of up to $1\,\mathrm{km\,s^{-1}}$ (relative to those documented in Table 1) might be required to generate specific PT conditions.

Electron microscopy. We studied the composition and mineralogy of matrix, opaque assemblages and chondrules. A JEOL JXA-8530 field emission gun 'Hyperprobe' was used to collect quantitative X-ray elemental maps and point analyses for Na, Mg, Al, Si, P, S, K, Ca, Fe, Ni and Co from a thin section using a 0.26-µm beam. Regions of matrix with distinct average chemical composition were identified. Broad-beam analyses of these regions were also obtained. These regions were also examined by FE-SEM to establish porosity, grain distribution, size and morphology.

TEM sections were prepared using a FEI Quanta 200 3D dual beam FIB workstation. FIB-TEM sections were examined using a JEOL 2010F field-emission gun TEM/scanning TEM equipped with an ultrathin window X-ray EDS. Four FIB sections were extracted from areas where the bulk composition and/or the porosity drastically changed. In addition, four other FIB sections from opaque assemblages were available from a previous study. Acquisition and processing of digital TEM images were conducted using GATAN's Digital Micrograph imaging software. Quantification of EDS data was performed using the Cliff–Lorimer thin-film approximation. EDS spectra were collected in the scanning TEM configuration, using a 0.6-nm probe diameter. A counting time of 100 s was used to minimize the effects of beam damage and drift, while retaining meaningful counting statistics. After collecting each data point, each grain was examined using bright-field imaging. If the grain was not stable under the EDS probe for the 100-s duration of the analysis (that is, analysis resulted in a hole in the sample), the resulting analysis was not included in the data set. These analyses were excluded because it was unclear whether there was a beam overlap between the phase of interest and phase(s) lying underneath. Compositional heterogeneities were correlated with mineralogical markers, such as disequilibrium mineralogical assemblages, temperature-related crystallographic transitions, thermal decomposition of mineral phases and extent of solid solutions, to determine the scale of temperature heterogeneity.

FE-EPMA was carried out at the Regional Microanalytical and Imaging Center at Fayetteville State University. FIB and TEM analyses were conducted at the Material Characterization Laboratory at Pennsylvania State University.

References

1. Stöffler, D., Keil, K. & Scott, E. R. D. Shock metamorphism of ordinary chondrites. *Geochim. Cosmochim. Acta* **55**, 3845–3867 (1991).
2. Scott, E. R. D., Keil, K. & Stöffler, D. Shock metamorphism of carbonaceous chondrites. *Geochim. Cosmochim. Acta* **56**, 4281–4293 (1992).
3. Sharp, T. G. & DeCarli, P. S. in *Meteorites and the Early Solar System II* (eds Lauretta, D. S. & McSween, Jr. H. Y.) (Univ. Arizona Press, 653–677, 2006).
4. Kieffer, S. W., Phakey, P. P. & Christie, J. M. Shock processes in porous quartzite: transmission electron microscope observations and theory. *Contrib. Mineral. Petrol.* **59**, 41–93 (1976).
5. Blum, J. Grain growth and coagulation. *Astrophys. Dust ASP Conf. Ser.* **309**, 369–391 (2004).
6. Ormel, C. W., Cuzzi, J. N. & Tielens, A. G. G. M. Co-accretion of chondrules and dust in the solar nebula. *Astrophys. J.* **679**, 1588–1610 (2008).
7. Bland, P. A. *et al.* Earliest rock fabric formed in the Solar System preserved in a chondrule rim. *Nat. Geosci.* **4**, 244–247 (2011).
8. Beitz, E., Güttler, C., Nakamura, A. M., Tsuchiyama, A. & Blum, J. Experiments on the consolidation of chondrites and the formation of dense rims around chondrules. *Icarus* **225**, 558–569 (2013).
9. Topel-Schadt, J. & Müller, W. F. The submicroscopic structure of the unequilibrated ordinary chondrites Chainpur, Mezo-Madaras amd Tieschitz: a transmission electron-microscope study. *Earth Planet. Sci. Lett.* **74**, 1–12 (1985).
10. Abreu, N. M., Bland, P. A. & Rietmeijer, F. J. M. Effects of shock metamorphism on the matrix of CR chondrites: GRA 06100. *Lunar Planet. Sci. Conf.* **45** abstract #2753 (2014).
11. Davison, T. M., Ciesla, F. J. & Collins, G. S. Post-impact thermal evolution of porous planetesimals. *Geochim. Cosmochim. Acta* **95**, 252–269 (2012).
12. Consolmagno, G. J., Britt, D. T. & Stoll, C. P. The porosities of ordinary chondrites: models and interpretation. *Meteorit. Planet. Sci.* **33**, 1221–1229 (1998).
13. Sasso, M. R. *et al.* Incompletely compacted equilibrated ordinary chondrites. *Meteorit. Planet. Sci.* **44**, 1743–1753 (2009).
14. Macke, R. J., Consolmagno, G. J. & Britt, D. T. Density, porosity, and magnetic susceptibility of carbonaceous chondrites. *Meteorit. Planet. Sci.* **46**, 1842–1862 (2011).
15. Watt, L. E., Bland, P. A., Prior, D. J. & Russell, S. S. Fabric analysis of Allende matrix using EBSD. *Meteorit. Planet. Sci.* **41**, 989–1001 (2006).
16. Weidenschilling, S. J. & Cuzzi, J. N. in *Meteorites and the Early Solar System II* (eds Lauretta, D. S. & McSween, Jr. H. Y.) (Univ. Arizona Press, 473–485, 2006).
17. Amsden, A. A., Ruppel, H. M. & Hirt, C. W. *SALE: Simplified ALE Computer Program for Fluid Flow at all Speeds, LA-8095* (Los Alamos National Laboratory, pp 101, 1980).
18. Collins, G. S., Melosh, H. J. & Ivanov, B. A. Modeling damage and deformation in impact simulations. *Meteorit. Planet. Sci.* **39**, 217–231 (2004).
19. Wünnemann, K., Collins, G. S. & Melosh, H. J. A strain-based porosity model for use in hydrocode simulations of impacts and implications for transient crater growth in porous targets. *Icarus* **180**, 514–527 (2006).
20. Benz, W., Cameron, A. G. W. & Melosh, H. J. The origin of the Moon and the single-impact hypothesis III. *Icarus* **81**, 113–131 (1989).
21. Brookshaw, L. in *Working Paper Series SC-MC-9813* (Faculty of Sciences, Univ. Southern Queensland, 1998).
22. Leroux, H. Microstructural shock signatures of major minerals in meteorites. *Eur. J. Mineral* **13**, 53–272 (2001).
23. Tomeoka, K., Yamahana, Y. & Sekine, T. Experimental shock metamorphism of the Murchison CM carbonaceous chondrite. *Geochim. Cosmochim. Acta* **63**, 3683–3703 (1999).
24. Brigatti, M. F. & Poppi, L. Natural and monoionic aliettite: hydration and dehydration states. *Clay Min.* **22**, 187–197 (1987).
25. Anovitz, L. M. & Essene, E. J. Phase equilibria in the system $CaCO_3$-$MgCO_3$-$FeCO_3$. *J. Petrol.* **28**, 389–415 (1986).
26. Schmid-Beurmann, P., Cemic, L. & Knitter, St. Crystal chemical properties of synthetic lazulite-scorzalite solid-solution series. *Phys. Chem. Miner.* **26**, 496–505 (1999).
27. Kimura, M., Grossman, J. N. & Weisberg, M. K. Fe-Ni metal and sulfide minerals in CM chondrites: an indicator for thermal history. *Meteorit. Planet. Sci.* **46**, 431–442 (2011).
28. Pierazzo, E. & Melosh, H. J. Melt production in oblique impacts. *Icarus* **145**, 252–261 (2000).
29. Cooper, R. F. & Kohlstedt, D. L. Sintering of olivine and olivine-basalt aggregates. *Phys. Chem. Miner.* **11**, 5–16 (1984).
30. Weiss, B. P., Gattacceca, J., Stanley, S., Rochette, P. & Christensen, U. R. Paleomagnetic records of meteorites and early planetesimal differentiation. *Space Sci. Rev.* **152**, 341–390 (2010).
31. Crawford, D. A. & Schultz, P. H. Electromagnetic properties of impact-generated plasma, vapour and debris. *Int. J. Imp. Eng.* **23**, 169–180 (1999).
32. Kojima, T., Lauretta, D. S. & Buseck, P. R. Accretion, dispersal, and reaccumulation of the Bishunpur (LL3.1) brecciated chondrite: evidence from troilite-silicate-metal inclusions and chondrule rims. *Geochim. Cosmochim. Acta* **67**, 3065–3078 (2003).
33. Brenker, F. E. & Krot, A. N. Late-stage, high-temperature processing in the Allende meteorite: record from Ca,Fe-rich silicate rims around dark inclusions. *Am. Miner.* **89**, 1280–1289 (2004).
34. Brenker, F. E., Palme, H. & Klerner, S. Evidence for solar nebula signatures in the matrix of the Allende meteorite. *Earth Planet. Sci. Lett.* **178**, 185–194 (2000).
35. Müller, W. F. Transmission electron microscopy study of some minerals of the meteorite Allende. *9th Lunar Planet. Sci. Conf.* 609–610 (1978).
36. Caillet, C., MacPherson, G. J. & Zinner, E. K. Petrologic and Al-Mg isotopic clues to the accretion of two refractory inclusions onto the Leoville parent body: One was hot, the other wasn't. *Geochim. Cosmochim. Acta* **57**, 4725–4743 (1993).
37. Rubin, A. E. Postshock annealing and postannealing shock in equilibrated ordinary chondrites: Implications for the thermal and shock histories of chondritic asteroids. *Geochim. Cosmochim. Acta* **68**, 673–689 (2004).
38. Amelin, Y. *et al.* U-Pb chronology of the Solar System's oldest solids with variable $^{238}U/^{235}U$. *Earth Planet. Sci. Lett.* **300**, 343–350 (2010).
39. Connelly, J. N. *et al.* The absolute chronology and thermal processing of solids in the solar protoplanetary disk. *Science* **338**, 651–655 (2012).
40. Connelly, J. N., Amelin, Y., Krot, A. N. & Bizarro, M. Chronology of the solar system's oldest solids. *Astrophys. J.* **675**, L121–L124 (2008).
41. Amelin, Y., Krot, A. N., Hutcheon, I. D. & Ulyanov, A. A. Lead isotopic ages of chondrules and calcium-aluminum-rich inclusions. *Science* **297**, 1678–1683 (2002).
42. Villeneuve, J., Chaussidon, M. & Libourel, G. Homogeneous distribution of ^{26}Al in the Solar System and the Mg isotopic composition of chondrules. *Science* **325**, 985–988 (2009).
43. Doyle, P. M., Nagashima, K., Jogo, K. & Krot, A. N. ^{53}Mn-^{53}Cr chronometry reveals secondary fayalite in Asuka 881317 (CV3) and MacAlpine Hills 88107 (CO/CM-like) formed 4-5Ma after CV CAIs. *Lunar Planet. Sci. Conf.* **44** abstract #1793 (2013).
44. Doyle, P. M., Krot, A. N., Nagashima, K., Dobrica, E. & Brearley, A. J. Manganese-chromium ages of aqueous alteration of unequilibrated ordinary chondrites. *Lunar Planet. Sci. Conf.* **45** abstract #1726 (2014).
45. Walsh, K. J., Morbidelli, A., Raymond, S. N., O'Brien, D. P. & Mandell, A. M. A low mass for Mars from Jupiter's early gas-driven migration. *Nature* **475**, 206–209 (2011).
46. Walsh, K. J., Morbidelli, A., Raymond, S. N., O'Brien, D. P. & Mandell, A. M. Populating the asteroid belt from two parent source regions due to the migration of giant planets – "The Grand Tack". *Meteorit. Planet. Sci.* **47**, 1941–1947 (2012).
47. Davison, T. M., O'Brien, D. P., Ciesla, F. J. & Collins, G. S. The early impact histories of meteorite parent bodies. *Meteorit. Planet. Sci.* **48**, 1894–1918 (2013).
48. Collins, G. S., Melosh, H. J. & Wünnemann, K. Improvements to the epsilon-alpha compaction model for simulating impacts into high-porosity solar system objects. *Int. J. Impact Eng.* **38**, 434–439 (2011).
49. Benson, D. J., Nesterenko, V. F., Jonsdottir, F. & Meyers, M. A. Quasistatic and dynamic regimes of granular material deformation under impulse loading. *J. Mech. Phys. Solids* **45**, 1955–1999 (1997).
50. Zhuang, S. M., Ravichandran, G. & Grady, D. E. W. An experimental investigation of shock wave propagation in periodically layered composites. *J. Mech. Phys. Solids* **51**, 245–265 (2003).
51. Trott, W. M., Baer, M. R., Castaneda, J. N., Chhabildas, L. C. & Asay, J. R. Investigation of the mesoscopic scale response of low-density pressings of granular sugar under impact. *J. Appl. Phys.* **101**, 024917–024921 (2007).

52. Güldemeister, N., Wünnemann, K., Durr, N. & Hiermaier, S. Propagation of impact-induced shock waves in porous sandstone using mesoscale modeling. *Meteorit. Planet. Sci.* **48,** 115–133 (2013).

53. Davison, T. M., Collins, G. S. & Ciesla, F. J. Numerical modeling of heating in porous planetesimal collisions. *Icarus* **208,** 468–481 (2010).

54. Melosh, H. J. A hydrocode equation of state for SiO₂. *Meteorit. Planet. Sci.* **42,** 2079–2098 (2007).

55. Hicks, D. G. *et al.* Dissociation of liquid silica at high pressures and temperatures. *Phys. Rev. Lett.* **97,** 025502 (2002).

56. Kraus, R. G. *et al.* Shock vaporization of silica and the thermodynamics of planetary impact events. *J. Geophys. Res.* **117,** E004082 (2012).

57. Wünnemann, K., Collins, G. S. & Osinski, G. R. Numerical modeling of impact melt production in porous rocks. *Earth Planet. Sci. Lett.* **269,** 529–538 (2008).

Acknowledgements

This work was funded by the Australian Research Council via their Laureate Fellowship programme, the UK Science and Technology Facilities Council, the Royal Society and the UK Natural Environment Research Council. We thank Yuri Amelin and Jürgen Blum for useful discussions, and Anton Kearsley for Renazzo imagery. We also gratefully acknowledge the iSALE developers (www.isale-code.de), in particular Nicole Güldemeister.

Author contributions

P.A.B. and G.S.C. originated the study and collaborated to explore potential implications, iterating towards a model appropriate for simulating chondrite compaction, and generating outputs that offered predictions against the chondrite record. P.A.B. led interpretation of the meteorite record and comparison with numerical results and wrote much of the paper. G.S.C. designed the mesoscale numerical approach described in this work. T.M.D. performed much of the numerical modelling and model analysis. N.M.A. performed combined microscopy and micro-analytical study, producing the isotherm map of GRA 06100. F.J.C. contributed intellectual input on both numerical modelling and the meteorite record. A.R.M. and J.M. contributed to models involving transient heating, and detail on implications for palaeomagnetism.

Additional information

Competing financial interests: The authors declare no competing financial interests.

28

An asteroidal origin for water in the Moon

Jessica J. Barnes[1], David A. Kring[2], Romain Tartèse[1,3], Ian A. Franchi[1], Mahesh Anand[1,4] & Sara S. Russell[4]

The Apollo-derived tenet of an anhydrous Moon has been contested following measurement of water in several lunar samples that require water to be present in the lunar interior. However, significant uncertainties exist regarding the flux, sources and timing of water delivery to the Moon. Here we address those fundamental issues by constraining the mass of water accreted to the Moon and modelling the relative proportions of asteroidal and cometary sources for water that are consistent with measured isotopic compositions of lunar samples. We determine that a combination of carbonaceous chondrite-type materials were responsible for the majority of water (and nitrogen) delivered to the Earth–Moon system. Crucially, we conclude that comets containing water enriched in deuterium contributed significantly <20% of the water in the Moon. Therefore, our work places important constraints on the types of objects impacting the Moon ~4.5–4.3 billion years ago and on the origin of water in the inner Solar System.

[1] Department of Physical Sciences, The Open University, Walton Hall, Milton Keynes MK7 6AA, UK. [2] Lunar and Planetary Institute, 3600 Bay Area Boulevard, Houston, Texas 77058, USA. [3] Institut de Minéralogie, de Physique des Matériaux et de Cosmochimie (IMPMC), Muséum National d'Histoire Naturelle, Sorbonne Universités, CNRS, UMPC & IRD, Paris 75005, France. [4] Earth Sciences Department, Natural History Museum, Cromwell Road, London SW7 5BD, UK. Correspondence and requests for materials should be addressed to J.J.B. (email: jessica.barnes@open.ac.uk).

Lunar sample studies have focused on the measurement of H (reported as equivalent amounts of OH or H_2O) in lunar volcanic glass beads[1-3], melt inclusions[1,2,4] and apatite in lunar basalts and highlands samples[5-8]. From this compilation of work the bulk silicate Moon (BSM) is estimated to contain between ~ 10 and 300 p.p.m. H_2O[4,9,10]. A few of these studies reported the identification of a primordial H-isotopic signature[2,7,8], corresponding to Earth-like δD values between ~ -200 and $\sim +180$‰ (where $\delta D = [(D/H)_{sample}/(D/H)_{SMOW}-1] \times 1,000$; SMOW: standard mean ocean water). Therefore, the similar nature of water (the term water is hereafter used to refer to H_2O equivalent) in the Earth and Moon suggests that water locked up inside of the Moon was either inherited directly from the proto-Earth or delivered to the Earth–Moon system shortly after the formation of the Moon (Fig. 1) through impacting carbonaceous chondrite (CC)-type asteroids, since bulk D/H ratios for most CC-types[11] are consistent with those of terrestrial[12,13] and lunar water (Fig. 2).

Models for the differentiation of the Moon invoke the presence of a lunar magma ocean (LMO) and place the lifetime of this LMO to between ~ 10 and 200 Ma after the Moon-forming event (as reviewed by Elkins-Tanton et al.[14]), which necessitates that volatile accretion occurred relatively early in the Moon's geological history (Fig. 1). The final dregs of the LMO, enriched in incompatible elements such as potassium (K), rare-earth-elements (REEs) and phosphorous (P)—collectively termed urKREEP—should have contained the vast majority of the lunar water, as OH would have been incompatible in the crystallizing mantle cumulates and the crust[10]. Lunar samples with a significant KREEP component, such as Mg-suite rocks and KREEP basalts[7,8], are characterized by H-isotopic compositions that are similar to those estimated for the mantle source regions of picritic glasses and mare basalts[2,3,6], suggesting that the crystallization of the LMO did not result in significant fractionation of the H isotopes of water. To retain volatiles in the LMO, a conductive lid is required, to have prevented significant loss of water via degassing of H-bearing species, and is consistent with the formation of an early anorthosite crust a few kilometres thick in the first few thousand years of the Moon's history[14]. It is likely that the lunar crust then grew by continuous underplating of more anorthositic materials over a period of between ~ 10 and ~ 200 million years until the LMO completely solidified, leading to the mantle cumulate overturn (see review by Elkins-Tanton et al.[14]). This time period between lunar accretion and the overturn of mantle cumulates permits any water (and other

volatiles) accreted to the LMO to be incorporated into the mantle source regions of mare basalts and picritic glasses ($> \sim 100$ km deep[15]), and the urKREEP liquid provided that impactors were able to breach the lunar crust. Crystallization of an insulating crust likely started only after $\sim 80\%$ LMO crystallization[14]. During its initial growth the crust was likely hot and tenuous and perhaps even re-melted if tidal heating played an important role[14], meaning not only that impactors could have reached the LMO melt pool but also that many early impact structures could have been erased during crustal underplating and modification. Therefore, volatile delivery through the earliest lunar crust into the LMO could have balanced out any potential loss of volatiles (accreted or indigenous) by degassing (before 80% solidification).

Studies that have modelled the flux of impactors to the Moon posit that it was likely that the time period between lunar accretion and differentiation received relatively fewer impact events than during the late heavy bombardment (LHB) (for example, Morbidelli et al.[16]), but importantly this was not a quiescent time. Indeed, there are ~ 12 preserved Pre-Nectarian lunar basins that have excavations depths around or greater than 34–45 km (ref. 17), which corresponds to the current estimates for lunar crustal thickness based on observations by the GRAIL spacecraft[18]. This demonstrates that large impacts occurred early in the Moon's history (albeit infrequently compared with the LHB epoch[19]). In summary, the time interval for water delivery to the LMO, hereafter referred to as the late accretion window 'LAW', is conceptually restricted to between time zero (Moon formation) and up to ~ 200 Ma thereafter. Importantly, the bulk of lunar water was delivered during LAW, which is distinct from a much younger basin-forming epoch on the Moon that is commonly referred to as the LHB, lunar cataclysm, or inner Solar System cataclysm[16,20-22].

It has been suggested that water could have been accreted together with the Moon-forming material in the so-called 'cold start'[9]. Whilst it is plausible that some water may have been retained since the Moon's formation[23], it is highly unlikely that all of BSM water (up to 300 p.p.m. H_2O) was accreted in this manner and survived LMO processing. Therefore, we envisage two scenarios that can account for the bulk water inventory of the Moon, the first that water was accreted solely during the LAW and no water was inherited from the Moon-forming material (scenario 1), and the second that a portion (we consider up to 25%) of the BSM water was accreted with Moon-forming materials and was then supplemented by water delivery during the LAW (scenario 2).

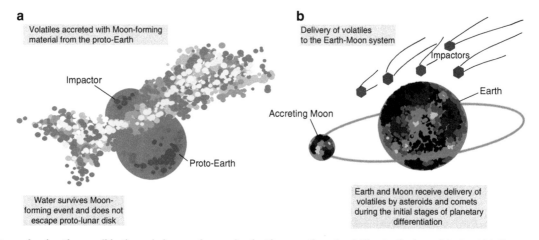

a

Volatiles accreted with Moon-forming material from the proto-Earth

Impactor

Proto-Earth

Water survives Moon-forming event and does not escape proto-lunar disk

b

Delivery of volatiles to the Earth-Moon system

Impactors

Earth

Accreting Moon

Earth and Moon receive delivery of volatiles by asteroids and comets during the initial stages of planetary differentiation

Figure 1 | Cartoon showing the possible time windows and scenarios for the accretion of volatiles to the lunar interior. Volatiles were accreted to the Moon during its formation (**a**) and/or continuously delivered by impacting bodies during the ca. 10–200 million years of crystallization of the lunar magma ocean (**b**). This graphic is not to scale.

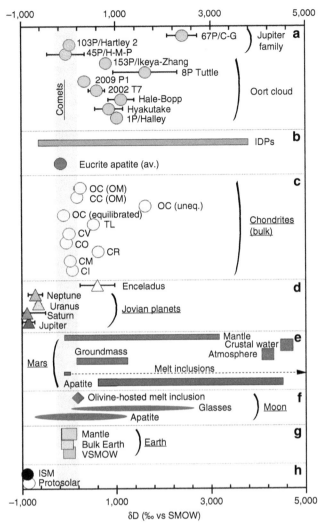

Figure 2 | Hydrogen isotope signatures for different objects in the Solar System. The grey bar indicates the range of δD values predicted for the lunar interior based on the previous studies of water and its H-isotopic composition in lunar samples[2,3,6-8]. (**a**) The H-isotopic compositions of comets, where the data for comets 1P/Halley[60], Hyakutake[61], Hale-Bopp[62], C/2002 T7 (ref. 63), C/2009 P1 (ref. 64), 8P/Tuttle[65], and 153P/Ikeya-Zhang[66], 45P/Honda-Mrkos-Pajdusakova[57], 103P/Hartley 2 (ref. 56) and 67P/Churyumov-Gerasimenko[58]. (**b**) The average H-isotopic composition of apatite grains from eucrites[47], and the range in values for interplanetary dust particles (IDPs)[67,68]. (**c**) The bulk H-isotope data for bulk Tagish Lake (TL) and other carbonaceous and ordinary chondrites (CC and OC, respectively)[11,55], carbonaceous and ordinary chondrite hydroxyl and organic matter[54]. (**d**) Hydrogen isotope data for Enceladus[69] and the jovian planets[70-72]. (**e**) The H-isotopic compositions of martian apatite[73-75], martian melt inclusions[46,74], martian meteorite groundmass[46], martian atmosphere[76] and martian crust and mantle[46,73-75]. (**f**) Data for lunar apatite[5-8,33,41,77], lunar picritic glasses[2,3] and lunar olivine-hosted melt-inclusions in picritic glass beads[2]. (**g**) The range in H-isotopic compositions of H_2O on Earth[12,13,78]. (**h**) The H-isotopic composition of Proto-solar and interstellar medium[42]. For carbonaceous chondrites, CI-, CM-, CV-, CO- and CR- refer to the different groups, named according to one prominent meteorite of the group, respectively Ivuna, Mighei, Vigorano, Ornans and Renazzo. OC stands for ordinary chondrites. Error bars indicate measured analytical uncertainties, please see original references for more information (Encleadus, Jovian planets, and comets error bars, 1 s.d., and Eucrite data, 2 s.d.).

In this study we utilize estimates of the water content of the BSM[4,9,10] and its H-isotopic composition[2,3,6-8] together with the current database for the bulk water contents of chondritic meteorites (Table 1, see Methods section), to determine the amount of water added to the early Moon by impacting objects. We consider a range of possible accretion scenarios during LAW to investigate the implications for the mass of water accreted and populations of accreting objects. This is constrained by the estimates for the proportion of lunar mass that was likely added post-lunar core-formation from studies of highly siderophile elements (HSEs) in lunar basalts (up to 0.02% lunar mass[24,25]) and highly volatile elements in mare basalts and picritic glasses (up to 0.4% lunar mass[4,9]). Our results indicate that carbonaceous chondrite types of impactors were the most important contributors of water to the lunar interior and that comets (with heavy water, Fig. 2) played a very small role.

Results

Mass of asteroidal materials added during the LAW. Based on the BSM water estimates of McCubbin *et al.*[10] ($\sim$10–100 p.p.m. H_2O), and Chen *et al.*[4] and Hauri *et al.*[9] ($\sim$100–300 p.p.m. H_2O), we have calculated the amount of material supplied by chondritic impactors to accrete $\sim$10, 100 and 300 p.p.m. bulk H_2O for LMO depths of 400 and 1,000 km (to match the various LMO depths estimated in previous studies[3,14,26,27]), and have recast these estimates into the context of how much lunar mass each would represent (Table 2 and Fig. 3). The types of carbonaceous chondrites are defined in the caption accompanying Fig. 2.

In scenario 1, for example, between 3.94×10^{16} kg ($\sim$1 p.p.m. H_2O) and 2.05×10^{18} kg H_2O ($\sim$52 p.p.m. H_2O) could have been accreted by CI-type CCs into a 400-km-deep LMO to remain below the addition threshold of <0.02% lunar mass (Table 1). In fact, using this <0.02% lunar mass threshold prohibits the addition of 300 p.p.m. H_2O by asteroids alone (Table 2). On the other hand, the accretion of up to 0.4% lunar mass in some cases, allows for the addition of H_2O in amounts greater than the high-end of BSM H_2O estimates (Table 2).

Importantly, because equilibrated ordinary (OC) and enstatite chondrites (EC) contain very little water (Table 1) the mass of such types of chondritic material added to accrete even 10 p.p.m. H_2O greatly exceeds the mass constraints imposed by HSE and highly volatile element abundances (Table 2). As a consequence, although ECs are considered an ideal geochemical match for the BSM and the bulk silicate Earth, and Os isotope signatures may suggest a large contribution of OC- and EC-type objects in materials accreted to the Earth–Moon system during late accretion[28,29], it is clear that these types of objects are unlikely to have contributed significant amounts of H_2O to the lunar interior. The accretion during the LAW of CC material is consistent with the impact of CC-type asteroids during the LHB[20]. HSE data have been interpreted to mean that EC and OC impactors also played an important role during the LHB[30,31], which may reflect a change in impactor populations over the first $\sim$500 million years of the Moon's evolution and, thus, dynamical mixing of material in the Solar System. The same data have also been interpreted, however, to represent a mixture of OC and CC impactors together with differentiated iron impactors[29-32], in which case CC asteroids remained an important source of impactors.

Assessing the contribution of cometary water during LAW. The first scenario in this study assumes that the Moon accreted completely dry or that any primordial water was completely lost very early in the Moon's history, and that late-accretion delivered all of the water in the BSM. Considering a BSM of 100 p.p.m. H_2O (the most recent BSM water estimates overlapping at this

Table 1 | Data used in mass balance calculations and mixing models.

Chondrite type	Bulk δD(‰)	s.d.(‰)	Bulk H_2O(wt.%)	s.d.(wt%)	Bulk $\delta^{15}N$(‰)	s.d.(‰)	Bulk N content(wt%)	s.d.(wt%)
CI	80	7	14.05	0.10	47	11	0.31	0.19
CM	30	62	11.57	0.17	32	12	0.11	0.02
CO*	−50	4	4.91	0.09	−9	13	0.05	0.04
CV*	14	8	2.03	0.02	−6	18	0.03	0.02
CR	630	24	4.87	1.44	150	44	0.14	0.07
OC uneq*	1,616	19	1.11	—	—	—	—	—
OC eq	−120	73	0.06	0.07	—	—	—	—
EC	−130	—	<0.05	—	—	—	—	—

CC, carbonaceous chondrite; EC, enstatite chondrites; OC, ordinary chondrites.
Data for bulk H_2O and bulk H-isotopic compositions of CC and unequilibrated OC meteorites (weighted averages from ref. 11), equilibrated OCs and ECs from ref. 55, and average bulk N and bulk N-isotopic composition[53,59] of chondritic materials. s.d. is the standard deviation of literature data for given chondrite types.
*Single data points from Alexander et al.[11] uncertainties represent analytical errors.

Table 2 | Results of mass balance calculations for the late addition of water to the Moon assuming either a 400- or 1,000-km-deep LMO.

Chondrite type	10 p.p.m. H_2O BSM (CC, kg)	%LM added	100 p.p.m. H_2O BSM (CC, kg)	%LM added	300 p.p.m. H_2O BSM (CC, kg)	%LM added	Max. H_2O for 0.02% LM (p.p.m.)	Max. H_2O for 0.4% LM (p.p.m.)
400-km-deep LMO								
CI	2.8E+18	<0.01	2.8E+19	0.04	8.4E+19	0.12	52	1,030*
CM	3.4E+18	<0.01	3.4E+19	0.05	1.0E+20	0.14	42	850*
CO	8.0E+18	0.01	8.0E+19	0.11	2.4E+20	0.33	18	360*
CV	1.9E+19	0.03	2.0E+20	0.27	5.8E+20	0.81	7	150
CR	8.1E+18	0.01	8.1E+19	0.11	2.4E+20	0.34	18	358*
EC	7.9E+20	1.09	7.9E+21	10.88	2.4E+22	32.63	<1	<4
OC uneq.	3.6E+19	0.05	3.6E+20	0.49	1.1E+21	1.47	4	81
OC eq.	6.6E+20	0.91	6.6E+21	9.11	2.0E+22	27.32	<1	<5
1,000-km-deep LMO								
CI	4.8E+18	0.01	4.8E+19	0.07	1.4E+20	0.20	30	608*
CM	5.8E+18	0.01	5.8E+19	0.08	1.7E+20	0.24	25	500*
CO	1.4E+19	0.02	1.4E+20	0.19	4.1E+20	0.56	10	213
CV	3.3E+19	0.05	3.3E+20	0.46	9.9E+20	1.37	4	87
CR	1.4E+19	0.02	1.4E+20	0.19	4.1E+20	0.57	10	211
EC	1.3E+21	1.85	1.3E+22	18.47	4.0E+22	55.41	<1	<3
OC uneq.	6.0E+19	0.08	6.1E+20	0.83	1.8E+21	2.50	<3	48
OC eq.	1.1E+21	1.55	1.1E+22	15.46	3.4E+22	46.39	<1	<3

CC, carbonaceous chondrite; EC, enstatite chondrites; HSE, highly siderophile element; HVE, highly volatile element; LMO, lunar magma ocean; OC, ordinary chondrites.
Mass of chondritic material, and corresponding percentage (%) of lunar mass (LM), required for the addition of 10, 100, and 300 p.p.m. H_2O in 400- and 1,000km-deep LMO during the late accretion window (LAW) for different types of carbonaceous, enstatite and ordinary chondrites. Also included is the maximum amount of water that would be added to the Moon for accretion of 0.02% lunar mass (adhering to HSE abundance constraints[24,25]) and for up to 0.4% lunar mass (adhering to the upper limit defined by HVEs[4,9]) assuming scenario 1.
*Indicates cases where accretion of chondritic material results in H_2O abundances higher than BSM H_2O estimates.

abundance[4,9,10]) and a LMO depth of 400 km (see Methods section), a simple two-component mixing model (equation 1) permits a number of combinations for the mixing of water from bulk CCs (CI, CO, CV, CM type) and deuterium-depleted Kuiper belt (DDK) comets that all result in bulk BSM water δD values between ∼ −50 and ∼ +80‰ (Fig. 4), consistent with existing isotopic limits[2,3,6–8] (see Methods).

$$D_R = \frac{((H_2O_t \times f\,H_2O_A) \times D_A) + ((H_2O_t \times f\,H_2O_B) \times D_B)}{H_2O_t} \quad (1)$$

Where; D_R is the resultant D/H ratio, H_2O_t refers to the total mass of water added (kg), fH_2O is the proportion (fraction of total) of water accreted by a given type of impactor denoted with an A or B, characterized by D/H ratios D_A or D_B, respectively. The resultant isotopic ratios were subsequently converted into delta notation to ease comparisons.

However, because cometary water, excluding comets 103P and 45P, has significantly elevated δD values compared with chondrites (see Methods section), the amount of cometary water that can be added to the lunar interior is far more restricted. In

such cases, ∼2% H_2O from Oort cloud comets can be mixed with 98% H_2O derived from CI-chondrites, or ∼1% H_2O is allowed from D-enriched Kuiper belt (DEK) comets (Table 3). Mixing of cometary water with predominantly CV-, CM- or CO-type CCs, respectively, also result in minor cometary contributions of ∼6% for DEK comets and ∼15% from Oort cloud-like objects (Fig. 4 and Table 3).

In scenario 2, we consider that the Moon accreted with between 1 and 25% of BSM H_2O, with an initial δD value of −200‰ (refs 2,3,6–8) and reached a final BSM water content of 100 p.p.m. H_2O (in agreement with the constraints imposed by modelling of Elkins-Tanton and Grove[26]) (equation (2)). Note that the lower the initial δD value is, the higher the maximum possible amount of H_2O added from a cometary source.

$$D_R = \frac{(H_2O_i \times D_i) + ((H_2O_{LA} \times f\,H_2O_A) \times D_A) + ((H_2O_{LA} \times f\,H_2O_B) \times D_B)}{H_2O_t} \quad (2)$$

Where; H_2O_i refers to the initial mass of water present in the LMO (kg), which varied between 1% and 25% of the total amount of water (H_2O_t), D_i refers to the initial H-isotopic composition of the water in the LMO, H_2O_{LA} refers to the mass of water added during LAW (kg), fH_2O is the proportion (fraction of amount added during LAW) of water accreted by a given type of impactor denoted with A or B, respectively, and all other parameters are the same as in equation (1).

Mixing of water from CI-type CCs with water from CO, CM, CV-type CCs and DDK comets in different proportions can yield final δD values $< +100‰$ (Table 3), satisfying isotopic constraints[2,3,6–8]. If deuterium-enriched comets were involved in the addition of water during LAW then up to $\sim 14\%$ H_2O could have originated from Oort cloud-type comets and $\sim 5\%$ H_2O from DEK comets if the remaining water is assumed to come from CI-type chondrites (Fig. 5a), proportions that increase to ~ 7 and $\sim 19\%$ H_2O originating from DEK and Oort cloud comets, respectively, if the remaining water came from CM-type chondrites (Fig. 5b and Table 3). Significantly, in the case where CO-type CCs dominated the impactor population (Fig. 5c) up to $\sim 27\%$ H_2O could have been contributed by Oort cloud-like comets and up to $\sim 10\%$ H_2O from DEK comets (Table 3). In addition, we considered that the LMO underwent some degassing[26,33] before prior to late addition (see Methods section) and found that the results limit the amount of initial water present before late addition (assuming degassed H-isotopic

composition of $\sim +800‰$ (ref. 8)), which further restricts the amount of water contributed by D-rich comets (Table 4).

Discussion

To a first order, a dominantly asteroidal source of water accreted during the LAW is similar to the dominant source inferred for the subsequent basin-forming epoch of the Moon, based on geochemical[22,34] and mineralogical[20] markers, implying that asteroids and not comets dominated the impactor population hitting the Moon during its first 500 million years of geological history[35]. Although the source of objects delivered to the Earth–Moon system may have largely remained the same, the dynamical delivery could have been much different, as the basin-forming epoch appears to have been characterized by asteroids with higher impact velocities compared with those during the early accretional epoch (LAW) (Bottke et al.[19] and references therein). We note that the accretion of CC-type material during

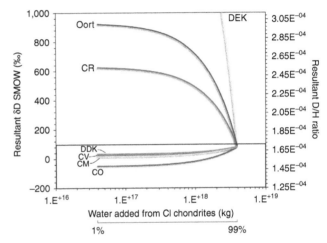

Figure 4 | An example of a two-component mixing model for scenario 1 considering that CI type CCs were dominant during the LAW. This model assumes 100 p.p.m. H_2O in BSM, equivalent to 3.94×10^{18} kg of H_2O for a 400-km-deep LMO. The plot shows the resultant δD value and D/H ratios of the water mixture versus the amount of water (kg) supplied by CI-type carbonaceous chondrites as an example (results in Table 3). The bar underneath the x-axis shows how the mass of water is related to % of water mixed. Only final H-isotopic compositions below $+100‰$ (denoted by the blue box) are acceptable within the constraints of the model (see Methods). Where: CI-type carbonaceous chondrites (CCs) are Ivuna-like, CV are Vigorano-like CO are Ornans-like, CM are Mighei-like and CR are Renazzo-like, respectively. Note that it only takes a couple of per cent contribution of water from Oort or DEK comets to produce H-isotope compositions outside of the model limits. DDK, deuterium-depleted Kuiper belt comets; DEK, deuterium-enriched Kuiper belt comets; Oort, average H-isotopic composition of Oort cloud comets.

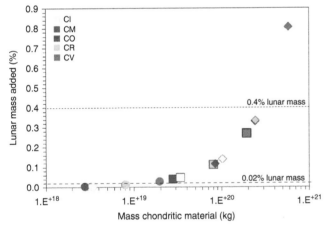

Figure 3 | Mass of chondritic material added during late accretion compatible with BSM water estimates. This figure shows the mass of the different types of chondritic material accreted (kg) to the Moon to add 10 (circles), 100 (squares) and 300 p.p.m. H_2O (diamonds), respectively, to a 400-km-deep LMO, and the corresponding amount in terms of lunar mass. The mass constraints imposed by HSE (0.02% lunar mass added) and highly volatile element abundances (up to 0.4% lunar mass added) are indicated by the dashed lines. Where: CI-type carbonaceous chondrites (CCs) are Ivuna-like, CO are Ornans-like, CV are Vigorano-like, CM are Mighei-like and CR are Renazzo-like, respectively.

Table 3 | Results from mixing models of scenario 1 and 2.

Major CC type	Scenario 1		Scenario 2			
	% H_2O from DEK comets	% H_2O from Oort-like comets	Max. % H_2O from DEK comets LL initial	Max. % H_2O from DEK comets UL initial	Max. % H_2O from Oort-like comets LL initial	Max. % H_2O from Oort-like comets UL initial
CI	<1	2	1	5	3	14
CM	3	8	3	7	8	19
CO	6	15	6	10	16	27
CV	<4	9	4	8	10	20

Max., maximum; LL initial., lower limit (1% BSM H_2O) of H_2O initially in lunar magma ocean; UL initial., upper limit (25% BSM H_2O) of H_2O initially in the lunar magma ocean.
Showing the maximum proportions (%) of water permitted from deuterium-enriched Kuiper belt (DEK) and Oort cloud comets assuming the rest of lunar water was delivered by CC-type asteroids.

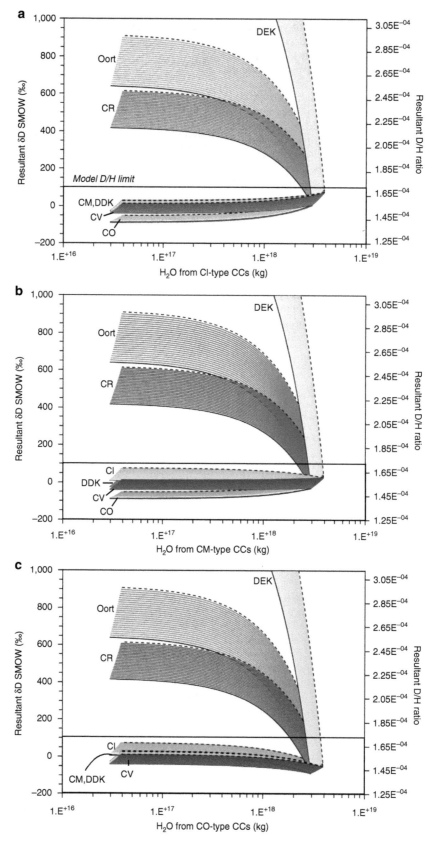

Figure 5 | Two-component mixing models for scenario 2 considering that CI or CM or CO-type CCs dominated the impactor population during the LAW. The resultant δD value of each water mixture (‰) versus the amount of water (kg) supplied by (**a**) CI-, (**b**) CM- and (**c**) CO-type carbonaceous chondrites. Table 3 also gives results from mixes with CV-type CCs. This model assumes that the LMO initially contained between 1% (dashed black lines) and 25% (solid black lines) of the BSM water (100 p.p.m. H_2O) with a δD value of -200‰. Only final H-isotopic compositions $< +100$‰ are acceptable within the model constraints (denoted by the black boxes). Where: CI-type carbonaceous chondrites (CCs) are Ivuna-like, CO are Ornans-like, CV are Vigorano-like, CM are Mighei-like and CR are Renazzo-like, respectively.

the LAW, that is required to account for the lunar H-isotope and water abundances in the lunar interior, is distinct from a late veneer dominated by EC- and OC-type objects that has been inferred from Os isotope studies of terrestrial rocks[28,29]. This dichotomy could imply that either the largest impactors that preferentially impacted the Earth due to stochastic accretion[19] were OC- and EC-type objects, while the population of smaller impactors included abundant CC objects that preferentially impacted the Moon, or that while the accreting objects had OC- and EC-type Os isotope signatures, they had CC-type water abundances and H-isotope signatures, consistent with other evidence that support the presence of planetesimal-types that are not represented in modern meteorite falls[36].

On the basis of combined N isotopes and C/N ratios, the lunar interior seems to be best matched with CO-type CCs[37]. In terms of N, we can estimate the amount of N delivered to the LMO based on our calculations of the mass of chondritic material added during LAW for each chondrite type (Table 2, restricted to consideration of material added to achieve a BSM with ~ 100 p.p.m. H_2O), and the average N contents in the different chondrite types (Table 1). Results yield N contents for the BSM of ~ 0.95 to 2.21 p.p.m. for LMO depths of 400 and 1,000 km (Table 5), which is consistent with an average of ~ 1 p.p.m. N in the lunar mantle as estimated by Füri et al.[38] If minimal fractionation of $^{15}N/^{14}N$ occurred during basalt ascent and eruption[38,39], then the N-isotopic composition measured in mare basalts of between ~ -2 and $+27$‰ (refs 37,38) should reflect the isotopic composition of the lunar mantle source regions from which the basalts were derived. If all of the N in the LMO was added during LAW (scenario 1) then mixing of CI-, CM-, CO- and CV-type CCs provide suitable matches for the origin of lunar N (Methods section and Table 1) and restrict cometary contribution of N to as much as ~ 5% (Table 5). This is comparable to scenario 2, where considering the lowest mare basalt $\delta^{15}N$ value of -2‰ (ref. 37) to represent the N-isotopic signature of N accreted with Moon-forming materials before LAW, then up to ~ 6% N could have been delivered by comets (Table 5). Still using scenario 2 but considering that the isotopic

signature of N accreted when the Moon formed was consistent with that of the primordial Earth mantle, which is thought to be as low as ~ -40 ‰ (reviewed by Füri and Marty[40]), increases the possible cometary contribution to a maximum of ~ 8% of total lunar interior N.

Therefore, it appears that in order to add the appropriate amounts of H_2O and N to the lunar interior during LAW, while keeping the H- and N-isotopic compositions in line with sample measurements, requires the contemporaneous accretion of the various types of CCs (mostly CO, CI, CM and CV), together with $< \sim 20$% cometary input.

The very-low apparently indigenous lunar δD values of ~ -600‰ reported by Robinson et al.[41] are quite anomalous. In fact, such low values have yet to be observed in terrestrial rocks, the lowest δD values being ~ -220‰ for Earth's depleted upper mantle[13]. Therefore, such low values reported for an intrusive lunar rock[41] could potentially indicate the presence of a proto-solar component (δD value of ~ -900‰ (ref. 42)) in the lunar mantle, in a similar way that has been used to explain Earth's noble gas record[43,44] and recent data from primitive terrestrial basalts[13]. If this source is indeed required to explain the extremely low H-isotopic compositions observed[41], then this signature must have been added to the Moon from either material accreted from the proto-Earth or by the impactor involved in the formation of the Moon. Alternatively, such a low δD value could have arisen from contamination of a water-poor magma by solar wind-derived H present in the lunar regolith during magma ascent and emplacement.

The mantles of the early-Earth[13,45] the Moon[2,3,6–8] and Mars[46] all seem to be characterized by water with remarkably similar H-isotopic compositions, suggesting that the same types of accreting objects as those modelled here for the Moon delivered a vast majority of the water to the rocky planets in the inner Solar System. Likewise, isotopic analyses of differentiated meteorites[47] suggest that the interior of the eucrite parent body, probably the asteroid 4-Vesta, contained water characterized by H-isotopic signatures indistinguishable to that of the terrestrial planetary objects in the inner Solar System (Fig. 2).

Regarding the timing of volatile accretion to the Moon, major constraints are imposed by the time interval between lunar accretion and solidification of the LMO and overturn of mantle cumulates (~ 10–200 Myr after Moon formation[14]) since there is water in the mantle source regions of the mare basalts[4,6] and picritic glasses[1–3,9]. It is possible that some water with Earth-like H-isotopic characteristics may have been accreted with the Moon-forming material, to explain the H-isotopic composition of water in the source regions for the picritic glasses (that were likely deep enough in the lunar mantle to have escaped processes such as degassing of H_2 (ref. 48)). To account for the H-isotopic composition of the primordial lunar mantle and the urKREEP reservoir together, it is necessary that any initial lunar water was complemented by the continuous accretion of water from CC-type objects to the LMO during LAW. Incidentally, late

Table 4 | Results from the modelling for scenario 2 considering that the water in the LMO before the late addition had undergone significant H_2 degassing.

Main CC Type	Max. H_2O in LMO before late accretion (p.p.m.)*	Max. % H_2O added by Oort comets	Max. % H_2O added by DEK comets
CI	3	2	0
CM	9	7	3
CO	17	15	6
CV	11	9	4

LMO, lunar magma ocean; max., maximum.
*Assuming a BSM of ~ 100 p.p.m. H_2O.

Table 5 | Results of modelling the amount of N in the lunar interior and the permitted proportion of cometary N added during late accretion.

Main CC type	N content of 400/1,000-km-deep LMO (p.p.m.)	Scenario 1: max. % N from comets	Scenario 2: max. % N from comets initial $\delta^{15}N = -2$ or -40‰, respectively
CI	2.21	0.0	0, 0
CO	1.01	~ 5.0	6, 8
CM	0.95	0.0	0, 2.5
CV	1.48	4.5	6, ~ 8

LMO, lunar magma ocean; max., maximum. N contents based on BSM of 100 p.p.m. H_2O.

accretion of material to the Earth–Moon system did not only affect their volatile inventories[49] but could also explain the recently identified difference between the bulk silicate Earth and BSM in terms of W isotopes[50,51].

In summary, the lunar interior is characterized by an average bulk water content of the order of ~100 p.p.m. based on samples studies[4,9,10] and we propose that the majority of this water was delivered during the *ca.* 10–200 million year-long LMO phase of lunar differentiation, with the possibility that some of this water was inherited proto-Earth-derived materials. This late accretion was dominated by water-rich carbonaceous chondrite asteroids, on the basis of H and N data, likely comprising a mixture of CO-, CI-, CM- and possibly CV-type CCs, with a minor contribution of water from deuterium-rich Oort cloud or Kuiper belt comets (much less than 20% of the total water in the BSM). Our work, therefore, confirms that the inner Solar System received a similar flux of asteroidal and cometary material for much of its early history from 4.5 billion years (this work and Dauphas *et al.*[52]) to at least 3.9 Gyr (ref. 20) (LHB).

Methods

Mass balance calculations. For mass balance calculations, we considered a lunar radius of 1,737.4 km, a mantle density of 3,300 kg m^{-3}, and varied the depth of the LMO from 400 to 1,000 km. We then calculated the mass of chondritic material added to the lunar interior that would result in BSM water contents of 10, 100 and 300 p.p.m. H_2O, and considered both the addition of up to 0.02% lunar mass[24,25] and up to 0.4% lunar mass[4,9]. Note that for a 400-km-deep LMO, a BSM with 10 p.p.m. H_2O equates to 3.94×10^{17} kg H_2O, 100 p.p.m. equates to 3.94×10^{18} kg H_2O, and 300 p.p.m. equates to 1.18×10^{19} kg H_2O (6.70×10^{17} kg H_2O, 6.70×10^{18} kg H_2O, and 2.01×10^{19} kg H_2O, for BSM H_2O contents of 10, 100 and 300 p.p.m., respectively, for a 1,000-km-deep LMO).

Data for the average water contents of different carbonaceous, ordinary and enstatite chondrites that were used in these calculations are shown in Table 1. For carbonaceous chondrites, CI-, CM-, CV-, CO- and CR- refer to the different groups, named according to one prominent meteorite of the group, respectively Ivuna, Mighei, Vigorano, Ornans and Renazzo. OC stands for ordinary chondrites. The results of these calculations, presented in Table 2, allowed us to identify suitable impactors for the modelling presented below. For example, to add 10 p.p.m. H_2O to the BSM through the accretion of enstatite chondrites alone implies the addition of >1% lunar mass for both 400- and 1,000-km-deep LMO, and, therefore, ECs were not deemed important carriers of water to the lunar interior. The same was also true for equilibrated ordinary chondrites (Table 2). Un-equilibrated ordinary chondrites were also not considered further on the basis of them having prohibitively high H-isotopic compositions (Table 1) for the mixing calculations. Note that all of the scenarios in this work consider that water was efficiently delivered to the Moon regardless of if it originated in asteroids or comets, that is, we have not attempted to calculate the proportions of volatiles lost during impact into the crystallizing LMO.

Mixing models. In the mixing models, literature data for the water abundances and D/H ratios of bulk carbonaceous chondrite meteorites were utilized. Carbonaceous, ordinary and enstatite chondrite meteorites are characterized by δD values mostly between ~ −200 and +300‰ (refs 11,53–55). There is also a possibility that the asteroid parent bodies to the CCs may have once contained ice[11]. If the asteroids that delivered water to the Moon contained ice, in addition to their bulk water contents hosted in phyllosilicates and organic matter compounds, then less amounts of chondritic material than those given in Table 2 would have been accreted to the Moon to account for the bulk inventory of water in the lunar interior. Until ice on asteroids is identified and characterized, it remains difficult to quantify how much ice, if any, could have contributed to the mixing of different H-reservoirs within the Moon. Comets 103P/Hartley 2 (ref. 56) and 45P/Honda-Mrkos-Pajdusakova[57] contain water characterized by δD values between ~ +30 and +400‰, whereas other comets studied (for example, 8P/Tuttle and 67P/Churyumov-Gerasimenko) contain water distinctly enriched in D, with δD values > +900‰ as reviewed by Hartogh *et al.*[56] In the modelling, the δD value for water in comet 67P/Churyumov-Gerasimenko (+2,400‰ (ref. 58)) was used to represent DEK comets, the δD value for water in Oort-cloud comets was set to ~ +930‰ (see Hartogh *et al.*[56] for a review) , and the water δD for DDK belt comets was given the same isotopic composition as comet 103P (+34‰ (ref. 56)).

The modelling for scenario 1 assumes no water in the lunar interior before lunar core-formation, and follows equation (1). We used the mid-range estimate for water content for the BSM of ~100 p.p.m. H_2O (refs 4,9,10) (equivalent to 3.94×10^{18} kg of H_2O for a 400-km-deep LMO) and imposed an upper limit on the resultant δD values after mixing of < +100‰ to remain consistent with H-isotopic data obtained for lunar samples[2,3,6–8]. Only chondrites that were likely

to have delivered significant amounts of water to the Moon without adding unrealistic amounts of chondritic material (Table 2) to the LMO were considered in the mixing calculations. Each chondrite-type (CI, CM, CO, CV) was considered in turn as being the most dominant chondrite in the impactor population (Table 3). CR-type CCs were not considered as being a significant part of the impactor population since the bulk δD value (Table 1) for such meteorites is significantly elevated with respect to the expected H-isotopic composition of the lunar mantle. In each case, the model extends from having 99% of the accreted water originating from CI-, CM-, CO- or CV-type CCs to having 99% of the water accreted from another type of chondrite or comet.

The two-component mixing models applicable to scenario 2 follow the same constraints as scenario 1 except that they assume that the Moon accreted with some water and late accretion added any remaining water needed to meet the BSM estimate of 100 p.p.m. H_2O. These calculations follow equation (2) and consider an initial amount of water varying between 1% and 25% of the BSM estimate (that is, 1–25 p.p.m. for a bulk BSM water content of 100 p.p.m. H_2O) characterized by a δD value of − 200‰ (refs 2,3,6–8). The model is also adapted to consider each type of CC (CI, CO, CM or CV) as being the dominant source (Table 3).

We also considered that water initially dissolved in the molten LMO might have experienced degassing through loss of H_2, in a similar way as has been proposed to explain for example the elevated H-isotopic compositions of water (present as OH) measured in apatite in some lunar basalts[6]. Such degassing would have resulted in any remaining water in the LMO before late addition having an elevated δD signature of ~ +800‰ (following the work of Tartèse *et al.*[8]). Therefore, we utilized equation (2) and allowed the model for scenario 2 to start with water having a δD signature of ~ +800‰ before the late addition of volatiles. All other parameters were kept the same (described above) and from this we were able to model the proportion of water that was initially in the LMO prior to late addition since the model is still constrained to a final H-isotopic composition of < + 100‰ and to constrain the maximum contribution made by comets to the water inventory of the Moon assuming an initial H_2O content of 1 p.p.m. prior to late addition (Table 4).

Bulk nitrogen content of the Moon and mixing calculations. Using the mass of each type of CC (CI, CO, CM and CV) accreted to an initially dry Moon during LAW (scenario 1) to end up with a BSM containing 100 p.p.m. H_2O (Table 2), the average N contents for the various types of carbonaceous chondrites (Table 1, data sources[53,59]), and assuming LMO depths of 400 or 1,000 km, we have estimated the amounts of N added to an initially N-free LMO, that is, the bulk N content of the BSM (Table 5). Note that in some cases (e.g., CV CCs) the mass of chondritic material added, in order to accrete 100 p.p.m. H_2O, exceeds the mass balance constraints imposed by HSEs. We then employed the mixing models described above for H_2O to calculate the acceptable proportion of N contributed by comets (with δ^{15}N values ~ +700‰ as a lower limit, see Füri and Marty[40] and references therein) for scenario 1 (Table 5) and scenario 2 (Table 5), where it was assumed that the initial N-isotopic compositions of N were − 2‰ (ref. 37) and − 40‰ (Füri and Marty[40] and references therein) in the same way as was done for H_2O (Table 3). Notice that like for H_2O, CR-type chondrites are unsuitable carriers of N due to their elevated δ^{15}N values (Table 1).

References

1. Hauri, E. H., Weinreich, T., Saal, A. E., Rutherford, M. C. & Van Orman, J. A. High pre-eruptive water contents preserved in lunar melt inclusions. *Science* **333**, 213–215 (2011).

2. Saal, A. E., Hauri, E. H., Van Orman, J. A. & Rutherford, M. J. Hydrogen isotopes in lunar volcanic glasses and melt inclusions reveal a carbonaceous chondrite heritage. *Science* **340**, 1317–1320 (2013).

3. Füri, E., Deloule, E., Gurenko, A. & Marty, B. New evidence for chondritic lunar water from combined D/H and noble gas analyses of single Apollo 17 volcanic glasses. *Icarus* **229**, 109–120 (2014).

4. Chen, Y. *et al.* Water, fluorine, and sulfur concentrations in the lunar mantle. *Earth Planet. Sci. Lett.* **427**, 37–46 (2015).

5. Greenwood, J. P. *et al.* Hydrogen isotope ratios in lunar rocks indicate delivery of cometary water to the Moon. *Nat. Geosci.* **4**, 79–82 (2011).

6. Tartèse, R. *et al.* The abundance, distribution, and isotopic composition of hydrogen in the Moon as revealed by basaltic lunar samples: Implications for the volatile inventory of the Moon. *Geochim. Cosmochim. Acta* **122**, 58–74 (2013).

7. Barnes, J. J. *et al.* The origin of water in the primitive Moon as revealed by the lunar highlands samples. *Earth Planet. Sci. Lett.* **390**, 244–252 (2014).

8. Tartèse, R. *et al.* Apatites in lunar KREEP basalts: the missing link to understanding the H isotope systematics of the Moon. *Geology* **42**, 363–366 (2014).

9. Hauri, E. H., Saal, A. E., Rutherford, M. J. & Van Orman, J. A. Water in the Moon's interior: Truth and consequences. *Earth Planet. Sci. Lett.* **409**, 252–264 (2015).

10. McCubbin, F. M. *et al.* Magmatic volatiles (H,C, N, F, S, Cl) in the lunar mantle, crust, and regolith: abundances, distributions, processes, and reservoirs. *Am. Mineral.* **100** (2015).

11. Alexander, C. M. O. *et al.* The provenances of asteroids, and their contributions to the volatile inventories of the terrestrial planets. *Science* **337**, 721–723 (2012).

12. Lécuyer, C. & Gillet, P. The hydrogen isotope composition of seawater and the global water cycle. *Chem. Geol.* **145**, 249–261 (1998).

13. Hallis, L. J. *et al.* Evidence for primordial water in Earth's deep mantle. *Science* **350**, 6262 (2015).

14. Elkins-Tanton, L. T., Burgess, S. & Yin, Q. Z. The lunar magma ocean: Reconciling the solidification process with lunar petrology and geochronology. *Earth Planet. Sci. Lett.* **304**, 326–336 (2011).

15. Grove, T. L. & Krawczynski, M. J. Lunar mare volcanism: Where did the magmas come from? *Elements* **5**, 29–34 (2009).

16. Morbidelli, A., Marchi, S., Bottke, W. F. & Kring, D. A. A sawtooth-like timeline for the first billion years of lunar bombardment. *Earth Planet. Sci. Lett.* **355-356**, 144–151 (2012).

17. Potter, R. W. K., Kring, D. A. & Collins, G. S. Scaling of basin-sized impacts and the influence of target temperature. *Geol. Soc. America. Spec. Papers* **518**, SPE518–06 (2015).

18. Wieczorek, M. A. *et al.* The crust of the Moon as seen by GRAIL. *Science* **339**, 671–675 (2013).

19. Bottke, W. F., Walker, R. J., Day, J. M. D., Nesvorny, D. & Elkins-Tanton, L. Stochastic late accretion to Earth, the Moon, and Mars. *Science* **330**, 1527–1530 (2010).

20. Joy, K. H. *et al.* Direct detection of projectile relics from the end of the lunar basin-forming epoch. *Science* **336**, 1426–1429 (2012).

21. Marchi, S., Bottke, W. F., Kring, D. A. & Morbidelli, A. The onset of the lunar cataclysm as recorded in its ancient crater populations. *Earth Planet. Sci. Lett.* **325-326**, 27–38 (2012).

22. Kring, D. A. & Cohen, B. A. Cataclysmic bombardment throughout the inner Solar System 3.9–4.0 Ga. *J. Geophys. Res.* **107** (2002).

23. Nakajima, M. & Stevenson, D. J. in *45th Lunar and Planetary Science Conference*, 2770, (2014).

24. Day, J. M. D., Pearson, D. G. & Taylor, L. A. Highly siderophile element constraints on accretion and differentiation of the Earth-Moon system. *Science* **315**, 217–219 (2007).

25. Day, J. M. D. & Walker, R. J. Highly siderophile element depletion in the Moon. *Earth Planet. Sci. Lett.* **423**, 114–124 (2015).

26. Elkins-Tanton, L. T. & Grove, T. L. Water (hydrogen) in the lunar mantle: Results from petrology and magma ocean modeling. *Earth Planet. Sci. Lett.* **307**, 173–179 (2011).

27. Tartèse, R. & Anand, M. Late delivery of chondritic hydrogen into the lunar mantle: Insights from mare basalts. *Earth Planet. Sci. Lett.* **361**, 480–486 (2013).

28. Meisel, T., Walker, R. J. & Morgan, J. W. The osmium isotopic composition of the Earth's primitive upper mantle. *Nature* **383**, 517–520 (1996).

29. Walker, R. J. *et al.* In search of late-stage planetary building blocks. *Chem. Geol.* **411**, 125–142 (2015).

30. Puchtel, I. S., Walker, R. J., James, O. B. & Kring, D. A. Osmium isotope and highly siderophile element systematics of lunar impact melt breccias: implications for the late accretion history of the Moon and Earth. *Geochim. Cosmochim. Acta* **72**, 3022–3042 (2008).

31. Liu, J., Sharp, M., Ash, R. D., Kring, D. A. & Walker, R. J. (2015Diverse impactors in Apollo 15 and 16 impact melt rocks: Evidence from osmium isotopes and highly siderophile elements. *Geochim. Cosmochim. Acta* **155**, 122–153.

32. Fischer-Gödde, M. & Becker, H. Osmium isotope and highly siderophile element constraints on ages and nature of meteoritic components in ancient lunar impact rocks. *Geochim. Cosmochim. Acta* **77**, 135–156 (2012).

33. Boyce, J. W. *et al.* The chlorine isotope fingerprint of the lunar magma ocean. *Sci. Adv.* **1**, e1500380 (2015).

34. Steele, A. *et al.* Graphite in an Apollo 17 impact melt breccia. *Science* **329**, 51 (2010).

35. Morbidelli, A. *et al.* Source regions and timescales for the delivery of water to the Earth. *Meteorit. Planet. Sci.* **35**, 1309–1320 (2000).

36. Fischer-Gödde, M., Burkhardt, C., Kruijer, T. S. & Kleine, T. Ru isotope heterogeneity in the solar protoplanetary disk. *Geochim. Cosmochim. Acta* **168**, 151–171 (2015).

37. Mortimer, J., Verchovsky, A. B., Anand, M., Gilmour, I. & Pillinger, C. T. Simultaneous analysis of abundance and isotopic composition of nitrogen, carbon, and noble gases in lunar basalts: Insights into interior and surface processes on the Moon. *Icarus* **255**, 3–17 (2015).

38. Füri, E., Barry, P. H., Taylor, L. A. & Marty, B. Indigenous nitrogen in the Moon: constraints from coupled nitrogen–noble gas analyses of mare basalts. *Earth Planet. Sci. Lett.* **431**, 195–205 (2015).

39. Cartigny, P., Jendrzejewski, N., Pineau, F., Petit, E. & Javoy, M. Volatile (C, N, Ar) variability in MORB and the respective roles of mantle source

heterogeneity and degassing: the case of the Southwest Indian Ridge. *Earth Planet. Sci. Lett.* **194**, 241–257 (2001).

40. Füri, E. & Marty, B. Nitrogen isotope variations in the Solar System. *Nat. Geosci.* **8**, 515–522 (2015).

41. Robinson, K. L. *et al.* in *45th Lunar and Planetary Science Conference*, 1607, (2014).

42. Geiss, J. & Gloecker, G. Abundance of deuterium and helium in the protosolar cloud. *Sp. Sci. Rev* **84**, 239–250 (1998).

43. Halliday, A. N. The origins of volatiles in the terrestrial planets. *Geochim. Cosmochim. Acta* **105**, 146–171 (2013).

44. Marty, B. The origins and concentrations of water, carbon, nitrogen and noble gases on Earth. *Earth Planet. Sci. Lett.* **313-314**, 56–66 (2012).

45. Sharp, Z. D., McCubbin, F. M. & Shearer, C. K. A hydrogen-based oxidation mechanism relevant to planetary formation. *Earth Planet. Sci. Lett.* **380**, 88–97 (2013).

46. Usui, T., Alexander, C. M. O. D., Wang, J., Simon, J. I. & Jones, J. H. Origin of water and mantle-crust interactions on Mars inferred from hydrogen isotopes and volatile element abundances of olivine-hosted melt inclusions in primitive shergottites. *Earth Planet. Sci. Lett.* **357-358**, 119–129 (2012).

47. Sarafian, A. R., Nielsen, S. G., Marschall, H. R., McCubbin, F. M. & Monteleone, B. D. Early accretion of water in the inner solar system from a carbonaceous chondrite-like source. *Science* **346**, 623–626 (2014).

48. Hirschmann, M. M., Withers, A. C., Ardia, P. & Foley, N. T. Solubility of molecular hydrogen in silicate melts and consequences for volatile evolution of terrestrual planets. *Earth Planet. Sci. Lett.* **345-348**, 38–48 (2012).

49. Albarede, F. *et al.* Asteroidal impacts and the origin of terrestrial and lunar volatiles. *Icarus* **222**, 44–52 (2013).

50. Touboul, M., Puchtel, I. S. & Walker, R. J. Tungsten isotopic evidence for disproportional late accretion to the Earth and Moon. *Nature* **520**, 530–533 (2015).

51. Kruijer, T. S., Kleine, T., Fischer-Gödde, M. & Sprung, P. Lunar tungsten isotopic evidence for the late veneer. *Nature* **520**, 534–537 (2015).

52. Dauphas, N., Robert, F. & Marty, B. The late asteroidal and cometary bombardment of Earth as recorded in water deuterium to protium ratio. *Icarus* **148**, 508–512 (2000).

53. Kerridge, J. F. Carbon, hydrogen and nitrogen in carbonaceous chondrites: abundances and isotopic compositions in bulk samples. *Geochim. Cosmochim. Acta* **49**, 1707–1714 (1985).

54. Robert, F. Water and organic matter D/H ratios in the solar system: a record of an early irradiation of the nebula? *Planet. Space Sci.* **50**, 1227–1234 (2002).

55. Robert, F., Javoy, M., Halbout, J., Dimon, B. & Merlivat, L. Hydrogen isotope abundances in the solar system. Part II: Meteorites with terrestrial-like ratio. *Geochim. Cosmochim. Acta* **51**, 1807–1822 (1987).

56. Hartogh, P. *et al.* Ocean-like water in the Jupiter-family comet 103P/Hartley 2. *Nature* **478**, 218–220 (2011).

57. Lis, D. C. *et al.* A Herschel study of D/H in water in the Jupiter-Family comet 45P/Honda-Mrkos-Pajdušáková and prospects for D/H measurements with CCAT. *Astrophys. J.* **774**, L3 (2013).

58. Altwegg, K. *et al.* 67P/Churyumov-Gerasimenko, a Jupiter family comet with a high D/H ratio. *Science* **347**, 3–6 (2015).

59. Pearson, V. K., Sephton, M. A., Franchi, I. A., Gibson, J. M. & Gilmour, I. Carbon and nitrogen in carbonaceous chondrites: Elemental abundances and stable isotopic compositions. *Meteorit. Planet. Sci.* **1918**, 1899–1918 (2006).

60. Balsiger, H., Altwegg, K. & Geiss, J. D/H and $^{18}O/^{16}O$ ratio in the hydronium ion and in neutral water from in situ ion measurements in comet Halley. *J. Geophys. Res.* **100**, 5827–5834 (1995).

61. Bockelée-Morvan, D. *et al.* Deuterated water in comet C/1996 B2 (Hyakutake) and its implication for the origin of comets. *Icarus* **133**, 147–162 (1998).

62. Meier, R. *et al.* A determination of the HDO/H2O ratio in comet C/1995 O1 (Hale-Bopp). *Science* **279**, 842–844 (1998).

63. Hutsémekers, D., Manfroid, J., Jehin, E., Zucconi, J.-M. & Arpigny, C. The OH/^{18}OH and OD/OH isotope ratios in comet C/2002 T7 (LINEAR). *Astron. Astrophys.* **490**, L31–L43 (2008).

64. Bockelée-Morvan, D. *et al.* Herschel measurements of the D/H and $^{16}O/^{18}O$ ratios in water in the Oort-cloud comet C/2009 P1 (Garradd). *Astron. Astrophys.* **544**, L15 (2012).

65. Villanueva, G. *et al.* A sensitive search for deuterated water in comet 8P/Tuttle. *Astrophys. J.* **690**, L5–L9 (2009).

66. Biver, N. *et al.* Radio wavelength molecular observations of comets C/1999 T1 (Mc Naught-Hartley), C/2001 A2 (LINEAR) and 153/P Ikeya-Zhang. *Astron. Astrophys.* **449**, 1255–1270 (2006).

67. Aléon, J., Engrand, C., Robert, F. & Chaussidon, M. Clues to the origin of interplanetary dust particles from the isotopic study of their hydrogen-bearing phases. *Geochim. Cosmochim. Acta* **65**, 4399–4412 (2001).

68. Floss, C. *et al.* Identification of isotopically primitive interplanetary dust particles: a NanoSIMS isotopic imaging study. *Geochim. Cosmochim. Acta* **70**, 2371–2399 (2006).

69. Waite, J. H. *et al.* Liquid water on Enceladus from observations of ammonia and 40Ar in the plume. *Nature* **460,** 487–490 (2009).

70. Lellouch, E. *et al.* The deuterium abundance in Jupiter and Saturn from ISO-SWS observations. *Astron. Astrophys.* **670,** 610–622 (2001).

71. Feuchtgruber, H. *et al.* Detection of HD in the atmospheres of Uranus and Neptune: a new determination of the D/H ratio. *Astron. Astrophys.* **341,** L17–L21 (1999).

72. Lellouch, E. *et al.* First results of Herschel-PACS observations of Neptune. *Astron. Astrophys.* **518,** L152 (2010).

73. Greenwood, J. P., Itoh, S., Sakamoto, N., Vicenzi, E. P. & Yurimoto, H. Hydrogen isotope evidence for loss of water from Mars through time. *Geophys. Res. Lett.* **35,** 1–5 (2008).

74. Hu, S. *et al.* NanoSIMS analyses of apatite and melt inclusions in the GRV 020090 Martian meteorite: hydrogen isotope evidence for recent past underground hydrothermal activity on Mars. *Geochim. Cosmochim. Acta* **140,** 321–333 (2014).

75. Boctor, N. Z., Alexander, C. M. O. D., Wang, J. & Hauri, E. The sources of water in Martian meteorites: Clues from hydrogen isotopes. *Geochim. Cosmochim. Acta* **67,** 3971–3989 (2003).

76. Bjoraker, G. L., Mumma, M. J. & Larson, H. P. Isotopic Abundance Ratios for Hydrogen and Oxygen in the Martian Atmosphere. in *21st Annual DPS Meeting* 991 (1989).

77. Barnes, J. J. *et al.* Accurate and precise measurements of the D/H ratio and hydroxyl content in lunar apatites using NanoSIMS. *Chem. Geol.* **337-338,** 48–55 (2013).

78. McKeegan, K. D. & Leshin, L. A. Stable isotope variations in extraterrestrial materials. *Rev. Mineral. Geochem.* **43,** 279–318 (2001).

Acknowledgements

STFC is thanked for a PhD studentship to J.J.B. and for research grants to M.A. (grant number ST/I001298/1 and ST/L000776/1). This research was supported in part by NASA Solar System Exploration Research Virtual Institute contract NNA14AB07A (David A. Kring, PI). We thank the reviewers for their insightful comments that have certainly enhanced the quality of this paper. Lunar and Planetary Institute contribution 1909.

Author contributions

J.J.B. performed all calculations and modelling and prepared the first draft of the manuscript. D.A.K, J.J.B. and M.A. moulded the project. All authors were involved in the discussion of ideas and preparation of the manuscript.

Additional information

Competing financial interests: The authors declare no competing financial interests.

29

Earth-like aqueous debris-flow activity on Mars at high orbital obliquity in the last million years

T. de Haas[1], E. Hauber[2], S.J. Conway[3], H. van Steijn[1], A. Johnsson[4] & M.G. Kleinhans[1]

Liquid water is currently extremely rare on Mars, but was more abundant during periods of high obliquity in the last few millions of years. This is testified by the widespread occurrence of mid-latitude gullies: small catchment-fan systems. However, there are no direct estimates of the amount and frequency of liquid water generation during these periods. Here we determine debris-flow size, frequency and associated water volumes in Istok crater, and show that debris flows occurred at Earth-like frequencies during high-obliquity periods in the last million years on Mars. Results further imply that local accumulations of snow/ice within gullies were much more voluminous than currently predicted; melting must have yielded centimetres of liquid water in catchments; and recent aqueous activity in some mid-latitude craters was much more frequent than previously anticipated.

[1] Faculty of Geosciences, Universiteit Utrecht, Heidelberglaan 2, 3584 CS, Utrecht, The Netherlands. [2] Institute of Planetary Research, German Aerospace Center, Rutherfordstrasse 2, Berlin DE-12489, Germany. [3] Department of Physical Sciences, Open University, Walton Hall, Milton Keynes MK7 6AA, UK. [4] Department of Earth Sciences, University of Gothenburg, Gothenburg SE-405 30, Sweden. Correspondence and requests for materials should be addressed to T.d.H. (email: t.dehaas@uu.nl).

At present Mars is very cold and dry and its thin atmosphere makes liquid water at its surface exceptionally rare[1,2]. However, climatic conditions differed during periods of high-orbital obliquity in the last few millions of years[3-6]. In these periods liquid water was probably more abundant, as testified by the presence of numerous mid-latitude gullies, which are small catchment-fan systems[7-12] (Fig. 1). During high-obliquity periods on Mars (>30°), increased polar summer insolation enhances polar ice sublimation, which increases atmospheric water content and amplifies circulation, leading to a more intense water cycle[4,5]. Precipitation of snow and ice is thought to become widespread in the mid-latitudes (from the poles to ~30° N and S), and in the high mountains in lower-latitude regions[4-6], leading to extensive glaciation[4,5]. Snow/ice probably melted during high-obliquity periods in favourable locations, forming thousands of gullies in the mid-latitudes[7-9]. Evidence of water-free sediment flows[13-15], debris flows[8,10] and fluvial flows[9,11] has been identified, which result in a large morphological diversity of gullies on Mars. The morphology and morphometry of many gullies imply that they are formed by liquid water[9-12], whilst other gullies are morphologically active today, probably driven mainly by CO_2 frost, suggesting a water-free continuing present-day activity[13,14]. Some aqueous gullies formed in the last few millions of years[10,16,17]. As such, they are the youngest record of liquid water and extensive aqueous activity on the surface of Mars, and therefore of critical importance in resolving the planet's recent hydrologic and climatic history. Obliquity on Mars has varied between 15° and 35° in the last 5 Myr, in cycles of approximately 120 Kyr[18]. The obliquity threshold for snow and ice transfer from the poles to lower latitudes is estimated at 30°[3], whereas the threshold for melting and associated morphological activity is probably higher but unknown[19].

Key questions that remain unanswered are how much water could potentially melt during these high-obliquity periods? And how frequent was the aqueous activity within the gullies?

Here, we address these questions by quantifying debris-flow size, frequency and associated liquid water content on Mars, in the very young Istok crater in Aonia Terra (Fig. 1) (formed 0.1–1 Myr ago; best-fit age: ~0.19 Ma; 45.11° S; 274.2° E)[10]. These analyses show that local accumulations of snow/ice within gullies were in the order of centimetres to decimetres during periods of high obliquity in the last Myr. Melting of this snow/ice must have yielded centimetres of liquid water in the gully catchments to produce the observed debris-flow volumes. Moreover, debris flows were much more frequent than previously anticipated and occurred at Earth-like frequencies in Istok crater at high obliquity.

Results

Study crater. The pole-facing slope of Istok crater hosts a bajada, a series of coalescing fans, with abundant debris-flow deposits[10], which are among the best preserved found on Mars to date. Although morphometric analyses suggest that many gullies are formed by debris flows[12], evidence thereof is generally absent on gully-fan surfaces. This is probably caused by post-depositional reworking of gully-fan surfaces by weathering and erosion[20] or emplacement of latitude-dependent mantle deposits[10] (LDM; a smooth, metres-thick deposit comprising layers of dust and ice that extends from the poles to the mid-latitudes[21,22]). The unusually pristine debris-flow deposits in Istok crater therefore make it the best, and only, crater wherein detailed quantitative analyses using debris-flow volumes can be performed today.

In Istok crater, contribution of fluvial and water-free sediment flows to the fan surface morphology on the bajada appears to be very minor. We base this on the absence of bright and dark deposits and 'fingering' depositional lobes, which are both associated with dry sediment flows on Mars[13-15], the abundance of well-developed leveed channels that have to date not been observed in recent dry flows on Mars; the remarkably

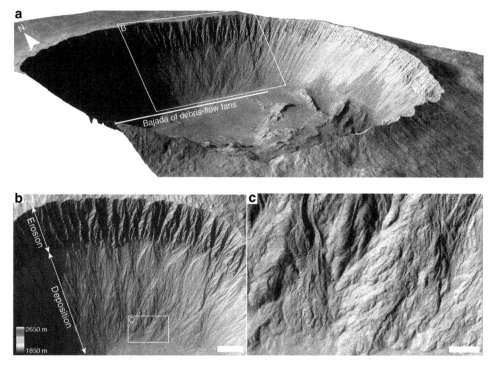

Figure 1 | Istok crater. (**a**) Bajada of remarkably pristine debris-flow fans on the pole-facing slope (45.11° S; 274.2° E). The bajada length is ~4 km. (**b**) Eroding alcoves supply sediments to the downslope bajada of fans. Scale bar, 250 m wide. (**c**) The fans are composed of debris-flow deposits, as testified by the widespread occurrence of paired levees, distinct depositional lobes and embedded boulders[10]. Scale bar, 50 m wide.

good morphological and textural agreement with terrestrial debris flows[10]; and the 8° to 20° depositional slopes, typical for non-cohesive terrestrial debris flows[10]. In contrast, the landforms on the crater slopes with non-polar azimuths appear unrelated to liquid water[10]. This landform disparity further supports debris-flow formation by insolation-driven melting of snow/ice, because melting is hypothesized to mainly occur on pole-facing slopes at high obliquity in the mid-latitudes[8,9]. In contrast to many other gullies on Mars, the source of water seems to be unrelated to the LDM. This is testified by the presence of highly brecciated alcoves hosting many boulders, solely exposing bedrock and the abundance of metre-sized boulders on the depositional fans. The absence of LDM in the gullies is further supported by the lack of landforms associated with the LDM, such as polygonally patterned ground and moraine-like ridges[10]. The absence of LDM suggests that the debris flows in Istok crater formed from top-down melting of relatively pure (that is, little dust) snow packs[10], and therefore inferences from debris-flow volumes, directly relate back to snowfall amounts and climate.

Debris-flow volume and frequency. Debris flows are high-concentration mixtures of solid particles and water that move as a single-phase high-density flow. Non-cohesive debris flows contain ∼20–60% water by volume[23–25]. They form deposits with paired levees and distinct depositional lobes that often incorporate large boulders. We use the distinct morphology of these deposits to estimate individual debris-flow volumes from a High-Resolution Imaging Science Experiment (HiRISE) Digital

Elevation Model (DEM) with a sampling distance of 1 m. Estimated individual debris-flow volumes roughly range from 400 to 5,100 m³ (Fig. 2a–c; Table 1) and are similar to those in unconfined terrestrial debris-flow systems[26–28] (Fig. 3).

We estimated the total number of debris flows by comparing the volume of a single, modal-sized, debris flow to the total volume of sediment eroded from the catchments. Here, an alcove is defined as a single source area divided by ridges, and a catchment as a set of alcoves that together feed a similar part of the bajada (Supplementary Fig. 1). In total, around 28,000 modal-sized debris flows were needed to form the entire bajada and ∼1,900 debris flows originated from each catchment (Supplementary Table 1). From this we calculated the cumulative time above a specific obliquity threshold for melting and then determined the debris-flow frequency within the gullies, expressed as their return period[26]. Debris-flow return periods ranged between 4 and 15 years on the bajada, and 64–221 years in the catchments for a conservative obliquity threshold for melting of 30°. A melting threshold of 35°[19] implies return periods of 0.2–0.8 year for the bajada and 3–12 years for the catchments (Fig. 2d,e; Supplementary Table 1).

Liquid water volumes. Using the known range of water concentrations of terrestrial debris flows in combination with the measured debris-flow volumes in Istok crater, we can make an estimation of the amount of liquid water required for each flow. The associated liquid water volume yields a minimum estimate of snow/ice deposition and subsequent melting within the alcoves.

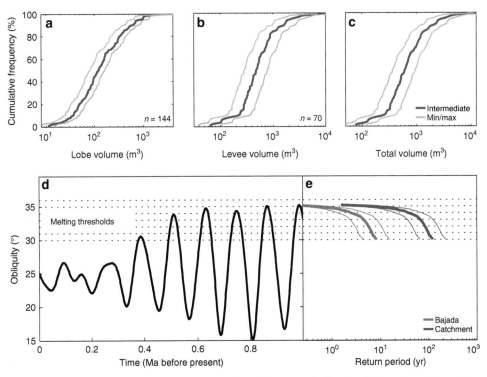

Figure 2 | Debris-flow return periods and size in Istok crater. (**a**) Cumulative frequency distribution of lobe volume. The minimum, maximum and intermediate estimates are based on a triangular, rectangular and trapezoidal-shaped lobe model, respectively (Supplementary Fig. 2). (**b**) Cumulative frequency distribution of levee volume, estimated by assuming paired levees of half the bajada length (900 m; 800–1,100 m range). The minimum estimate is based on triangular-shaped paired levees of 400 m long, the maximum estimate on rectangular-shaped paired levees of 550 m long, and the intermediate estimate on trapezoidal-shaped levees of 450 m long. (**c**) Cumulative frequency distribution of total debris-flow volume (lobe and levee volume combined). (**d**) Obliquity in the last Myr on Mars[18], and potential thresholds for melting on mid-latitude pole-facing crater walls. (**e**) Debris-flow return periods on the bajada and per catchment. The intermediate estimate (thick line) is calculated from the intermediate-estimate debris-flow size and best-estimate catchment size. The minimum and maximum estimates are calculated from the largest debris-flow size and smallest catchment volume and the smallest debris-flow size and maximum catchment volume, respectively. See Supplementary Tables 1–5 for raw data.

Table 1 | Minimum amounts of liquid water required for the generation of debris flows.

Debris-flow size	Debris-flow volume (m³)	Water:sediment ratio 0.2		Water:sediment ratio 0.6	
		Water volume (m³)	Water in alcove (mm)	Water volume (m³)	Water in alcove (mm)
Modal	605 (368–950)	121 (74–190)	3.0 (1.8–4.7)	363 (221–570)	9.0 (5.5–14.1)
95% largest	3,307 (2,031–5,101)	661 (406–1,020)	16.4 (10.0–25.2)	1,984 (1,219–3,061)	49.1 (30.1–75.7)

Error margins expressed as minimum and maximum values within brackets (see Fig. 2a–c and Supplementary Tables 3 and 4 for raw debris-flow volume data, and Supplementary Table 2 for raw alcove volume data).

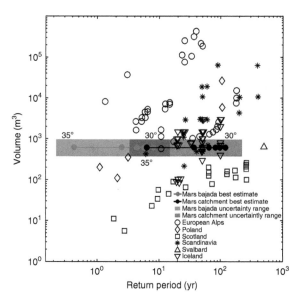

Figure 3 | Debris-flow volumes and return periods in Istok crater and examples from Earth. The return periods on Mars are clearly within the range of return periods observed in temperate to polar regions on Earth regardless of the uncertainty in debris-flow volume, return periods per bajada or individual catchment and obliquity thresholds for melting between 30° and 35°. Data from the European Alps, Poland, Scotland, Scandinavia and Svalbard are from Van Steijn[26] and references therein, additional data for the Alps from Helsen et al.[27], and Icelandic data from Decaulne and Sæmundsson[28].

Between 3 and 9 mm of liquid water uniformly spread over an average-sized alcove is required for the formation of modal-sized debris flows, and 16–50 mm of liquid water is required for the formation of large, 95 percentile-sized debris flows (Table 1). The actual thickness of the snow/ice layer must have been much larger because of the porosity of the snowpack, potential sublimation and evaporation losses, and the fact that uniform melting over an entire alcove will generally not occur[19]. On the other hand, snowdrift might have led to larger accumulations of snow in the alcoves than was originally emplaced[29]. Potential infiltration losses are likely negligible on Mars where a frozen permafrost layer acts as an aquiclude[11]. Therefore, we estimate that centimetres to decimetres of snow must have accumulated in the alcoves to form the observed debris-flow deposits.

Discussion

The surprisingly short debris-flow return periods at high orbital obliquity in Istok crater are very similar to those in various environments on Earth[26–28] (Fig. 3). Moreover, they are similar or even shorter than in terrestrial environments that are climatologically comparable to the dry and cold polar desert of Mars. In one of the driest regions on Earth, the hyperarid

Atacama Desert, debris-flow return periods along the coast of northern Chile and southern Peru range between 40 and 2,400 years[30] and are probably higher further inland where precipitation is even less frequent. In the periglacial, polar semi-desert of Svalbard, which displays many periglacial landforms similar to Mars[31], debris-flow return periods ranging between 80 and 500 years were found[32]. These results suggest that pole-facing mid-latitude crater walls on Mars, at least in Istok crater, were extremely active environments with Earth-like debris-flow activity during high-obliquity periods in the last few million years. Moreover, the debris-flow return periods during high obliquity are similar to present-day return periods of dry, CO_2-aided, sediment flows in some of the currently most active gullies on Mars[14]. The debris-flow return periods in Istok crater imply that these gullies were among the most active aqueous gullies on Mars in the last Myr. They are the first estimated return periods of aqueous activity for Martian gullies. Yet, it is not unlikely that similar activity has occurred on other sites in the past.

Generally, inferences drawn from Global Climate Models (GCMs) suggest that annual atmospheric precipitation of snow/ice at 35° obliquity does not exceed a spatially averaged 10 mm per year[6,33], and precipitation is even less at lower obliquity. However, recently Madeleine et al.[5] incorporated the effect of radiatively active water-ice clouds in their Global Climate Model, resulting in annual snow/ice accumulations of ~10 cm in the mid-latitudes at 35° obliquity, which corresponds well with the snow accumulation inferred from debris-flow volumes at Istok crater. Yet, although snowdrift causes larger accumulations of snow within the alcoves and snow might accumulate over multiple years, most snow is thought to sublimate[6,19]. Moreover, Williams et al.[19] estimate that only a very small amount, in the order of 1 mm per year, of liquid water was produced by melting on pole-facing mid-latitude crater walls out of a 5-cm-thick snowpack at high obliquity. Kite et al.[34] estimate that melting over a Mars-year produces 9 mm of liquid water in total. Clearly, these models do not explain the amounts of liquid water needed for the formation of the debris flows in Istok crater. This implies that melting of snow/ice must locally have been much larger than currently predicted by most climate models, and supports the recent improvements to the climate models by Madeleine et al.[5].

We conclude that debris flows occurred at Earth-like frequencies in Istok crater during high-obliquity periods in the last million years on Mars. Although this required much more atmospheric deposition and subsequent melting than is currently predicted by climate models, these findings fit well into the emerging view of a much more dynamic recent and present Mars than anticipated only a few years ago[1,14]. Mars was long presumed to be a hyperarid environment dominated by wind since the onset of the Amazonian period, 3 Gyr ago. However, pristine debris-flow deposits in Istok crater provide compelling evidence for very active aqueous environments on pole-facing slopes in the mid-latitudes during high obliquity in the last million years. The surprisingly large amount of liquid water on these slopes means we should revise our understanding of Mars' recent climate, but also points to more habitable recent environments than previously predicted.

Methods

Production of the digital elevation model. The DEM used for the extraction of alcove and debris-flow volumes was constructed using the methods described by Kirk et al.[35], from HiRISE stereo images PSP_006837_1345 and PSP_007127_1345. The ground sampling distance of the DEM is 1 m. The vertical precision of the DEM can be estimated based on viewing geometry and pixel scale. The stereo convergence angle of the HiRISE images is 20.1°, the largest spatial resolution of the two images is 0.258 m, and assuming 1/5 pixel correlations yields a vertical precision of $0.258/5/\tan(20.1) = 0.13\,\text{m}^{35}$.

Extraction of alcove and debris-flow volumes. The volume of material eroded from the alcoves was determined from the DEM. Following Conway and Balme[22] we assume that the top of the alcove crests represent the initial pre-gully surface. In reality, alcove crests will also be eroded, and thus our assumption yields a lower bound estimate. The eroded volume was derived by subtracting the original from the pre-gully surface. Error propagation calculations by Conway and Balme[22] suggest that such volume estimates are accurate within 15%. Because a bajada is composed of a series of coalescing fans, it is impossible to directly determine total bajada volume given the uncertainty of the pre-gully crater profile. However, as (1) the material eroded from the alcoves consists of pure bedrock, unrelated to the LDM, and (2) the composition and rheology of debris flows prevent significant escape of material after deposition, the gullies are probably closed systems wherein the amount of sediment eroded from the alcoves is approximately equal to the amount of material deposited on the bajada, as quantitatively demonstrated by Conway and Balme[22]. We therefore approximate the total bajada volume by calculating the total amount of material eroded from the alcoves. As such, we neglect the potential input of material by rockfalls to the bajada. However, the abundance of debris-flow deposits on the bajada suggests that the volume transfer by rockfalls is minor compared with the transfer by debris flows.

Individual debris-flow volumes were determined from the orthorectified image and from the DEM. We measured width, length and height of 144 clearly resolvable lobes and width and height of 70 levees (Supplementary Fig. 1; Supplementary Tables 3 and 4), which we subsequently combined into debris-flow volume based on percentiles (Fig. 2a–c). Width and length were measured using the orthorectified image, and height using the DEM. We combined percentiles of lobe and levee size to obtain debris-flow volumes, because it was impossible to directly determine debris-flow volumes, as the debris flows on the bajada are strongly amalgamated. We avoided measurement of lobes that were largely buried by subsequent debris flows to prevent an under-prediction of lobe volume (see Supplementary Fig. 2 for an example of the delineation of lobes). It is impossible to determine whether the debris flows in Istok crater formed one or multiple lobes, based on remote sensing data only. However, on Earth it is much more common for debris flows to form one lobe rather than multiple lobes[27,32,36] and there is no model suggesting that forking should occur more frequently under Martian conditions. Therefore, we assume that each debris flow formed one depositional lobe. This might underestimate the volume of very large debris flows, which can bifurcate and form multiple lobes, but is probably a good representation of the modal-sized debris flow and therefore debris-flow return periods. To account for errors associated with the estimation of the debris-flow cross-section, levee and lobe volumes were calculated by assuming a triangular (minimum estimate), rectangular (maximum estimate) and a trapezoidal cross-section (intermediate estimate) (Supplementary Fig. 3). Total levee volume for the modal-sized debris flows was calculated by assuming paired levees of half the bajada length (900 m; 800–1,100 m range), where we used the mean, minimum and maximum length for the calculation of the intermediate, minimum and maximum volume estimates, respectively. This is a rough estimation, but as accurate direct measurement of modal debris-flow length is very ambiguous half the bajada length was chosen as a parsimonious approximation. This assumption will lead to an underestimation of the volume for relatively long debris flows and an overestimation for relatively short debris flows, but is a good approximation of the modal-sized debris flow and therefore the return periods.

Estimation of debris-flow return periods. The debris-flow return periods were calculated by dividing the cumulative time above an obliquity threshold (Fig. 2d)[18,19] since the formation of Istok crater, by an estimate of the total number of debris flows on the bajada and per catchment. We conservatively estimate return periods by using the maximum age of Istok crater (formed 1 Ma)[10]. The number of debris flows was calculated by dividing the volume of sediment eroded from the alcoves by the volume of the modal-sized debris flow that formed the bajada. As we can only observe, and therefore measure, the volume of surficial debris flows, we assume that volumes remained quasi-static over time. Approximately 45% of the pole-facing crater wall hosting the fans is covered in shadows, and hence debris flows were not accurately discernible (Fig. 1). Therefore, we calculated total volume

on the bajada by extrapolating the volume on the sun-illuminated part to the shadowed part by assuming similar geometry. Return periods were calculated for the entire bajada and per average-sized catchment and for a range of obliquities exceeding 30°.

Robustness of results. Our calculations are based on a number of assumptions. However, this is inevitable for the current analysis that can only be based on remote sensing data. For this reason we conservatively estimated possible errors and propagated these through our calculations. Moreover, we conservatively calculated alcove volumes and used the maximum estimated age of the host crater as an overestimate of the gully-system age. Even then, the uncertainty in age range (up to a factor of 5: 0.2–1.0 Myr) is larger than the error associated with volume calculations. For example, if we adopt the extreme scenario that every debris flow formed two lobes and that we underestimated lobe volume by a factor of 2 (for example, by underestimating runout length or missing part of the lobe volume due to partial burial), this results in an error of a factor of 4 in lobe volumes and associated return periods and liquid water volumes. As such, this error is smaller than the error range associated with crater age uncertainty.

Second, our conclusions on debris-flow return periods are largely insensitive to the assumptions. If we adopt an extreme scenario where we overestimate or underestimate debris-flow volume by a factor of 4, the number of debris flows that formed the bajada, and the return periods also change by a factor of 4. However, compared with the large range of debris-flow return periods observed on Earth this is not significantly different (Fig. 3). Neither does a factor 4 difference in liquid water and snow accumulation volumes change the result that centimetres to decimetres of snow were required for the formation of debris flows. Therefore, the conclusions are robust.

References

1. McEwen, A. S. et al. Recurring slope lineae in equatorial regions of Mars. *Nat. Geosci.* **7**, 53–58 (2014).
2. Jakosky, B. M. & Phillips, R. J. Mars' volatile and climate history. *Nature* **412**, 237–244 (2001).
3. Head, J. W., Mustard, J. F., Kreslavsky, M. A., Milliken, R. E. & Marchant, D. R. Recent ice ages on Mars. *Nature* **426**, 797–802 (2003).
4. Forget, F., Haberle, R. M., Montmessin, F., Levrard, B. & Head, J. W. Formation of glaciers on Mars by atmospheric precipitation at high obliquity. *Science* **311**, 368–371 (2006).
5. Madeleine, J.-B. et al. Recent ice ages on Mars: the role of radiatively active clouds and cloud microphysics. *Geophys. Res. Lett.* **41**, doi:10.1002/2014GL059861 (2014).
6. Madeleine, J. B. et al. Amazonian northern mid-latitude glaciation on Mars: a proposed climate scenario. *Icarus* **203**, 390–405 (2009).
7. Malin, M. C. & Edgett, K. S. Evidence for recent groundwater seepage and surface runoff on Mars. *Science* **288**, 2330–2335 (2000).
8. Costard, F., Forget, F., Mangold, N. & Peulvast, J. P. Formation of recent Martian debris flows by melting of near-surface ground ice at high obliquity. *Science* **295**, 110–113 (2002).
9. Dickson, J. L. & Head, J. W. The formation and evolution of youthful gullies on Mars: Gullies as the late-stage phase of Mars' most recent ice age. *Icarus* **204**, 63–86 (2009).
10. Johnsson, A., Reiss, D., Hauber, E., Hiesinger, H. & Zanetti, M. Evidence for very recent melt-water and debris flow activity in gullies in a young mid-latitude crater on Mars. *Icarus* **235**, 37–54 (2014).
11. Heldmann, J. L. et al. Formation of Martian gullies by the action of liquid water flowing under current Martian environmental conditions. *J. Geophys. Res.-Planet* **110**, doi:10.1029/2004JE002261 (2005).
12. Conway, S. J. et al. The indication of Martian gully formation processes by slope-area analysis. *Geological Society, London, Special publications* **356**, 171–201 (2011).
13. Dundas, C. M., McEwen, A. S., Diniega, S., Byrne, S. & Martinez-Alonso, S. New and recent gully activity on Mars as seen by HiRISE. *Geophys. Res. Lett.* **37**, L07202 (2010).
14. Dundas, C. M., Diniega, S. & McEwen, A. S. Long-term monitoring of martian gully formation and evolution with MRO/HiRISE. *Icarus* **251**, 244–263 (2014).
15. Pelletier, J. D., Kolb, K. J., McEwen, A. S. & Kirk, R. L. Recent bright gully deposits on Mars: Wet or dry flow? *Geology* **36**, 211–214 (2008).
16. Reiss, D., Van Gasselt, S., Neukum, G. & Jaumann, R. Absolute dune ages and implications for the time of formation of gullies in Nirgal Vallis, Mars. *J. Geophys. Res. Planet* **109** (2004).
17. Schon, S. C., Head, J. W. & Fassett, C. I. Unique chronostratigraphic marker in depositional fan stratigraphy on Mars: Evidence for ca. 1.25 Ma gully activity and surficial meltwater origin. *Geology* **37**, 207–210 (2009).
18. Laskar, J. et al. Long term evolution and chaotic diffusion of the insolation quantities of Mars. *Icarus* **170**, 343–364 (2004).
19. Williams, K., Toon, O., Heldmann, J. & Mellon, M. Ancient melting of mid-latitude snowpacks on mars as a water source for gullies. *Icarus* **200**, 418–425 (2009).

20. De Haas, T., Hauber, E. & Kleinhans, M. G. Local late Amazonian boulder breakdown and denudation rate on Mars. *Geophys. Res. Lett.* **40,** 3527–3531 (2013).

21. Mustard, J. F., Cooper, C. D. & Rifkin, M. K. Evidence for recent climate change on Mars from the identification of youthful near-surface ground ice. *Nature* **412,** 411–414 (2001).

22. Conway, S. J. & Balme, M. R. Decametre-thick remnant glacial ice deposits on Mars. *Geophys. Res. Lett.* **41,** 5402–5409 (2014).

23. Iverson, R. M. The physics of debris flows. *Rev. Geophys.* **35,** 245–296 (1997).

24. Costa, J. E. in *Flood Geomorphology.* (eds Baker, V. R. *et al.*) Ch 7 (Wiley-Intersciences, 1988).

25. Pierson, T. C. 'Distinguishing between debris flows and floods from field evidence in small watersheds'. *USGS Fact Sheet 2004-3142* (2005).

26. Van Steijn, H. Debris-flow magnitude-frequency relationships for mountainous regions of Central and Northwest Europe. *Geomorphology* **15,** 259–273 (1996).

27. Helsen, M. M., Koop, P. J. M. & Van Steijn, H. Magnitude–frequency relationship for debris flows on the fan of the Chalance torrent, Valgaudemar (French Alps). *Earth Surf. Proc. Land* **27,** 1299–1307 (2002).

28. Decaulne, A. & Sæmundsson, T. Debris-flow characteristics in the Gleidarhjalli area, northwestern Iceland. *Debris-flow hazards mitigation: mechanics, prediction, and assessment* **2,** 1107–1118 (2003).

29. Christiansen, H. H. Nivation forms and processes in unconsolidated sediments. *NE Greenland. Earth Surf. Proc. Land* **23,** 751–760 (1998).

30. Vargas, G., Rutllant, J. & Ortlieb, L. ENSO tropical–extratropical climate teleconnections and mechanisms for Holocene debris flows along the hyperarid coast of western South America (17–24S). *Earth Planet. Sci. Lett.* **249,** 467–483 (2006).

31. Hauber, E. *et al.* Landscape evolution in Martian mid-latitude regions: insights from analogous periglacial landforms in Svalbard. *Geological Society, London, Special Publications* **356,** 111–131 (2011).

32. André, M. F. Frequency of debris flows and slush avalanches in Spitsbergen: a tentative evaluation from lichenometry. *Pol. Pol. Res.* **11,** 345–363 (1990).

33. Mischna, M. A., Richardson, M. I., Wilson, R. J. & McCleese, D. J. On the orbital forcing of Martian water and CO2 cycles: a general circulation model study with simplified volatile schemes. *J. Geophys. Res. Planet* **108,** 5062 (2003).

34. Kite, E. S., Halevy, I., Kahre, M. A., Wolff, M. J. & Manga, M. Seasonal melting and the formation of sedimentary rocks on Mars, with predictions for the Gale Crater mound. *Icarus* **223,** 181–210 (2013).

35. Kirk, R. L. *et al.* Ultrahigh resolution topographic mapping of Mars with MRO HiRISE stereo images: Meter-scale slopes of candidate Phoenix landing sites. *J. Geophys. Res.-Planet* **113,** E00A24 (2008).

36. Suwa, H. & Okuda, S. Deposition of debris flows on a fan surface, Mt. Yakedake, Japan. *Z. Geomorphol. Suppl.* **46,** 79–101 (1983).

Acknowledgements

This work is supported by the Netherlands Organisation for Scientific Research (NWO) and the Netherlands Space Office (NSO) (grant ALW-GO-PL17-2012 to MGK). EH was partly supported by the Helmholtz Association through the research alliance 'Planetary Evolution and Life'. SJC is funded by a Leverhulme Trust Grant RPG-397. AJ was supported by the Swedish National Space Board (SNSB) (grant 2012-R). We acknowledge Dennis Reiss and Kim Cohen for valuable discussions on the manuscript.

Author contributions

All authors contributed significantly to this work. T.d.H. designed the study, collected and analysed the data and prepared the manuscript. S.J.C. made the DEM and contributed to data analysis and manuscript preparation. M.G.K. and E.H. contributed to data analysis and manuscript preparation. H.V.S. and A.J. contributed to manuscript preparation.

Additional information

Permissions

All chapters in this book were first published in NC, by Nature Publishing Group; hereby published with permission under the Creative Commons Attribution License or equivalent. Every chapter published in this book has been scrutinized by our experts. Their significance has been extensively debated. The topics covered herein carry significant findings which will fuel the growth of the discipline. They may even be implemented as practical applications or may be referred to as a beginning point for another development.

The contributors of this book come from diverse backgrounds, making this book a truly international effort. This book will bring forth new frontiers with its revolutionizing research information and detailed analysis of the nascent developments around the world.

We would like to thank all the contributing authors for lending their expertise to make the book truly unique. They have played a crucial role in the development of this book. Without their invaluable contributions this book wouldn't have been possible. They have made vital efforts to compile up to date information on the varied aspects of this subject to make this book a valuable addition to the collection of many professionals and students.

This book was conceptualized with the vision of imparting up-to-date information and advanced data in this field. To ensure the same, a matchless editorial board was set up. Every individual on the board went through rigorous rounds of assessment to prove their worth. After which they invested a large part of their time researching and compiling the most relevant data for our readers.

The editorial board has been involved in producing this book since its inception. They have spent rigorous hours researching and exploring the diverse topics which have resulted in the successful publishing of this book. They have passed on their knowledge of decades through this book. To expedite this challenging task, the publisher supported the team at every step. A small team of assistant editors was also appointed to further simplify the editing procedure and attain best results for the readers.

Apart from the editorial board, the designing team has also invested a significant amount of their time in understanding the subject and creating the most relevant covers. They scrutinized every image to scout for the most suitable representation of the subject and create an appropriate cover for the book.

The publishing team has been an ardent support to the editorial, designing and production team. Their endless efforts to recruit the best for this project, has resulted in the accomplishment of this book. They are a veteran in the field of academics and their pool of knowledge is as vast as their experience in printing. Their expertise and guidance has proved useful at every step. Their uncompromising quality standards have made this book an exceptional effort. Their encouragement from time to time has been an inspiration for everyone.

The publisher and the editorial board hope that this book will prove to be a valuable piece of knowledge for researchers, students, practitioners and scholars across the globe.

List of Contributors

Antoine Lucas
Équipe de sismologie, Institut de Physique du Globe de Paris, Sorbone Paris Cité, Université Paris Diderot, UMR 7154 CNRS, 1 rue Jussieu, 75238 Paris cedex 05, France
Division of Geological and Planetary Sciences, California Institute of Technology, 1200 E. California Blvd, Pasadena, California 95125, USA
Laboratoire Astrophysique, Instrumentation et Modélisation (AIM), CNRS-UMR 7158, Université Paris-Diderot, CEA-Saclay, Gif-sur-Yvette 91191, France

Jean Paul Ampuero
Division of Geological and Planetary Sciences, California Institute of Technology, 1200 E. California Blvd, Pasadena, California 95125, USA

Anne Mangeney
Équipe de sismologie, Institut de Physique du Globe de Paris, Sorbone Paris Cité, Université Paris Diderot, UMR 7154 CNRS, 1 rue Jussieu, 75238 Paris cedex 05, France
Équipe ANGE INRIA, Laboratoire Jacques-Louis Lions, UPMC Paris 6, 4 place Jussieu, Case 187 75252 Paris cedex 05, France

David M. Kass
Jet Propulsion Laboratory, California Institute of Technology, Pasadena, California 91109, USA

Renyu Hu, Bethany L. Ehlmann and Yuk L. Yung
Jet Propulsion Laboratory, California Institute of Technology, Pasadena, California 91109, USA
Division of Geological and Planetary Sciences, California Institute of Technology, Pasadena, California 91125, USA

Shiva Kavosi and Joachim Raeder
Department of Physics and Space Science Center, University of New Hampshire, 8 College Road, Durham, New Hampshire 03824, USA

Asmaa Boujibar
Laboratoire Magmas et Volcans, Université Blaise Pascal, CNRS UMR-6524, 5 rue Kessler, 63000 Clermont-Ferrand, France
Astromaterials Research and Exploration Science, NASA Johnson Space Center, 2101 Nasa Parkway, Houston, Texas 77058, USA

Denis Andrault, Nathalie Bolfan-Casanova, Mohamed Ali Bouhifd and Julien Monteux
Laboratoire Magmas et Volcans, Université Blaise Pascal, CNRS UMR-6524, 5 rue Kessler, 63000 Clermont-Ferrand, France

Boris Laurent, Mathieu Roskosz, Hugues Leroux, Christophe Depecker and Jean-Marc Lefebvre
UMET, Université Lille 1, CNRS UMR 8207, Villeneuve d'Ascq F-59655, France

Laurent Remusat and François Robert
IMPMC, CNRS UMR 7590, Sorbonne Universités, Université Pierre et Marie Curie, IRD, Muséum National d'Histoire Naturelle, CP 52, 57 rue Cuvier, Paris 75231, France

Hervé Vezin
LASIR, Université de Lille 1, CNRS UMR 8516, Villeneuve d'Ascq F-59655, France

Nicolas Nuns
Institut M.E. Chevreul, Université de Lille 1, CNRS, FR 2638, Villeneuve d'Ascq F-59655, France

Gerardo Dominguez
Department of Physics, California State University, San Marcos, San Marcos, California 92096-0001, USA
Department of Chemistry and Biochemistry, University of California, San Diego, La Jolla, California 92093, USA

Mark Thiemens
Department of Chemistry and Biochemistry, University of California, San Diego, La Jolla, California 92093, USA

D.N. Basov, A.S. Mcleod and P. Kelly
Department of Physics, University of California, San Diego, La Jolla, California 92093, USA

Zack Gainsforth and Andrew Westphal
Space Sciences Laboratory, University of California, Berkeley, Berkeley, California 94720, USA

Hans A. Bechtel
Advanced Light Source, Lawrence Berkeley National Laboratory, Berkeley, California 94720, USA

Fritz Keilmann
Ludwig-Maximilians-Universität and Center for Nanoscience, 80539 München, Germany

Xi Zhang
Department of Earth and Planetary Sciences, University of California Santa Cruz, Santa Cruz, California 95064, USA

Robert A. West
Jet Propulsion Laboratory, California Institute of Technology, 4800 Oak Grove Drive, Pasadena, California 91109, USA

Patrick G.J. Irwin
Atmospheric, Oceanic and Planetary Physics, University of Oxford, Clarendon Laboratory, Parks Road, Oxford OX1 3PU, UK

Conor A. Nixon
NASA Goddard Space Flight Center, Greenbelt, Maryland 20771, USA

Yuk L. Yung
Division of Geological and Planetary Sciences, California Institute of Technology, Pasadena, California 91125, USA

Tímea Szabó
Department of Earth and Environmental Science, University of Pennsylvania, 251 Hayden Hall, 240 South 33rd Street, Philadelphia, Pennsylvania 19104, USA
Department of Mechanics, Materials and Structures, Budapest University of Technology and Economics, Mu+egyetem rkp. 1-3. K261, Budapest 1111, Hungary

Douglas J. Jerolmack
Department of Earth and Environmental Science, University of Pennsylvania, 251 Hayden Hall, 240 South 33rd Street, Philadelphia, Pennsylvania 19104, USA

Gábor Domokos
Department of Mechanics, Materials and Structures, Budapest University of Technology and Economics, Mu+egyetem rkp. 1-3. K261, Budapest 1111, Hungary

John P. Grotzinger
Division of Geological and Planetary Sciences, California Institute of Technology, 1200 East California Boulevard, Pasadena, California 91125, USA

Hiroki Ando and Takeshi Imamura
Institute of Space and Astronautical Science, Japan Aerospace Exploration Agency, Sagamihara, Kanagawa 252-0222, Japan

Norihiko Sugimoto
Research and Education Center for Natural Sciences, Department of Physics, Keio University, Yokohama, Kanagawa 223-8521, Japan

Masahiro Takagi
Faculty of Science, Kyoto Sangyo University, Kita-ku, Kyoto 603-8555, Japan

Hiroki Kashimura
Japan Agency for Marine-Earth Science and Technology, Yokohama, Kanagawa 236-0001, Japan

Yoshihisa Matsuda
Department of Astronomy and Earth Sciences, Tokyo Gakugei University, Koganei, Tokyo 184-8501, Japan

A.V. Artemyev
LPC2E/CNRS, 3A, Avenue de la Recherche Scientifique, 45071 Orleans Cedex 2, France
Space Research Institute (IKI) 117997, 84/32 Profsoyuznaya Street, Moscow, Russia. (A.V.A.)

V.V. Krasnoselskikh
LPC2E/CNRS, 3A, Avenue de la Recherche Scientifique, 45071 Orleans Cedex 2, France

O.V. Agapitov
Space Sciences Laboratory, University of California, 7 GaussWay, Berkeley, California 94720, USA
Astronomy and Space Physics Department, National Taras Shevchenko University of Kiev, 2 Glushkova Street, 03222 Kiev, Ukraine (O.V.A.)

F.S. Mozer
Space Sciences Laboratory, University of California, 7 GaussWay, Berkeley, California 94720, USA

D. Mourenas
CEA, DAM, DIF, F-91297 Arpajon Cedex, France

Kellie T. Wall
School of the Environment, Washington State University, Webster Physical Science Building, Room 1228, Pullman, Washington 99164, USA

Michael C. Rowe and Jennifer D. Eccles
School of Environment, University of Auckland, Commerce A Building, Private Bag 92019, Auckland 1142, New Zealand

Ben S. Ellis
Institute of Geochemistry and Petrology, ETH Zurich, 8092 Zurich, Switzerland

Mariek E. Schmidt
Department of Earth Sciences, Brock University, 500 Glenridge Avenue, Saint Catharines, Ontario, Canada L2S 3A1

Max Popp
Max Planck Institute for Meteorology, Bundesstrasse 53, Hamburg 20146, Germany

Program in Atmospheric and Oceanic Sciences, Princeton University, 300 Forrestal Road, Sayre Hall, Princeton, New Jersey 08544, USA
NOAA's Geophysical Fluid Dynamics Laboratory, Princeton, New Jersey, USA

Hauke Schmidt and Jochem Marotzke
Max Planck Institute for Meteorology, Bundesstrasse 53, Hamburg 20146, Germany

Patricia M. Doyle and Kaori Jogo
Hawai'i Institute of Geophysics and Planetology, University of Hawai'i at Ma%noa, Pacific Ocean Science & Technology (POST) Building, 1680 East-West Road, Honolulu, Hawai'i 96822, USA
University of Hawai'i NASA Astrobiology Institute, Honolulu, Hawai'i 96822, USA

Kazuhide Nagashima
Hawai'i Institute of Geophysics and Planetology, University of Hawai'i at Ma%noa, Pacific Ocean Science & Technology (POST) Building, 1680 East-West Road, Honolulu, Hawai'i 96822, USA

Alexander N. Krot
Hawai'i Institute of Geophysics and Planetology, University of Hawai'i at Ma%noa, Pacific Ocean Science & Technology (POST) Building, 1680 East-West Road, Honolulu, Hawai'i 96822, USA
University of Hawai'i NASA Astrobiology Institute, Honolulu, Hawai'i 96822, USA

Shigeru Wakita
Center for Computational Astrophysics, National Astronomical Observatory of Japan, 2-21-1 Osawa, Mitaka, Tokyo 181-8588, Japan

Fred J. Ciesla
Department of the Geophysical Sciences, University of Chicago, 5734 South Ellis Avenue, Chicago, Illinois 60637, USA

Ian D. Hutcheon
Glenn Seaborg Institute, Lawrence Livermore National Laboratory, L-231, Livermore, California 94551, USA

Péter Németh
Institute of Materials and Environmental Chemistry, Research Centre for Natural Sciences, Hungarian Academy of Sciences, Budapest 1117, Hungary
Department of Chemistry and Biochemistry, Arizona State University, Tempe, Arizona 85287-1604, USA

Peter R. Buseck
Department of Chemistry and Biochemistry, Arizona State University, Tempe, Arizona 85287-1604, USA
School of Earth and Space Exploration, Arizona State University, Tempe, Arizona 85287-1404, USA

Laurence A.J. Garvie
Center for Meteorite Studies, Arizona State University, Tempe, Arizona 85287-6004, USA
School of Earth and Space Exploration, Arizona State University, Tempe, Arizona 85287-1404, USA

Toshihiro Aoki
LeRoy Eyring Center for Solid State Science, Arizona State University, Tempe, Arizona 85287-1704, USA

Natalia Dubrovinskaia
Material Physics and Technology at Extreme Conditions, Laboratory of Crystallography, University of Bayreuth, Bayreuth D-95440, Germany

Leonid Dubrovinsky
Bayerisches Geoinstitut, Universita¨t Bayreuth, Bayreuth D-95440, Germany

Yasuhito Sekine
Department of Earth and Planetary Science, University of Tokyo, Bunkyo 113-0033, Japan

Takazo Shibuya
Laboratory of Ocean-Earth Life Evolution Research, Japan Agency for Marine-Earth Science and Technology, Yokosuka 237-0061, Japan
Research and Development Center for Submarine Resources / Project Team for Next- Generation Technology for Ocean Resources Exploration, Japan Agency for Marine-Earth Science and Technology, Yokosuka 237-0061, Japan

Katsuhiko Suzuki and Yuka Masaki
Research and Development Center for Submarine Resources / Project Team for Next-Generation Technology for Ocean Resources Exploration, Japan Agency for Marine-Earth Science and Technology, Yokosuka 237-0061, Japan

Frank Postberg
Institut für Geowissenschaften, Universita¨t Heidelberg, Heidelberg 69120, Germany
Institut für Raumfahrtsysteme, Universita¨t Stuttgart, Stuttgart 70569, Germany

Hsiang-Wen Hsu
Laboratory for Atmospheric and Space Physics, University of Colorado, Boulder, Colorado 80303, USA

Tatsu Kuwatani
Department of Solid Earth Geochemistry, Japan Agency for Marine-Earth Science and Technology, Yokosuka 237-0061, Japan

Shogo Tachibana and Megumi Mori
Department of Natural History Science, Hokkaido University, Sapporo 060-0810, Japan

Peng K. Hong
The University Museum, University of Tokyo, Bunkyo 113-0033, Japan

Motoko Yoshizaki
Department of Earth and Planetary Science, Tokyo Institute of Technology, Meguro 152-8551, Japan

Sin-iti Sirono
Graduate School of Environmental Science, Nagoya University, Nagoya 464-8601, Japan

Martha-Cary Eppes and Stephen Abernathy
Department of Geography and Earth Sciences, University of North Carolina at Charlotte, Charlotte, North Carolina 28223, USA

Beibei Zhou and Andrew Willis
Department of Electrical and Computing Engineering, University of North Carolina at Charlotte, Charlotte North Carolina 28223, USA

Jamie Molaro
Lunar and Planetary Laboratory, University of Arizona, Tucson, Arizona 85721, USA

B. Schmitz
Astrogeobiology Laboratory, Department of Physics, Lund University, 221 00 Lund, Sweden
Hawai'i Institute of Geophysics and Planetology, University of Hawai'i at Manoa, Honolulu, Hawaii 96822, USA

M. Tassinari
Astrogeobiology Laboratory, Department of Physics, Lund University, 221 00 Lund, Sweden

C.E. Caplan and G.R. Huss
Hawai'i Institute of Geophysics and Planetology, University of Hawai'i at Manoa, Honolulu, Hawaii 96822, USA

Q.-Z. Yin
Department of Earth and Planetary Sciences, University of California at Davis, Davis, California 95616, USA

G.W. Evatt, M.J. Coughlan and I.D. Abrahams
School of Mathematics, University of Manchester, Manchester M13 9PL, UK

K.H. Joy
The School of Earth, Atmospheric and Environmental Sciences, University of Manchester, Manchester M13 9PL, UK

A.R.D. Smedley and P.J. Connolly
Centre for Atmospheric Science, The School of Earth, Atmospheric and Environmental Sciences, University of Manchester, Manchester M13 9PL, UK

Timothy D. Glotch and Jessica A. Arnold
Department of Geosciences, Stony Brook University, Stony Brook, New York 11794-2100, USA

Joshua L. Bandfield
Space Science Institute, 4750 Walnut St #205, Boulder, Colorado 80301, USA

Paul G. Lucey
Hawaii Institute of Geophysics and Planetology, University of Hawaii, Honolulu, Hawaii 96822, USA

Paul O. Hayne and Benjamin T. Greenhagen
Jet Propulsion Laboratory, M/S 183-301, 4800 Oak Grove Drive, Pasadena, California 91109, USA

Rebecca R. Ghent
Department of Earth Sciences, University of Toronto, Toronto, Ontario, Canada M5S 3B1

David A. Paige
University of California Los Angeles, Box 951567, Los Angeles, California 90095-1567, USA

Chizu Kato
Institut de Physique du Globe de Paris, Université Paris Diderot, CNRS UMR 7154, Paris 75005, France
Department of Earth and Planetary Sciences and McDonnell Center for Space Sciences, Washington University in St Louis, St Louis, Missouri 63130, USA

Maria C. Valdes
Department of Earth and Planetary Sciences and McDonnell Center for Space Sciences, Washington University in St Louis, St Louis, Missouri 63130, USA

Frederic Moynier
Institut de Physique du Globe de Paris, Université Paris Diderot, CNRS UMR 7154, Paris 75005, France
Institut Universitaire de France, Paris 75005, France

Jasmeet K. Dhaliwal and James M.D. Day
Geosciences Research Division, Scripps Institution of Oceanography, La Jolla, California 92093-0244, USA

Zongcheng Ling
Shandong Provincial Key Laboratory of Optical Astronomy and Solar-Terrestrial Environment, Institute of Space Sciences, Shandong University, Weihai 264209, China

Department of Earth & Planetary Sciences and McDonnell Center for the Space Sciences,Washington University, St Louis, Missouri 63130, USA
Key Laboratory of Lunar and Deep Space Exploration, National Astronomical Observatories, Chinese Academy of Sciences, Beijing 100012, China

Bradley L. Jolliff and AlianWang
Shandong Provincial Key Laboratory of Optical Astronomy and Solar-Terrestrial Environment, Institute of Space Sciences, Shandong University, Weihai 264209, China
Department of Earth & Planetary Sciences and McDonnell Center for the Space Sciences,Washington University, St Louis, Missouri 63130, USA

Chunlai Li, Jianjun Liu and Xin Ren
Key Laboratory of Lunar and Deep Space Exploration, National Astronomical Observatories, Chinese Academy of Sciences, Beijing 100012, China.

Jianzhong Liu
Institute of Geochemistry, Chinese Academy of Sciences, Guiyang 550002, China

Long Xiao
Planetary Science Institute, School of Earth Sciences, China University of Geosciences, Wuhan 430074, China

Wenxi Peng, Huanyu Wang and Xingzhu Cui
Institute of High Energy Physics, Chinese Academy of Sciences, Beijing 100049, China

Zhiping He and Jianyu Wang
Key Laboratory of Space Active Opto-Electronics Technology, Shanghai Institute of Technical Physics, Chinese Academy of Science, Shanghai 200083, China

Joseph R. Dwyer
Department of Physics and Space Sciences, Florida Institute of Technology, 150 West University Boulevard, Melbourne, Florida 32901, USA
Department of Physics, University of New Hampshire, Durham, New Hampshire 03824, USA

Hamid K. Rassoul, Ningyu Liu and Nicholas Spiva
Department of Physics and Space Sciences, Florida Institute of Technology, 150 West University Boulevard, Melbourne, Florida 32901, USA

Dwayne Free
Space Coast Intelligent Solutions, Melbourne, Florida 32934, USA

Steven A. Cummer
Department of Electrical and Computer Engineering, Duke University, Durham, North Carolina 27708, USA

H. Kjeldsen, A.B. Justesen, V. Silva Aguirre, V. Van Eylen, C. Vang, T. Arentoft, T.L. J. Christensen-Dalsgaard, R. Handberg and B. Tingley
Stellar Astrophysics Centre (SAC), Department of Physics and Astronomy, Aarhus University, Ny Munkegade 120, DK-8000 Aarhus C, Denmark

M.S. Lundkvist
Stellar Astrophysics Centre (SAC), Department of Physics and Astronomy, Aarhus University, Ny Munkegade 120, DK-8000 Aarhus C, Denmark
Zentrum für Astronomie der Universität Heidelberg, Landessternwarte, Königstuhl 12, 69117 Heidelberg, Germany

Campante, W.J. Chaplin, Y.P. Elsworth, M.N. Lund, A. Miglio and G.R. Davies
Stellar Astrophysics Centre (SAC), Department of Physics and Astronomy, Aarhus University, Ny Munkegade 120, DK-8000 Aarhus C, Denmark
School of Physics and Astronomy, University of Birmingham, Birmingham B15 2TT, UK

S. Basu
Department of Astronomy, Yale University, New Haven, Connecticut 06511, USA

D. Stello, S. Albrecht, D. Huber and T.R. Bedding
Stellar Astrophysics Centre (SAC), Department of Physics and Astronomy, Aarhus University, Ny Munkegade 120, DK-8000 Aarhus C, Denmark
Sydney Institute for Astronomy (SIfA), School of Physics, University of Sydney, New SouthWales 2006, Australia

C. Karoff
Stellar Astrophysics Centre (SAC), Department of Physics and Astronomy, Aarhus University, Ny Munkegade 120, DK-8000 Aarhus C, Denmark
Department of Geoscience, Aarhus University, Høegh-Guldbergs Gade 2, DK-8000 Aarhus C, Denmark

J.F. Rowe
NASA Ames Research Center, Moffett Field, California 94035, USA
SETI Institute, Mountain View, California 94043, USA

T. Barclay
NASA Ames Research Center, Moffett Field, California 94035, USA
Bay Area Environmental Research Institute, 596 1st Street West, Sonoma, California 95476, USA

R.L. Gilliland
Center for Exoplanets and Habitable Worlds, The Pennsylvania State University, 525 Davey Lab, University Park, Pennsylvania 16802, USA

S. Hekker
Stellar Astrophysics Centre (SAC), Department of Physics and Astronomy, Aarhus University, Ny Munkegade 120, DK-8000 Aarhus C, Denmark
Max Planck Institute for Solar System Research, D-37077 Göttingen, Germany

S.D. Kawaler
Department of Physics and Astronomy, Iowa State University, Ames, Iowa 50011, USA

T.S. Metcalfe
Stellar Astrophysics Centre (SAC), Department of Physics and Astronomy, Aarhus University, Ny Munkegade 120, DK-8000 Aarhus C, Denmark
Space Science Institute, Boulder, Colorado 80301, USA

T.R. White
Stellar Astrophysics Centre (SAC), Department of Physics and Astronomy, Aarhus University, Ny Munkegade 120, DK-8000 Aarhus C, Denmark
Institut für Astrophysik, Georg-August-Universität Göttingen, Friedrich-Hund-Platz 1, D37077 Göttingen, Germany

Michael A. Balikhin, Simon N. Walker and Keith H. Yearby
Department of Automatic Control and Systems Engineering, University of Sheffield, Mappin Street, Sheffield S1 3JD, UK

Yuri Y. Shprits
Department of Earth Planetary and Space Sciences, UCLA, 595 Charles Young Drive East, Box 951567, Los Angeles, California 90095-1567, USA
Department of Earth Atmospheric and Planetary Sciences, MIT, 77 Massachusetts Avenue, Cambridge, Massachusetts 02139-4307, USA

Benjamin Weiss
Department of Earth Atmospheric and Planetary Sciences, MIT, 77 Massachusetts Avenue, Cambridge, Massachusetts 02139-4307, USA

Lunjin Chen
W.B. Hanson Center for Space Sciences, Department of Physics, The University of Texas at Dallas, 800 West Campbell Road, Richardson, Texas 75080-3021, USA

Nicole Cornilleau-Wehrlin
LPP, CNRS, École Polytechnique, Palaiseau 91128, France
LESIA, Observatoire de Paris, Section de Meudon, 5, Place Jules Janssen, Meudon 92195, France

Iannis Dandouras
CNRS, IRAP, 9, Avenue du Colonel Roche, Toulouse BP 44346-31028, France
UPS-OMP, IRAP, 14, Avenue Edouard Belin, Toulouse 31400, France

Ondrej Santolik
Department of Space Physics, Institute of Atmospheric Physics ASCR, Bocni II/1401, 14131 Praha 4, Czech Republic
Faculty of Mathematics and Physics, Charles University in Prague, V Holesovickach 2, 18000 Praha 8, Czech Republic

Christopher Carr
Blackett Laboratory, Imperial College London, South Kensington Campus, London SW7 2AZ, UK

Michael A. Balikhin, Simon N. Walker and Keith H. Yearby
Department of Automatic Control and Systems Engineering, University of Sheffield, Mappin Street, Sheffield S1 3JD, UK

Yuri Y. Shprits
Department of Earth Planetary and Space Sciences, UCLA, 595 Charles Young Drive East, Box 951567, Los Angeles, California 90095-1567, USA
Department of Earth Atmospheric and Planetary Sciences, MIT, 77 Massachusetts Avenue, Cambridge, Massachusetts 02139-4307, USA

Benjamin Weiss
Department of Earth Atmospheric and Planetary Sciences, MIT, 77 Massachusetts Avenue, Cambridge, Massachusetts 02139-4307, USA

Lunjin Chen
W.B. Hanson Center for Space Sciences, Department of Physics, The University of Texas at Dallas, 800 West Campbell Road, Richardson, Texas 75080-3021, USA

Nicole Cornilleau-Wehrlin
LPP, CNRS, École Polytechnique, Palaiseau 91128, France
LESIA, Observatoire de Paris, Section de Meudon, 5, Place Jules Janssen, Meudon 92195, France

Iannis Dandouras
CNRS, IRAP, 9, Avenue du Colonel Roche, Toulouse BP 44346-31028, France
UPS-OMP, IRAP, 14, Avenue Edouard Belin, Toulouse 31400, France

Ondrej Santolik
Department of Space Physics, Institute of Atmospheric Physics ASCR, Bocni II/1401, 14131 Praha 4, Czech Republic
Faculty of Mathematics and Physics, Charles University in Prague, V Holesovickach 2, 18000 Praha 8, Czech Republic

Christopher Carr
Blackett Laboratory, Imperial College London, South Kensington Campus, London SW7 2AZ, UK

Yoichi Nakajima
Materials Dynamics Laboratory, RIKEN SPring-8 Center, RIKEN, Hyogo 679-5148, Japan

Alfred Q.R. Baron
Materials Dynamics Laboratory, RIKEN SPring-8 Center, RIKEN, Hyogo 679-5148, Japan
Research and Utilization Division, SPring-8, Japan Synchrotron Radiation Research Institute, Hyogo 679-5198, Japan

Saori Imada
Department of Earth and Planetary Sciences, Tokyo Institute of Technology, Tokyo 152-8551, Japan
Earth-Life Science Institute, Tokyo Institute of Technology, Tokyo 152-8550, Japan

Kei Hirose
Earth-Life Science Institute, Tokyo Institute of Technology, Tokyo 152-8550, Japan
Laboratory of Ocean-Earth Life Evolution Research, Japan Agency for Marine-Earth Science and Technology, Kanagawa 237-0061, Japan

Shigehiko Tateno
Earth-Life Science Institute, Tokyo Institute of Technology, Tokyo 152-8550, Japan
Institute for Study of the Earth's Interior, Okayama University, Tottori 682-0193, Japan

Tetsuya Komabayashi
Department of Earth and Planetary Sciences, Tokyo Institute of Technology, Tokyo 152-8551, Japan
School of GeoSciences and Centre for Science at Extreme Conditions, University of Edinburgh, Edinburgh EH9 3FE, UK

Haruka Ozawa
Laboratory of Ocean-Earth Life Evolution Research, Japan Agency for Marine-Earth Science and Technology, Kanagawa 237-0061, Japan
Institute for Study of the Earth's Interior, Okayama University, Tottori 682-0193, Japan

Satoshi Tsutsui
Research and Utilization Division, SPring-8, Japan Synchrotron Radiation Research Institute, Hyogo 679-5198, Japan

Yasuhiro Kuwayama
Geodynamics Research Center, Ehime University, Ehime 790-8577, Japan

Derek W.T. Jackson
School of Environmental Sciences, Ulster University, Coleraine BT52 1SA, UK

Mary C. Bourke
Department of Geography, Trinity College Dublin, Dublin D2, Ireland

Thomas A.G. Smyth
School of the Environment, Earth Sciences Building, Flinders University, GPO Box 2100, Adelaide, South Australia 5001, Australia

P.A. Bland
Department of Applied Geology, Curtin University, GPO Box U1987, Perth, Western Australia 6845, Australia

G.S. Collins, T.M. Davison, A.R. Muxworthy and J. Moore
Impacts & Astromaterials Research Centre (IARC), Department of Earth Science & Engineering, Imperial College London, South Kensington Campus, London SW7 2AZ, UK

N.M. Abreu
Earth Science Program, Pennsylvania State University — Du Bois Campus, Du Bois, Pennsylvania 15801, USA

F.J. Ciesla
Department of Geophysical Science, University of Chicago, 5734 South Ellis Avenue, Chicago, Illinois 60430, USA

Jessica J. Barnes and Ian A. Franchi
Department of Physical Sciences, The Open University, Walton Hall, Milton Keynes MK7 6AA, UK

David A. Kring
Lunar and Planetary Institute, 3600 Bay Area Boulevard, Houston, Texas 77058, USA

Romain Tartèse
Department of Physical Sciences, The Open University, Walton Hall, Milton Keynes MK7 6AA, UK
Institut de Minéralogie, de Physique des Matériaux et de Cosmochimie (IMPMC), Muséum National d'Histoire Naturelle, Sorbonne Universités, CNRS, UMPC & IRD, Paris 75005, France

Mahesh Anand
Department of Physical Sciences, The Open University, Walton Hall, Milton Keynes MK7 6AA, UK
Earth Sciences Department, Natural History Museum, Cromwell Road, LondonSW7 5BD, UK

Sara S. Russell
Earth Sciences Department, Natural History Museum, Cromwell Road, LondonSW7 5BD, UK

T. de Haas, H. van Steijn and M.G. Kleinhans
Faculty of Geosciences, Universiteit Utrecht, Heidelberglaan 2, 3584 CS, Utrecht, The Netherlands

E. Hauber
Institute of Planetary Research, German Aerospace Center, Rutherfordstrasse 2, Berlin DE-12489, Germany

S.J. Conway
Department of Physical Sciences, Open University, Walton Hall, Milton Keynes MK7 6AA, UK

A. Johnsson
Department of Earth Sciences, University of Gothenburg, Gothenburg SE-405 30, Sweden

Index